Actuarial Finance

Actuarial Finance

Derivatives, Quantitative Models and Risk Management

Mathieu Boudreault and Jean-François Renaud
Université du Québec à Montréal, Canada

WILEY

This edition first published 2019

Registered Office
John Wiley & Sons, Inc., 111 River Street, Hoboken, NJ 07030, USA

Editorial Office
111 River Street, Hoboken, NJ 07030, USA

For details of our global editorial offices, customer services, and more information about Wiley products visit us at www.wiley.com.

Wiley also publishes its books in a variety of electronic formats and by print-on-demand. Some content that appears in standard print versions of this book may not be available in other formats.

Library of Congress Cataloging-in-Publication data is available for this book.

9781119137009 (hardback)

Cover Design: Wiley
Cover Image: © precinbe/iStock.com

Set in 10/12pt WarnockPro by Aptara Inc., New Delhi, India

Printed in the United States of America.

V10008650_031119

À Joëlle, Justine, Raphaël et Thomas
À Julie, Étienne et Anne-Sophie

Contents

Acknowledgments

We are deeply grateful to our families for their unconditional support.

Special thanks to Geneviève Gauthier, Carole Bernard and Christiane Lemieux for their impact on our first years of teaching this material. Many thanks to colleagues from both academia and industry who took some of their precious time to review one or several chapters of earlier versions of the book: (in alphabetical order) Maciej Augustyniak, Jean-François Bégin, Bruno Gagnon, Ron Gebhardtsbauer, Jonathan Grégoire, Anne MacKay, Clarence Simard, Tommy Thomassin and Marie-Claude Vachon. Your comments were greatly appreciated and helped improve our book.

We also wish to thank the students who have attended our lectures for their feedback. This book is for you.

Finally, warm thanks to Jean-Mathieu Potvin for turning our handwritten plots into beautiful TikZ figures, to Christel Jackson Book for carefully reviewing each and every example and exercise, and to Adel Benlagra who also contributed to solve and revise exercises in the first half of the book. All remaining errors and typos are ours.

Preface

What is actuarial finance?

The work of actuaries has been strongly impacted by the deregulation of financial markets and the financial innovation that followed. Nowadays, there are embedded options in many life insurance and annuity products. Moreover, the use of weather and catastrophe derivatives in P&C insurance, to manage losses from earthquakes, hurricanes, floods and extreme weather, is constantly increasing, while longevity derivatives have been created to manage longevity risk on pools of pensioners.

As a consequence, in recent years *actuarial finance* has become an emerging field at the crossing of actuarial science and mathematical finance, i.e. when both actuarial and financial risks have to be taken into account. Despite its common roots with modern financial mathematics, actuarial finance has its own challenges due to:

- the very long-term nature (several years or decades) of insurance liabilities;
- the presence of mortality risk and other contingencies;
- the structure and regulations of the insurance and pension markets.

Therefore, it is now widely recognized that actuaries should have a basic knowledge of:

- derivatives and other assets to manage actuarial and financial risks;
- embedded options in actuarial liabilities.

A book written by and for actuaries

As we are writing this book for the *typical* actuarial science (undergraduate) student or practitioner, our main goal is to find an appropriate balance between the level of mathematics and finance in the presentation of *classical* financial models such as the binomial tree and the Black-Scholes-Merton model. Given the growing complexity of many actuarial liabilities and their ties to financial markets, we also felt the need for a book with actuarial applications at the forefront.

Therefore, this book has been motivated, designed and written for actuaries (and by actuaries). This means that we spend more time on and put more energy into what matters to actuaries, i.e. topics of particular relevance for actuaries are introduced at a more accessible level and analyzed in greater depth. For example, in this book, we:

- compare pricing in insurance and financial markets;
- present credit default swaps which are largely used by insurance companies to manage credit risk and we compare them with term life insurance;

- discuss event-triggered derivatives such as weather, catastrophe and longevity derivatives and how they can be used for risk management;
- introduce equity-linked insurance and annuities (EIAs, VAs), relate them to common derivatives and describe how to manage mortality for these products;
- put more emphasis on replication in the (two-period and n-period) binomial model and in the Black-Scholes-Merton model to better grasp the origins of the so-called actuarial/real-world and risk-neutral probabilities (or the (in)famous $\mathbb{P}$ and $\mathbb{Q}$ measures);
- implement and rebalance dynamic hedging strategies over several periods;
- introduce pricing and replication in incomplete markets and analyze the impact of market incompleteness on insurance and risk management;
- derive the Black-Scholes formula using simple limiting arguments rather than using stochastic calculus;
- implement Monte Carlo methods with variance reduction techniques to price simple and exotic derivatives;
- present immunization techniques alongside Greeks-based hedging;
- cover in details how to delta-gamma/rho/vega hedge a liability and how to rebalance periodically a hedging portfolio.

Moreover, the book contains many actuarial applications and examples to illustrate the relevance of various topics to actuaries.

To meet the above-mentioned objectives, we assume the reader has basic skills in calculus (differentiation and integration of functions), probability (at the level of the Society of Actuaries' Exam P), interest theory (time value of money) and, ideally, a basic understanding of elementary stochastic processes such as random walks.

Structure of the book

This book is targeted mainly toward undergraduate students in actuarial science and practitioners with an actuarial science background seeking a solid but yet accessible introduction to the quantitative aspects of *modern finance*. It covers pricing, replication and risk management of derivatives and actuarial liabilities, which are of paramount importance to actuaries in areas such as asset-and-liability management, liability-driven investments, banking, etc.

The book is divided into three parts:

Part 1: Introduction to actuarial finance
Part 2: Binomial and trinomial tree models
Part 3: Black-Scholes-Merton model

In each chapter, on top of the presented material, there is a set of specific objectives, numerical examples, a point-form summary and end-of-chapter exercises. Additional complementary information, such as historic notes or mathematical details, is presented in boxes. Finally, *warning icons* appear in the margin when a given topic, a concept or a detail deserves extra care or thought.

The content of the book is as follows.

Part 1: Introduction to actuarial finance

After an introductory chapter that puts into perspective the work of actuaries in the financial world, more standard chapters on financial securities, forwards and futures, swaps and options follow. Then, the next two chapters are devoted to the *engineering* of derivatives payoffs and liabilities, i.e. to the analysis of the structure of payoffs/liabilities, which is at the core of *no-arbitrage pricing*. Finally, a whole chapter describes insurance products bearing financial risk, namely *equity-linked insurance and annuities* (ELIAs).

In **Chapter 1 – Actuaries and their environment**, we put into context the role of the actuary in an insurance company or a pension plan. We explain how to differentiate between (actuarial) liabilities and (financial) assets, and between financial and insurance markets. We describe the various insurance policies and financial securities available and we compare actuarial and financial risks, short- and long-term risks, and diversifiable and systematic risks. Finally, we analyze various risk management methods for systematic risks.

In **Chapter 2 – Financial markets and their securities**, we provide an introduction to financial markets and financial securities, especially stocks, bonds and derivatives. We present the term structure of interest rates, we calculate the present and future value of cash flows and we explain the impact of dividends on stock prices. We also explain how actuaries can use derivatives and why pricing in the financial market is different from pricing in the insurance market. Finally, we look at price inconsistencies and how to create arbitrage opportunities.

In **Chapter 3 – Forwards and futures**, we provide an introduction to forwards and futures. We look at situations where forward contracts and futures contracts can be used to manage risks, and we explain the difference between a forward contract and a futures contract. We explain how to replicate the cash flows of forward contracts and calculate the forward price of stocks and of foreign currencies. Finally, we describe the margin balance on long and short positions of futures contracts.

In **Chapter 4 – Swaps**, we provide an introduction to swaps with an emphasis on those used in the insurance industry, namely interest rate swaps, currency swaps and credit default swaps. We present their characteristics, explain their cash flows and compute their values.

In **Chapter 5 – Options**, we give an introduction to standard options. We explain the differences between options to buy (call) and options to sell (put), as well as the difference between options and forward contracts. We explain when an option is used for hedging/risk management or for speculating purposes. Finally, we describe various investment strategies using options.

In **Chapter 6 – Engineering basic options**, we want to understand how to build and relate simple payoffs and then use no-arbitrage arguments to derive parity relationships. We see how to use simple mathematical functions to design simple payoffs and relate basic options and how to create synthetic versions of basic options, including binary options and gap options. Finally, we analyze when American options should be (early-)exercised.

In **Chapter 7 – Engineering advanced derivatives**, we provide an introduction to exotic/path-dependent options and event-triggered derivatives. We describe the payoff of various derivatives including barrier, Asian, lookback and exchange options, as well as weather, catastrophe and longevity derivatives. We explain why complex derivatives exist and how they can be used. Finally, we show how to use no-arbitrage arguments to identify relationships between the prices of some of these derivatives.

In **Chapter 8 – Equity-linked insurance and annuities**, we give an introduction to a large class of insurance products known as equity-linked insurance and annuities. First, we present relationships and differences between ELIAs and other derivatives. Then, after defining three indexing methods, we show how to compute the benefit(s) of typical guarantees included in ELIAs. We explain how equity-indexed annuities and variable annuities are funded and analyze the losses tied to these products. Finally, we explain how mortality is accounted for when risk managing ELIAs.

Part 2: Binomial and trinomial tree models

In the second part, we focus on the binomial tree model and the trinomial tree model, two discrete-time market models for the replication, hedging and pricing of financial derivatives and equity-linked products. By keeping the level of mathematics low, the binomial model allows greater emphasis on replication (as opposed to pricing), a concept of paramount importance for actuaries in asset and liability management. The intuition gained from the binomial model will be used repeatedly in the Black-Scholes-Merton model. Finally, as *market incompleteness* is a crucial concept in insurance markets, the trinomial tree model is treated in a chapter of its own. This model is simple and yet powerful enough to illustrate the idea of market incompleteness and its consequences for hedging and pricing.

In **Chapter 9 – One-period binomial tree model**, we first describe the basic assets available and we identify the assumptions on which this model is based. Then, we explain how to build a one-period binomial tree. Most of the chapter is devoted to the pricing of derivatives by replication of their payoff, from which we obtain risk-neutral pricing formulas.

In **Chapter 10 – Two-period binomial tree model**, we consider again the replication and the pricing of options and other derivatives, but now in a two-step tree. First, we explain how to build a two-period binomial tree using three one-period binomial trees. Then, we build dynamic replicating strategies to price options, from which we obtain risk-neutral pricing formulas. Finally, we determine how to price options in more complex situations: path-dependent options, options on assets that pay dollar dividends, variable annuities or stochastic interest rates.

In **Chapter 11 – Multi-period binomial tree model**, we see how to build a general binomial tree. We relate the asset price observed at a given time step to the binomial distribution and we highlight the differences between simple options and path-dependent options. We explain how to set up the dynamic replicating strategy to price an option and then obtain risk-neutral pricing formulas in a multi-period setup.

In **Chapter 12 – Further topics in the binomial tree model**, we extend the binomial tree to more realistic situations. We determine replicating portfolios and derive risk-neutral formulas for American-style options, options on stocks paying continuous dividends, currency options and futures options.

In **Chapter 13 – Market incompleteness and one-period trinomial tree models**, we define *market incompleteness* and we present the one-period trinomial tree model. We build sub-replicating and super-replicating portfolios for derivatives and explain that, in incomplete markets, there is a range of prices that prevent arbitrage opportunities. Then, we derive bounds on the admissible risk-neutral probabilities and relate the resulting prices to sub- and super-replicating portfolios in a one-period trinomial tree model. Second, we show how to replicate derivatives if the model has three traded assets. Also, we examine the risk management implications of ignoring possible outcomes when replicating a derivative. Finally, we analyze how actuaries cope with the incompleteness of insurance markets.

Part 3: Black-Scholes-Merton model

The third and last part of the book is devoted to the Black-Scholes-Merton model, the *famous* Black-Scholes formula and its applications in insurance. Both the model and its main formula are presented without the use of *stochastic calculus*; justifications are provided mainly by using the detailed work done previously in the binomial model and taking the appropriate limits. For the sake of completeness, a more classical treatment with stochastic calculus is also presented in two *starred* chapters, which can be skipped. In the last chapters, we apply generalizations of the Black-Scholes formula to price more advanced derivatives and equity-linked products, we provide an introduction to simulation methods and, finally, we present several sensitivity-based hedging strategies for equity risk, interest rate risk and volatility risk.

In **Chapter 14 – Brownian motion**, we provide the necessary background on Brownian motion to understand the Black-Scholes-Merton model and how to price and manage (hedge) options in that model. We also focus on simulation and estimation of this process, which are very important in practice. First, we provide an introduction to the lognormal distribution and compute truncated expectations and the stop-loss transform of a lognormally distributed random variable. Then, we define standard Brownian motion as the limit of random walks and present its basic properties. Linear and geometric Brownian motions are defined as transformations of standard Brownian motion. Finally, we show how to simulate standard, linear and geometric Brownian motions to generate scenarios, and how to estimate a geometric Brownian motion from a given data set.

In **Chapter 15 – Introduction to stochastic calculus*****, we provide a heuristic introduction to stochastic calculus based on Brownian motion by defining Ito's stochastic integral and stochastic differential equations (SDEs). First, we define stochastic integrals and look at their basic properties, including the computations of the mean and variance of a given stochastic integral. Then, we show how to apply Ito's lemma in simple situations. Next, we explain how a stochastic process can be the solution to a stochastic differential equation. Finally, we study the SDEs for linear and geometric Brownian motions, the Ornstein-Uhlenbeck process and the square-root process, and understand the role played by their coefficients.

In **Chapter 16 – Introduction to the Black-Scholes-Merton model**, we lay the foundations of the famous *Black-Scholes-Merton (BSM) market model* and we provide a heuristic approach to the Black-Scholes formula. More specifically, we present the main assumptions of the Black-Scholes-Merton model, including the dynamics of the risk-free and risky assets, and connect the Black-Scholes-Merton model to the binomial model. We explain the difference between real-world (actuarial) and risk-neutral probabilities. Then, we compute call and put options prices with the Black-Scholes formula and price simple derivatives using risk-neutral probabilities. Also, we analyze the impact of various determinants of the call or put option price. Finally, we derive replicating portfolios for simple derivatives and show how to implement a delta-hedging strategy over several periods.

In **Chapter 17 – Rigorous derivations of the Black-Scholes formula*****, we provide a more advanced treatment of the BSM model. More precisely, we provide two *rigorous* derivations of the Black-Scholes formula using either partial differential equations (PDEs) or changes of probability measures. In the first part, we define PDEs and show a link with diffusion processes as given by the Feynman-Kač formula. Then, we derive and solve the Black-Scholes PDE for simple payoffs and we show how to price and replicate simple derivatives. In the second part, we explain the effect of changing the probability measure on random variables and on Brownian motions (Girsanov theorem). Then, we compute the price of simple and exotic derivatives using the risk-neutral probability measure.

In **Chapter 18 – Applications and extensions of the Black-Scholes formula**, we analyze the pricing of options and other derivatives such as options on dividend-paying assets, currency options and futures options, but also insurance products such as investment guarantees, equity-indexed annuities and variable annuities, as well as exotic options (Asian, lookback and barrier options). Also, we explain how to compute the break-even participation rate or annual fee for common equity-linked insurance and annuities.

In **Chapter 19 – Simulation methods**, we apply simulation techniques to compute approximations of the no-arbitrage price of derivatives under the BSM model. As the price of most complex derivatives does not have a closed-form expression, we illustrate the techniques by pricing simple and path-dependent derivatives with crude Monte Carlo methods. Then, we describe three variance reduction techniques, namely stratified sampling, antithetic and control variates, to accelerate convergence of the price estimator.

In **Chapter 20 – Hedging strategies in practice**, we analyze various risk management practices, mostly hedging strategies used for interest rate risk and equity risk management. First, we apply cash-flow matching or replication to manage interest rate risk and equity risk. Then, we define the so-called *Greeks*. We explain how Taylor series expansions can be used for risk management purposes and highlight the similarities between duration-(convexity) matching and delta(-gamma) hedging. We show how to implement delta(-gamma) hedging, delta-rho hedging and delta-vega hedging to assets and liabilities sensitive to changes in both the underlying asset price and the other corresponding financial quantity. Finally, we compute the new hedging portfolio (rebalancing) as conditions in the market evolve.

Further reading

This book strives to find a balance between actuarial science, finance and mathematics. As such, the reader looking for additional information should find the following references useful.

For readers seeking a more advanced treatment of stochastic calculus and/or mathematical finance, important references include (in alphabetical order): Baxter & Rennie [1], Björk [2], Cvitanic & Zapatero [3], Lamberton & Lapeyre [4], Mikosch [5], Musiela & Rutkowski [6] and both volumes of Shreve, [7] and [8].

From a finance and/or business perspective: Boyle & Boyle [9], Hull [10], McDonald [11] and Wilmott [12].

On simulation methods: Devroye [13] and Glasserman [14].

Finally, for more details on ELIAs, there are two key references: Hardy [15] and Kalberer & Ravindran [16].

Part I

Introduction to actuarial finance

1

Actuaries and their environment

Actuaries are professionals using scientific and business methods to quantify and manage risks. They mostly work for insurance companies, pension plans and other social security systems. Actuarial professionals may hold different positions in an organization and perform several functions, such as pricing, reserving, capital determination, risk and asset management. For example, actuaries:

- determine the cost of protections found in insurance policies and pension plans;
- analyze how the financial commitments arising from insurance policies and pension plans will affect the solvency of the organization;
- calculate the appropriate amount of money to set aside to ensure the solvency of the insurance company or the pension plan;
- find and manage investments that will help meet the company's short-term and long-term financial commitments.

Therefore, actuaries:

- deal with assets and liabilities of insurance companies and pension plans;
- play an active role in insurance and financial markets;
- need to manage several types of risks whether they occur over the short or the long term, and whether they are systematic or diversifiable.

The main objective of this chapter is to put into context the role of the actuary in an insurance company or a pension plan. The specific objectives are to:

- differentiate between (actuarial) liabilities and (financial) assets;
- differentiate between financial and insurance markets;
- describe the various insurance policies and financial securities available in the markets;
- compare actuarial and financial risks, short- and long-term risks, diversifiable and systematic risks;
- analyze various risk management methods for systematic risks.

1.1 Key concepts

In this section, we define several general concepts to better understand the challenges faced by the actuary.

Actuarial Finance: Derivatives, Quantitative Models and Risk Management, First Edition.
Mathieu Boudreault and Jean-François Renaud.

1.1.1 What is insurance?

Insurance is an instrument designed to protect against a (potential) financial loss. Formally, it is a risk transfer mechanism whereby an individual or an organization pays a premium to another entity to protect against a loss due to the occurrence of an adverse event. Insurance in a broad sense thus includes typical insurance policies (life, homeowner's), pension plans and other social security systems.

Example 1.1.1 *Homeowner's insurance*
In a basic homeowner's insurance policy, the insurance company commits to repairing or rebuilding the house, and to buying new furniture, if a fire occurs. Therefore, the homeowner has transferred the *fire risk* to the insurance company. In exchange, the owner agrees to pay a fixed monthly fee, i.e. an insurance premium. ■

Instead of taking the risk of making one large and random payment, if for example a fire destroys their house, individuals are willing to make small and fixed payments to an insurance company (or a pension plan sponsor) in exchange for some protection.

Example 1.1.2 *Pension plan*
A pension plan can also be viewed as a risk transfer mechanism set up by an employer, known as the pension plan sponsor, for its employees in order to provide them with revenue during retirement. Each employee faces the risk of having insufficient savings for retirement because one cannot predict investment returns, nor when one will die. The pension plan thus provides protection against these risks and is funded by contributions, similar to fees or premiums, jointly paid by the employer and its employees. ■

1.1.2 Actuarial liabilities and financial assets

As a result of selling insurance protection, an insurance company or a pension plan sponsor:

- receives premiums or contributions that are invested in the financial markets;
- reimburses claims and/or pays out death and survival benefits.

Therefore, insurance companies and pension plans have important *assets* and *liabilities*. Generally speaking, an **asset** is what you *own* and a **liability** is what you *owe* while the **equity** is the difference between the two, i.e. your net worth. We have the fundamental accounting equation:

$$\text{Assets} = \text{Liabilities} + \text{Equity}. \tag{1.1.1}$$

Examples of assets for ordinary people range from a bank account (and other savings) to a house or a car. Typical financial assets for an insurance company mostly consist of investments such as stocks, bonds and derivatives.

Common liabilities for most people are loans, e.g. a student loan, and a mortgage, which is a loan on an asset provided as collateral. Liabilities of an insurance company are contractual obligations tied to the insurance policies sold to its policyholders. For pension plans, the benefits promised to the participants constitute the single most important liability. To differentiate between liabilities tied to insurance contracts and pension plans from other types of liabilities, such as loans or accounts payable, we will refer to insurance and pension obligations as *actuarial liabilities*.

Accounting and regulatory environment

The role of accounting is to make sure that financial information is disclosed to investors in a trustworthy and consistent manner. Therefore, accounting bodies are responsible for determining rules and assumptions to value the assets and liabilities of companies. Insurance companies and pension plans are no exception and, when reporting the value of their assets and liabilities, they must abide by a set of rules and assumptions specific to them. The accounting environment is driven by what is known as *generally accepted accounting principles* (GAAP).

Regulators closely monitor financial institutions (including insurance companies) to assure their solvency and therefore protect the public's interests. Like accounting bodies, they set up assumptions and design methods to value assets and liabilities and are therefore more conservative by nature. Each jurisdiction has a regulatory body depending upon where these companies are constituted.

1.1.3 Actuarial functions

As the actuary is involved in the management of financial assets and actuarial liabilities, typical actuarial functions include:[1]

- **Pricing**: computing the cost of an insurance protection (or of a specific pension plan design), designing policies and protections that are best for the customer (or employee) and the company (or pension plan sponsor), participating in the price determination.
- **Valuation**: given the current market conditions (interest rates, returns on financial markets) and a set of assumptions, computing the current value of actuarial liabilities, i.e. of contractual obligations. Valuation is useful to determine the amount of money *to reserve*, an actuarial function known as *reserving*, and to determine capital requirements.
- **Investments**: designing investment strategies and finding financial assets to mitigate the risks related to the organization's actuarial obligations. Not as common for property and casualty (P&C) insurance companies.

Example 1.1.3 *Pricing in P&C insurance*
The process of finding the appropriate premium for a property and casualty insurance policy is often known as *ratemaking*. Actuaries use a large history of claims to better understand the risks. They use factors such as age, sex, characteristics of the car or house, etc. to determine the appropriate premium (rate) for a specific policyholder. P&C actuaries are also involved in the design of insurance policies through determining deductibles, limits and exclusions. ■

Example 1.1.4 *Valuation in life insurance*
Interest rates affect the value of life insurance policies. The valuation actuary determines the reserve reflecting the company's current mortality experience and the level of interest rates. If, in the future, interest rates go down significantly, the actuary will have to increase the reserve. ■

1 Actuarial functions can be further divided and made specific to each organization. This list is not meant to be exhaustive, it is an overview of the most important responsibilities of the actuary.

Valuation of actuarial liabilities
Valuation of actuarial liabilities depends upon the body interested in such information and therefore assumptions and valuation techniques will vary accordingly. There are GAAP reserves (accounting purpose), statutory reserves (regulatory purpose, capital requirements), tax reserves (tax purposes) and reserves guided by actuarial practice (e.g. Actuarial Standards Board).

Example 1.1.5 *Liability-driven investments*
One of the roles of the investment actuary is e.g. to find the appropriate allocation between stocks and bonds to meet actuarial obligations in the future. This is known as a **liability driven investment (LDI)** strategy. The actuary can also manage the investment portfolio on a day-to-day basis or advise the pension plan administrators on appropriate investment strategies. ∎

As time passes, interest rates and financial returns will evolve and so will the company's claim and/or mortality experience. Therefore, insurance companies and pension plans will make gains or suffer losses. To mitigate the impact of important losses, the actuarial obligations (liabilities) and the investment portfolio (assets) should be aligned, as much as possible. The process of managing assets and liabilities together is known as **asset and liability management (ALM)**. It is a key concept and it will be discussed throughout the book under the wording *risk management*, *replication* and *hedging*.

1.2 Insurance and financial markets

A market is a system in which individuals or organizations can buy and sell goods. In particular, a **financial market** is where financial securities, currencies and commodities are traded. Similarly, an **insurance market** is where insurance policies are sold. Therefore, each actuarial function requires knowledge of the insurance market, the financial market, or both. We do not discuss pension plans as they are set up by an employer and an employee obviously cannot *shop for* her pension plan.

The market for financial securities differs significantly from the market for insurance policies. This has an important impact on how we should price and manage the products sold in each of these markets. This section highlights key aspects of insurance and financial markets.

1.2.1 Insurance market

To enter (buy) an insurance policy, regulations require individuals to have an *insurable interest*. If you have an accident with a car you own, you suffer losses and therefore you have an insurable interest in that car. You obviously also have an insurable interest in yourself due to the temporary or permanent injuries you might suffer from an accident. But unless you can prove it, you do not have an insurable interest in the life of your neighbor.

The insurance market is tightly regulated. It is composed of competing chartered insurance companies (registered to a government agency) which are the sole sellers of insurance policies. Individuals and organizations buying insurance policies are not allowed to sell them. If you want to get rid of your insurance policy, it is not permitted to (re-)sell it to another individual or company. Insurance policies are not *tradable* assets. The best a policyholder can do is to terminate the contract and pay the penalty.

Typical insurance policies sold in the insurance market are:

- Life insurance: a fixed or random benefit is paid, upon the death of the policyholder, to a beneficiary. Classes of policies are permanent and term insurance, universal life, etc.
- Annuity: contract providing a stream of income until death or until a predetermined date. Classes of policies are life or term-certain annuities, fixed and variable annuities, etc.
- General insurance: protection covering specific hazards (fire, accident, flood, etc.) to an owned property (car, house, etc.). Includes car insurance, homeowner's insurance, etc.
- Health and disability insurance: protection covering the costs of treatments, hospitalization, doctors and/or the loss of income following an injury, disease, etc.
- Group insurance: life, health and disability insurance sold to a group of employees.

In practice, policies are sold directly by the insurance company or through a *broker*, i.e. an intermediary between an individual and an insurance company. Even the broker is not allowed to buy back insurance policies from policyholders.

1.2.2 Financial market

The financial market is composed of thousands of knowledgable investors, ranging from individuals to institutional investors, such as investment banks, insurance companies and pension plans. In between, there are *market makers*, i.e. intermediaries selling to investors willing to buy and buying from investors willing to sell.

The financial market is not as tightly regulated as the insurance market. Consequently, investors are allowed to buy and sell securities quite easily. In practice, however, only large institutional investors can easily access the breadth of securities available.

The range of goods traded over the financial market is extremely large. There are securities, i.e. financial assets (such as stocks, bonds and derivatives), commodities (such as aluminum, wheat, gold) and currencies. Even carbon emissions are now traded on specially-designed markets.

Formally, a (financial) **security** represents a legal agreement between two parties. It can be traded between investors. The most common financial securities traded on the financial market are:

- Bond: type of loan issued by an entity such as a corporation or government;
- Stock: share of ownership of a corporation;
- Derivative: financial instrument whose value is derived from the price of another security or from a contingent event. Examples include futures and forwards, swaps and options.

These financial assets will be discussed in more detail in Chapter 2 for bonds and stocks, in Chapter 3 for forwards and futures, in Chapter 4 for swaps, in Chapter 5 for options and in Chapter 7 for other derivatives.

Other important assets are exchanged on the financial market:

- Commodity: tangible or non-tangible good such as crude oil, gold, coal, aluminum, copper, wheat, electricity (non-tangible).
- Currency: money issued by the government of a given country.

Securitization
Securitization is a process by which non-tradable assets or financial products, such as mortgages, car loans and credit cards, are aggregated to build a tradable security.

Mortgage-backed securities (MBS) are securities whose cash flows depend on a pool of residential or commercial mortgages. For example, there are special types of bonds based upon the principal or interest payments of the underlying loans. Life insurers and pension plans invest in MBS to manage interest rate risk.

Securitization has led to complex structured products such as **collateralized debt obligations (CDOs)**. In CDOs, low-quality (subprime) mortgages and MBS were repackaged into other securities. The collapse of the housing market in the U.S. in 2007–2009 and the financial crisis that followed have confirmed the difficulty of assessing the risk of these complex securities.

1.2.3 Insurance is a derivative

According to its definition, a **derivative** is a financial instrument whose value is derived *from a contingent event.* Therefore, insurance can be viewed as a derivative based on the occurrence (or not) of a risk. For example, life insurance is a derivative based on the life of an individual.

Example 1.2.1 *Homeowner's insurance*
You buy a house for $300,000. The building itself and your belongings are worth $200,000. In the event of a fire, the value of your home would drop considerably. Your homeowner's insurance policy will restore your home's value to what it was prior to the fire by providing enough money to repair or even rebuild your house, buy new furnitures and clothes, etc. It can therefore be viewed as a derivative contingent upon the occurrence of a fire and paying the amount of damages, up to a value of $200,000. ■

Therefore, insurance companies sell *actuarial derivatives*, i.e. insurance, to individuals in the insurance market, similar to investment banks selling financial derivatives to investors in the financial market. But, as discussed above, there are many aspects in which the insurance market and the financial market differ.

There are two additional differences between insurance policies and typical financial derivatives. Insurance policies have long maturities: decades for life insurance and pension plans, 1–2 years for general insurance, whereas most financial derivatives mature within a few months. Moreover, insurance policies are paid for over the life of the contract with periodic premiums whereas financial derivatives usually require a single premium paid up-front.

In conclusion, pricing *actuarial derivatives* in the insurance market is different from pricing derivatives in the financial market. We will come back to this topic in Chapter 2.

1.3 Actuarial and financial risks

Insurance companies and pension plans face various types of risks due to the nature of their obligations. In this section, we will divide risks according to two criteria, whether it is a short-term or a long-term risk, and whether it is actuarial or financial.

Long-term risks are found mostly in life insurance and pension plans, as they are arising from commitments taking place over decades.

- Mortality risk: uncertainty as to when a person or a group of persons will die affects the timing of the benefit payment and, in the case of annuities, the number of payments as well.
- Time value of money: uncertainty as to the level of future interest rates and the returns on investment portfolios affects the *time value of money*.

- Longevity risk: the overall improvement of life expectancy, for the general population or sub-groups, makes it difficult to predict how much longer people will live in the future and therefore has a mostly negative impact on pension plans.
- Equity-linked death and living benefits: for universal life policies and equity-linked insurance and annuities, the benefits are themselves random and mostly tied to stock market returns.

Short-term risks are typically covered by P&C insurance policies whose maturities are usually 1 or 2 years. In this case, the short-term risk can be decomposed according to its:

- Frequency: uncertainty as to the occurrence or not of the adverse event;
- Severity: uncertainty as to how much the loss will be, if the event occurs.

For example, the number of car accidents over the covered period refers to the frequency while the amount of losses of these accidents refers to the severity. The same reasoning would apply for other adverse events such as fire, theft, vandalism, etc. Losses resulting from natural hazards such as earthquakes, hurricanes, floods, etc., included in homeowner's insurance, are also in this category.

In light of the above discussion on long-term and short-term risks, we can now define what is an actuarial risk and what is a financial risk.

- **Financial risk** is uncertainty arising from the movements in the level of economic and financial variables such as interest rates, stock market returns, foreign exchange rates, commodity prices, etc.
- **Actuarial risk** is uncertainty arising from the occurrence, the timing and the amount of losses tied to adverse events such as death, disease, fire, theft, vandalism, earthquakes, hurricanes, etc.

Therefore, life insurance companies and pension plans deal with both financial and actuarial risks that are mainly long-term in nature, whereas P&C insurers mostly manage actuarial risks that are short-term in nature.

1.4 Diversifiable and systematic risks

Insurance policies and pension plan agreements are contracted simultaneously with hundreds or even thousands of people. The obligations toward each policyholder or employee are usually aggregated and managed as a whole. Not all risks behave the same way once pooled together. The purpose of this section is to further classify actuarial and financial risks to determine how they should be managed.

Some risks are said to be *diversifiable* as opposed to *systematic*. In what follows, we will illustrate the difference between the two using the following proverb:

> "Do not put all your eggs in the same basket."

1.4.1 Illustrative example

A **systematic risk** affects many or most individuals, if not all. For example, if we have 100 eggs to carry from point A to point B, then we can make the decision to put them all in a single basket and have one carrier bring them to their destination. In this case, if the carrier drops the basket, many or all eggs may break. However, if the basket is not dropped, they will all be intact. This extreme example is not that far from reality: there are many real-life examples of (close to) systematic risks such as the risk of a stock market crash or the risk of a natural catastrophe in a given region.

At the exact opposite, a **non-systematic risk** or a **diversifiable risk** for one individual does not affect the same risk for another individual. For example, if we have instead 100 people each carrying a basket containing one egg, then it is nearly impossible to break all 100 eggs or to bring them all safely to their destination. We can reasonably expect that only a few eggs will be broken, as some clumsy carriers will drop their basket. In this case, the risk of dropping eggs is *diversified away* over 100 carriers. We can of course imagine other situations such as having 25 people each carrying a basket with four eggs. There is also a wide range of real-life risks that are (close to being) diversifiable that we will discuss later in this section. The benefit of **diversification** is to reduce the uncertainty of the aggregate outcome. It is at the core of actuarial science.

1.4.2 Independence

One important assumption above was the *independence* between risks: the clumsiness of an egg carrier does not affect the skills of other egg carriers. In general, risks are independent if they do not influence each other.[2] In particular, independent risks do not depend on a common *source of risk*.

Example 1.4.1 *Mortality risk*
In most cases, death or survival of an individual does not affect the death or survival of other individuals. Therefore, mortality risk is usually assumed to be independent from one individual to another, except in the following circumstances:

- Epidemics, wars, etc.: those are events that can cause the death of many people in the same period and thus create statistical dependence. Death from these causes is usually excluded in life insurance policies.
- Family members: spouses often have similar living habits (eating habits, physical activities, etc.) and death can be explained by similar factors. There is also the well-known *broken heart syndrome* where the death of one of the spouses can accelerate death of the other. ■

Example 1.4.2 *Car accidents*
Reasons for a car accident are not necessarily related to those of another accident. Most accidents are caused by individual factors: driver distraction or tiredness, speeding, driving while impaired, local weather, mechanical failure, etc. Like many other types of claims in P&C insurance, the risks associated with car accidents are assumed to be independent between policyholders. ■

Example 1.4.3 *Natural hazards*
In a given region, say a state or a small country, all insureds might be exposed to a natural hazard such as an earthquake or a hurricane. Therefore, losses resulting from these natural catastrophes are not independent. ■

2 Mathematically speaking, this is known as statistical dependence between random variables.

1.4.3 Framework

In the rest of this section, we will compare diversifiable and systematic risks with (copies of) a *generic* die. Consider a die with six possible outcomes:

⚀ ⚁ ⚂ ⚃ ⚄ ⚅

The die is assumed to be well balanced, i.e. outcomes are equally likely. In other words, each outcome has a probability of $\frac{1}{6}$ of appearing on any given throw of the die.

More precisely, we will denote by X the result of a throw, i.e. the number of points appearing on the face of the die (⚀ ⚁ ⚂ ⚃ ⚄ ⚅). Said differently, $X = 1$ when the outcome is ⚀, $X = 2$ when the result is ⚁, and so on. Clearly, X is a random variable uniformly distributed on $\{1, 2, \dots, 6\}$. We can easily verify that

$$\mathbb{E}[X] = \frac{1}{6}(1 + 2 + 3 + 4 + 5 + 6) = \frac{7}{2} = 3.5$$

and

$$\begin{aligned}\mathbb{V}\mathrm{ar}(X) &= \frac{1}{6}((1 - 3.5)^2 + (2 - 3.5)^2 + (3 - 3.5)^2 + (4 - 3.5)^2 + (5 - 3.5)^2 + (6 - 3.5)^2) \\ &= \frac{35}{12} = 2.91\overline{6}.\end{aligned}$$

1.4.4 Diversifiable risks

Understanding diversifiable risks and their impact in insurance can be illustrated with dice. Suppose there are n policyholders each throwing their own die. Each die behaves as the generic one described above. More precisely, let X_i be the result for policyholder/die number $i = 1, 2, \dots, n$. As we also assume that the throws do not influence each other, the random variables $X_1, X_2, \dots, X_n$ are independent and identically distributed. In particular, for each $i = 1, 2, \dots, n$, we have

$$\mathbb{E}[X_i] = 3.5 \quad \text{and} \quad \mathbb{V}\mathrm{ar}(X_i) = 2.91\overline{6}.$$

Let us now look at the average of the n throws:

$$\overline{X}_n = \frac{1}{n}\sum_{i=1}^{n} X_i.$$

Under the above assumptions, we know that

$$\mathbb{E}[\overline{X}_n] = 3.5 \quad \text{and} \quad \mathbb{V}\mathrm{ar}(\overline{X}_n) = \frac{2.91\overline{6}}{n}.$$

This means that, no matter how many die we consider, the expectation of $\overline{X}_n$ will always be 3.5. Yet the variance of $\overline{X}_n$ will decrease if the number of die increases, which means that the possible values of $\overline{X}_n$ will be more and more *concentrated* around 3.5. In fact, according to the law of large numbers, if n is large enough, then

$$\overline{X}_n \approx 3.5,$$

i.e. the average of the n throws will be almost equal to the expected result of a single throw, with a *very high* probability.

For the sake of illustration, assume that the i-th policyholder faces a loss amount of X_i for a given event. Of course, this is an over-simplified situation as the loss can only take the values $1, 2, \ldots,$ or 6 (e.g. hundreds or thousands of dollars). For each policyholder, the loss amount X_i is unknown at the beginning of the year or, said differently, it is random.

From the perspective of each policyholder, it is a risky situation. Indeed, no one knows in advance whether the loss they will suffer will be zero, small or large. Consequently, a risk-averse person would prefer to transfer this risk to an insurance company in exchange for the payment of a non-random premium.

From the point of view of the insurance company, if the number of policyholders is large enough, then the realization $\overline{X}_n$ of its average loss over the whole portfolio will be close to 3.50, no matter which policyholder suffers large or small losses. By aggregating a large number of individual risks, the insurance company has diversified away the risk of suffering large losses. This is how *pure diversification* works for the benefit of individuals, and for the company as well. In this setting, the insurer could charge each policyholder the value of this *pure premium* of 3.50 and break even,[3] with a very high probability.

Diversification is the cornerstone of traditional insurance and, as we have seen, it is based upon the law of large numbers. It does not really matter who exactly claims a small or a large loss, what matters is the number of insureds in the portfolio so that $\overline{X}_n$ approaches 3.50 as much as possible. As highlighted in Examples 1.4.1 and 1.4.2, typical actuarial risks such as mortality, fire, theft, vandalism, etc. are independent and as such are considered diversifiable risks.

However, in practice, there is a lot of heterogeneity in insurance portfolios, i.e. policyholders represent different risks with different probability distributions. For example, in example 1.4.1, a 55-year-old male smoker does not represent the same mortality risk as a 25-year-old non-smoker female. Even if those two lives can be considered independent, they are not identically distributed. Similarly, for car insurance risk as in example 1.4.2, the probability of a car accident in any given year depends on individual risk factors such as age and sex of the driver. Again, even if the independence assumption is reasonable, car accident risk is not identically distributed from one insured to the other.

Insurance companies use several characteristics to distinguish between individuals. In fact, they use buckets of insureds, in which policyholders represent similar risks. Those sub-portfolios are known as *risk classes*. Even if there are fewer policyholders in a given risk class, in many cases the diversification principle described above still applies within a risk class. Whenever the number of policies within a class is not large enough, margins for adverse deviations are added to the premium.

Finally, it might not always be possible to add more people in a portfolio. Take the case of a pension plan. The capability of the pension plan to diversify mortality risk depends on the size of the plan, which in turn depends on the number of employees of the sponsor. Therefore, some significant mortality risk might remain in the portfolio.

1.4.5 Systematic risks

When dealing with diversifiable risks, if we have more policyholders (assumed to be independent and representing similar risks), then the diversification benefit is stronger, i.e. aggregated losses are closer to the expected value. Things are drastically different for *systematic risks*.

In this direction, let us look at an entirely different setup. Suppose now that we draw a single die and that the random variables $X_1, X_2, \ldots, X_n$ are all linked to this unique throw. More

3 Note that we are not taking into account the other expenses of the company. In practice, the true premium charged to the policyholder will be larger than the pure premium.

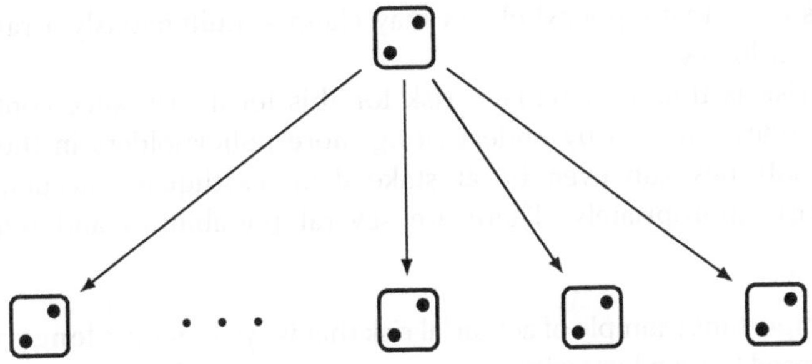

Figure 1.1 Systematic risk

precisely, let X be the result of this throw. Then, we set $X_1 = X_2 = \ldots = X_n = X$. This is illustrated in Figure 1.1.

In this situation, the average of the n throws is

$$\overline{X}_n = \frac{1}{n}\sum_{i=1}^{n} X_i = \frac{1}{n} \times nX = X.$$

The average number of points is always (in all scenarios) equal to X, no matter how small or large n is. This means that $\overline{X}_n$ has a uniform distribution over $\{1, 2, \ldots, 6\}$. This is completely different from the situation in Section 1.4.4.

Still interpreting X_i as the amount of loss for the i-th policyholder, we deduce that there is absolutely no diversification benefit from pooling risks together. Indeed, even if n is large, $\overline{X}_n$ still has a uniform distribution over $\{1, 2, \ldots, 6\}$. Sometimes, it will be equal to 2, with probability 1/6, while in other scenarios it will be equal to 6, with the same probability. It will not get closer to 3.5, with a high probability, even for a very large value of n. In this case, adding more policies to the (sub-)portfolio will not generate diversification as risks are systematic.

In the portfolio of policyholders of an insurance company or in a pension plan, the most important systematic risk is usually financial risk. Indeed, all premiums and employee contributions are managed and invested in the financial market with different kinds of investments. The returns earned on the company's or plan's investments are the same for all policyholders and employees. Therefore, uncertainty tied to the time value of money and thus the returns required to meet the company's commitments are systematic risks. Having more insureds or employees will not diminish the overall risk.

Example 1.4.4 *Interest rates*

For many years after the financial crisis of 2007–2009, interest rates have reached levels close to zero in many industrialized countries, including the United States, Canada, Germany, France and Japan. This has significantly increased the present value of cash flows tied to life insurance and annuities (including pension), increasing in turn required premiums and pension contributions of millions of savers.

Because the interest rate level is common to all contracts, again the insurance company clearly does not benefit from underwriting more policies. *Interest rate risk* is therefore a systematic risk. ■

Example 1.4.5 *Natural catastrophes*

A local Californian P&C insurer offers earthquake risk protection in its homeowner's insurance policies. If an earthquake occurs, it is exposed to an important systematic risk.

Indeed, in this case, many policyholders may claim simultaneously a random amount according to their losses.

Earthquake risk is thus a systematic risk for this local insurance company and the insurer clearly does not gain by underwriting more policyholders in the same region. The insurer's solvency can even be at stake if an earthquake occurs, so it should manage this risk appropriately. There are several possibilities and reinsurance is a popular one. ■

Another very important example of actuarial risk that is systematic is **longevity risk**. Reasons that explain increased human longevity are common to all: overall quality of the healthcare system, medical research, better living habits, etc. The uncertainty as to how many years humans will live in the long run is a systematic risk that life insurance companies and pension plans need to bear.

1.4.6 Partially diversifiable risks

In reality, most risks fall somewhere between being purely diversifiable or purely systematic. In these situations, there are benefits to diversification but they will be limited. To illustrate how partially diversifiable risks behave, we will use dice once more.

Suppose we have $n + 1$ dice. Let Y_i be the number of points on the throw of the i-th die for $i = 1, 2, \dots, n$ and let Z be the number of points on the extra die. Then, define:

$$X_i := Y_i + Z$$

for $i = 1, 2, \dots, n$. In other words, X_i is the sum of an individual component Y_i and a common element Z. Clearly, the X_is are not independent random variables, but they have the same probability distribution.

Increasing the number n in the portfolio, we can diversify the individual components away but not the *common shock* Z. If Z is small (large), it will be small (large) for all Xs, potentially generating small (large) losses for all Xs. Figure 1.2 shows for example that when $Z = 6$, then

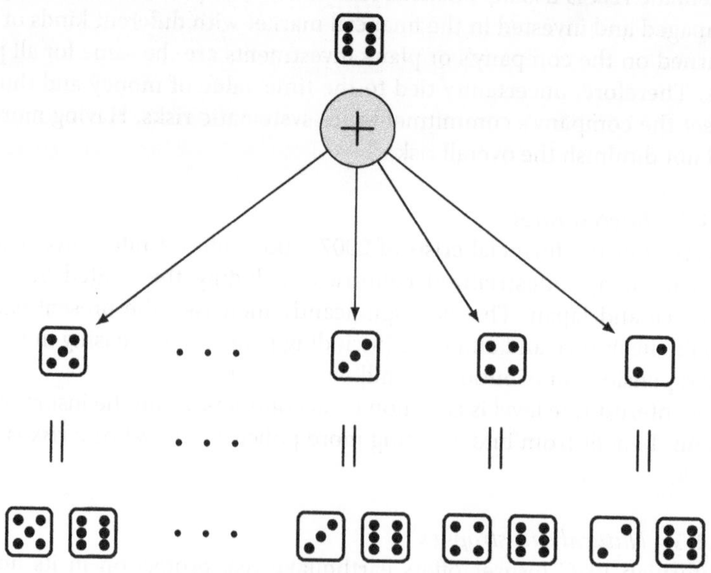

Figure 1.2 Partially diversifiable risks

most Xs will be relatively large: we have $X_i \geq 7$. The factors that are common to many risks (Z in this illustration) are known as the *systematic components* whereas the individual parts are known as the *diversifiable* or *idiosyncratic components*.

Let us now analyze which actuarial and financial risks are partially diversifiable. Risks tied to natural hazards can be systematic or partially diversifiable depending on the insurer's ability to geographically diversify its risks. As discussed in Example 1.4.5, a company that operates in a U.S. state does not have the same diversification capabilities as a large global insurer. Natural hazards are better diversified globally as earthquake risk in for example California is not related to earthquake risk in the Middle East. The same applies for hurricanes in the North Atlantic (East Coast of the U.S.) and West Pacific (typhoons in Asia).

Returns in a portfolio of stocks are also an example of a partially diversifiable risk. Stock returns are affected by macroeconomic (e.g. economic growth), country- and sector-specific factors. For example, oil companies, auto makers and software companies can be affected by the general growth of the economy but the crude oil price will not affect oil drilling companies in the same way as auto makers, and certainly will have a negligible impact on most tech firms. Again, due to the presence of common factors explaining stock returns, there is an important limit to diversification benefits we can achieve by increasing the number of stocks in a portfolio.

1.5 Risk management approaches

Depending on whether an actuarial or a financial risk is diversifiable, partially diversifiable or systematic, risk management will not rely upon the same tools. This section mostly describes risk management of systematic risks as this will be a recurrent topic in this book.

As discussed, whenever a risk is diversifiable or partially diversifiable, a solution to reduce the overall aggregate risk is to diversify over additional (independent) individuals or geographically over additional countries and continents. This applies to traditional insurance policies such as term life insurance, car insurance, etc. Tools from short-term (non-life) and long-term (life) actuarial mathematics are used to price, reserve and manage these risks on a day-to-day basis.

But how do we manage systematic risks? Whenever interest rates, stock prices (and indices), exchange rates, etc. affect the value of actuarial obligations for all or most policyholders, the (investment) actuary uses one of the following two strategies:

- Trade in the financial markets to reproduce the (behavior of) cash flows of liabilities. This approach is generally costly but there is very small *insolvency risk*. This is also known as *replication* or *hedging*.
- Manage a balanced portfolio of financial securities with the objective of increasing returns, thus lowering the time value of money. However, it comes with the risk that investments are insufficient to meet obligations. Portfolio managers focus on reducing this last risk while trying to maximize returns.

Both approaches are illustrated in the next example.

Example 1.5.1 *Life insurance*
Suppose an insurance company has issued the following life insurance contract to a very large number, say n, of individuals:

- It pays \$100 in 1 year if the individual survives;
- It pays \$105 in 1 year (to the beneficiaries) if the individual dies within the year.

Based upon mortality tables and other experience data, the actuary has determined that the expected loss per policy is \$101. Following the law of large numbers, the average loss, which is a random quantity, will be *close to* \$101 if it is a large portfolio and if we assume that individual mortality risks are independent and identically distributed. We will now illustrate how the actuary can manage the systematic risk tied to the time value of money (investment returns).

The aggregated financial commitments of the insurer, in 1 year from now, will be close to $n \times 101$. The first strategy would be to find securities in the financial market that allow to lock in a value of $n \times 101$ in a year, with (almost) certainty. Assume a risk-free zero-coupon bond, trading for \$97, will pay \$100 in a year. Therefore, buying $101n/100$ zero-coupon bonds for a total cost of

$$\frac{101n}{100} \times 97 = 97.97n$$

perfectly matches the liability of the insurer.

If the investment actuary believes she can earn a return of 6% over the year, then she could instead set aside

$$\frac{101n}{1.06} = 95.2830n$$

and invest it in the financial market. However, if the returns earned over the year are less than 6%, then the invested amount will not be enough to cover future benefits. This is why a conservative discount rate is usually assumed to make sure the company meets its financial commitments. ■

As mentioned above, another way of mitigating the impact of systematic risks for an insurance company is to use reinsurance. **Reinsurance** can be viewed as insurance for insurers. Reinsurance contracts specify the risks that are covered and how losses are distributed between the insurer and the reinsurer (proportional, stop loss, excess of loss). Just like individuals with typical insurance policies, insurers need to pay regular premiums to the reinsurer in exchange for protection.

One case where reinsurance works very well is natural hazards. For example, one can think of earthquake risk being systematic for a given region but being highly diversifiable at the international level. Reinsurance can also be used to limit longevity risk for life insurance companies and pension plans. Reinsurance companies diversify internationally and across many risks as well. For example, earthquake and longevity risks can be considered independent.

Finally, recent financial innovations, such as *longevity derivatives* or *catastrophe derivatives* (see Chapter 7, Section 7.2), are providing additional tools to manage these two systematic risks.

1.6 Summary

Actuary

- Professional using scientific and business methods to quantify and manage risks.
- Works for insurance companies, pension plans and other social security systems.
- Typical tasks include:
  - Determine the cost of protections;
  - Analyze how financial commitments affect solvency;
  - Calculate how much to set aside (reserve, capital requirements);
  - Find and manage investments to meet financial commitments.

Key concepts

- Insurance is a risk transfer mechanism.
- Assets: what you own.
- Liabilities: what you owe.
- Financial assets: securities, investments.
- Actuarial liabilities: commitments from insurance policies and pension plans.
- Fundamental accounting equation: Assets = Liabilities + Equity.

Insurance and financial markets

- The insurance market is tightly regulated: insurable interest, insurance policies are not tradable assets.
- Main financial securities: bonds, stocks, derivatives, commodities and currencies.
- The financial market is not very regulated: investors can almost freely buy and sell financial securities.

Classifications of risks from an actuarial perspective

- Actuarial risk vs. financial risk:
  - Actuarial risk: uncertainty arising from the occurrence, timing and loss tied to adverse events.
  - Financial risk: uncertainty arising from economic and financial variables.
- Short-term risk vs. long-term risk:
  - Short-term risk: mostly in P&C insurance policies.
  - Long-term risk: mostly in life insurance policies and pension plans.
- Diversifiable risk vs. systematic risk:
  - Diversifiable risk: independent of other risks, i.e. not affected by nor affecting other risks.
  - Systematic risk: common to many or most individuals or investors.
  - Partially diversifiable risk: partly diversifiable, partly systematic (most common type of risk).

Risk management

- Diversification: reduces (some of) the uncertainty of aggregated (partially) diversifiable risks by adding more risks.
- Hedging, replication: trade in the financial market to reproduce, as much as possible, cash flows of liabilities.
- Investment strategy, speculation: trade in the financial market to increase returns and lower the time value of money.

1.7 Exercises

1.1 For each of the following events, explain its impact on the assets and liabilities of a pension plan:

(a) The stock market drops by 7%.

(b) The Treasury rate increases by 1%.

(c) A hurricane hits a region, affecting 10% of your portfolio.

(d) A great medical discovery cures two of the deadliest cancers.

(e) A massive heat wave increases mortality by 3%.

(f) The economy enters into a recession and many corporations go bankrupt.

Repeat the exercise for a life insurance company, a P&C insurance company and an investment firm.

1.2 According to each partial description, determine whether the corresponding security is a stock, a bond/loan or a derivative.

(a) Difference between the current stock price and its mean over the last week, only if this difference is positive.

(b) 2% of the voting rights in the management of ABC inc.

(c) U.S. Treasury promises to pay 1.5% annually on \$100.

(d) \$1 if rain accumulates to more than 50mm during the week of August 1st at LAX airport.

(e) \$1 if more than 95% of a reference population survives.

1.3 For each of the following situations, identify all diversifiable, systematic and partially diversifiable risks.

(a) You lend \$10 each to 100 different people working for the same company. You want to assess the risk that some people will not pay you back. Repeat the experiment if the 100 people work for 100 different firms but in the same city/country.

(b) You work for an insurer selling health insurance policies that reimburse hospital fees and loss of income in case the policyholder gets cancer.

(c) Autonomous and semi-autonomous cars are vulnerable to problems in their respective softwares. You sell policies to individual drivers (policyholders) covering damage in case of a car accident (including fully and semi-autonomous cars). What happens in this insurance market if there are only 3 or there are 20 major autonomous automakers?

1.4 Your insurance company assumes the risks tied to $X_1, X_2, \dots, X_n$. For each of the following situations:

(a) Analyze the distribution of $X_1, X_2, \dots, X_n$ and determine whether the risks are diversifiable, systematic or partially diversifiable. Explain why.

(b) Describe mathematically the behavior of $\overline{X}_n = \frac{1}{n}\sum_{i=1}^{n} X_i$.

(c) Determine the behavior of $\overline{X}_n$ when $n \to \infty$ and compare the results with your intuition in (a).

Situation # 1 Let $X_i|\Theta$ be represented by an exponential distribution with mean Θ. Suppose that Θ is itself a random variable with mean μ and standard deviation σ. Given Θ, the X_i are independent.

Situation # 2 Let $X_i|\Theta_i$ be represented by an exponential distribution with mean Θ_i. Suppose that $\Theta_i, i = 1, 2, \dots, n$ are independent and all distributed as another random variable Θ with mean μ and standard deviation σ. Given Θ_i, the X_i are independent.

Situation # 3 Let $Y_i, i = 1, 2, \dots, n$ and Z be independent and identically distributed random variables. Suppose that $X_i = \min(Y_i, Z)$ for each i.

1.5 A customer purchases a special kind of life insurance policy. The policyholder first invests \$100 in an investment portfolio composed of 20 stocks. The insurance company adjusts the death and maturity benefits according to the returns of that portfolio, subject to a minimum return. The insurer sells such a policy to 10,000 policyholders. From a risk management point of view, what is the impact of increasing/decreasing the number of stocks in the portfolio? And what is the impact of increasing/decreasing the number of policyholders?

(a) For the policyholder.

(b) For the life insurance company.

1.6 Let the probability distribution of a set of risks $X_i, i = 1, 2, \ldots, n$ be described by the following joint cumulative distribution function:

$$F_{X_1,X_2,\ldots,X_n}(x_1, x_2, \ldots, x_n) = \exp\left[-\left(\sum_{k=1}^{n}(-\log F_{X_i}(x_i))^{\theta}\right)^{1/\theta}\right]$$

where $\theta \geq 1$.

Are these risks diversifiable, systematic or partially diversifiable:

(a) if $\theta = 1$;

(b) if $\theta > 1$, but is relatively small;

(c) if θ is very large, converging to infinity?

First, consider the case $n = 2$, and then consider the general case.

2

Financial markets and their securities

As described in Chapter 1, insurance companies and pension plans collect premiums and contributions in exchange for *protection*. These premiums are then invested in the financial market, in securities such as stocks, bonds and derivatives. Moreover, life insurers offer complex policies, in which living and death benefits are sometimes indexed to the returns of an investment portfolio. Consequently, the financial market and financial securities have an impact on actuarial liabilities, so a basic understanding is necessary.

The main objective of this chapter is to provide an introduction to the financial market and financial securities, especially stocks, bonds and derivatives, from the point of view of an actuary. The specific objectives are to:

- understand the term structure of interest rates and the ties with bond prices;
- calculate the present and future value of cash flows using spot rates and forward rates;
- understand the impact of dividends on stock prices and how dividends can be reinvested;
- determine how actuaries can use derivatives;
- explain why pricing in the financial market is different from pricing in the insurance market;
- understand basic financial terminology related to trading;
- identify price inconsistencies and create arbitrage opportunities.

It is important to note that this chapter is not meant to be an exhaustive analysis of stocks, bonds or derivatives and their market-specific characteristics. This chapter covers stocks and bonds at a level that is sufficient for the rest of this book, and it provides a brief high-level overview of derivatives.

2.1 Bonds and interest rates

A **bond** is a type of loan issued by an entity, such as a corporation or a government, and traded on the financial market. With bonds, both small and institutional investors can lend money. Bonds are part of a broader class of investments known as *fixed-income securities.*

2.1.1 Characteristics

A bond is characterized by:

1) its issuer, i.e. the borrower;
2) its interest schedule, i.e. when interest is paid over time;
3) its maturity, i.e. when the principal, or face value, is paid back to the lender;
4) additional protections, for the borrower and/or the lender.

Let us discuss each of these items.

Actuarial Finance: Derivatives, Quantitative Models and Risk Management, First Edition.
Mathieu Boudreault and Jean-François Renaud.

Table 2.1 Outstanding U.S. bond market debt (billions of $US) as of the end of 2015

Type	Amount	Percentage
Municipal	3,796	9.9%
Treasury	13,192	34.5%
Mortgage-backed securities (MBS)	8,759	22.9%
Corporate debt	8,157	21.4%
Federal agency securities	1,995	5.2%
Money markets	941	2.5%
Other asset-backed securities (ABS)	1,362	3.6%
Total	**38,203**	**100%**

Source: Securities Industry and Financial Markets Association (SIFMA).

Bonds are issued by companies (corporate bonds), cities (municipal bonds), public utilities, states and provinces, countries, etc. Bonds issued by national governments are called Treasuries, if held by local investors, or sovereign bonds, if held by foreign investors. Table 2.1 shows the size of the bond market in each of these categories.

There are two types of bonds: coupon bonds and zero-coupon bonds. A **coupon bond** is a bond where interest is paid periodically (twice a year normally) and principal is paid back at maturity. The interest payments are known as **coupons**. In a **zero-coupon bond**, both interest and principal are paid back at maturity.

Maturities of bonds range from a couple of months to several years, up to 15, 20 or even 30 years. Long-maturity bonds are popular investments for life insurance companies and pension plans because they can be used as part of an ALM strategy to match long-term actuarial liabilities.

However, bonds are exposed to **credit risk**, i.e. the loss a bondholder will suffer if the bond issuer is unable to fully pay coupons and/or principal. This event is known as *default*. The safest bonds are often called risk-free bonds (e.g. Treasuries), risk-free in the sense that default risk is negligible. The resulting yield is known as the **risk-free rate**. Insurance companies and pension plans need to invest in highly-rated bonds to meet solvency requirements set by regulators. To protect against credit risk, many investors, including insurance companies, use *credit default swaps (CDS)* (see Section 4.4).

U.S. Treasuries

U.S. Treasuries are a very important part of the fixed-income securities market with about 13 trillions worth of Treasuries outstanding in the market (as of December 2015, see Table 2.1). Treasuries have different maturities: 3 and 6 months and 1, 2, 3, 5, 7, 10, 15 and 30 years. Depending on the maturity, Treasuries have different names: notes, bills or bonds. A Treasury bill is a zero-coupon bond with a maturity of less than 1 year, a note is a coupon bond with a maturity of 10 years or less whereas a Treasury bond has a maturity of more than 10 years.

Bonds have additional features protecting either the issuer or the holder. A bond may be **callable (redeemable)**, meaning that the issuer can buy the bond back from the investor on specific dates. When interest rates go down, it may become attractive for a bond issuer to refinance the bond at a lower interest rate, just like individuals refinance their mortgages in this situation. In the event of a bankruptcy leading to a liquidation of all the company's assets, (corporate) bondholders have priority over shareholders, i.e. those who own stocks of the company (see Section 2.2). This is also known as the **absolute priority rule.**

2.1.2 Basics of bond pricing

Without loss of generality, suppose a coupon bond pays a total of n fixed coupons, of \$$c$ each, at the following dates: $t_1 < t_2 < \cdots < t_n$. Assume the bond matures at time $T = t_n$ and that it has a principal (or face value) of \$$F$. In what follows, we ignore credit risk.

Then, the initial bond price B_0 can be written as

$$B_0 = \sum_{i=1}^{n} \frac{c}{(1+y)^{t_i}} + \frac{F}{(1+y)^T} \tag{2.1.1}$$

where y is the corresponding **bond yield,** or **yield to maturity** (YTM). Usually, the bond yield y is compounded annually (or semi-annually), which means that the t_is are also annual (or semi-annual). Of course, the bond yield y depends on the coupon amount c and the bond maturity time T. Finally, in the bond markets, c and y are usually such that $B_0 = F$, i.e. the bond is commonly priced *at par* at inception.

Example 2.1.1 *Bond price*
Let us find the initial price of a 10-year bond with semi-annual coupons of \$3 and a face value of \$100, if the bond yield is 7.5% and is compounded semi-annually.

For this bond, we have $T = 10$, $n = 20$, $c = 3$ and $t_1 = 0.5, t_2 = 1, \ldots, t_{20} = 10$, and also $y = 0.075/2 = 0.0375$. So, using formula (2.1.1), we get

$$B_0 = \sum_{i=1}^{20} \frac{3}{1.0375^{t_i}} + \frac{100}{1.0375^{10}} = 89.57784684.$$

■

Note that a zero-coupon bond is obtained when $n = c = 0$, in which case formula (2.1.1) simplifies to

$$B_0 = \frac{F}{(1+y)^T}. \tag{2.1.2}$$

Again, the zero-coupon bond yield y depends on the maturity time T.

2.1.3 Term structure of interest rates

The **term structure of interest rates** is the relationship between the maturity of a loan, or an investment, and the *annual* interest rate underlying that loan. The idea is that investing for 5 years should earn a different *annual* interest rate than investing for 1 year. This is illustrated in Figure 2.1, where it is shown that a short-maturity loan will earn a lower annual interest rate than a long-maturity loan.

Let us first illustrate the idea of the term structure of interest rates with three examples.

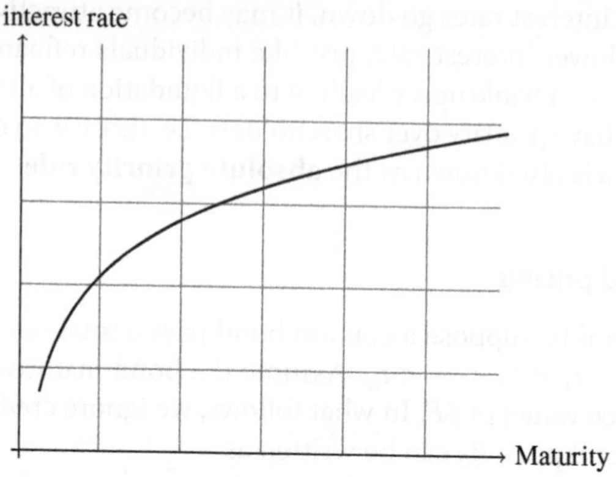

Figure 2.1 Typical term structure of interest rate

Example 2.1.2 *Simple loan*
Bank ABC offers simple loans where principal and interest are only paid back at maturity. It advertises the following rates:

Maturity (term)	Rate
1 year	3.8%
2 years	4.1%
3 years	5.0%
4 years	5.7%
5 years	6.5%

Therefore, you can borrow \$1000 at an annual rate of 5% for 3 years. The principal and interest are paid back to the lender at maturity. This payment will be $1000 \times (1.05)^3 = 1157.63$.

Suppose there is another borrower who can only afford to pay back \$5000 in 5 years. How much can she borrow right now using these simple loans? Five-year loans credit interest at an annual rate of 6.5% and therefore the amount she can borrow is given by (2.1.2):

$$\frac{5000}{1.065^5} = 3649.40.$$

■

Example 2.1.3 *Bullet guaranteed investment contracts (GICs)*
Bullet GICs are savings products sold by insurance companies where principal and interest are paid to the investor at maturity. One insurer offers the following rates for its bullet GICs:

Maturity (term)	Rate
1 year	0.5%
3 years	1.0%
5 years	1.5%
7 years	2.0%
10 years	2.5%
15 years	3.0%

Therefore, an investor can accumulate savings over 10 years at a rate of 2.5% per year. Thus, $1000 accumulates to

$$1000 \times (1.025)^{10} = \$1280.08.$$

Similarly, another investor that requires principal and interest after 3 years will earn an annual rate of 1% per annum. ■

Example 2.1.4 *Zero-coupon bonds*
The following zero-coupon bonds are traded in the financial market. Prices as of today are quoted such that $100 (principal and interest) is paid back at maturity.

Maturity (term)	Price
1 year	99.00
2 years	97.50
3 years	96.00
4 years	94.50
5 years	92.00
10 years	84.00

For example, in this case, the 5-year yield rate can be obtained with equation (2.1.2):

$$y = \left(\frac{F}{B_0}\right)^{1/T} - 1 = \left(\frac{100}{92}\right)^{\frac{1}{5}} - 1 = 1.68\%.$$

Therefore, someone investing $92 for 5 years in this zero-coupon bond will earn an annual return of 1.68%. ■

Zero-coupon bond prices, with face value equal to $F = \$1$, can serve as *discount factors.* Indeed, in this case, since equation (2.1.2) becomes

$$B_0 = \frac{1}{(1+y)^T},$$

we see that multiplying by B_0 is equivalent to discounting over a period of length T.

This is illustrated in the next example.

Example 2.1.5 *Zero-coupon bonds (continued)*
Let us now illustrate how zero-coupon bond prices can be used as discount factors, using the context of example 2.1.4. What is the price of a 3-year bond paying annual coupons of 6%?

First, let us normalize the prices as follows:

Maturity (term)	Discount factor
1 year	0.99
2 years	0.975
3 years	0.96
4 years	0.945
5 years	0.92
10 years	0.84

Now, we can discount any cash flow occurring $1, 2, \dots, 5$ years from now. Therefore, for the coupon bond price, we have

$$B_0 = 6 \times 0.99 + 6 \times 0.975 + 106 \times 0.96 = 113.55,$$

i.e. the *market-consistent price* of this 3-year bond is \$113.55. ■

2.1.3.1 Spot rates

The T-**year spot rate** is the annual interest rate agreed upon today, at time 0, for a loan or an investment starting immediately, whose capital and interest are paid only at maturity, at time T. Mathematically, it is denoted as r_0^T. The **spot rate curve** or **yield curve** is the relationship between the spot rates and the maturity of the underlying loans, i.e. the relationship between T and r_0^T. If the interest rate depicted in Figure 2.1 is r_0^T, as a function of T, then the graph is a typical increasing spot rate curve or yield curve.

Therefore, we can rewrite the bond price formula of equation (2.1.1) using spot rates instead of the corresponding yield rate and get

$$B_0 = \sum_{i=1}^{n} \frac{c}{\left(1 + r_0^{t_i}\right)^{t_i}} + \frac{F}{\left(1 + r_0^T\right)^T}.$$

The underlying idea is simple: for a cash flow occurring at time t_i, such as a coupon payment, we discount using the spot rate that applies between time 0 and time t_i, i.e. the rate $r_0^{t_i}$. Consequently, we see that there is a strong relationship between the yield and the spot rates associated with a given bond.

Clearly, for a zero-coupon bond, we have the following pricing formula, related to (2.1.2):

$$B_0 = \frac{F}{\left(1 + r_0^T\right)^T}. \tag{2.1.3}$$

It is possible to trade a single payment of a coupon bond. This process is known as **stripping** and can also be used to determine the price of a zero-coupon bond from coupon bonds. As a result, we can also recover the spot rate curve, as in the following example.

Example 2.1.6 *Stripping a bond*
Suppose the following coupon bonds, with annual coupons and face value $F = 100$, are traded with prices given by:

Maturity	Coupon	Price
1	n/a	97.09
2	6%	103.83
3	8%	109.80

Let us find the price of the 2-year and 3-year zero-coupon bonds and spot rates.

Using the pricing formula in (2.1.3), we know that $r_0^T = (F/B_0)^{1/T} - 1$. Then, the 1-year rate is $r_0^1 = (100/97.09) - 1 = 0.03$.

The 2-year coupon bond price is 103.83 and it can be represented as follows:

$$103.83 = 6 \times 0.9709 + 106 \times x,$$

where x is the 2-year discount factor in this market. We find that $x = 0.9246$ and thus the 2-year zero-coupon bond price is 92.46. Then, from (2.1.3), we deduce that the 2-year rate is $r_0^2 = 0.04$.

Finally, the 3-year coupon bond price is 109.80 and we have

$$109.80 = 8 \times 0.9709 + 8 \times 0.9246 + 108 \times x,$$

where x is now the 3-year discount factor. We find that $x = 0.8763$ and thus the corresponding 3-year zero-coupon bond price is 87.63. Again, from (2.1.3), we deduce that the 3-year rate is $r_0^3 = 0.045$. ■

It is important to note that there exists a spot rate curve for each type of bond or other fixed-income security. There is a spot rate curve for Treasuries, a spot rate curve for the corporate bonds of a company, a spot rate curve for the municipal bonds issued by a city, etc. It suffices to have a set of prices and maturities, and then we can deduce the underlying spot rates or discount factors, and thus the spot rate curve.

Figure 2.2 shows the yield curve for U.S. Treasuries as of December 31st, 2015. It is a common assumption, as in Figure 2.2, to use linear interpolation for maturities occurring in between two bond maturities.

Typical curves are increasing, but in the course of history, we have also observed decreasing (inverted) and hump-shape curves. Figure 2.3 illustrates the three typical shapes of interest rate term structure we usually observe.

2.1.3.2 Forward rates

A **forward rate** is an annual interest rate agreed upon today, at time 0, for a loan or an investment starting some time in the future, say at time T_1, and whose capital and interest are paid at a later maturity time, say at time $T_2 > T_1$. Mathematically, the $[T_1, T_2]$-forward rate is denoted by $f_0^{T_1,T_2}$. It is the forward rate applicable over the time interval $[T_1, T_2]$. It can be related to the T_1-year spot rate and the T_2-year spot rate, as follows: \$1 accumulated over $[0, T_2]$ at the T_2-year rate is equivalent to \$1 accumulated over the interval $[0, T_1]$ at the T_1-year rate and further accumulated over the interval $[T_1, T_2]$ at the $[T_1, T_2]$-forward rate. Mathematically, on an annual basis, we have

$$\left(1 + r_0^{T_2}\right)^{T_2} = \left(1 + r_0^{T_1}\right)^{T_1} \times \left(1 + f_0^{T_1,T_2}\right)^{T_2 - T_1}. \tag{2.1.4}$$

This is also illustrated in Figure 2.4.

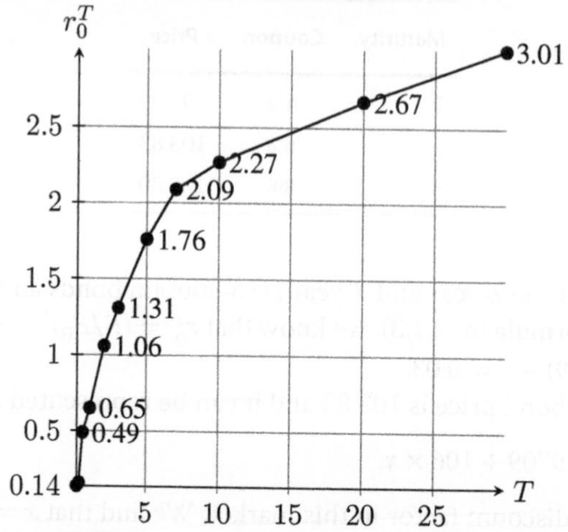

Figure 2.2 U.S. Treasury yield curve as of December 31st, 2015. Data from Federal Reserve Economic Database (FRED) (Constant Maturity Treasury). Rates are linearly interpolated between available maturities

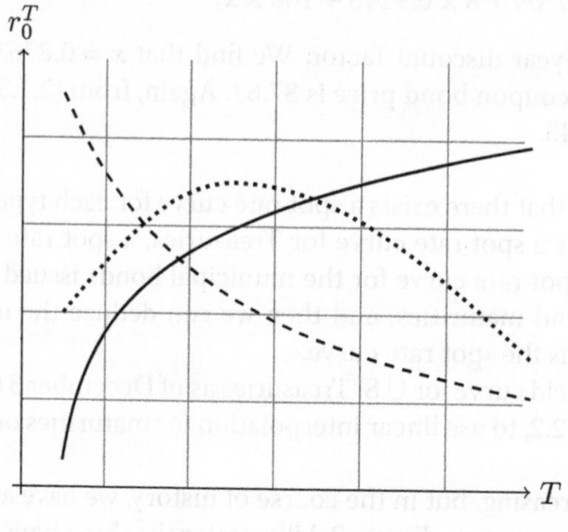

Figure 2.3 Different shapes of the term structure of interest rates

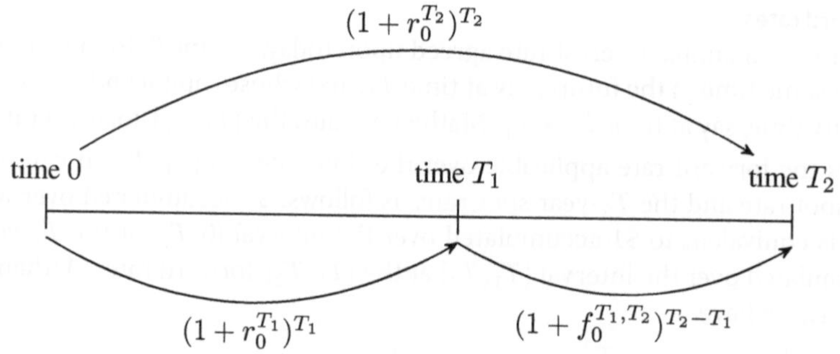

Figure 2.4 Relationship between spot and forward rates

Example 2.1.7 *Bullet GICs (continued)*
In the context of example 2.1.3, let us determine the forward rate applicable between year 5 and year 10. Using the relationship in (2.1.4), we need to solve for $f_0^{5,10}$ in

$$1.025^{10} = (1.015)^5 \times \left(1 + f_0^{5,10}\right)^5$$

and we immediately find $f_0^{5,10} = 3.51\%$. ■

Example 2.1.8 *Zero-coupon bonds (continued)*
In the context of example 2.1.4, let us determine the forward rate applicable between year 1 and year 3. Using (2.1.4), we need to solve for $f_0^{1,3}$ in

$$0.96 = 0.99 \times \frac{1}{\left(1 + f_0^{1,3}\right)^2}$$

and we immediately find $f_0^{1,3} = 1.55\%$. ■

⚠ It is very important to understand that the forward rate $f_0^{T_1,T_2}$ is not equivalent to the $(T_2 - T_1)$-year spot rate that will prevail at time T_1. This is simply because $f_0^{T_1,T_2}$ is determined today, using information available today, while the spot rate $r_{T_1}^{T_2}$ is the rate that will apply in the future, at time T_1, for a loan starting and contracted at time T_1 and maturing at time $T_1 + T_2$. With the information available today, at time 0, the rate $r_{T_1}^{T_2}$ is still unknown, so it is a random variable, while $f_0^{T_1,T_2}$ is a known value.

Determinants of the term structure

For a given maturity, why is the interest rate 3% rather than 1% or 10%? Why is the term structure increasing or decreasing? The first question addresses the general level of interest rates, whereas the second question is concerned with the general shape of the term structure.

Interest in a broad sense is a compensation to convince an investor to spend later rather than spending now. Such compensations, often known as *risk premiums*, need to take into consideration various aspects:

- Real interest rate: lending prevents you from generating wealth yourself. The real rate of interest compensates for this loss.
- Inflation: spending later exposes you to inflation, which is the general increase of the price of goods.
- Credit risk: lending also means bearing the risk of the borrower's insolvency.
- Liquidity risk: by lending, you assume the risk that you might be unable to convert your investment into cash at a later date.

Explaining the shape of the yield curve is more complicated as it requires economic theories about how consumers and investors allocate money over time. Theories explaining increasing and decreasing term structures are: *expectations theory* and *segmentation theory*.

2.2 Stocks

From the point of view of a corporation, issuing a corporate bond is equivalent to having a loan with several lenders, i.e. the bondholders. An alternative to raise money and finance growth of a company is to sell shares of the equity of the firm (see equation (1.1.1)). This is known as **(capital) stock**.

Stocks entitle the investor, known in this case as a **shareholder**, to a fraction of ownership of the company. And, as an owner, the shareholder participates in periodic meetings and votes on important issues regarding the operations of the company, such as election of Board members.

Historically, life insurance companies and pension plans have invested in stocks to achieve higher returns over the long term.

Stock indices

A stock index is the value of an *artificial* portfolio of stocks, chosen for example to represent a sector, within the larger financial market. It is a subset of the entire spectrum of stocks available. Indices are maintained by financial information companies or large institutional investors.

The most common examples in the U.S. are the Dow Jones Industrial Average (DJIA) and the S&P 500 index. For example, the DJIA 30 index is composed of 30 of the most important industrial firms. As of August 2017, companies such as Apple, Coca-Cola, GE, IBM, Nike and Wal-Mart were part of this index.

The companies that constitute an index can evolve over time. In 2008, following the financial crisis, AIG and General Motors were withdrawn from the DJIA 30. Only GE has been in the DJIA since its inception in 1896.

Stock prices vary widely over time depending on investors' perception of the corporation's future growth. Throughout this book, we will represent by S_t the price of a share of stock at time t, where time $t = 0$ represents *today*, while any other time t, for $t > 0$, is *in the future*. The evolution of the price of a given stock will be represented by $S = \{S_t, t \geq 0\}$.

We will not perform any *stock price analyses* or try to find their values as we did above for bonds. Instead, we will use various stock price models, such as the *binomial model*, the *trinomial model* and the *Black-Scholes-Merton model*, to represent the behavior of stock prices. In other words, $S = \{S_t, t \geq 0\}$ will be a stochastic process whose dynamics is generated by the chosen model.

2.2.1 Dividends

The profits of a corporation are usually shared (or divided) periodically among shareholders in the form of regular payments. These payments are known as **dividends**. Dividends are paid as:

- a fixed *amount*, for example \$0.50 per quarter;
- a fixed *percentage* of the stock price, e.g. 1% of the stock price.

In the latter case, 1% is known as the **dividend yield**. Not all shareholders are entitled to dividends in the same way; it depends on the type of shares held by the investor.

The return on a stock has two components:

- capital gain or loss, which is the variation in the value of an investment over some time horizon $[0, T]$, i.e. $S_T - S_0$;
- dividends.

Future capital gain and dividends form the basis of many *stock valuation models* used by financial analysts to determine whether a stock should be bought or sold (e.g. Gordon & Shapiro's method, also known as the dividend discount model). As mentioned above, this is not a topic we will consider any further.

On the day a dividend is paid, the price of the stock goes down by the amount of the dividend. This is simply because the cash used to pay the dividends is an outflow that directly reduces the value of the firm. However, from the point of view of the stockholder, there is no effect: the price drop is exactly compensated by the dividend received.

In the markets, the dividend payment date is known as the **ex-dividend date** whereas the stock price immediately after the dividend payment is known as the **ex-dividend price**. Thus, if a dividend D is paid at time t, which becomes the ex-dividend date, then the ex-dividend price is

$$S_t = S_{t-} - D$$

whether D is fixed or proportional to the stock price. Here, the notation $t-$ means an *instant*, think of a millisecond, before time t.

Example 2.2.1 *Impact of dividends for a shareholder*
On November 18th a share of ABC Inc. was trading at \$45 at the close of the stock market. The next day, on November 19th, the exchange opened at 9:30 AM and the share of ABC Inc. then traded at \$45.50. A dividend of \$1 was scheduled to be paid at 10:00 AM and the last quoted price before the payment of the dividend was \$45.25.

Mathematically, let t represent November 19th at 10:00 AM precisely, whereas $t-$ can be one millisecond before 10:00 AM. Therefore, November 19th is the ex-dividend date for that dividend payment, since a dividend was paid on that day. The stock price before the dividend payment is $S_{t-} = 45.25$, and with a dividend of $D = 1$, the ex-dividend price $S_t = 45.25 - 1 = 44.25$.

Assume now that, at the end of the day, the stock price closed at \$44.75. The variation in the stock price from November 18th to November 19th (close to close) is $44.75 - 45 = -0.25$. This is a negative capital gain, or a loss, of 25 cents. However, for the stockholder that was entitled to the dividend, the total return was composed of the capital loss and the dividend, and therefore the profit was $44.75 - 45 + 1 = 0.75$. ■

2.2.2 Reinvesting dividends

It is also possible to further increase the potential return by reinvesting dividends in the stock itself or in other assets. In this book, we will consider only two cases:

1) Dividends are paid discretely and reinvested at the risk-free rate.
2) Dividends are paid as a fixed proportion of the stock price (fixed dividend yield) and constantly reinvested in the stock.

Those are instances where reinvestment yields easy and closed-form solutions, for example in the pricing of forwards and options (see Chapters 3 and 5).

At time 0, buying one share of stock for S_0 may entitle the holder to a stream of dividends between time 0 and time T.

First, let us begin by assuming that dividends are paid discretely and let D_T denote the accumulated (at the risk-free rate) time-T value of all dividends paid during the period $[0, T]$. At the end of the period, i.e. at time T, the investor will hold a value of

$$S_T + D_T.$$

Second, let us assume instead that dividends are constantly reinvested in the stock, i.e. whenever a dividend is received, we buy additional shares of the stock. Let γ be the annualized dividend yield and suppose that every $1/n$-th year dividends are paid. In other words, dividends are paid at time $1/n$, $2/n$, etc.

At time $1/n$, the stock will be worth $S_{1/n}$, which is an unknown quantity at time 0. Said differently, $S_{1/n}$ is a random variable. At time $1/n$, the dividend will be worth $\gamma/n \times S_{1/n}$. Therefore, the total value of the investment will be $(1+\gamma/n) \times S_{1/n}$, including the share and the dividend payment, which is equivalent to owning $1+\gamma/n$ shares of the stock. So, we buy γ/n extra shares.

At time $2/n$, the dividend per share will be $S_{2/n} \times \gamma/n$. Since we hold $1+\gamma/n$ shares, the total value of the investment, i.e. capital and the dividend, will be

$$\underbrace{(1+\gamma/n)}_{\#\text{shares}} \times \underbrace{S_{2/n}}_{\text{stock price}} + \underbrace{(1+\gamma/n)}_{\#\text{shares}} \times \underbrace{S_{2/n} \times \gamma/n}_{\text{dividend}} = S_{2/n} \times (1+\gamma/n)^2.$$

We buy $(1+\gamma/n) \times \gamma/n$ extra shares with the total dividend payment, which means we now hold

$$(1+\gamma/n) + (1+\gamma/n) \times \gamma/n = 1 + 2\gamma/n + (\gamma/n)^2 = (1+\gamma/n)^2$$

shares.

At time $3/n$, the process is similar and we get

$$\underbrace{(1+\gamma/n)^2}_{\#\text{shares}} \times \underbrace{S_{3/n}}_{\text{stock price}} + \underbrace{(1+\gamma/n)^2}_{\#\text{shares}} \times \underbrace{S_{3/n} \times \gamma/n}_{\text{dividend}} = S_{3/n} \times (1+\gamma/n)^3.$$

We buy more shares and then the investment is equivalent to owning $(1+\gamma/n)^3$ shares.

Of course, we can continue the analysis for other future dividend payments. We can easily see the *compounding effect*: reinvesting dividends has the same effect as increasing the number of shares invested in the stock. Therefore, after 1 year with n such dividend payments, the value of the investment will go from S_0 to

$$S_1 \times (1+\gamma/n)^n,$$

which is a random value.

If dividends were paid continuously, which is equivalent to taking the limit when $n \to \infty$, then we would get a (random) value of

$$S_1 e^{\gamma}$$

at time 1, i.e. after 1 year. Therefore, over a time horizon $[0, T]$, a share of stock worth S_0 at time 0 yields a value of $S_T e^{\gamma T}$ at time T.

2.3 Derivatives

There is another family of financial securities widely used by institutional investors which are known as (financial) derivatives. **Derivatives** are financial instruments whose value is derived from either:

1) the (history of) prices of another security, known as the **underlying asset**;
2) a contingent event.

The objective of this section is to give an overview of the most common derivatives.

2.3.1 Types of derivatives

Here are the most common examples of derivatives depending on the value of another security or financial variable:

- **Futures contract** and **forward contract**: a contract that *engages* one party to buy (and the other to sell) a stock, a commodity or a currency some time in the future for a price determined today. More on futures contracts and forward contracts in Chapter 3.
- **Swap**: an agreement to exchange specific cash flows during a fixed period of time. The most popular swaps are:
  - interest rate swaps: to exchange fixed interest rate for floating, or vice versa;
  - currency swaps: to exchange fixed/floating interest on a loan in a foreign currency or vice versa;
  - credit default swaps: to cede credit risk exposure for a fixed periodic premium;
  - commodity swap: to fix today the price paid for delivery of the commodity on several dates.

  More on swaps in Chapter 4.
- **Option**: a contract that provides the *opportunity* for one party to buy or sell (to the other party), for example a stock or a commodity, some time in the future for a price determined today. More on options in Chapter 5.
- **Exotic option**: an option whose payoff is a more complex function of the underlying asset price. Includes Asian, barrier and lookback options. More on exotic options in Chapter 7.

Event-triggered derivatives are securities whose value depends on the occurrence of specific events:

- Weather derivative: a contract whose payoff depends upon weather variables such as temperature, precipitation or wind, observed at a fixed location during a specified interval of time.
- Catastrophe derivative: a contract whose payoff depends upon the occurrence and strength of a given natural catastrophe observed at a fixed location and time.
- Longevity derivative: a contract whose payoff depends upon the number of survivors in a cohort of individuals.

More on event-triggered derivatives in Chapter 7.

Following Section 1.2.3, we can say that traditional insurance (e.g. homeowner's insurance, life insurance) is a sort of event-triggered derivative sold on the insurance market. Some insurance policies and pension plans now include death and living benefits that are tied to the returns of a reference portfolio of assets (e.g. universal life insurance, variable annuities). They have similarities with both types of derivatives and a detailed treatment of *equity-linked insurance or annuity* is provided in Chapter 8.

2.3.2 Uses of derivatives

A derivative holder will win or lose depending on the direction taken by (the price of) the underlying asset or the occurrence or not of the contingent event. Whenever this gain is used to counterbalance the exposure to a risk or the loss associated with a liability, we say the derivative is used for risk management, for asset and liability management or for hedging purposes. In the opposite case, i.e. whenever the gain (or loss) does not offset a loss (or gain), we say the derivative is held for speculative reasons (a *bet* in certain situations).

For example:

1) In 6 months, your company will have to buy 1 ton of aluminum whose price fluctuates a lot. You buy call options on aluminum to fix today the maximum price your company will have to pay in 6 months. These options are used here for risk management.
2) Whenever interest rates go down, the value of actuarial liabilities goes up. An insurance company could enter into an interest rate swap that profits when interest rates decrease. In this case, the swap is used for hedging purposes because the profit from the swap offsets the loss resulting from an increase in the actuarial liabilities.
3) You believe the stock price will go up in 3 months. You buy stock options to benefit from such an increase. The option is then used for speculation.
4) A hurricane derivative pays for losses in excess of $1 billion in a specific region if a hurricane does occur. As an insurer selling hurricane coverage in your homeowner's policies, buying this hurricane derivative would be a way to manage the hurricane risk.
5) A Wall Street investor seeks to diversify its asset portfolio and sells hurricane derivatives: the investor does it for speculative reasons.

In our context, speculation should not have a negative meaning. It is viewed in opposition to hedging. There are many other reasons to buy or sell derivatives, including:

- Arbitrage: derivatives can be used to profit from price inconsistencies in financial markets. More on this in Section 2.5.
- Accounting and regulations: accounting and regulatory bodies recognize the use of derivatives to hedge risk exposures. Accounting bodies allow to reduce the value of liabilities that are hedged with derivatives whereas regulators can provide capital relief in similar situations.

Insurance companies and pension plans generally use derivatives for ALM purposes. They are mostly used to offset the random fluctuations of their actuarial liabilities. Use of derivatives by insurance companies is documented regularly by the National Association of Insurance Commissioners (NAIC) in the U.S.[1] Key results from its latest poll indicate:

- 95% of amounts invested in derivatives is for ALM.
- About 95% of the insurance industry exposure to derivatives comes from life insurers.
- About 67% of derivatives are used to manage interest rate risk whereas 25% of amounts are used for equity (stock) risk management.
- Swaps and options are the most popular instruments, with 49% and 45% of assets invested respectively.

Large institutional investors such as investment banks, insurance companies and pension plans have very complex assets and liabilities to deal with. The decision to hedge or not involves additional factors that managers need to weigh in their decision. When hedging is done appropriately, it has the advantage of improving the solvency of the company. This is a primary motivation for hedging as an insolvent or riskier company loses the confidence of its customers, shareholders, etc. In some cases, hedging can also be required by law and enforced by regulators.

1 NAIC and Center for Insurance Policy and Research, Capital Markets Special Report 2015.

2.4 Structure of financial markets

This section gives an overview of how and where securities such as stocks, bonds and derivatives are traded in the financial market. We will avoid most *market specifics* and focus on the general ideas needed for the upcoming chapters.

2.4.1 Overview of markets

Securities, such as stocks, bonds and derivatives, are launched usually on a **primary market**. Then, these securities can change hands and trade between thousands of investors on the **secondary market**. The secondary market, or what is usually known as *the* financial market, is of two types: exchange-traded markets and over-the-counter (OTC) markets.

In an **exchange-traded market**, or just an **exchange**, securities are standardized and trades need to follow a set of regulations. This is to protect investors, on both sides of the transaction, to make sure obligations binding in the contract are respected. An exchange can be a physical location, such as a building, or it can be virtual, through an electronic communication network. Stocks, options, futures, commodities are usually traded on such markets.

The most known stock exchanges are the New York Stock Exchange (NYSE), the NASDAQ (National Association of Securities Dealers Automated Quotations), London Stock Exchange (LSE), Toronto Stock Exchange (TSX), etc. Commodities, futures and options are mostly traded on the Chicago Board of Trade (CBOT), Chicago Mercantile Exchange (CME), New York Mercantile Exchange (NYMEX), London Metal Exchange (LME), Montréal Exchange (MX).

An **over-the-counter market** is usually set up by a network of dealers. A company that wants to enter into a contract may contact a broker or intermediary whose primary concern will be to find another party to complete the agreement. The contract between both parties can be entirely customized to fit the needs of both sides. Forwards and swaps are *traded OTC*. Although not customizable, bonds are considered being traded OTC as well.

For popular contracts, such as forwards and swaps, the International Swaps and Derivatives Association (ISDA) oversees transactions. Because OTC markets lack regulations of exchanges, contracts have less protection in the case of one party failing to meet its obligations. Prices are also rarely published or maintained, making an OTC market much less transparent. Table 2.2 shows the size of OTC markets as of December 2014.

Finally, the **foreign exchange (forex or FX) market** is where currencies are traded. Currencies are not traded on a formal physical exchange (in a building) but rather through a network of dealers and financial institutions. It is the largest financial market, with over $5 trillion traded daily. Currency derivatives are also traded in this market.

2.4.2 Trading and financial positions

For a trade to take place, there must be a buyer and a seller, who both agree on a price. The **bid price** of an asset is the highest price that a buyer is willing to pay whereas the **ask price** or **offer price** is the smallest price at which the seller is willing to sell that same asset.

The simplest example is probably with currencies. For example, if you are a U.S. citizen and if you are traveling to Europe, you might want *to buy* a few Euros beforehand. In this case, the security is the Euro and typically the price to pay to obtain Euros will be different than that *to sell* Euros. For example, on a given day, maybe you will have to pay 1.21 U.S. dollars (ask price) to get one Euro, while you will receive only 1.17 U.S. dollars (bid price) in exchange for one Euro.

Table 2.2 Amounts outstanding of OTC derivatives by risk category in billions of $US as of the end of the year 2014

Instrument	Amounts
Foreign exchange contracts	**75 879**
Forwards and forex swaps	37 076
Currency swaps	24 204
Options	14 600
Interest rate contracts	**505 454**
Forward rate agreements	80 836
Interest rate swaps	381 028
Options	43 591
Equity-linked contracts	**7 940**
Forwards and swaps	2 495
Options	5 445
Commodity contracts	**1 868**
Gold	300
Other commodities	1 568
Forwards and swaps	1 053
Options	515
Credit default swaps	**16 399**
Single-name instruments	9 041
Multi-name instruments	7 358
Index products	6 747
Unallocated	**22 609**
Total contracts	**630 149**

Source: Bank for International Settlements, Derivatives Statistics, Table 19.

Here is another example.

Example 2.4.1 *Bid and ask prices*
On December 1st, a dealer displays the price of a share of ABC Inc. as follows: Bid: 52.25. Ask: 52.75. On that day, if you buy 100 shares of the stock of ABC Inc. from this dealer, you will have to pay $52.75 per share, for a total of $5275.

Suppose that, as of December 15th, the price of the same stock has dropped as follows: Bid: 49.75. Ask: 50.50. If you sell 100 shares to this dealer, then you will receive $49.75 per share, for a total of $4975 on the sale.

It is interesting to notice that if we buy and immediately sell the 100 shares on December 1st, we lose 50 cents per share, i.e. $50. A similar transaction would have cost $75 on December 15th. This loss on an immediate buy and sell operation corresponds to the profit made by the dealer. ■

The **bid-ask spread** is the difference between the bid and ask prices. This profit compensates the broker for finding a buyer or a seller at the other end of the transaction, or the dealer for entering the other side of the trade herself. The more frequently a stock trades, the smaller the bid-ask spread, because it is easier to close the transaction (thus lowering the required compensation). Therefore, the bid-ask spread is also an indication of how liquid a security is, i.e. how easily we can sell that security and convert it into cash.

Unless stated otherwise, when we talk about the price of a security, we implicitly assume that delivery occurs immediately (in a matter of milliseconds through computer systems). This price is also known as the **spot price**. But whenever the price of an asset is agreed upon today and to be delivered only later in the future, then this price is known as the **forward price** of this asset; it is also called the **futures price** in some markets. These are similar to spot and forward interest rates, as seen in Section 2.1.3. We will come back to forward and futures prices in Chapter 3.

The quantity of a security owned or owed is known as a **financial position**. Whenever an investor owns (owes) a security, we say that she has a **long** (**short**) position in this instrument. Normally, when you sell securities, you have to own them first, unless you perform a short sale. A **short sale** is the act of selling a security that you do not yet own. The idea is similar to borrowing a stock (with a fee) that you immediately sell to make a profit on the sale. The short seller also agrees to *give back* the security at a predetermined date.

Example 2.4.2 *Long and short positions*
On December 1st, Mr Brown believes the price of a share of ABC Inc. will increase whereas Mrs McFly thinks it will decrease. They respectively enter into a long position and a short position, for one share of this stock. The dealer charges a 0.5% fee for short selling a stock. The prices are the same as in example 2.4.1.

Mr Brown buys the stock at \$52.75 on December 1st and sells the stock at \$49.75 on December 15th, for a net loss of \$3. Mrs McFly short sells the stock at \$52.25 and pays a fee of \$0.26. On December 15th, she closes her short position by buying a stock at \$50.50. The net profit is $52.25 - 0.26 - 50.50 = \$1.49$. Therefore, Mrs McFly has profited from the transaction since the stock price dropped in the two-week period. ■

To assure the short seller will be able to meet its obligations, i.e. buying back the security, an exchange normally requires a cash deposit in an account known as a **margin**. The margin earns interest, reducing the net cost of the short sale.

2.4.3 Market frictions

In the microstructure of financial markets, in addition to the bid-ask spread, there are other constraints or limits to trading, generally known as **market frictions**. Examples include:

- transaction costs: fees required to enter into a trade, like a commission. Include fixed fees per trade and variable fees that depend on the size of the transaction (volume). Those are generally excluded from the bid-ask spread;
- taxes: capital gains, dividends and other investment income are often taxable at the state/province and country level;
- volume of trading: minimum or maximum transaction size, only blocks of securities can be traded (e.g. 100 shares per block), etc.;
- short selling: often regulated, it requires the maintenance of a margin and additional fees.

Despite these market frictions, the financial market is relatively frictionless for large institutional investors. Transaction costs generally decrease with trading volume so larger institutions have very low *relative* transaction costs. The bid-ask spread is generally much smaller for large investors than for small investors. The constraints on the volume of trading have a negligible impact for institutional investors as they trade billions of dollars every day.

As we have seen for bonds, investors are also subject to credit risk, also known in general as **default risk** or **counterparty risk**. Default occurs when the other side of the transaction (the counterparty) cannot meet its obligations. Credit risk is negligible for many institutional investors.

As a result, let us define a **frictionless market** as a theoretical market whose microstructure is such that there is/are:

- no bid-ask spread, i.e. there is only one price, for a given security, at a given time;
- no transaction costs, no taxes, etc.;
- no constraints on the number of securities we can buy or sell (fractional quantities are allowed);
- no limits on borrowing and short-selling, which in some sense is assuming agents have infinite wealth;
- no credit or default risk.

Even if default risk is not a market friction, it is usually included in the above definition of a frictionless market.

In summary, in a frictionless market, we can buy or sell any amount of any security, for a single price, without constraint and at no extra cost. We can also borrow and lend at the risk-free rate as there is no credit risk. The frictionless market assumption holds approximately true for large investors, such as investment banks and other institutional investors like insurance companies and pension plans, mutual funds, etc.

2.5 Mispricing and arbitrage opportunities

Just like markets for common goods, such as food, cars, smartphones, etc., the insurance and financial markets are subject to the forces of supply and demand. But because financial securities and insurance policies are often easily ***substitutable***, or at least more liquid than most goods, financial and insurance markets should also be free of any mispriced securities. Let us look at various situations where we might believe, rightfully or not, that a good or a security is mispriced or that securities are inconsistently priced with respect to each other.

Example 2.5.1 *Mispriced cars*
You want to buy a new car.

- If, for exactly the same make and model, two dealers advertise different prices, then prices are inconsistent.
- A car from brand A has different features than car B but they sell for identical prices. You *prefer* features of car A and thus believe it is a better deal. Cars are not easily substitutable because it is impossible to find two identical cars from different automakers (this is true for most consumer goods). Thus, what is mispriced for you might not be for someone else. ■

Example 2.5.2 *Mispriced stock*
One share of ABC inc. stock trades for \$50. You have made a thorough analysis of ABC inc.'s fundamentals and truly believe the stock is underpriced. Again, what is mispricing to you might not be the case to another investor. ■

Example 2.5.3 *Inconsistently priced insurance policies*
You shop for homeowner's insurance and call two similarly trustworthy companies to get a quote. For exactly the same protection, you find that the premium is different. Prices appear to be inconsistent in this insurance market. ■

Example 2.5.4 *Inconsistently priced derivatives*
Let us look at a fictitious weather derivative market where \$10 will be paid, in a week, if the maximum temperature attained over the week is over x degrees (denoted by $M \geq x$).

Seller	Condition	Payoff	Selling price
A	$M \geq 90$	10	5.00
B	$M \geq 90$	10	4.50
C	$M \geq 95$	10	4.50

These derivatives, sold by the three investment banks, are very similar but there are some inconsistencies between their prices. Companies A and B have issued exactly the same product for a different price. Moreover, for \$4.50 you can buy two similar derivatives, from companies B and C, but the threshold temperatures are different (90 or 95 degrees). Company C's derivative is less likely to pay and is thus overpriced compared with Company B's derivative. ■

2.5.1 Taking advantage of price inconsistencies

Rational investors and well-informed customers will examine the information available and will attempt to exploit pricing inconsistencies as much as they possibly can and are allowed to. However, the extent to which a rational investor can exploit mispricing depends on the substitutability of the goods involved and on market frictions.

Example 2.5.5 *Exploiting mispricing of new cars*
We continue example 2.5.1 by analyzing the first situation, i.e. two identical cars are sold at different prices by two dealers. It is well known that a large depreciation applies if a new car is immediately resold on the market or to a car dealer. Hence, buying and immediately reselling the car entails a large loss. Therefore, exploiting mispricing in this market is close to impossible and the best you can do is buy the car from the dealer advertising it at the lowest price. ■

As we have seen in Chapter 1, because of the insurable interest, the insurance market operates only one way, i.e. insurers sell to customers. Clearly, the insurance market is not frictionless, which is a major impediment to exploiting inconsistently priced insurance policies.

Example 2.5.6 *Exploiting price inconsistencies related to insurance policies*
We continue example 2.5.3. We know it is not possible for an individual to sell an overpriced insurance policy. Therefore, if a rational customer believes both contracts are easily substitutable (she believes one company or the other is as trustworthy), then she should choose the company offering the cheapest premium for the same protection. ■

Apparent mispricing in insurance markets appear because:

- each company's experience (data) is different and therefore actuaries value the same or similar risks differently;
- some companies have explicit preferences for some market segments (young/old, risk-averse people, etc.) and charge premiums accordingly;
- the contracts are in fact slightly different (additional minor clauses that cannot be found in other companies).

2.5.2 Arbitrage opportunities and no-arbitrage pricing

In financial markets, price inconsistencies between two securities, as opposed to mispricing of one security, lead to *arbitrage opportunities*, a concept of paramount importance when studying actuarial and financial derivatives, and leading to the *no-arbitrage pricing principle*.

The idea is pretty simple: there is an **arbitrage opportunity** if we can take a financial position bearing no risk and generating only non-negative cash flows. In other words, there is no cash outflow and there is even a *chance* to win some money. This means that in some scenarios, we will break even, while in other scenarios, we will make a profit.

Example 2.5.7 *Exploiting mispriced stock*
We concluded Example 2.5.2 by saying the stock was not necessarily mispriced. Indeed, each investor has its own assessment of the future profitability of ABC inc. and therefore can come up with different conclusions as to how the stock is priced on the market. If you believe you are right, the best you can do is buy the stock and hope your expectations will realize. It is still a risky position and therefore clearly not an arbitrage opportunity: in some scenarios, you will make a profit, and in others, you will not. ■

(!) Price inconsistencies between two securities do not refer to flawed financial analyses, i.e. assessing that a stock is under/over-priced or that the interest rates are not what we think they should be, but rather arise from an *anomaly* between the price of two (or more) financial assets.

Consequently, to construct arbitrage opportunities, we need first to find pricing inconsistencies between two or more financial products. Then, the simplest strategy is to buy the under-priced security and sell the over-priced one.

Example 2.5.8 *Arbitrage opportunities in a weather derivatives market*
Let us continue Example 2.5.4. Assume the market is frictionless, which means in particular that any investor can buy or sell the derivatives at the quoted prices.

Derivative	Condition	Payoff	Bid/Ask prices
A	$M \geq 90$	10	5.00
B	$M \geq 90$	10	4.50
C	$M \geq 95$	10	4.50

We identified earlier two price inconsistencies that we can exploit to create arbitrage opportunities.

1) It is probably easier to exploit the inconsistency between derivative A and derivative B: it suffices to short sell derivative A and buy derivative B. You lock in a profit

of 50 cents at inception. It remains to make sure the investor cannot lose from this trade.

A week from now, if the temperature is below 90 degrees, then you owe nothing on the short derivative and receive nothing from derivative B. If the max temperature is over 90 degrees, then the payment you receive from B covers the payment you have to make to A. In both scenarios, your (net) cash flows are equal to zero.

In conclusion, you have locked in a profit of 50 cents at inception whereas you owe or receive absolutely nothing a week later. This is an arbitrage opportunity, i.e. a profit at no risk.

2) Derivative B seems like a *better deal* than derivative C because the scenarios in which derivative C pays are included in those for which derivative B pays. In other words, there seems to be a price inconsistency. Therefore, you will create an arbitrage if you sell derivative C and buy derivative B. The cost at inception, for taking this position, is zero.

   If the maximum temperature is below 90 degrees, then you owe zero with derivative C and you receive zero from derivative B. If the maximum temperature is between 90 and 95 degrees, then you receive 10 from derivative B and owe zero on derivative C. Finally, if the maximum temperature is 95 degrees and over, then you owe 10 with derivative C and will receive 10 from derivative B, for a net cash flow of zero.

   Overall, for an initial cost of zero, you will receive 10 if the maximum temperature is between 90 and 95 degrees and nothing otherwise. This is an arbitrage opportunity: we cannot lose money and there is a *chance* to win some. ■

Price inconsistencies can also arise when trying to replicate the cash flows of one security using other assets available in the market, as shown in the next example.

Example 2.5.9 *Arbitrage between (risk-free) bonds*
In a simple market, three bonds are traded, each with a face value of $100:

- a 1-year zero-coupon bond currently trades at $90;
- a 2-year zero-coupon bond currently trades at $81;
- a 2-year coupon bond with annual coupons of 7% currently trades at $95.

It might not be clear at first, but there is a price inconsistency to benefit from by setting up an arbitrage strategy in this market. First, recall that a coupon bond can be replicated with a portfolio of zero-coupon bonds. Thus, we will check whether the price of the coupon bond reflects this.

The cash flows of the coupon bond are: $7 at time 1 and $107 at time 2. If we ignore market frictions, to exactly replicate the cash flows of the coupon bond, we need 0.07 unit of the 1-year zero-coupon bond and 1.07 unit of the 2-year zero-coupon bond. The total cost is

$$0.07 \times 90 + 1.07 \times 81 = 92.97.$$

To be *consistent* with the two zero-coupon bonds, that coupon bond should trade for $92.97. At $95, it is over-priced. There is an arbitrage opportunity.

Indeed, as we are in a frictionless market, we can sell the coupon bond and receive $95, and then buy the replicating portfolio of zero-coupon bonds, i.e. 0.07 unit of the 1-year zero-coupon bond and 1.07 unit of the 2-year zero-coupon bond, for a cost of $92.97. We lock in a profit of \$2.03 at time 0. In 1 year and 2 years from now, the cash flows will match exactly. The short coupon bond outflows (−7, −107), at time 1 and time 2 respectively, are matching both zero-coupon bonds' inflows (+7, +107).

This strategy is an arbitrage: positive cash flow at inception, offsetting cash flows at time 1 and time 2. ■

In the previous example, we can expect that more than one investor would exploit this inconsistency, so that bond prices would reach eventually their *relative consistent values*. In general, price inconsistencies in the financial markets tend to disappear very quickly.

In practice, there might be impediments to exploit price inconsistencies. Depending on the size of the institutional investor, transaction costs and limits on short sales may limit the possibilities to create arbitrage opportunities.

In the financial markets, there are firms whose main purpose is to look for price inconsistencies and exploit them. They are known as *arbitrageurs*. Nowadays, with the very high number of traders in the markets and sophisticated computing infrastructure that conclude trades within milliseconds, arbitrage opportunities are very rare but might exist in a very small time frame.

This leads us to the basis of one fundamental pricing principle, known as the *no-arbitrage pricing principle*. The **no-arbitrage price** or the **no-arbitrage value** of a security is a consistent price relative to the prices of other already available securities. It will also be called the **fair price** of this security.

If a security can be replicated using a portfolio made of (some of) the other securities, then the no-arbitrage price is simply the cost for setting up the replicating portfolio. In the previous example, the no-arbitrage price of the coupon bond is \$92.97.

Arbitrage in practice and the story of Long-Term Capital Management

Companies also have a somewhat looser definition of an arbitrage opportunity and are willing to bear a small risk of loss. Assessing the loss probability is very arbitrary and it is easy to neglect some risks, as was the case for Long-Term Capital Management (LTCM).

LTCM was a hedge fund founded by John Meriwether whose board of directors included Myron S. Scholes and Robert C. Merton, famously known for the Black-Scholes-Merton model (see Chapter 16). One of their investment strategies involved exploiting arbitrage opportunities in the fixed-income markets and LTCM was very successful for some time, from its foundation in 1994.

But in 1998, when Russia defaulted on its debt, LTCM's main investment strategy failed and the hedge fund was bailed out by a \$4 billion plan supervised by the Federal Reserve. LTCM was forced to close down in 2000 in the aftermath of the bailout plan.

2.6 Summary

Bond characteristics

- Loan made by investors to an organization and traded in the financial market.
- Interest schedule: regular coupons (coupon bond) or interest accumulated until maturity (zero-coupon bond).
- Maturity: couple of months to several years. Their long maturities make them interesting to insurance companies and pension plans to match long-duration liabilities.
- Issuer: corporations, cities, public utilities, governments (and central banks), etc.

Bond pricing

- Fixed coupons c paid at times $t_1, t_2, \ldots, t_n$ where $T = t_n$ is the maturity.
- Principal or face value of F.
- Yield to maturity of y.
- Initial bond price is

$$B_0 = \sum_{i=1}^{n} \frac{c}{(1+y)^{t_i}} + \frac{F}{(1+y)^T}.$$

Term structure of interest rates

- Spot rate r_0^T: rate agreed upon today for a loan/investment starting today (time 0) and maturing at time T.
- Forward rate $f_0^{T_1,T_2}$: rate agreed upon today for a loan/investment starting at time T_1 and maturing at time T_2.
- Link between spot and forward rates:

$$\left(1+r_0^{T_2}\right)^{T_2} = \left(1+r_0^{T_1}\right)^{T_1} \times \left(1+f_0^{T_1,T_2}\right)^{T_2-T_1}.$$

Stocks

- Share of ownership of a corporation; may provide voting rights and/or dividends.
- Dividends: profits shared between shareholders.
- Ex-dividend date: date of dividend payment.
- Ex-dividend price: stock price immediately after the dividend payment.
- Reinvesting dividends:
  - Fixed discrete dividends reinvested at the risk-free rate. Buying one share at time 0 yields $S_T + D_T$ at time T.
  - Fixed proportional dividends (dividend yield) constantly reinvested in the stock. Buying one share at time 0 yields $S_T(1+\gamma/n)^{nT}$ or $S_T e^{\gamma T}$ at time T.

Types of derivatives

- Value is derived from an underlying asset.
- Value is derived from the occurrence or not of a contingent event.

Main uses of derivatives

- Risk management, hedging, ALM: gain (loss) from derivative position is used to offset loss (gain) from a risk exposure or actuarial liability.
- Investment, speculation: gain (loss) from derivative position is used to increase profits, at the risk of suffering important losses.

Financial market

- Exchange-traded markets: securities are standardized and trades are regulated. Mostly for stocks, options, futures and commodities.
- Over-the-counter markets: customized products and trades but much less transparent market. Mostly for forwards and swaps (and also bonds).
- Foreign exchange market.

Trading and financial positions

- Bid price: highest price a buyer is willing to pay.
- Ask price: lowest price at which a seller is willing to sell.

- Bid-ask spread: difference between bid and ask prices.
- Long position: quantity/value owned/bought in a security.
- Short position: quantity/value owed/sold in a security.
- Short sale: selling a security we do not yet own.

Frictionless market

- No bid-ask spread, i.e. there is only one price at any given time.
- No transaction costs, no taxes, etc.
- No constraints on the number of securities we can buy or sell (fractional quantities are allowed).
- No limits on borrowing and short-selling.
- No credit or default risk.

Pricing inconsistencies and arbitrage opportunities

- Pricing inconsistencies: incoherence between the prices of two or more goods, financial securities, insurance policies, etc.
- Taking advantage of pricing inconsistencies: depends on substitutability, constraints on buying and selling, market regulations, etc.
- Financial securities and insurance policies are highly substitutable.
- Financial market: easy to buy and sell securities.
- Insurance market: one-way market, insurers sell policies to customers; customers can only choose company with lowest premium.
- Arbitrage opportunity:
  - Situation where we cannot lose money (no risk of loss) and there is a chance to win some.
  - No injection of capital and non-negative final cash flows.
  - Comes from a pricing inconsistency, not a mispricing or flawed financial analysis.
- No-arbitrage price of a security: price which is consistent with the prices of other securities on the market.

2.7 Exercises

2.1 The following table shows current zero-coupon bond prices.

Maturity (term)	Price
1 year	99.01
2 years	97.07
3 years	93.54
4 years	88.85
5 years	82.19

(a) Calculate the T-year spot rates for $T = 1, 2, \ldots, 5$.
(b) Draw the spot rate curve in this market. What is the shape of the yield curve?
(c) What is the forward rate applicable between time 2 and time 4?
(d) One year later, you observe the following zero-coupon bond prices.

Maturity (term)	Price
1 year	97.56
2 years	95.18
3 years	92.86
4 years	90.60
5 years	88.39

What happened to the shape of the yield curve?

2.2 As of today, the yield curve is such that:

- 6-month rate: 2%
- 1-year rate: 2.25%
- 1.5-year rate: 2.4%
- 2-year rate: 2.5%

Assume that all rates are compounded semi-annually.

(a) What is the 1-year forward rate applicable in the second year (compounded semi-annually)?

(b) What is the current price of a 2-year coupon bond paying \$1.25 semi-annually having a face value of \$100?

2.3 You are given the following annual coupon bonds:

Maturity	Coupon	Price
1	3%	101.980
2	4%	103.922
3	2%	98.620
4	5%	107.329

Compute the entire term structure of interest rates (spot rates).

2.4 You are given the following zero-coupon bond prices (same as question 2.1).

Maturity (term)	Price
1 year	99.01
2 years	97.07
3 years	93.54
4 years	88.85
5 years	82.19

Your company owes \$5 million, to be paid in 20 months from now.

(a) Using linear interpolation on spot rates, compute the present value of the latter debt (assume interest is compounded annually).

(b) Using linear interpolation on bond prices, compute the present value of the latter debt (assume interest is compounded annually).

2.5 Borrower A has a 25% chance of defaulting on a 5-year loan whereas borrower B has a 5% chance of going bankrupt within that same time frame. If a borrower defaults, then interest is lost and only 50% of capital is recovered. Otherwise, capital and interest are paid at maturity. If the 5-year risk-free rate is 2%, find the compensation for credit risk that the lender should add to the risk-free rate such that on average, the lender has the same amount as investing at the risk-free rate. Do that calculation for both borrowers.

2.6 A share of stock currently trades for \$52 and assume the dividend yield is 2% paid continuously. If the stock price 1 year from now is \$55, calculate the accumulated value over 1 year of \$5200 invested in that stock when dividends are continuously reinvested in the stock.

2.7 On May 1st, you observe the following stock price from a dealer:
Bid: \$23.52 Ask \$24.17
Suppose that on May 31st, the dealer advertises the following stock prices:
Bid: \$21.78 Ask \$22.41

(a) You buy 100 shares of stock from that dealer on May 1st that you sell on May 31st to that dealer. Calculate your profit or loss.

(b) You short-sell 100 shares of stock to that dealer on May 1st that you buy back on May 31st from that dealer. A margin of 50% of your investment is required on which you earn 2% interest per annum compounded monthly. Calculate your profit or loss.

2.8 Suppose that the market in example 2.1.4 is frictionless.

(a) What is the forward rate determined today for a loan beginning at $t = 3$ and ending at $t = 5$?

(b) Determine how to trade today in the bonds available in the market to exactly lock in the return found in (a) over the time interval $[3, 5]$. Your initial investment at $t = 0$ should be 0.

2.9 You are given the following zero-coupon bond prices (same as question 2.1):

Maturity (term)	Price
1 year	99.01
2 years	97.07
3 years	93.54

(a) Find the no-arbitrage price of a 3-year coupon bond with annual coupons of 1.5%.

(b) Suppose a 3-year coupon bond with annual coupons of 1.5% currently trades for par. Exploit this arbitrage opportunity and explain each step.

(c) In (b), how should you adjust quantities bought or sold to have a zero cost at time 0? What are the future cash flows then?

2.10 In example 2.5.9, suppose instead the coupon bond trades at \$92.50 and there is a transaction cost of 0.50% on securities bought and sold from a given market maker. Would you attempt to exploit this arbitrage opportunity? Explain your answer.

2.11 In a frictionless market, two assets are available:

- a risk-free (zero-coupon) bond with initial value of 100 that will be worth 105 a year from now;
- a stock with initial value of 100 that will be worth 105 or 110 a year from now.

Determine whether there is an arbitrage opportunity in this market and if this is the case, explain in detail how you can exploit such an opportunity.

3

Forwards and futures

In many situations, it is good risk management to fix today the price at which an asset will be purchased in the future. Here are two examples:

- Consider a company regularly buying gold as an input for its business, e.g. that company could transform gold to make jewelry. This company is clearly exposed to the (random) fluctuations of the gold price. More specifically, it would suffer from an increase in the price. In this case, it may be a good idea to enter into a financial agreement that fixes the price of gold today for delivery in the future. Therefore, the cost of jewelry made over the next few weeks/months can be known in advance.
- An insurance company based in the U.S. but also doing business in Canada is exposed to changes in the currency rate between the U.S. dollar (USD) and the Canadian dollar (CAD). It could enter into a financial agreement to fix today the exchange rate between these currencies that will apply in the future.

Forward contracts and futures contracts can be used specifically for this purpose: fix today the price of a good to be bought in the future. They are also commonly known as *forwards* and *futures*, without the word contract or agreement attached to them.

The role of this chapter is to provide an introduction to forwards and futures. The specific objectives are to:

- recognize situations where forward contracts and futures contracts can be used to manage risks;
- understand the difference between a forward contract and a futures contract;
- replicate the cash flows of forward contracts written on stocks or on foreign currencies;
- calculate the forward price of stocks and of foreign currencies;
- calculate the margin balance on long and short positions of futures contracts.

3.1 Framework

This section lays the foundations of forwards and futures for the rest of the chapter. The content applies regardless of the underlying asset.

3.1.1 Terminology

A **forward contract** or a **futures contract** is a contract that engages one party to buy (and the other to sell) an asset some time in the future for a price determined today at inception of the

Actuarial Finance: Derivatives, Quantitative Models and Risk Management, First Edition.
Mathieu Boudreault and Jean-François Renaud.

contract. That time in the future is known as the expiration, **maturity** or **delivery date** whereas the said price is known as the **delivery price**.

The most important difference between forwards and futures is that futures are standardized and exchange-traded. Exchanges require investors to hold money aside to protect both sides of the transaction from default risk. This is usually known as the process of *marking-to-market*. This difference aside, forwards and futures serve the same purpose: fixing today the price at which an asset will be bought in the future.

Forward contracts and futures contracts also differ from *spot contracts*:

- A spot contract is agreed upon today between a buyer and a seller, the asset is paid for and delivered (almost) immediately.
- A forward (futures) contract is also agreed upon today but the asset will be paid for and delivered at maturity of the forward (futures) contract.

Therefore, if you need to acquire the asset, say in 3 months, then instead of waiting 3 months and buying the asset on the spot market and paying a price viewed as random today, the forward contract fixes that price today.

To simplify the presentation of this chapter, we will start by looking at forward contracts and we will come back to futures contracts in Section 3.5.

3.1.2 Notation

Let us begin with the following standard notation:

- K is the delivery price;
- T is the maturity date.

We say that delivery of the underlying asset will occur at time T or that the forward contract will *mature at time T*.

Entering the long position of a forward contract, or said differently *buying a forward contract*, does not usually require an initial payment/premium. However, to avoid arbitrage opportunities, this restriction will have an impact on the *right value* of K. For now, we will consider that K can take any value and that entering a forward contract might require, or not, an upfront payment. We will come back to this issue later when we discuss the *forward price*.

The cash flows of a forward contract are:

- At inception (at time 0), both parties agree on K (and on T) and an up-front premium might be required (paid by the forward buyer (long position) to the forward seller (short position)), or vice versa.
- At maturity (at time T), the investor with the long position pays K and receives the underlying asset worth S_T from the investor with the short position, no matter what the realization of the random variable S_T is. This is illustrated in Figure 3.1.

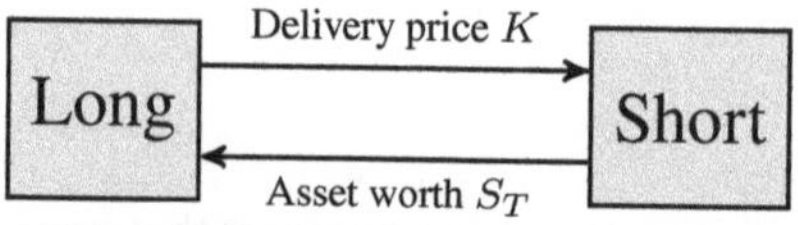

Figure 3.1 Cash flows of a forward contract at maturity

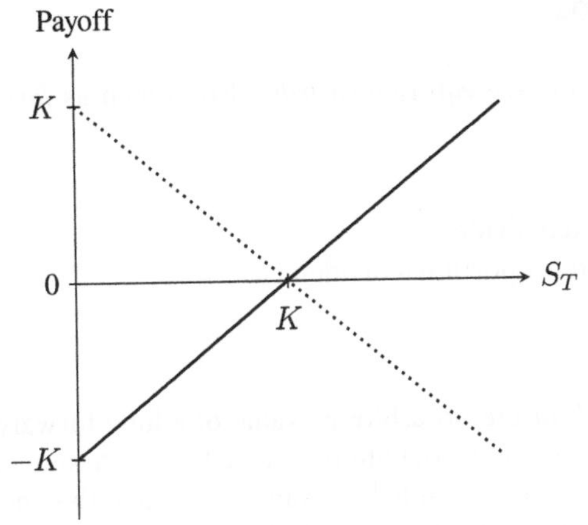

Figure 3.2 Payoff of a long forward (continuous line) and a short forward (dotted line)

3.1.3 Payoff

In general, the payoff of a derivative corresponds to its value at maturity. Usually, it is the net amount received by the investor with the long position. For a forward contract, the payoff can be positive (gain), negative (loss) or zero. Mathematically, the payoff of a **long forward** is given by

$$S_T - K \tag{3.1.1}$$

which is positive if $S_T > K$, but is negative otherwise. Therefore, the payoff of a **short forward** is given by

$$-(S_T - K) = K - S_T.$$

The payoff of a long or a short position in a forward contract is illustrated in Figure 3.2 as a function of the terminal value of the underlying asset, i.e. as a function of S_T.

(!) No matter the perspective (either long or short position), the payoff of a forward contract is *random*. Indeed, as of today, it is impossible to determine whether it is beneficial for either party to fix today the buying or selling price of the asset.

Example 3.1.1 *Payoff of a forward on a stock*
A stock currently trades at \$65. You intend to buy this stock 3 months from now and you would like to fix the buying price today. Therefore, you choose to enter the long position in a forward contract with a delivery price of \$67 and a maturity date in 3 months. Assume it costs nothing at inception to enter this forward contract.

Let us determine the gain/loss of the long forward if the stock price reaches \$70 or if it remains at \$65 in 3 months.

At time 0, there are no cash flows. At maturity, in the first scenario, you have to buy an asset worth \$70 for only \$67. Therefore, the gain is $70 - 67 = 3$. In the second scenario, you have to buy an asset worth \$65 for \$67, which is a gain of $65 - 67 = -2$. In other words, in this second scenario, it is a loss of \$2. ■

Just like a long (resp. short) position in an asset, the value of a long (resp. short) position in a forward contract written on this same asset increases (resp. decreases) whenever the price of the underlying asset increases (resp. decreases).

3.2 Equity forwards

In this section, we will analyze *equity forwards*, also known as *forwards on stocks*. We will consider stocks:

1) not paying dividends;
2) paying discrete and fixed dividends;
3) paying continuous and proportional dividends.

3.2.1 Pricing

Our first objective is to find the no-arbitrage value of a long forward contract with delivery date T and delivery price K. Let us denote this value by V_0^T. Assume a bond or bank account earning $r > 0$ (continuously compounded) is available and that the underlying financial market is frictionless – this will be the case for the rest of the book.

Just as a coupon bond can be replicated by a portfolio of zero-coupon bonds (see Section 2.5.2), we will replicate (the payoff of) a forward contract but using only two assets: the underlying stock and a zero-coupon bond.

There are two components in the payoff of a long forward contract and each component requires a specific investment at time 0 to be replicated:

- At time T, we will have to pay K. This is equivalent to a debt of K maturing at time T. To replicate this component, it suffices to borrow, at time 0, the present value of K, i.e. borrow Ke^{-rT}. This is equivalent to selling a bond with face value K and maturity T.
- At time T, we will receive an asset worth S_T. To replicate this component, it suffices to buy, at time 0, the stock for a price of S_0.

In summary, we have obtained the following *static* replication strategy for a long forward contract: at time 0, we take a

- short position in a bond, i.e. we receive Ke^{-rT}, with maturity time T;
- long position in the stock, i.e. we buy one unit of the stock for S_0 and we hold on to it until time T.

These trades are summarized in the following table.

Position	time 0	time T
Long stock	S_0	S_T
Short bond or loan	$-Ke^{-rT}$	$-K$
TOTAL	$S_0 - Ke^{-rT}$	$S_T - K$

This strategy is also called a (long) **synthetic forward**. In other words, by buying a stock and borrowing at the risk-free rate (shorting a bond), we create a portfolio replicating the cash flows of a long forward contract. This means that in any possible scenario, i.e. for all possible values of the random variable S_T, the cash flows of the portfolio and the forward contract coincide. By the no-arbitrage principle, we can say that the time-0 value V_0^T of a long forward contract with delivery date T and delivery price of K must be equal to the cost of setting up this portfolio. As the initial cost of this portfolio is clearly equal to $S_0 - Ke^{-rT}$, we have obtained the following pricing formula:

$$V_0^T = S_0 - Ke^{-rT}. \tag{3.2.1}$$

This is the up-front premium required to enter the long position in such a forward contract.

Similarly, shorting the stock and lending at the risk-free rate, we can verify that the no-arbitrage time-0 value of the short forward contract is equal to

$$-V_0^T = Ke^{-rT} - S_0.$$

Example 3.2.1 *Value of a forward contract*
The stock of company ABC inc. currently trades at \$54 and you want to enter into a 3-month forward contract on that stock with a delivery price of \$54.25. If the annual risk-free interest rate is continuously compounded at 4%, determine the current value of the long forward contract.

We have $S_0 = 54$, $T = 0.25$, $K = 54.25$ and $r = 0.04$. Using equation (3.2.1), the time-0 value of the long position of this forward agreement is

$$V_0^T = 54 - 54.25e^{-0.04\times0.25} = 0.29.$$

Therefore, you would need to pay 29 cents at inception to enter the long position. ■

Example 3.2.2 *Arbitrage opportunities related to a forward agreement*
Suppose now that the forward contract of example 3.2.1 can be entered into at no cost. Let us describe the arbitrage opportunity and explain how to exploit this inconsistency.

We know that the initial value of the forward contract should be 0.29, so if it can be entered at no cost then it is underpriced for the buyer and there is an arbitrage opportunity. We can exploit this situation by taking the long position in the forward (at no cost) and by short-selling the synthetic forward. At time 0, this will create an excess of 29 cents, but at maturity all positions will offset each other.

Let us describe how this works. At time 0, you enter the long forward position at no cost, you short-sell the stock for \$54 and you invest $54.25e^{-0.04\times0.25} = 53.71$ at the risk-free rate. This yields an immediate positive cash flow of 0.29 which can be invested at the risk-free rate. At maturity time T, your long position in the forward contract means that you have to buy a share of stock for \$54.25, which can be done by using the risk-free investment now worth \$54.25, thanks to accumulation of interest. Then, you use this share of stock to close your initial short-sell.

In conclusion, at time T, all positions are covered (there is no possibility of loss) and you still have the accumulated value of \$0.29. ■

Using a similar methodology, the value of a forward contract can also be replicated at any time t, between inception and maturity. Suppose you enter the long position in a forward contract today (at time 0). Let us foresee ourselves in the future: time goes by and, at time $t > 0$, you want to determine the no-arbitrage value V_t^T of your position. Of course, as V_t^T is a value in the future, it is a random quantity.

It turns out that, as we did above, buying the stock (at time t) and taking a loan at the risk-free rate (at time t) will also replicate the cash flows of the long forward position at maturity:

Position	time t	time T
Long stock	S_t	S_T
Short bond or loan	$-Ke^{-r(T-t)}$	$-K$
TOTAL	$S_t - Ke^{-r(T-t)}$	$S_T - K$

Therefore, the value at time t of a long forward contract maturing at time T is simply

$$V_t^T = S_t - Ke^{-r(T-t)}, \tag{3.2.2}$$

which is, as expected, a random variable since S_t is itself a random variable.

The value of the forward contract thus depends on the spot price S_t of the underlying stock and the remaining time to maturity $T - t$. Depending on the value of the stock at time t, i.e. the realization of S_t, the long forward contract will have a positive or a negative value.

It is interesting to note that equation (3.2.2) is in fact valid for any time $t \in [0, T]$. Indeed, if we set $t = 0$, then we see that it is a generalization of the time-0 value obtained in equation (3.2.1). And if we set $t = T$, then we recover the payoff of a long forward as obtained in equation (3.1.1). Consequently, equation (3.2.2) encompasses the latter two equations.

Example 3.2.3 *Value of a forward*
Today, you enter the long position in a costless forward contract on a stock with delivery price \$25 and a maturity date in 3 months. If the risk-free rate is 3%, let us determine what will be the value of the long position 2 months from now.

Using equation (3.2.2), we get

$$V_{2/12}^{3/12} = S_{2/12} - Ke^{-r(3/12-2/12)} = S_{2/12} - 25e^{-0.03/12}.$$

Now, consider the following scenario: the stock price will be at \$26 in 2 months, i.e. $S_{2/12} = 26$. Then, the value of your position in the forward contract will be

$$V_{2/12}^{3/12} = 26 - 25e^{-0.03/12} = 1.06242194.$$

If instead we observe the scenario that the stock price will be at \$24 in 2 months, then the corresponding value of your position in the forward contract will be

$$V_{2/12}^{3/12} = 24 - 25e^{-0.03/12} = -0.93757806.$$

■

3.2.2 Forward price

It is a market convention that forward contracts should be costless at inception, i.e. no up-front premium should be required. Therefore, following equation (3.2.1), we must have

$$S_0 - Ke^{-rT} = 0,$$

so the only *fair* delivery price, i.e. not introducing an arbitrage opportunity, is $K = S_0e^{rT}$. This unique delivery price is known as the **forward price** of the underlying stock. As we can see, it depends on the maturity date considered.

In conclusion, the *T*-**forward price of a stock** is the only delivery price such that the corresponding forward contract requires no up-front premium at inception. It is usually denoted by F_0^T, so we have the following definition:

$$F_0^T = S_0 e^{rT}. \tag{3.2.3}$$

A few words of caution about the terminology:

- *Value* of a forward contract: the value of a forward contract represents the cost for replicating the payoff $S_T - K$. It also represents the amount received (or paid) if the position in the forward were to be liquidated (or offset).
- Forward *price*: it is the delivery price for which a forward contract can be entered at no cost.

Example 3.2.4 *Forward price*
A stock currently trades at $37 and the annual interest rate is 1%. Determine the 3-month forward price of this stock.

Using equation (3.2.3), the forward price of this stock is simply

$$F_0^{0.25} = 37e^{0.01\times 0.25} = 37.09.$$

Entering today a forward contract on this stock with a delivery price of $K = 37.09$ does not require an up-front payment. ■

Example 3.2.5 *Another arbitrage opportunity with a forward contract*
A stock currently trades at $84 and the interest rate is 1.25%. The 6-month forward price of this stock is quoted at $85. Let us construct an arbitrage strategy.

From equation (3.2.3), the forward price should be equal to

$$F_0^{0.5} = 84e^{0.0125\times 0.5} = 84.53.$$

In some sense, the forward price quoted in the market is *too high* since we can build a zero-premium portfolio, namely a synthetic long forward contract, which has a *better* payoff.

Indeed, at time 0, if you short the forward contract and if you borrow $84 to buy one share of the stock, then the overall initial cost for setting up this strategy is 0. Then, at maturity, you have to deliver the stock as part of your obligation in the forward agreement for $85. Also, you pay back the loan, i.e. you give back $84e^{0.0125\times 0.5} = 84.53$, leaving an extra $85 - 84.53 = 0.47$. This is an arbitrage opportunity.

The details are in the following table.

Position	time 0	time T
Short forward	0	$-(S_T - 85)$
Short bond or loan	−84	−84.53
Long stock	S_0	S_T
TOTAL	0	0.47

Clearly, this does not depend on the realization of the random variable S_T. ■

Finally, we can also define the *T*-**forward stock price at time** t, i.e. the forward price that will prevail at time t for the purchase and delivery of a share of stock at time T.

Using equation (3.2.2), which provides the time-t value of a long forward, this forward price should be

$$F_t^T = S_t e^{r(T-t)}, \tag{3.2.4}$$

for any $0 \leq t \leq T$. For validation purposes, if we set $t = 0$ in equation (3.2.4), then we see that it is a generalization of the T-forward stock price obtained in equation (3.2.3), and if t increases to T, then we see that F_t^T gets closer and closer to S_T. At the limit, we have $F_T^T = S_T$. We can interpret this last relationship as if we issue a forward maturing in a few seconds, then the forward price should be the price at which it is actually/about to be traded.

Note that in the notation F_t^T the parameter T is the maturity date of the forward contract, it is not the time to maturity (which is here $T - t$), except of course if $t = 0$ in which case the maturity and the time to maturity coincide.

Example 3.2.6 *Evolution of the forward price*
A 3-month forward contract is issued today. The current stock price is \$34 and the risk-free rate is 6%. Consider the following: the stock price increases to \$37 1 month later and decreases to \$35 1 month prior to maturity. Let us calculate the forward price after each month in this scenario.

We have $T = 0.25 = \frac{3}{12}$, $r = 0.06$ and $S_0 = 34$. Also, in the specified scenario, we have $S_{1/12} = 37$ and $S_{2/12} = 35$. We make use of equation (3.2.4) so that

$$F_0^{0.25} = S_0 e^{0.06\times(3/12-0)} = 34.5138442$$
$$F_{1/12}^{0.25} = S_{1/12} e^{0.06\times(3/12-1/12)} = 37.37185618$$
$$F_{2/12}^{0.25} = S_{2/12} e^{0.06\times(3/12-2/12)} = 35.17543823.$$

■

Forward price and expected spot price

We know from Chapter 2 that investing in a bond, i.e. lending money, entails some risk such as credit risk. Stocks also entail an important level of risk as the future capital gain on the stock is unknown.

For both bonds and stocks, investors require a risk premium. This compensation is generally paid as an extra return over what is known as the risk-free rate (such as the Treasury rate). In other words, investing in stocks should yield a higher return *on average* than investing in a risk-free bond.

Suppose you have \$$S_0$ to invest: you can either buy a share of stock or invest at the risk-free rate. We know that investing S_0 at the risk-free rate yields

$$S_0 e^{rT} = F_0^T \tag{3.2.5}$$

and, therefore, based upon the above risk-return argument, the expected spot price should be higher than the forward price.

In practice, it is difficult to assess the expected stock price from market participants because additional (modeling) assumptions would be needed. When we present the binomial tree and Black-Scholes-Merton models later in the book, we will take a second look at this argument and determine the implications on the stock market.

Determining the relationship between the expected spot price and its forward price is more difficult to address for assets other than stocks.

3.2.3 Discrete and fixed dividends

For a stock paying dividends, holding a share is different from receiving it later (as in the long position of a forward contract), as in the former case the shareholder is entitled to the stream of dividends. Consequently, dividends paid between inception and maturity of a forward contract will have an impact on the pricing of this forward contract and the forward price of the stock.

Now, let us assume that dividends paid by a given stock are fixed (known in advance), paid periodically and reinvested at the risk-free rate. Let us denote by D_0 (resp. D_T) the present (resp. future) value of all dividends to be paid on the stock between time 0 and time T. Under our assumptions, D_0 and D_T are deterministic quantities related to each other by the following equation:

$$D_0 = e^{-rT} D_T.$$

As in Section 3.2.1, the objective is to replicate $S_T - K$. To do so, at time 0, we now need to take the following positions:

- a long position in the stock;
- a loan with principal $Ke^{-rT} + D_0$.

Clearly, D_0 is the adjustment to the loan to account for future dividends paid between time 0 and time T.

The cash flows of this replicating portfolio, also known as a (long) synthetic forward, are:

Position	**time 0**	**time T**
Long stock	S_0	$S_T + D_T$
Short bond or loan	$-Ke^{-rT} - D_0$	$-K - D_T$
TOTAL	$S_0 - Ke^{-rT} - D_0$	$S_T - K$

We see that the loan is larger than in the no-dividend case as dividends are an income.

Since we have replicated the payoff of the forward contract, the initial value of this forward contract is given by

$$V_0^T = S_0 - D_0 - Ke^{-rT}.$$

The information about the size and the timing of the dividends is hidden in the value of D_0.

The T-forward price F_0^T of a stock paying fixed dividends is then given by the value K such that

$$S_0 - D_0 - Ke^{-rT} = 0.$$

We easily get that

$$F_0^T = e^{rT}(S_0 - D_0) = S_0 e^{rT} - D_T. \tag{3.2.6}$$

Example 3.2.7 *Forward price of a stock paying quarterly dividends*
A stock currently trades at \$38 and quarterly dividends of 25 cents are due 1 month and 4 months from now. Let us find the 3-month forward price of this stock if the risk-free rate is 2%.

We need to accumulate one dividend over 2 months so that $D_T = 0.25e^{0.02\times\frac{2}{12}} = 0.2508$. Using equation (3.2.6), the 3-month forward price is simply given by

$$F_0^{0.25} = 38e^{0.02\times\frac{3}{12}} - 0.2508 = 37.94.$$

■

Example 3.2.8 *Forward price of a stock paying semi-annual dividends*
A stock currently trades at \$67 and a semi-annual dividend of \$2 is due sometime before the maturity of a forward contract. If the forward contract matures in 4 months and the corresponding forward price is 65.6583, let us find when the dividend is due given that the risk-free rate is 3%.

Here, we have $D_T = 2e^{0.03\times\frac{m}{12}}$, where m is the number of months between the dividend payment and the maturity of the forward contract. Using equation (3.2.6), we have the following relationship:

$$65.6583 = 67e^{0.03\times\frac{4}{12}} - 2e^{0.03\times\frac{m}{12}}.$$

We find that $m = 3$. Hence the dividend is due in 1 month from now. ∎

3.2.4 Continuous and proportional dividends

Now, let us assume that dividends are paid continuously as a fraction of the value of the stock (fixed dividend yield) and continuously reinvested in the stock. Let us denote by γ the corresponding dividend yield/rate on that stock. In the case of a stock index, γ is interpreted as the average dividend yield on the stocks that compose the index.

Recall from Section 2.2.1 that buying one share for S_0 at time 0, and continuously reinvesting dividends, will accumulate to a value of $S_T e^{\gamma T}$ at time T. This means that, at time T, we will hold $e^{\gamma T}$ units of this stock.

Again, holding a forward on a stock is different from holding a stock as the latter is entitled to dividends. Therefore, replicating the payoff of a forward by purchasing a share of stock and borrowing the delivery price at the risk-free rate will yield more than $S_T - K$. Dividends require the replicating portfolio to be adjusted accordingly.

Hence, we take the following positions: at time 0,

- buy $e^{-\gamma T}$ shares of the stock;
- take a loan with principal Ke^{-rT}.

The cash flows of this replicating portfolio, also known as a (long) synthetic forward, are:

Position	time 0	time T
Long stock	$e^{-\gamma T}S_0$	S_T
Short bond or loan	$-Ke^{-rT}$	$-K$
TOTAL	$S_0e^{-\gamma T} - Ke^{-rT}$	$S_T - K$

Because dividends are reinvested in the stock and S_0 accumulates to $S_T e^{\gamma T}$, we simply need a lesser number of shares to replicate S_T. Indeed, purchasing $e^{-\gamma T}$ shares of stock and reinvesting dividends will accumulate to S_T, i.e. one share at maturity.

At maturity, we have indeed replicated the cash flows and we see that the value of the forward contract at inception is

$$V_0^T = S_0e^{-\gamma T} - Ke^{-rT}, \tag{3.2.7}$$

and thus the T-forward stock price is

$$F_0^T = S_0e^{(r-\gamma)T}. \tag{3.2.8}$$

Example 3.2.9 *Forward price of a stock with dividends continuously reinvested*
A stock currently trades at \$104 and dividends of 1% are continuously reinvested in the stock. If the interest rate is 2%, let us find the 3-month forward price of that stock.

Using equation (3.2.8), we find

$$F_0^{0.25} = 104e^{(0.02-0.01)\times 0.25} = 104.26.$$ ■

Example 3.2.10 *Forward price of a stock with dividends continuously reinvested*
A stock currently trades at \$45 and dividends of 100γ% are continuously reinvested in the stock. If the interest rate is 3% and the 6-month forward price of that stock is \$45.51, let us find the dividend yield γ.

Using equation (3.2.8), we have

$$F_0^{0.5} = 45e^{(0.03-\gamma)\times 0.5} = 45.51.$$

We find $\gamma = 0.75\%$. ■

3.3 Currency forwards

To exchange goods from one country to another, it is necessary to buy or sell currencies on the foreign exchange (forex) market, also known as the currency market. The price of one currency (e.g. one Euro) with respect to another currency (e.g. the U.S. dollar) is known in the market as the **foreign exchange rate** or **currency rate**.

Like many other assets, it is possible to fix today the price for the purchase (or sale) of a foreign currency sometime in the future. This contract is known as a **currency forward**. But before looking at currency forwards and the forward exchange rate, we need to provide some background on currencies.

3.3.1 Background

Without loss of generality, we will consider the U.S. dollar as the **domestic currency** and any other currency as a **foreign currency**. The following example illustrates how to buy and sell Euros on the spot market.

Example 3.3.1 *Converting U.S. dollars and Euros*
As of year-end, one USD could buy 0.915 Euros (EUR). This is also expressed as an exchange rate of 0.915 EUR *per* USD or 0.915 EUR/USD.[1] Let us determine the value in Euros of \$10,000 and, similarly, let us determine the value in U.S. dollars of 2000€.

We have

$$10000 \text{ USD} \times 0.915 \frac{\text{EUR}}{\text{USD}} = 9150 \text{ EUR}$$

i.e. 10000 USD can buy 9150 EUR. Notice how we can cancel out the USD units on the left-hand side of the equation. Borrowing the vocabulary from financial markets, we need 10000 USD *to buy* 9150 EUR.

1 On many websites, the exchange rate *units* are expressed differently. We have chosen this approach, which we think is more intuitive.

Conversely,

$$2000 \text{ EUR} \times \left(0.915 \frac{\text{EUR}}{\text{USD}}\right)^{-1} = 2000 \text{ EUR} \times 1.09289617 \frac{\text{USD}}{\text{EUR}}$$
$$= 2185.79 \text{ USD}.$$

We are now *selling* 2000 EUR for a price of 2185.79 USD. ■

As the (spot) exchange rate evolves randomly over time, a currency forward can be very useful for risk management purposes to fix the foreign exchange rate on future cash flows. This is shown in the next example.

Example 3.3.2 *Currency forwards for risk management*
Your company, based in the U.S., will receive 1 million CAD in 3 months and then will have to convert it into USD. A 3-month currency forward contract written on Canadian dollars is available for a delivery price of 0.77 U.S. dollar per Canadian dollar. Let us explain how your company can manage this currency risk.

There are (at least) two possibilities:

1) Do nothing and wait, i.e. convert 1 million CAD in 3 months when it is received.
2) Enter the long position of this currency forward contract to fix the price at 770,000 USD, for 1 million CAD, to be purchased in 3 months from now.

In the first case, the exchange rate we will observe in 3 months may as well be 0.70 USD/CAD or 0.80 USD/CAD, making 1 million Canadian dollars worth either 700,000 USD or 800,000 USD. Of course, it could be an even wider range. In the second case, the value of 1 CAD being locked at 0.77 USD, your company will receive 770,000 USD in exchange for 1 million Canadian dollars, no matter what the exchange rate is at that time. ■

3.3.2 Forward exchange rate

When we analyze currency forwards, we need to determine what it entails for an investor to buy one unit of a foreign currency. Just like cash earns interest at the (domestic) risk-free rate in a (domestic) bank account, holding a foreign currency should earn interest at the foreign risk-free rate in a foreign bank account. Moreover, the price of a foreign currency in U.S. dollars (exchange rate) is random and evolves over time. As a result, a unit of a foreign currency is a risky asset earning a continuous stream of income similar to continuous dividends.

Formally, let S_t be the time-t price (in U.S. dollars) for one unit of the foreign currency, say the Euro or the Canadian dollar. Therefore, S_t is the time-t (spot) exchange rate. As usual, the (domestic) risk-free rate is denoted by $r > 0$ whereas the foreign currency earns interest at a rate of $r_f > 0$, known as the foreign risk-free rate.

To determine the forward exchange rate, we need to first calculate the value of a currency forward with a fixed delivery price K, expressed in USD for one unit of the foreign currency. Viewing one unit of a foreign currency as a risky asset with a dividend yield of $\gamma = r_f$, then using the formula in (3.2.7) we deduce that the initial value of a currency forward is given by

$$V_0^T = S_0 e^{-r_f T} - K e^{-rT}.$$

Consequently, using the formula in (3.2.8), the T-**forward exchange rate** F_0^T, i.e. the forward price in USD for one unit of the foreign currency, is given by

$$F_0^T = S_0 e^{(r-r_f)T}. \tag{3.3.1}$$

It is the delivery price in a costless forward contract written on this foreign currency.

Example 3.3.3 *Arbitrage in foreign exchange markets*
The current USD/CAD (U.S. dollars per Canadian dollar) exchange rate is 0.70 and the 3-month forward exchange rate is 0.7012. Given that the U.S. Treasury 3-month rate is 0.1% and the Canadian Treasury 3-month rate is 0.25%, let us determine whether or not there exists an arbitrage opportunity in this situation.

As mentioned above, if we take the perspective of a U.S. company, we have $S_0 = 0.7$, $F_0^T = 0.7012$ whereas $r = 0.0010$ and $r_f = 0.0025$. We realize that

$$F_0^T = 0.7012 > 0.699737549 = 0.7 \times e^{(0.001-0.0025)\times 0.25} = S_0 \times e^{(r-\gamma)T}.$$

From equation (3.3.1), we know that it should be an equality. Therefore, there exists an arbitrage opportunity between USD and CAD: we should enter the short position in a forward contract on Canadian dollars and simultaneously buy a synthetic forward on Canadian dollars. ■

3.4 Commodity forwards

We mentioned earlier that holding a stock, as opposed to holding a forward contract on this stock, entitles the investor to a stream of dividend payments (if any). However, as many commodities are physical goods that need to be stored, such as gold, bushels of wheat, etc., owning such a *physical* commodity, as opposed to holding a forward on this commodity, often means that we have to pay for storage costs. Of course, these costs are avoided by the forward contract holder. In some sense, the storage costs act like negative dividend payments.

Consequently, valuation of commodity forwards must account for storage costs in a fashion similar to dividends on a stock, i.e.

$$F_0^T = S_0 e^{rT} + C_T = S_0 e^{(r+c)T},$$

where C_T represents the accumulated storage costs, which we assume is given by a proportional (to the commodity price) storage cost c.[2] This last equality should be reminiscent of formula (3.2.8).

Example 3.4.1 *Gold forward price*
An ounce of gold currently trades for \$1,200 and costs \$50 per month to store securely. Given that the risk-free rate is 4% and compounded continuously, let us compute the forward price (per ounce) of gold delivered 3 months from now.

The accumulated storage costs are

$$C_{3/12} = 50e^{2r/12} + 50e^{r/12} + 50 = 150.5014.$$

Therefore, the 3-month forward price is given by

$$F_0^{3/12} = 1200e^{3r/12} + C_{3/12} = 1362.5616.$$

■

More generally, the forward price of a commodity is defined in terms of what is known as the *cost of carry*. The **cost of carry** is simply $\xi = r + c - \gamma$, i.e. the level of interest rate necessary to finance the long position in the underlying asset (when replicating the forward contract), plus the storage costs (expressed as a percentage of the commodity value), minus any income earned

2 Despite the choice of notation, c has nothing to do here with coupon payments.

on the commodity (such as dividends) if any. Therefore, the forward price of a commodity can be written as $F_0^T = S_0 e^{\xi T}$.

In practice, valuation of commodity forwards is more complex as it depends on the type of commodity and whether it can be transformed. As a result, commodities are classified as being either an **investment asset**, i.e. held for investment reasons (such as gold), or a **consumption asset**, i.e. held for future transformation (such as oil). Forward contracts written on consumption assets have an additional return known as the **convenience yield** which compensates for the loss of additional income earned by transforming the asset.

3.5 Futures contracts

In a forward contract, we say that settlement takes place at maturity: an investor sells the underlying asset to her counterparty at a price known as the delivery price. Both parties are expected to fulfill their obligations, no matter what is the asset price at maturity.

Suppose both parties in a forward contract agree to exchange one share of a stock 3 months from now for a (delivery) price of \$100. If, 3 months later, the stock is worth \$50, then the long forward contract holder has to buy a stock worth \$50 for a price of \$100. This investor might be more inclined to back out from the agreement as the deal is very bad for her: in this scenario, she would prefer to buy the stock at its spot price of \$50 rather than at the delivery price of \$100.

Note that the short forward contract holder may be tempted to do the same if the stock price goes up to \$150 at maturity. Indeed, selling for \$100 an asset worth \$150 would generate an important loss.

As a result, we say that forward contracts entail **default risk**, i.e. the risk that one party fails to meet its contractual obligations. To reduce default risk on such transactions, one should trade *futures contracts* instead of forward contracts.

Essentially, a **futures contract** is a regulated and standardized version of a forward contract. It is traded on exchanges, i.e. exchange-traded markets, which require both parties to contribute to a **margin account**. This margin account is not managed by the investors. The purpose of the margin is to reduce default risk by periodically crediting gains and losses on the futures contract positions, a process known as **marking-to-market**; see the next subsection.

Main differences between forwards and futures

- Forward contracts are *customizable* and traded *over the counter* whereas futures contracts are *standardized* and *exchange-traded*.
- Futures contracts are marked to market using a margin.
- Futures contracts may have flexibility on the delivery dates rather than a single delivery date. For example, delivery could occur in October 2015 for a futures and on October 21st, 2015 for a similar forward.
- Most forward contracts are held until maturity so that delivery of the asset often occurs whereas most futures contracts are closed out before maturity and delivery rarely occurs.
- Just like stocks and other exchange-traded assets, futures contracts are subject to price limits, meaning that when the price drops by a large amount, transactions can be halted.

It is important to note that futures contracts are derivatives giving access to variations in the price of the underlying asset without actually holding this asset. Sometimes it is difficult, costly or even impossible to buy the underlying, for example if it is an index, while in other cases it is too expensive to handle/store, for example if it is a commodity such as aluminum, gold or oil. As a result, futures contracts are mostly used for risk management purposes, as well as for speculation.

3.5.1 Futures price

Like the forward price, today's **futures price** of an asset to be delivered at maturity time T is denoted by F_0^T. It is the delivery price at which a futures contract can be entered at no cost, for both the short and the long positions.

For example, if today is September 1st, 2015, if the underlying commodity is aluminum and if maturity is October 2015, then F_0^T refers to the futures price for *October 2015 aluminum*. Note that the maturity date is not a specific day of the month.

Like the forward price, the time-t **futures price** for delivery at time T is denoted by F_t^T. From today's point of view, i.e. at time 0, this is a value in the future so it is unknown. Again, it is the delivery price for the asset, which will prevail at time t, to enter a futures contract on this underlying asset at no cost.

3.5.2 Marking-to-market without interest

On the futures market, the futures price varies with time. On September 1st, 2015, the futures price for *October 2015 aluminum* was different from what it was 1 month earlier, i.e. on August 1st, 2015.

Let us fix an arbitrary underlying asset, which we do not specify explicitly, and let us fix a maturity date T. Also, for illustration purposes, let us fix two other dates t_1 and t_2 such that $0 < t_1 < t_2 < T$. In what follows, we take the point of view of a long position.

At time t_1, the futures price $F_{t_1}^T$ will be typically different from what it is today at time 0. Therefore, the *gain or loss* of a long futures position will be quoted at time t_1 as

$$F_{t_1}^T - F_0^T,$$

which is the change in the delivery price over that period of time. It is not a profit or loss that is immediately materialized.

In the scenario that, at time t_1, we observe a futures price such that $F_{t_1}^T > F_0^T$, then an investor who has entered at time 0 a long position in a T-futures contract will be in a *good position*. Indeed, for the same delivery date T, she will be allowed to buy the underlying asset for less than if she had waited until time t_1 to enter the long position in a futures contract. In other words, her position has now a strictly positive value, while she had entered in a position, with the same payoff, at no cost at time 0. However, if she decides to close her position, instead of waiting until maturity, she will leave with $F_{t_1}^T - F_0^T$; more details below.

In the scenario that, at time t_1, we observe a futures price such that $F_{t_1}^T < F_0^T$, the analysis is reversed.

At time t_2, for the time interval $[t_1, t_2]$, the *gain or loss* of a long T-futures will be quoted as

$$F_{t_2}^T - F_{t_1}^T.$$

And, finally, for the period from time t_2 to maturity time T, the *gain or loss* will be

$$S_T - F_{t_2}^T$$

since $F_T^T = S_T$.

Let us illustrate the marking-to-market process by assuming it takes place only at times t_1 and t_2 and at maturity:

- At time t_1, the futures price will be $F_{t_1}^T$. At that time, marking-to-market requires that the short position pays $F_{t_1}^T - F_0^T$ to the long position's margin account if this difference is positive. Or vice versa if the difference is negative.
- At time t_2, the futures price will be $F_{t_2}^T$. Again, at that time, if the difference $F_{t_2}^T - F_{t_1}^T$ is positive, this amount is debited from the short position's margin and then credited to the long position's margin. Or the other way around, if the difference is negative.
- At maturity, the asset price will be S_T. This time, the gain or loss over that last period is given by $S_T - F_{t_2}^T$. Again, this amount is debited and credited from the margins.

At the end of the process, the total payoff $S_T - F_0^T$ has been funded from deposits and withdrawals to and from each margin account: marking-to-market does not affect the final payoff of futures contract. In other words, holding a forward contract or a futures contract leads to the same payoff at maturity. This is easy to see from the following equality:

$$S_T - F_0^T = \underbrace{\left(F_{t_1}^T - F_0^T\right)}_{\text{Gain/Loss in the 1st period}} + \underbrace{\left(F_{t_2}^T - F_{t_1}^T\right)}_{\text{Gain/Loss in the 2nd period}} + \underbrace{\left(S_T - F_{t_2}^T\right)}_{\text{Gain/Loss in the last period}} .$$

It emphasizes that the gain or loss from the futures over the period $[0, T]$ can be decomposed as the sum of gains and losses over the sub-periods $[0, t_1]$, $[t_1, t_2]$ and $[t_2, T]$. Of course, we can consider more than just three periods. We would reach the same conclusion.

Instead of waiting until maturity to obtain the payoff $S_T - F_0^T$ as in a long forward contract, the marking-to-market process amortizes the changes in value on a periodic basis with margins held on both the long and the short positions. This is how it reduces default risk: an investor can default only by failing to deposit the required daily amount.

Example 3.5.1 *Marking-to-market*
Suppose two investors enter into a T-futures contract whose current futures price is \$100. Let us calculate the gains and losses for both the long and short positions, along with the margin balance, if the marking-to-market process is done on a monthly basis over the next 3 months.

In this case, $t_1 = 1/12$, $t_2 = 2/12$ and $T = 3/12$. Let us consider the following scenario for our computations: the futures price evolves to $F_{1/12}^{3/12} = 101$, $F_{2/12}^{3/12} = 99$ and $S_{3/12} = 98$.

		Long position		Short position	
t	Futures price	Gain/Loss	Margin	Gain/Loss	Margin
0	100		0		0
1/12	101	+1	1	−1	−1
2/12	99	−2	−1	+2	1
3/12	98	−1	−2	+1	2

At time t_1, the futures price has increased from 100 to 101. The investor having the long position has a gain of \$1 funded from the investor having the short position. At time

t_2, when the futures price drops by \$2, this amount is withdrawn from the long position's margin and deposited in the short futures holder's margin. At maturity, each investor's margin balance corresponds to $\pm(S_T - F_0^T)$, respectively. Overall, the futures price has decreased by \$2, since $S_T = 98$, which is a loss for the long position and a gain for the short position. ■

When a futures position is closed/cleared, the investor recovers its margin balance. More precisely, to close her long futures position prior to maturity, the investor needs to take immediately the opposite short futures position, for the same maturity.

3.5.3 Marking-to-market with interest

In practice, the margins earn interest at the risk-free rate so that deposits and withdrawals are effectively accumulated until maturity. Let us look at the impact of interest on the marking-to-market process, assuming again that it takes place at times t_1, t_2 and at maturity time T, and assuming also that interest is continuously compounded at an annual rate r:

- At time t_1, the futures price will be $F_{t_1}^T$. Marking-to-market requires that the short position pays $F_{t_1}^T - F_0^T$ to the long position's margin account if this difference is positive. The accumulated value of this deposit at maturity time T will then be

  $$\left(F_{t_1}^T - F_0^T\right) e^{r(T-t_1)}.$$

  If the difference is negative, then the operations are reversed, as before.
- At time t_2, the futures price will be $F_{t_2}^T$. Again, at that time, if the difference $F_{t_2}^T - F_{t_1}^T$ is positive, this amount is debited from the short position's margin and then credited to the long position's margin. Again, the accumulated value of this deposit at maturity time T will be

  $$\left(F_{t_2}^T - F_{t_1}^T\right) e^{r(T-t_2)}.$$

  Again, if the difference is negative, then the operations are reversed.
- At maturity, the gain or loss over the last period is $S_T - F_{t_2}^T$, it is debited and credited from the margins, but no accumulation of interest is possible as we are already at maturity.

The end balance of the long position's margin will always be different from $S_T - F_0^T$ because interest will *magnify* each periodic gain or loss. To see this, let us write down mathematically the margin balance at maturity:

$$\underbrace{\left(F_{t_1}^T - F_0^T\right)}_{\text{Gain/Loss in the 1st period}} \times \underbrace{e^{r(T-t_1)}}_{\text{Accumulated from } t_1 \text{ to } T} + \underbrace{\left(F_{t_2}^T - F_{t_1}^T\right)}_{\text{Gain/Loss in the 2nd period}} \times \underbrace{e^{r(T-t_2)}}_{\text{Accumulated from } t_2 \text{ to } T}$$
$$+ \underbrace{\left(S_T - F_{t_2}^T\right)}_{\text{Gain/Loss in the last period}}$$

Example 3.5.2 *Marking-to-market with interest*
Let us continue example 3.5.1 by considering interest on gains and losses. Assume for simplicity that \$1 accumulates to \$1.01 over a 1-month period. The margin balance at the end of each month is shown in the table.

		Long position		Short position	
t	Futures price	Gain/Loss	Margin	Gain/Loss	Margin
0	100		0		0
1/12	101	+1	1	−1	−1
2/12	99	−2	−0.99	+2	0.99
3/12	98	−1	−1.9999	+1	1.9999

Again, the periodic gain and loss is funded from each party's margin but now the margin earns interest. The end balance is obtained recursively as

$$\text{Balance at the end of the period} = \text{Balance at the beginning of the period} \times 1.01 + \text{Gain/Loss on the futures price}$$

or by accumulating each gain/loss with interest. In our case,

$$-1.9999 = 1 \times 1.01^2 - 2 \times 1.01 - 1.$$

Instead of a cash flow of $-(S_T - F_0^T) = 2$, the investor having the short position will gain 1.9999 instead. This is because the first period's loss was magnified by interest. ■

Example 3.5.3 *Marking-to-market with interest*
In the context of the preceding example, let us consider a different scenario. Suppose that the monthly futures prices are 100, 99, 102 and 101, respectively. Let us calculate the corresponding margin.

		Long position		Short position	
t	Futures price	Gain/Loss	Margin	Gain/Loss	Margin
0	100		0		0
1/12	99	−1	−1	+1	1
2/12	102	+3	1.99	−3	−1.99
3/12	101	−1	1.0099	+1	−1.0099

Rather than a profit of \$1 with a similar forward contract, the investor with a long position in this futures has a margin balance of \$1.0099 due to interest on the margin. Similarly, the investor with the short position will suffer a loss of 1.0099 rather than \$1. ■

Because the margin earns interest, the timing of gains and losses will have an unpredictable effect on the margin balance at maturity. It will be close but usually different from a margin not crediting interests. Of course, this difference will be amplified by the size of the position.

3.5.4 Equivalence of the futures and forward prices

We have just shown that if the margin earns interest, holding a futures contract or a forward contract will yield a different payoff at maturity (in most situations). How does this marking-to-market process affect the value of the futures price compared to the forward price? In fact, it has absolutely no impact.

Let us see why this is the case. Suppose you implement the following strategy:

- At inception, you enter the long position in e^{rt_1} units of a futures contract written on some asset and maturing at time T. Of course, this is done at no cost.
- At time t_1, the futures price will be $F_{t_1}^T$. Due to the marking-to-market process, you will receive an amount of

$$\underbrace{e^{rt_1}}_{\text{Qty}} \times \underbrace{\left(F_{t_1}^T - F_0^T\right)}_{\text{Gain/Loss in the 1st period}} .$$

Once we accumulate this until maturity, we simply get

$$e^{rT}\left(F_{t_1}^T - F_0^T\right) = \underbrace{e^{rt_1}}_{\text{Qty}} \times \underbrace{\left(F_{t_1}^T - F_0^T\right)}_{\text{Gain/Loss in the 1st period}} \times \underbrace{e^{r(T-t_1)}}_{\text{Accumulated from } t_1 \text{ to } T} .$$

Now we need to enter into additional T-futures contracts so that we hold a total of e^{rt_2} units instead: we enter, at no cost, into $e^{rt_2} - e^{rt_1}$ new T-futures.
- At time t_2, the futures price will be $F_{t_2}^T$. Due to the marking-to-market process, you will receive

$$\underbrace{e^{rt_2}}_{\text{Qty}} \times \underbrace{\left(F_{t_2}^T - F_{t_1}^T\right)}_{\text{Gain/Loss in the 2nd period}} .$$

Once we accumulate this until maturity, we simply get

$$e^{rT}\left(F_{t_2}^T - F_{t_1}^T\right) = \underbrace{e^{rt_2}}_{\text{Qty}} \times \underbrace{\left(F_{t_2}^T - F_{t_1}^T\right)}_{\text{Gain/Loss in the 2nd period}} \times \underbrace{e^{r(T-t_2)}}_{\text{Accumulated from } t_2 \text{ to } T} .$$

Now we need to enter into additional T-futures contracts so that we hold a total of e^{rT} units instead. Again, we can do this at no cost.
- At maturity, the asset price is S_T and due to marking-to-market, you will receive

$$\underbrace{e^{rT}}_{\text{Qty}} \times \underbrace{\left(S_T - F_{t_2}^T\right)}_{\text{Gain/Loss in the last period}} .$$

This does not need to be accumulated.

The final cash flow of this zero-cost strategy is simply

$$e^{rT}\left(F_{t_1}^T - F_0^T\right) + e^{rT}\left(F_{t_2}^T - F_{t_1}^T\right) + e^{rT}\left(S_T - F_{t_2}^T\right) = e^{rT}\left(S_T - F_0^T\right).$$

Combining this strategy with an investment of F_0^T at the risk-free rate will yield at maturity the following payoff:

$$e^{rT}\left(S_T - F_0^T\right) + e^{rT}F_0^T$$

where the first term is the strategy's payoff and the second term is the accumulated value from your investment at the risk-free rate. Consequently, the net payoff is

$$e^{rT}S_T = e^{rT}\left(S_T - F_0^T\right) + e^{rT}F_0^T,$$

for an initial investment of F_0^T.

Here is another way of generating the payoff $e^{rT}S_T$. For now, suppose that the forward price, with delivery at time T, is given by G_0^T. Let us implement the following *static* investment strategy:

- At inception, you invest an amount of G_0^T at the risk-free rate and enter the long position in e^{rT} units of a T-forward contract. So, your initial investment is G_0^T.
- There are no actions required between time 0 and time T.
- At maturity, you receive $e^{rT}G_0^T$ from your investment at the risk-free rate. Once combined with the forward contracts, the net payoff is $e^{rT}S_T$.

These two strategies have exactly the same payoff at maturity, in all possible scenarios, and both have net cash flows of zero between inception and maturity. To avoid arbitrage opportunities, they must have the same initial price. In other words, $G_0^T = F_0^T$. This means that the futures price and the forward price are equal.

Relationship between the forward price and the futures price

We have just demonstrated that the marking-to-market process does not affect the futures price when interest rates are non-random. In reality, interest rates are random and asset prices are themselves affected by the term structure of interest rates.

As a result, it is not easy to determine a precise relationship between the forward price and the futures price as it depends on the relationship between interest rates and the spot market. When that relationship is positive, i.e. when the interest rate tends to increase when the underlying asset increases as well (or vice versa), the long futures holder will make gains and these gains will be invested at a greater interest rate. Similarly, if the stock price drops, the interest rate will also drop and losses will be financed at a smaller interest rate. This makes futures generally more attractive than an otherwise equivalent forward contract.

Other factors may also affect futures prices differently than forward prices such as taxes, transaction costs, etc. It is generally well accepted that for small values of the maturity time T (small time to maturity), futures and forward prices can be considered equivalent. This is not necessarily the case for larger values of T (larger time to maturity).

3.5.5 Marking-to-market in practice

In the last subsections, we have purposely oversimplified the mechanics of margins to illustrate how marking-to-market works. There are, however, additional regulations in the futures market on how margins work to ensure obligations of the futures are met. As a result, investors having long or short positions in a futures both need to deposit an amount in a margin account when entering the contract. This is known as the **initial margin**.

Whenever the margin balance is too low, investors might need to deposit additional funds to make sure the margin remains over a minimum level. That minimum level is known as the **maintenance margin requirement** and the additional deposit required is known as a **margin call**.

Finally, marking-to-market is usually performed on a daily basis and, as stated previously, margin accounts earn interest. Therefore, the margin balance at the end of any period varies with:

1) gains/losses resulting from changes in the futures price;
2) margin calls;
3) interest.

Let us illustrate how these additional constraints affect the evolution of margins.

Example 3.5.4 *Margin for a long futures contract*
You are given the following information about the margin requirements set forth by a specific futures exchange:

- Initial margin: \$5 per contract.
- Maintenance margin: 70% of the initial margin.
- Margin calls are deposited at the beginning of the period.
- Marking-to-market is performed on a monthly basis.
- Interest credited is 12% compounded monthly (1% per month).

Now, consider the following scenario of futures prices for delivery in 5 months: at the end of each month, the futures price will be 101, 99, 98, 97 and 99.

The current futures price is $F_0^5 = 100$.

You enter the long position in 1,000 futures contracts. Let us calculate the evolution of the margin balance in this specific scenario.

As per the regulations of the exchange, the initial margin is \$5 per contract, so at inception \$5000 must be deposited in the margin. If the margin goes below level $5000 \times 0.7 = 3500$, then it will trigger a margin call to bring the margin back up to its initial level of \$5000. The table shows the evolution of the margin account in this scenario.

t	F_t^5	Initial balance	Margin call	Interest	Gain/Loss	End balance
0	100					5000
1	101	5000	0	50	1000	6050
2	99	6050	0	60.5	−2000	4110.5
3	98	4110.5	0	41.105	−1000	3151.605
4	97	3151.605	1848.395	50	−1000	4050
5	99	4050	0	40.5	2000	6090.5

Again, the mechanics of calculating the margin balance is similar to what we did before. For example, at the end of the first period, the balance is computed as follows:

$$\underbrace{5000}_{\text{initial margin}} + \underbrace{5000 \times 0.01}_{\text{acc. of interest}} + \underbrace{1000 \times (101 - 100)}_{\text{no. of units} \times \text{gain}} = 6050.$$

After three periods, the margin balance is below the margin requirement of \$3500. Therefore an additional deposit, i.e. a margin call, of $5000 - 3151.605 = 1848.395$ is immediately required to bring the margin balance back to \$5000. Interest will be computed on \$5000 during that next period, yielding an interest deposit of 50 at the end of the period.

The end balance of the margin is 6090.50, keeping in mind that you have deposited a total of 5000 + 1848.395 = 6848.395 in the margin. If you had instead entered into a long forward, you would have lost \$1 per contract, i.e. \$1000, instead of 6848.395 − 6090.50 = 757.895. Therefore, interests credited on the margin have absorbed part of the loss in this scenario. ■

Example 3.5.5 *Margin for a short futures contract*
In Example 3.5.4, assume you had taken instead the short position. Let us calculate the margin balance at the end of each month, using the same scenario. The margin balance is given in the table.

t	F_t^5	Initial balance	Margin call	Interest	Gain/Loss	End balance
0	100					5000
1	101	5000	0	50	−1000	4050
2	99	4050	0	40.5	2000	6090.5
3	98	6090.5	0	60.905	1000	7151.405
4	97	7151.405	0	71.51405	1000	8222.91905
5	99	8222.91905	0	82.2291905	−2000	6305.14824

The futures price did not increase too much and therefore no margin calls were necessary. The profit per contract is \$1 and, due to interests credited on the margin, the profit of 1305.14824 is greater than if you had entered into a short forward. ■

3.6 Summary

Forwards and futures

- Agreements engaging one party to buy (and the other to sell) the underlying asset in the future for a price paid upon delivery but determined at inception of the contract.
- Different from a spot contract, which is agreed upon today and where the asset is paid for and delivered immediately.

Important differences between forwards and futures

- Forwards are customized and traded over-the-counter.
- Futures are standardized and traded on exchanges.
- Futures: to protect both sides from default, a margin account is managed and regularly marked-to-market.

Notation and payoff

- Delivery price: K.
- Maturity date: T.
- Risk-free interest rate: r, continuously compounded.
- The underlying asset price S_0 is known.
- The underlying asset future time-t price S_t is random.
- Payoff of the long position: $S_T - K$.
- Payoff of the short position: $-(S_T - K) = K - S_T$.

Equity forwards: replication and initial value

- Synthetic long forward on a non-dividend-paying stock:
  - a long position in the stock: worth S_T at maturity;
  - a loan with principal Ke^{-rT}: reimburse K at maturity;
  - initial value: $S_0 - Ke^{-rT}$.
- Synthetic long forward on a stock paying discrete dividends:
  - a long position in the stock: worth $S_T + D_T$ at maturity;
  - a loan with principal $D_0 + Ke^{-rT}$: reimburse $K + D_T$ at maturity;
  - initial value: $S_0 - D_0 - Ke^{-rT}$.
- Synthetic long forward on a stock paying dividends at rate γ:
  - a long position in $e^{-\gamma T}$ units of the stock: worth S_T at maturity;
  - a loan with principal Ke^{-rT}: reimburse K at maturity;
  - initial value: $S_0e^{-\gamma T} - Ke^{-rT}$.

Equity forward price

- Forward price F_0^T: delivery price such that the forward contract costs nothing to enter (long or short) at inception.
- Non-dividend-paying stock: $F_0^T = S_0e^{rT}$.
- Stock paying discrete dividends: $F_0^T = (S_0 - D_0)e^{rT}$.
- Stock paying dividends at rate γ: $F_0^T = S_0e^{(r-\gamma)T}$.

Currency forwards

- Domestic currency: U.S. dollar.
- Domestic risk-free rate: r.
- Foreign currency: Euro, Canadian dollar, British pound, etc.
- Foreign risk-free rate: r_f.
- S_t is the time-t price (in U.S. dollars) for one unit of the foreign currency.
- K is the delivery price (in U.S. dollars) for one unit of the foreign currency.
- One unit of a foreign currency priced in U.S. dollars can be viewed as a risky asset earning a dividend yield of r_f.
- Initial value of a forward contract written on a foreign currency: $S_0e^{-r_fT} - Ke^{-rT}$.
- F_0^T is the forward price (in U.S. dollars) of one unit of the foreign currency also known as the forward exchange rate.
- $F_0^T = S_0e^{(r-r_f)T}$.

Commodity forwards

- Owning a commodity may entail storage costs: similar to a negative dividend on a stock.
- Accumulated storage costs: C_T.
- Proportional storage cost: c.
- Forward price: $F_0^T = S_0e^{rT} + C_T = S_0e^{(r+c)T}$.
- Cost of carry: $F_0^T = S_0e^{\xi T}$ with $\xi = r + c - \gamma$.
- Forward contracts on consumption assets, as opposed to investment assets, may require a convenience yield to compensate for the loss of income resulting from transforming the commodity.

Futures contracts and margin accounts

- Gains and losses are deposited in/withdrawn from a margin account.
- Initial margin: first deposit in the account.

- Margin call: additional deposit required whenever margin balance is below the minimum margin level.
- Margin account varies with periodic gains/losses, margin calls and interest.

3.7 Exercises

3.1 A 3-month zero-coupon bond currently trades at 98 (face value of 100) and a share of stock (that does not pay dividends) is worth $74.25.

(a) Compute the 3-month forward price of the stock.

(b) Describe how to replicate the cash flows of a 3-month forward at maturity.

3.2 You enter into a 6-month forward on a stock whose latest spot price is $23. The term structure of risk-free interest rates is flat at 3% (continuously compounded).

(a) What is the delivery price on that forward such that it is costless to enter at inception?

(b) Three months later, the stock (spot) price is $21 and the 3-month spot rate has risen to 4%. What is the value of the forward contract in (a) that you initiated 3 months ago?

(c) Suppose in (b) that you want to offset/liquidate your position. How much will you receive (or pay)?

3.3 Other methods to buy a stock:

(a) In a fully leveraged purchase of a stock, the stock is delivered immediately but paid for in the future (at time T). What amount should be paid at T for a leveraged purchase of a stock? Relate to the forward price.

(b) A prepaid forward is a forward contract for which the delivery price is paid today. How much should the investor pay today for a prepaid forward?

3.4 Synthetic securities:

(a) Using no-arbitrage arguments, show how to use a stock and a forward on a stock to replicate an investment in a risk-free bond (known as *synthetic* long bond position). Detail how the cash flows match at maturity and the initial cost of such a strategy.

(b) Using no-arbitrage arguments, show how to use a forward on a stock and a risk-free bond to replicate a long position in a stock (known as *synthetic* long stock position). Detail how the cash flows match at maturity and the initial cost of such a strategy.

3.5 You are given the following about a T-forward on a stock paying dividends:

- Fixed and known dividends D_i are expected to be paid at times $t_i, i = 1, 2, \dots, n$ with each $0 < t_i < T$.
- Dividends are reinvested at the risk-free rate.
- Term structure of risk-free rates flat at $100r\%$ (continuously compounded) and will remain so until T.

(a) If the current stock price is S_0, derive the current T-forward price of the stock at time 0 (F_0^T).

(b) If the stock price at t is S_t, derive the T-forward price of the stock at time t (F_t^T).

(c) Suppose now that dividends are paid continuously at a rate of γ and those dividends are continuously reinvested in the stock. Derive the T-forward price of the stock at time t (F_t^T).

3.6 A quarterly dividend of 45 cents is paid on a stock. If the current stock price is \$37 and the next dividend is expected 2 months from now, compute the 1-year forward price when the risk-free rate is constant at 3% (continuously compounded).

3.7 The dividend yield γ on a stock is 3% whereas the risk-free rate is 4% (both rates are continuously compounded). Assume dividends are continuously reinvested in the stock and the current stock price is \$50.

(a) Calculate the 6-month forward price.

(b) How many shares do you need to buy at time 0 to exactly replicate the payoff of a long forward contract?

3.8 A dealer advertises the following prices for a share of stock (with almost immediate delivery):

Bid : 50   Ask: 51.

If the 3-month rate is 2% (continuously compounded), then using no-arbitrage arguments, determine the long forward price and the short forward price for the stock. Assume the agreement matures in 3 months and the forward is initiated at no cost.

3.9 A food company needs 100 bushels of corn 1 year from now but unfortunately the price of corn on the spot market fluctuates a lot due to speculation. The current price per bushel is \$4.85 and you know it reached a record high of \$7.10 2 years ago. The 1-year forward price of corn is \$5.50 per bushel. Describe the implications of:

(a) waiting for a year to purchase corn on the spot market;

(b) immediately entering a 1-year forward contract.

3.10 Wheat is usually transformed to make cereal boxes sold to consumers. Storage costs are usually determined to be \$0.10 per bushel per year. Suppose the current spot price of wheat is \$4 per bushel and the risk-free rate is set at 3% compounded continually.

(a) Find the maximum 1-year forward price per bushel of wheat. Why is this a minimum forward price?

(b) If the forward price is found to be \$4.04, what is the convenience yield?

3.11 Your insurance company is headquartered in the U.S. and insures risks in both the U.S. and Germany (whose currency is the Euro). The German actuary tells you that the U.S. headquarters will need to pay claims in excess of premiums in the amount of €3 million 2 months from now. The U.S. headquarters wants to fix the cost of these claims in USD right now. The current exchange rate USD/EUR is 1.10, the 2-month forward exchange rate is 1.15 and the exchange rate is expected to vary a lot within the next 2 months. Describe the risk management implications of:

(a) waiting for 2 months to convert the foreign exchange rate on the spot market;

(b) immediately entering a 2-month currency forward contract.

3.12 In example 3.3.3, describe in details how you can exploit this arbitrage opportunity.

3.13 A 3-month futures is issued on the stock of ABC inc. Suppose the price of a share of ABC inc. evolves as follows over 3 months (time is in months): $S_0 = 45, S_1 = 47, S_2 = 48$, $S_3 = 49$. The margin requirement is 5% and a margin call is necessary whenever margin

drops below 70% of the initial balance. Interest is credited at the rate of 1.2% compounded monthly.

(a) Compute the 3-month futures price at times 0, 1, 2 and 3.

(b) Compute the margin balance at the end of every month for an investor that buys 1000 futures.

(c) Compute the margin balance at the end of every month for an investor that sells 1000 futures.

4

Swaps

A **swap** is an agreement between two parties to exchange cash flows at predetermined dates with contractual terms set at inception. A swap usually involves exchanging periodically *variable (random) payments*, whose value is based on a financial benchmark or an underlying asset (a stock, a bond, an index or some economic or financial quantity), for *fixed payments*. No premium is exchanged at inception. Figure 4.1 illustrates cash flows of a swap, where the words *variable*, *floating* and *risky* are used interchangeably.

Based upon this definition of a swap, common insurance policies can also be viewed as *swaps for ordinary people*. For example, a 2-year car insurance policy is similar to a swap since the policyholder makes a fixed monthly payment in exchange for variable/random amounts, i.e. the compensation for losses due to car accidents, theft or vandalism. In Figure 4.1, the policyholder is the investor in the left box whereas the insurance company is the investor in the right box.

Just like other financial derivatives, swaps are mostly used for risk management, i.e. to offset a risk exposure, or for speculation, i.e. to bet on/against the underlying asset. Risks that investors may want to swap are interest rate risk, currency risk, credit risk, etc. Examples of applications include:

1) When interest rates go down, the present value of actuarial liabilities can increase significantly. Insurance companies and pension plan sponsors can use *interest rate swaps* to mitigate the impact of a decrease in interest rates.
2) Your company uses gold to make jewelry and needs to regularly buy ounces of gold. To help fix those periodic cash outflows, your company can enter into a *gold swap* to fix the price it will pay for regular deliveries of gold over the next few months.
3) A company based in the U.S. receives a regular stream of Euros from its subsidiaries. Those cash flows need to be converted into U.S. dollars, but this is of course subject to fluctuations of the USD/EUR exchange rate. The company may enter into a *currency swap* to fix the value, in U.S. dollars, of cash inflows in Euros.
4) A bank has a large portfolio of fixed-rate loans, which is an asset for the bank. The investment manager believes interest rates will increase, so they enter into an *interest rate swap* to convert these fixed-rate loans into variable-rate loans. In this case, the swaps are used for speculative reasons.

As in the first three applications, insurance companies typically use swaps for risk management purposes. Table 4.1 shows the types of swaps used by insurers and their main purpose as of December 31st, 2014 according to the NAIC.

Even though swaps are traded over-the-counter, the legal aspects of swaps have been formalized by the International Swaps and Derivatives Association (ISDA).

Actuarial Finance: Derivatives, Quantitative Models and Risk Management, First Edition.
Mathieu Boudreault and Jean-François Renaud.

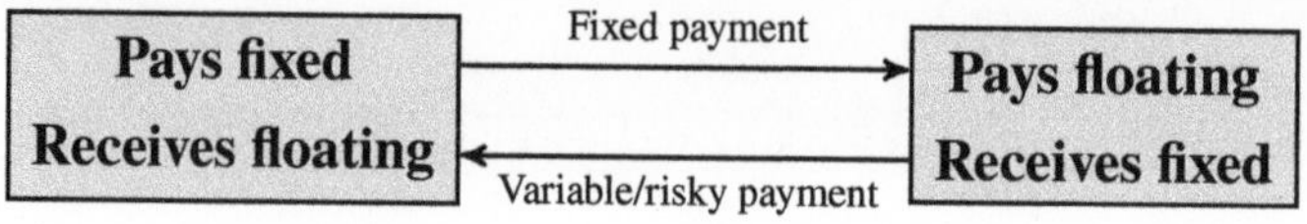

Figure 4.1 Periodic cash flows of a general swap

Table 4.1 Insurance industry swaps exposure by type and purpose/strategy in millions of US$ as of December 31st, 2014

Contract type	Hedging	Replication	Other	Total	%
Interest rate	807 108	9 123	2 132	818 363	83.45
Currency and FX	65 256	–	4 230	69 486	7.09
Credit default	11 323	22 641	966	34 930	3.56
Total return	38 758	275	863	39 897	4.07
Other	17 907	–	71	17 978	1.83
Total	940 353	32 040	8 263	980 655	100.00
Percentage	95.89	3.27	0.84	100.00	–

Source: Table 11, Capital Markets Special Reports, NAIC.

The role of this chapter is to provide an introduction to swaps with an emphasis on those used in the insurance industry, namely interest rate swaps, currency swaps and credit default swaps. For completeness, we will also discuss commodity swaps. The specific objectives are to:

- define the characteristics of interest rate, currency, credit default swaps and commodity swaps;
- calculate the cash flows of a swap;
- compute the value of a swap.

4.1 Framework

A swap is usually specified by:

- an underlying asset or financial variable;
- the dates of the payments;
- the maturity date, which is known as the **swap term** or **tenor**;
- how the payments are calculated.

The stream of predetermined fixed payments is known as the **fixed leg**, whereas the stream of variable payments is called the **variable leg** or **floating leg**.

Most swaps fall into the **fixed-for-floating** category, meaning that a variable (risky, random, floating) payment is exchanged periodically for a fixed amount.

In what follows, we will focus on the following swaps:

- **Interest rate swap**: exchange payments based on a variable/random interest rate for payments based on a fixed interest rate.
- **Currency swap**: exchange payments based on a given currency for payments based on another currency.

- **Credit default swap**: compensation, if the underlying bond defaults, in exchange for a fixed periodic payment/premium.
- **Commodity swap**: delivery of a commodity (whose price varies) at different dates, in exchange for a fixed periodic payment.

Interest rate swaps are by far the most popular swaps used by insurance companies to manage interest rate risk: 83% of all swaps are used to manage this risk only (see Table 4.1). In the entire OTC swap market, the amount of notional outstanding in interest swaps is about 400 trillion dollars (see Table 2.2).

The most popular types of swaps include the above-mentioned, as well as *equity swaps*, which is an exchange of payments based on the returns of a stock (or stock index) for payments based on an interest rate, and *total return swaps*, which is an exchange of an interest rate (fixed or floating) for the total return of a bond (including default).

Minor variations on standard swaps are *prepaid swaps* and *deferred swaps*. A *prepaid swap* is a regular swap whose fixed leg is paid as an up-front premium at inception, while a *deferred swap* is issued today but payments are exchanged later at a predetermined date.

4.2 Interest rate swaps

In an **interest rate swap**, parties exchange a variable interest rate, applied to a given amount, in exchange for a fixed interest rate, applied on the same amount. This amount on which the payments are calculated is known as the **notional** (principal, par value, face value). The *swaps of payments* occur at predetermined dates for a given period of time. Typically, payments are exchanged quarterly or semi-annually and interest rate swaps usually last from 2 to 15 years. Because their cash flows are similar, we will use the words *loans* and *bonds* interchangeably.

4.2.1 Fixed-rate and floating-rate loans

A **fixed-rate loan** is simple: the borrower receives the principal F at inception in exchange for future periodic interest payments of size c. The principal is repaid at maturity. No matter how the term structure of interest rates evolves over time, i.e. if it increases or decreases, all interest payments will be of size c. Therefore, for this type of loan, the interest rate risk is assumed by the lender. This is similar to a bond with fixed coupons.

In a **floating-rate loan**, the interest payments are variable because they are based on a variable interest rate. The borrower does not know in advance the payments she will have to make: when interest rates increase (resp. decrease), the borrower pays more (resp. less). Therefore, for this type of loan, the interest rate risk is assumed by the borrower. This is similar to a bond with variable coupons.

Example 4.2.1 *5-year floating-rate loan, renewable annually*
Suppose you borrow \$100 from your bank by entering into a 5-year floating-rate loan, renewable annually. Therefore, you are committed to make interest payments every year according to the 1-year rate observed upon each renewal. Principal must be repaid at maturity. Let us compute the cash flows for this loan.

At inception, your bank gives you \$100 and you observe that the current 1-year rate is 4%. Therefore, you will have to make a payment of \$4 at the end of the year.

At the end of this first year, you make your payment of \$4. Now, suppose that the 1-year rate is at 4.5%. Therefore, you will have to make a payment of \$4.50 at the end of the second year.

At the end of the second year, you make the payment of \$4.50. Now, suppose that the 1-year rate is at 3.7%. Therefore, you will have to make a payment of \$3.70 at the end of the third year.

The same reasoning applies for the remaining payments. ■

Recall from Section 2.1.3 that we use the notation r_t^T for the (annual and annually compounded) spot interest rate we will observe at time t, i.e. in t years from now, to invest/borrow from time t up to time $t + T$.

In the previous example, we considered a scenario where $r_0^1 = 0.04$, $r_1^1 = 0.045$ and $r_2^1 = 0.037$.

4.2.2 Cash flows

Now that we understand fixed-rate and floating-rate loans, let us look at the cash flows of a very simple interest rate swap.

Example 4.2.2 *Cash flows of a simplified interest rate swap*
Investors A and B enter into a 5-year interest rate swap agreement with a notional amount of \$100. Investor A agrees to pay investor B a fixed interest rate of 4% in exchange for the 1-year rate (from investor B). Interest rates are compounded annually and payments also occur annually. Let us compute the cash flows of this interest rate swap in a scenario where the 1-year rate evolves as follows:

Date	1-year rate
Today	3.8%
In 1 year	3.95%
In 2 years	4.12%
In 3 years	4.29%
In 4 years	3.73%
In 5 years	3.87%

Meanwhile, for the fixed-rate loan, we have:

Time	Amount
0	0
1	4
2	4
3	4
4	4
5	100+4

The floating leg of the swap behaves like a 5-year floating-rate loan, renewable annually. Each and every year, the interest payment made at the end of the year depends on the 1-year rate observed at the beginning of that same year, not before. At inception, the 1-year rates that will be observed are unknown. In the scenario considered above, the evolution of the 1-year rate gives us the following interpretation:

- a loan contracted at time 0 and maturing at time 1 is subject to an interest rate of $r_0^1 = 3.8\%$;
- a loan contracted at time 1 and maturing at time 2 will be subject to an interest rate of $r_1^1 = 3.95\%$;
- a loan contracted at time 2 and maturing at time 3 will is subject to an interest rate of $r_2^1 = 4.12\%$;
- etc.

Therefore, the cash flows are:

Time	Amount
0	0
1	3.80
2	3.95
3	4.12
4	4.29
5	100+3.73

Overall, the cash flows of the swap are:

Time	Fixed	Floating	Net
1	4	3.80	0.2
2	4	3.95	0.05
3	4	4.12	−0.12
4	4	4.29	−0.29
5	104	103.73	0.27

The column "Net" corresponds to the net amount received by the investor paying the floating rate. ■

Example 4.2.3 *Cash flows of a standard interest rate swap*
Let us now consider the case of a standard fixed-for-floating interest rate swap, on a principal of \$100, where interest rates are compounded semi-annually and payments occur every 6 months for 3 years. Investor A agrees to pay, to investor B, a fixed interest rate of 3.8% in exchange for a floating interest rate. We will illustrate how cash flows on both sides of the swap are computed.

For this swap, the floating interest rate is the 6-month rate. Mathematically, the rates used to compute the floating-rate interest payments are $r_t^{0.5}$ for $t = 0, 0.5, 1, 1.5, \ldots, 2.5$. All those rates, except for $r_0^{0.5}$, are unknown at inception.

Let us consider the following scenario and the resulting cash flows:

time t	$r_t^{0.5}$	Fixed	Floating	Net
0	0.04			
0.5	0.035	1.9	2	−0.1
1	0.037	1.9	1.75	0.15
1.5	0.042	1.9	1.85	0.05
2	0.045	1.9	2.1	−0.2
2.5	0.041	1.9	2.25	−0.35
3	0.038	1.9	2.05	−0.15

The fixed leg is easy to determine: every 6 months, a payment of \$1.90 should be made. This is due to the fixed 6-month rate of 1.9% = 3.8%/2 applied to \$100.

The first variable payment, to be made 6 months after inception, is already known: it corresponds to $100 \times \left(\frac{1}{2} \times r_0^{0.5}\right) = 2$, since the interest rate between time 0 and time 0.5 is $r_0^{0.5} = 0.04$. As for the rest of the variable leg, we need to look at the realizations (time series) of the floating rate in the chosen scenario. The second variable payment, due after 1 year, depends on the realization of $r_{0.5}^{0.5}$, which in this scenario is $r_{0.5}^{0.5} = 0.035$. Consequently, this second interest payment is $100 \times \frac{1}{2} \times r_{0.5}^{0.5} = 1.75$. The other cash flows are calculated in a similar fashion for the rest of the payment schedule. ■

Of course, payments are made only by the investor who is the net payer. In the above example and scenario, investor B would pay 0.1 to investor A at time 0.5 (6 months after inception), while investor A would pay 0.15 to investor B at time 1 (1 year after inception).

In most swaps, the principal upon which payments are calculated is not exchanged at maturity.

In most fixed-for-floating (or floating-for-floating) interest rate swaps, the floating payment is determined by the interest rate observed at the *beginning of the corresponding period*. For example, the floating payment to be made in 6 months from now is based on the current 6-month rate. This is similar to a loan contracted today and maturing in 6 months. Moreover, the principal/notional is not exchanged at maturity (it is the same on both sides anyway).

When entering a swap (no matter what is the underlying risk), both sides assume the default/credit risk of the counterparty. In an interest rate swap, where the notional is not exchanged at maturity, credit risk is much smaller than on conventional loans.

4.2.3 Valuation

Our approach to valuation of interest rate swaps is based on the fact that a long fixed-for-floating interest rate swap is equivalent to holding simultaneously:

1) a long position in a fixed-rate bond;
2) a short position in a variable-rate bond.

The short position in the swap is equivalent to holding the reversed positions in the bonds.

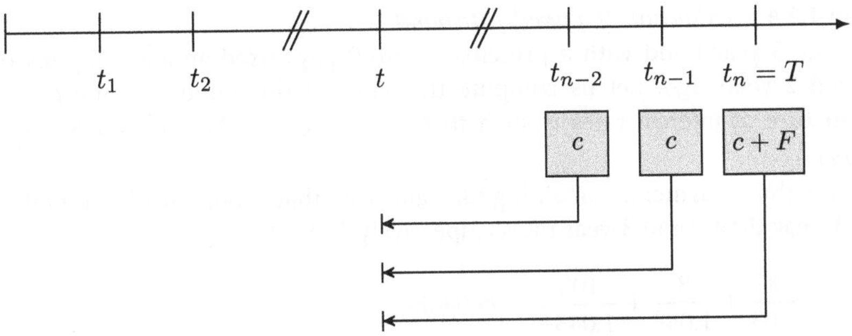

Figure 4.2 Valuation of a fixed-rate bond

Therefore, under the absence-of-arbitrage assumption, the value of the swap is equal to the *difference* between these two positions.[1]

4.2.3.1 Fixed-rate bonds

Assume the principal of the bond is F and its holder receives periodic payments/coupons of size c at dates $t_1, t_2, \dots, t_n = T$. Those payments are fixed and known in advance.

The principle underlying bond valuation is simple: future cash flows are discounted using the observed term structure of interest rates. Let B_t^{x} be the time-t value of a fixed-rate bond.[2]

We already know that the initial value is given by

$$B_0^{\text{x}} = \sum_{i=1}^{n} \frac{c}{\left(1 + r_0^{t_i}\right)^{t_i}} + \frac{F}{\left(1 + r_0^{t_n}\right)^{t_n}}. \tag{4.2.1}$$

This is the same formula as in equation (2.1.1) but using the notation B_0^{x} rather than B_0. It corresponds to the present value of cash flows occurring after time 0.

If $t = t_k$ (right after the k-th payment/coupon), for any $k = 1, 2, \dots, n-1$, then

$$B_{t_k}^{\text{x}} = \sum_{i=k+1}^{n} \frac{c}{\left(1 + r_{t_k}^{t_i - t_k}\right)^{t_i - t_k}} + \frac{F}{\left(1 + r_{t_k}^{t_n - t_k}\right)^{t_n - t_k}}$$

which corresponds to the present value of cash flows paid after t_k.

In general, for any time $t \in [0, t_n)$, we have

$$B_t^{\text{x}} = \sum_{i:t_i > t} \frac{c}{\left(1 + r_t^{t_i - t}\right)^{t_i - t}} + \frac{F}{\left(1 + r_t^{t_n - t}\right)^{t_n - t}}, \tag{4.2.2}$$

where the sum is taken over all is such that $t_i > t$. The formula in equation (4.2.2) is simply the present value of both coupons and principal to be paid immediately after time t.

This is also illustrated in Figure 4.2.

1 A fixed-for-floating interest rate swap can also be viewed as a portfolio of forward rate agreements (FRAs). Valuation based upon that perspective will not be covered in this book.

2 We use the superscript x as in fixed; we will use the superscript ℓ as in fℓoating.

Example 4.2.4 *Valuation of a fixed-rate bond*
A fixed-rate 5-year bond with a principal of \$100 pays fixed annual coupons of \$3. It was issued 2 years ago. Let us compute the value of this bond (at time $t = 2$) if the term structure of interest rates is such that $r_2^1 = 0.03$, $r_2^2 = 0.04$, $r_2^3 = 0.045$, $r_2^4 = 0.05$, $r_2^5 = 0.055$.

There are three payments remaining (3, 3 and 103) that should be discounted with the current 1-year, 2-year and 3-year rates respectively. We get

$$B_2^{\text{x}} = \frac{3}{1.03} + \frac{3}{1.04^2} + \frac{103}{1.045^3} = 95.9448.$$

■

4.2.3.2 Floating-rate bond

The fundamental bond valuation principle states that future cash flows should be discounted using the current interest rate term structure. In what follows, for simplicity, we will consider bonds paying coupons only on an annual basis, i.e. at times $1, 2, \ldots, n = T$. Again, we will start by finding the price at coupon dates and then deduce the value at any other dates.

The cash flows of a floating-rate bond are random since future 1-year rates are unknown. In fact, they are equal to $F \times r_k^1$, for each $k = 1, 2, \ldots, n$. To handle this difficulty, we will work backward from maturity to inception. Let us illustrate the idea with an n-year floating-rate loan with interest compounded annually.

After $n - 1$ years, immediately after the coupon payment, the remaining value of the bond is a cash flow of $F + Fr_{n-1}^1$ (principal + final interest payment) to be paid at time n, i.e. 1 year later. To compute the corresponding value of the bond at that time, we need to discount $F(1 + r_{n-1}^1)$ back from time n to time $n - 1$ using the 1-year rate observed at time $n - 1$. Letting B_t^{ℓ} be the time-t value of a floating-rate bond then, we have

$$B_{n-1}^{\ell} = \frac{F\left(1 + r_{n-1}^1\right)}{1 + r_{n-1}^1} = F.$$

In other words, the value of this floating-rate bond at time $n - 1$ is equal to the principal value.

The bond value at time $n - 2$ comes from the present value of both:

1) the bond value at time $n - 1$, which is given by $B_{n-1}^{\ell} = F$;
2) the interest payment due at time $n - 1$, which is given by $F \times r_{n-2}^1$.

Discounting these two components back from time $n - 1$ to $n - 2$ using the 1-year rate observed at time $n - 2$, we have

$$B_{n-2}^{\ell} = \frac{F\left(1 + r_{n-2}^1\right)}{1 + r_{n-2}^1} = F.$$

Again, the value of this floating-rate bond at time $n - 2$ is equal to the principal value.

In fact, it is easy to deduce that right after any coupon payment, we have

$$B_{k-1}^{\ell} = \frac{F\left(1 + r_{k-1}^1\right)}{1 + r_{k-1}^1} = F, \tag{4.2.3}$$

for each $k = 1, 2, \ldots, n$.

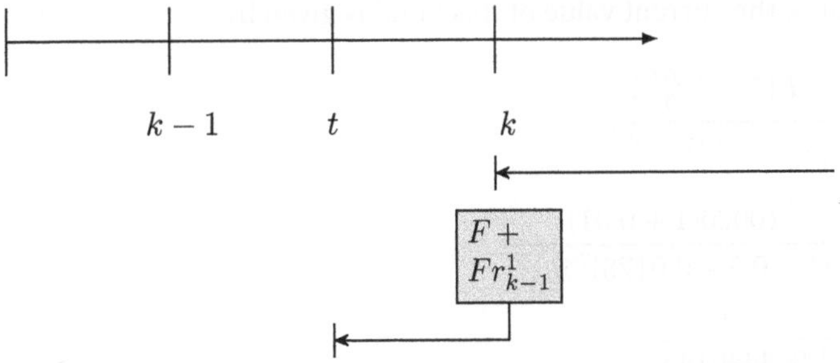

Figure 4.3 Valuation of a floating-rate bond

Now, let us determine the bond value between two coupon payments. Let $t \in [k-1, k)$, for $k = 1, 2, \ldots, n$. Since $B^{\ell}_k = F$ and an additional cash flow of $F \times r^1_{k-1}$ is due at time k, then

$$B^{\ell}_t = \frac{F\left(1 + r^1_{k-1}\right)}{\left(1 + r^{k-t}_t\right)^{k-t}}.$$

It corresponds to the present value of $F \times (1 + r^1_{k-1})$, which is paid at time k, discounted back from time k to time t using the $(k-t)$-rate observed at time t. This is also shown in Figure 4.3.

Example 4.2.5 *Valuation of a floating-rate bond with annual coupons*
A 10-year floating-rate bond of $10,000 was issued 3.5 years ago. Coupons are annual and calculated using the 1-year rate observed at the beginning of each year. Suppose that we observe the following term structure of interest rates: $r^{0.5}_{3.5} = 0.01$ and $r^1_{3.5} = 0.015$. Six months ago, the 1-year rate was $r^1_3 = 0.0125$. All rates are compounded annually. Find the value, at time $t = 3.5$, of this floating-rate bond.

In 6 months, after the coupon payment, the value of the bond will be equal to its principal, i.e. $10,000. The coupon payment due in 6 months is based upon the 1-year rate observed six months ago, i.e. 1.25%, so it is equal to $125. We discount this total value of $10,125, due in 6 months from now, using $r^{0.5}_{3.5} = 0.01$ and then we obtain

$$B^{\ell}_{3.5} = \frac{10125}{1.01^{0.5}} = 10074.75.$$ ■

If coupons/payments are paid on a semi-annual or a quarterly basis (and compounded accordingly), then the valuation methodology presented above is still valid, up to some minor modifications. Suppose interest is paid and compounded m times per year. At time $t = k/m$, for any $k = 1, 2, \ldots, mn$, that is right after a coupon payment, we still have

$$B^{\ell}_t = F.$$

To obtain the value of the bond between two coupon payments, we need to discount the next coupon and the principal value.

Let us illustrate this with an example.

Example 4.2.6 *Valuation of a floating-rate bond with semi-annual coupons*
A 3-year floating-rate bond with principal $F = 10,000$ was issued 3 months ago. Coupons are paid semi-annually and determined by using the 6-month rates. We observe $r^{0.5}_0 = 2\%$. Let us consider the following scenario: $r^{0.25}_{0.25} = 1.78\%$.

In this scenario, the current value of this bond is given by

$$\begin{aligned}B^{\ell}_{0.25} &= \frac{F\left(1+\frac{1}{2}r^{0.5}_{0}\right)}{\left(1+\frac{1}{2}r^{0.25}_{0.25}\right)^{2\times 0.25}}\\ &= \frac{10000(1+0.01)}{(1+0.5\times 0.0178)^{2\times 0.25}}\\ &= \frac{10100}{1.004440143}\\ &= 10055.3528.\end{aligned}$$

■

4.2.3.3 Valuation of a swap: the difference of two bonds

Using the above, we have that the time-t value of an interest swap for an investor:

- paying fixed and receiving floating is $V^{\ell}_t = B^{\ell}_t - B^{\times}_t$;
- paying floating and receiving fixed is $V^{\times}_t = B^{\times}_t - B^{\ell}_t$.

Example 4.2.7 *Valuation of an interest rate swap*
A 2-year interest rate swap was initiated 3 months ago. At that time, the 6-month rate was 1.14%. Coupons are semi-annual and the face value is $10,000. An investor has agreed to pay 2% per year whereas the counterparty has agreed to pay coupons based upon the observed 6-month rate. All rates are semi-annually compounded.

Assuming the current term structure of interest rates is given by $r^{T}_{0.25} = 0.01 + 0.004T$, let us find the current value of this swap, i.e. at time $t = 0.25$.

There are four cash flow dates remaining:

1) At time 0.5, which is 6 months after inception or 3 months from now.
2) At time 1, which is 1 year after inception or 9 months from now.
3) At time 1.5, which is 18 months after inception or 15 months from now.
4) At time 2, which is 2 years after inception or 21 months from now.

We need to identify the rates at which payments will be discounted:

1) Cash flow at time 0.5: 3-month rate observed at time 0.25.
2) Cash flow at time 1: 9-month rate observed at time 0.25.
3) Cash flow at time 1.5: 15-month (1.25-year) rate observed at time 0.25.
4) Cash flow at time 2: 21-month (1.75-year) rate observed at time 0.25.

The following table shows the corresponding rates and the present value of each fixed payment.

Maturity	Rate	Amount	PV
0.25	0.011	100	99.72613
0.75	0.013	100	99.03286
1.25	0.015	100	98.14934
1.75	0.017	10100	9805.183

For example, for the second fixed payment occurring at time 1 which is 9 months from now, we have $r_{0.25}^{0.75} = 0.01 + (0.004 \times 0.75) = 0.013$. The coupon value is $10000 \times (0.5 \times 0.02) = 100$ whereas the present value is

$$\frac{100}{(1 + (0.5 \times 0.013))^{2 \times 0.75}} = 99.03286224.$$

Therefore, summing up the elements in the last column, we get

$$B_{0.25}^{\mathrm{x}} = 99.726 + 99.033 + 98.149 + 9805.183 = 10102.09158.$$

On the other hand, the floating-rate payment due in 3 months is equal to

$$F \times \frac{r_0^{0.5}}{2} = 10000 \times \frac{0.0114}{2} = 57$$

and hence the value of the floating-rate bond at the next coupon date will be \$10,057. We discount this value at the current 3-month rate $r_{0.25}^{0.25} = 0.011$ (which is semi-annually compounded) and we get

$$B_{0.25}^{\ell} = \frac{10057}{(1 + (0.5 \times 0.011))^{2 \times 0.25}} = 10029.45681.$$

Therefore, the current value of this swap, for the fixed-rate receiver, is given by $V_{0.25}^{\mathrm{x}} = B_{0.25}^{\mathrm{x}} - B_{0.25}^{\ell} = 10102.09 - 10029.46 = 72.63$. Of course, we have a value of $B_{0.25}^{\ell} = -72.63$ for the floating-rate receiver. ■

4.2.4 Market specifics

At inception, it costs nothing to enter a swap. Therefore, the floating-rate bond should be worth the same as the fixed-rate bond, i.e. we should observe $B_0^{\ell} = B_0^{\mathrm{x}}$. Recall from equation (4.2.3) that a floating-rate bond is always worth F at inception and immediately after any coupon payment. Consequently, the fixed-rate bond should also be worth F at inception of the swap. Consequently, to avoid arbitrage opportunities, there is only one possible coupon rate for this fixed-rate bond. This coupon rate is known as the **swap rate**. Mathematically, the swap rate is the coupon rate such that $B_0^{\mathrm{x}} = F = B_0^{\ell}$, where an expression for B_0^{x} can be found in equation (4.2.1). This is how swaps are usually quoted.

Example 4.2.8 *Swap rate*
The current term structure of interest rates is such that $r_0^1 = 0.03$, $r_0^2 = 0.04$, $r_0^3 = 0.045$, $r_0^4 = 0.05$, $r_0^5 = 0.055$. Let us find the swap rate on a 5-year interest rate swap with a principal of \$100.

Recall that the swap rate is the coupon rate on the fixed-rate bond such that it trades at par. Using equation (4.2.1), we must solve for c in the following equality:

$$\begin{aligned} 100 &= c\left(\frac{1}{1.03} + \frac{1}{1.04^2} + \frac{1}{1.045^3} + \frac{1}{1.05^4} + \frac{1}{1.055^5}\right) + \frac{100}{1.055^5} \\ &= 4.359563432 \times c + 76.51343538. \end{aligned}$$

We find $c = 5.387366193$. Therefore, the swap rate is 5.39%. ■

Until now, we have not been specific about which floating interest rate to use. In the financial market, the floating rate most commonly used is the LIBOR rate, which is the London Interbank

Offered Rate. For example, the 6-month LIBOR rate is a common benchmark used to compute floating interest payments.

We have not specified either what interest rates should be used to discount cash flows from the swap. We know, however, they should be consistent with the benchmark used to compute floating interest payments. In this chapter, we will say it is obtained from a LIBOR *zero curve* which should not be confused with LIBOR *rates* (see below).

London Interbank Offered Rate (LIBOR)

The LIBOR is fixed by a council of banks doing business in London. It represents the interest rate an average bank from that council would need to pay for a short-term loan contracted from any of the other banks. The US LIBOR is the rate a U.S. bank would pay if it were to borrow from these leading banks.

The LIBOR rate is available for seven different maturities: 1 day (overnight), 1 week, 1 month, 2, 3, 6 and 12 months. Since most swaps have semi-annual payments, the 6-month LIBOR rate is thus often used to determine the periodic payment of the floating-rate leg of an interest rate swap.

The LIBOR zero curve is an interest rate term structure obtained from LIBOR rates (for maturities below 1 year) and Eurodollar futures for maturities between 1 year and 5 years.

Example 4.2.9 *Another swap rate*

Let us find the swap rate on a 2-year interest rate swap with semi-annual payments based upon the 6-month LIBOR. It should be noted the LIBOR zero curve is such that:

- 6-month rate: 3%
- 12-month rate: 3.66%
- 18-month rate: 4.04%
- 24-month rate: 4.26%.

All rates are compounded semi-annually.

We need to find c such that

$$100 = \frac{c}{1+\frac{1}{2}\times 0.03} + \frac{c}{\left(1+\frac{1}{2}\times 0.0366\right)^2} + \frac{c}{\left(1+\frac{1}{2}\times 0.0404\right)^3} + \frac{100+c}{\left(1+\frac{1}{2}\times 0.0426\right)^4}$$

and thus the semi-annual coupon is $c = 2.1217414$ or, said differently, the (semi-annual) swap rate is 4.2434828%. ■

Finally, swaps are commonly and formally initiated by an intermediary whose role is to either match two investors in a swap or play the role of the counterparty in the swap. The intermediary will incur costs and usually charge a fixed spread over the LIBOR rate, as shown in Figure 4.4.

Figure 4.4 Swap between two parties initiated with an intermediary such as a financial institution

4.3 Currency swaps

Generally speaking, a *swap on currencies* is an agreement between two parties to exchange future cash flows in different currencies. There are two commonly traded types of swaps on currencies: *cross-currency basis swaps* and *foreign exchange (FX) swaps*. In what follows, we will focus on the former type.

In a **cross-currency basis swap**, simply called a **currency swap** in what follows, both parties agree to exchange a principal and its interests in one currency for the same principal and its interests in another currency. There are three types of currency swaps:

- Fixed-for-fixed: the interest rates are fixed on both sides of the swap.
- Floating-for-floating: the interest rates are variable on both sides of the swap.
- Fixed-for-floating: the interest rate is fixed on one side and is floating on the other side.

Interests in each currency, either fixed or floating, are exchanged on both sides. Currency swaps are used by institutional investors and multinational companies to swap foreign-denominated loans or to lower borrowing costs abroad.

Example 4.3.1 *Constant flow of foreign revenues*
A U.S.-based insurance company also operates in Europe. Once every quarter, it will receive €1 million to be converted immediately into U.S. dollars. Let us explain how this American insurer can manage this foreign exchange rate risk over the next 12 months.

The company has three options:

1) If the investment manager believes the USD/EUR exchange rate will move favorably in the next couple of months, then she could simply wait, receive €1 million every quarter and then convert it using the spot exchange rate. In this case, the company is fully exposed to foreign exchange rate risk.
2) They could simply enter into four separate forward contracts, each to sell €1 million with maturities 3, 6, 9 and 12 months, respectively. This way, the company would fix the exchange rate for the next year.
3) They could also enter into a currency swap to pay €1 million every quarter and receive in exchange a certain amount in U.S. dollars. This amount would need to be determined. Depending on the agreement, this might also require an exchange of principal at inception and at maturity. ■

An *FX swap* is an agreement to buy (sell) one unit of a currency in exchange for selling (buying) it back at a predetermined date in the future. The price at which it is traded in the future is the forward exchange rate (as seen in Chapter 3). Therefore, a FX swap is a long (short) spot transaction combined with a short (long) forward transaction.

According to the Triennial Central Bank Survey conducted by the BIS in 2013, $2.2 trillion of FX swaps were traded daily compared with only $54 billion of currency swaps.[3]

4.3.1 Cash flows

The cash flows of a currency swap are similar to exchanging interest payments from two loans each denominated in a different currency. This is illustrated in the next example.

3 BIS, Foreign exchange turnover in April 2013: preliminary global results, Table 1.

Example 4.3.2 *Currency swap to lower borrowing costs*
A U.S.-based insurance company wants to start doing business in Europe, whereas a European reinsurance company wants to expand in the United States. The companies have access to the following rates.

On one hand, the American company may borrow:

- at 2% in USD from an American bank;
- at 4% in EUR from a European bank, which is a foreign bank for the company.

On the other hand, the European company may borrow:

- at 1% in EUR from a European bank;
- at 5% in USD from an American bank, which is a foreign bank for the company.

Today, both companies need 100 million U.S. dollars, or its equivalent in Euros according to the current exchange rate, i.e. 90 million Euros. We see it is much cheaper for each company to borrow from a domestic bank than from a foreign bank. Let us illustrate how the two companies can enter into a currency swap to lower their borrowing costs.

Indeed, if the companies enter into a fixed-for-fixed currency swap with a notional of 100 million USD, which is currently equal to 90 million EUR, and if each company borrows from its domestic bank, then they will both lower their costs.

At inception, the notional is exchanged in their respective currencies, i.e. the American company receives 90 million EUR, a loan worth 100 million USD, while the European company receives 100 million USD, a loan worth 90 million EUR. The net cash flow is zero.

Periodically, the European company will pay 1% on its loan, i.e. 0.9 million EUR. It receives the same amount through the swap from the U.S.-based company. Similarly, the American company will pay 2% on its loan, i.e. 2 million USD, but will receive the same amount through the swap from the Europe-based company.

At maturity, when the loans expire, the swap requires an exchange of notional and then each company can repay its respective loans.

The next two tables illustrate the cash flows for the U.S.-based company and for the European-based company, respectively (unit is 1 million):

Time	Local loan	Swap
Inception	+100 USD	−100 USD, +90 EUR
Periodic	−2 USD	+2 USD, −0.9 EUR
Maturity	−100 USD	+100 USD, −90 EUR

and

Time	Local loan	Swap
Inception	+90 EUR	−90 EUR, +100 USD
Periodic	−0.9 EUR	+0.9 EUR, −2 USD
Maturity	−90 EUR	+90 EUR, −100 USD

If the American company borrows from its local bank and enters into a currency swap, then this will be equivalent to a loan at 1% in EUR. If the European company borrows from its local bank and enters into a currency swap, then this will be equivalent to a loan at 2% in USD. ∎

It is worth mentioning that credit risk on a currency swap is larger than on an interest rate swap (which is on a single currency) as the notional is exchanged at maturity.

4.3.2 Valuation

Similar to interest rate swaps, it is possible to view a currency swap as a long (short) position in a locally-denominated bond (either fixed-rate or floating-rate) and a short (long) position in a foreign-denominated bond (either fixed-rate or floating-rate). Therefore, under the no-arbitrage assumption, pricing a currency swap is equivalent to pricing the *difference* of two bonds (or loans), each denominated in its own currency.[4] As a result, it suffices to value each bond in its local currency and convert the foreign-denominated bond using the current exchange rate.

Again, let B_t^d (d for domestic) be the time-t value of the locally-denominated bond (either fixed-rate or floating-rate), B_t^f (f for foreign) be the time-t value of the foreign-denominated bond (either fixed-rate or floating-rate), and S_t be the time-t exchange rate expressed as the number of local currency needed to buy one unit of the foreign currency.

For example, if the domestic currency is USD and the foreign currency is EUR, then B_t^f is the time-t price in EUR of the European bond. Consequently, $S_t B_t^f$ is the time-t value in USD of this European bond. Of course, B_t^d is the time-t price in USD of the American bond. It is important to note that B_t^d and B_t^f are computed using equation (4.2.2), or equation (4.2.3) (depending on whether interest is fixed or floating), in their respective currency.

Therefore, at time t, a currency swap where we receive the domestic interest and pay the foreign interest is worth

$$V_t^d = B_t^d - S_t B_t^f,$$

while a currency swap where we pay the domestic interest and receive the foreign interest is worth

$$V_t^f = S_t B_t^f - B_t^d.$$

Note that both values V_t^d and V_t^f are expressed in the domestic currency.

Example 4.3.3 *Valuation of a fixed-for-fixed currency swap*
The following fixed-for-fixed currency swap has been issued for some time:

- At inception, the USD and EUR were trading at par.
- Notional is 100.
- 5% in USD is exchanged for 4% in EUR.
- Maturity is in 3 years.
- Payments are annual.

---

4 A fixed-for-floating currency swap can also be viewed as a portfolio of currency forwards. Valuation based upon that perspective will not be further covered in this book.

The current market conditions are described as follows:

- U.S. LIBOR curve is flat at 4.5%.
- Euro LIBOR curve is flat at 4.7%.
- Current exchange rate is 0.90 USD/EUR.

All rates are compounded annually. Let us find the current value of this swap.

First note that as the exchange rate between the U.S. dollar and the Euro was at par at inception, hence the notional is 100 (USD and EUR) on both bonds.

We need to find the value of the 3-year bonds using the appropriate discount rates: the U.S.-(Euro-) denominated bond is discounted with the U.S.(Euro) LIBOR curve. We have

$$B_0^d = \sum_{k=1}^{3} \frac{5}{1.045^k} + \frac{100}{1.045^3} = 101.3744822 \text{ USD},$$

$$B_0^f = \sum_{k=1}^{3} \frac{4}{1.047^k} + \frac{100}{1.047^3} = 98.08295965 \text{ EUR}.$$

Therefore, this swap is worth

$$\left(0.9\frac{\text{USD}}{\text{EUR}} \times 98.08295965 \text{ EUR}\right) - 101.3744822 \text{ USD} = -13.09981852 \text{ USD},$$

from the point of view of a company paying U.S. dollars and receiving Euros. ■

4.4 Credit default swaps

A bond issuer defaults (on its obligation) whenever it is unable to make one or several coupon payments on that bond. The risk of occurrence of such an event, and the loss associated with it, are known as credit risk (see Chapter 2). Therefore, investing in a bond, or equivalently lending to a borrower, entails credit risk.

A **credit default swap** is a credit derivative similar to an insurance against default, compensating an investor losing because of a default event. Using the terminology for swaps, a CDS is an agreement between two parties to receive a (random) compensation in case of default in exchange for a fixed stream of payments. The **protection buyer** is the investor paying the fixed stream of cash flows, known as CDS premiums. The **protection seller** assumes the credit risk of the reference bond and agrees to compensate the protection buyer in case a pre-specified credit event occurs. The cash flows are illustrated in Figure 4.5.

CDS are offered with maturities (tenors) varying between 1 and 10 years, 5 years being the most popular. As with many other swaps, the ISDA formalizes the agreements. Credit default swaps are among the most popular credit derivatives traded in the market but lost much popularity after the financial crisis of 2007–2009. As of December 2014, the market size of CDS was about $16 trillion (BIS) but was almost four times larger in the months prior to the burst of the financial crisis (ISDA).

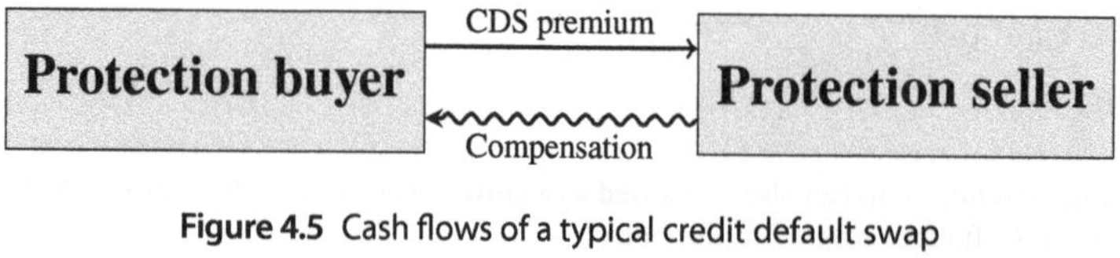

Figure 4.5 Cash flows of a typical credit default swap

As life insurance companies use corporate bonds to manage their long-term liabilities, they are exposed to credit risk. The financial crisis of 2007–2009 has shown that even investment-grade securities could default and, therefore, asset managers use CDS to protect against such insolvency. According to the NAIC, 93% of insurers who hedge credit risk on their assets use credit default swaps. However, only 4% of their total use of swaps involves CDS.[5]

4.4.1 Cash flows

Cash flows and valuation of CDS are based upon an underlying reference bond. There are two scenarios:

1) If the borrowing company does not default on the reference bond, which is the most likely scenario, the protection buyer will pay a semi-annual premium for the CDS and receive nothing in return. This is just like a conventional insurance policy where the *insured event* does not occur.
2) If the reference company does default prior to the maturity of the CDS, then the protection buyer receives a compensation and stops paying premiums. Compensation is tied to the market value of the reference bond upon default.

Whenever there is default, the buyer chooses between a physical or a cash settlement. In a physical settlement, the protection buyer sells the underlying reference bond to the protection seller at some given price, thus compensating the loss in value following default. Whenever the CDS is settled in cash, the difference between the market value of the defaulted bond and that given price is provided in cash to the protection buyer. This is illustrated in the next example.

Example 4.4.1 *Cash flows of a credit default swap*
A life insurance company holds 100M in a 20-year corporate bond issued by a highly-reputable company named ABC inc. However, the insurer fears that if default occurs on that bond, it will suffer important losses. It thus seeks default protection and enters into a CDS.

The CDS issuer sells a 5-year protection for a premium of 1% per year. The premium is paid at the end of each year and compensation in case of default is paid at year end. If default occurs, the protection buyer sells the defaulted bond to the protection seller at a price equivalent to a similar Treasury bond.

Let us illustrate the cash flows of the CDS if the bond issuer:

1) does not default; or,
2) defaults towards the end of the second year.

In the latter case, each defaulted bond is worth 50 whereas an equivalent 3-year Treasury bond trades at 94 (for a face value of 100).

In the most likely scenario, ABC inc. does not default over the next 5 years and the insurer pays $0.01 \times 100 = 1$ million per year for 5 years. In this scenario, it will receive nothing in return.

In the second scenario, the cash flows are as follows:

- Time 0: nothing;
- Time 1: payment of a premium of 1 million;
- Time 2: payment of a premium of 1 million, settlement of the CDS.

5 Source: Capital Markets Special Reports, Update on the Insurance Industry's Use of Derivatives and Exposure Trends 2015, NAIC.

There are two options for the settlement of the CDS:

- If it is settled physically, the investor (life insurer) delivers the defaulted bond to the protection seller who provides for 94 million in exchange.
- If it is settled in cash, the protection buyer receives 94 − 50 = 44 per 100 of face value, i.e. 44 million. The life insurer should also sell the underlying bond for 50 per 100 of face value and earn an extra 50 million, for a total of 94 million as well.

■

CDS are usually quoted in basis points (bps) where 100 bps corresponds to 1% of face value. The CDS premium is also known as the CDS spread and it is usually paid quarterly until maturity or default, whichever comes first.

CDS and the financial crisis

The CDS depicted above are written on typical bonds whose issuer could go bankrupt. However, CDS became famous during the financial crisis of 2007–2009 for CDS written on other securities that could potentially fail.

For example, American International Group (AIG), a major U.S. insurance company operating mostly in property and casualty insurance, expanded in the early 2000s in the CDS market. AIG sold CDS written on collateralized debt obligations themselves secured by residential mortgages (see "Securitization" in Chapter 2). When the residential market collapsed in 2008, billions of dollars worth of CDO also collapsed triggering payments on the CDS issued by AIG. Unfortunately, the reserves set aside by AIG to cover these events were insufficient. AIG was on the verge of declaring bankruptcy in 2008 when it was salvaged by the U.S. federal government.

The book and movie called "The Big Short" depict the story of several investors who bet on the collapse of the U.S. housing market. One of them asked several investment banks to issue CDS on mortgage bonds (written on residential mortgages) that would trigger a payment upon the failure of these bonds. In exchange for a low periodic premium, the CDS would provide a very high reward whenever the mortgage bonds would fail. Therefore, these CDS were used for speculation purposes.

4.4.2 Valuation

Valuation of a CDS is based upon a very important principle: holding the reference bond and a CDS on that reference bond should be a risk-free investment, or there would be arbitrage opportunities. Thus, we have:

$$\text{Reference bond} + \text{CDS on that bond} = \text{Risk-free (Treasury) bond.}$$

The following example illustrates this.

Example 4.4.2 *Valuation of a single-premium CDS*
Suppose there are two assets traded in the market:

- a risk-free zero-coupon bond, with $B_0 = 0.95$ and $B_T = 1$;
- a defaultable zero-coupon bond with $DB_0 = 0.93$ and random payoff DB_T.

Moreover:

- If the issuer of the defaultable bond does not default, then $DB_T = 1$;
- If it defaults, the creditor recovers $DB_T = 0.6$.

Let us find the (single) CDS premium to compensate the protection buyer in case the bond defaults.

Obviously, when the firm does not default, the return on the bond should be higher than a risk-free bond to compensate a risk-averse buyer for investing in that bond. That explains why $DB_0 < B_0$.

We want to issue a CDS that will compensate a bondholder in case of default. Therefore, the payoff of the CDS should be 0.4 when the firm defaults and 0 otherwise.

We know that the defaultable zero-coupon bond DB_T combined with the CDS should yield a risk-free position. Therefore, the up-front premium of the CDS should be 0.02, i.e. 0.95 – 0.93. Indeed, buying the defaultable bond and the CDS cost 0.95 in total and the portfolio provides a final payoff of 1, no matter whether the firm survives or not (1 + 0 vs. 0.6 + 0.4). Since the defaultable bond combined with the CDS yields the same payoffs as a risk-free bond in any scenario, they must trade at the same price to avoid arbitrage opportunities. ■

Computing the periodic CDS spread usually involves assessing default risk over time using a credit risk model setup under no arbitrage. This is beyond the scope of this book.

4.4.3 Comparing a CDS with an insurance policy

From its definition, it is easy to view a CDS as being similar to a term life insurance, but instead of paying a fixed or random benefit at death of an individual, the benefit is paid upon default of a reference company. This is possibly the only similarity between CDS and term insurance because there are many differences. Most of these differences stem from where CDS and insurance are sold as the financial market is much less regulated than the insurance market, as discussed in Chapter 1.

First, it is not required to have an *insurable interest* in the reference bond or company as is the case for insurance policies. Whereas an insurance policy can be viewed solely as a risk management instrument for an individual or a family, a CDS can be used for both speculation and risk management. As investors, life insurance companies use CDS for hedging purposes in a large proportion. In some sense, the life insurer has an *insurable interest* because, when investing in bonds, they lend money and are exposed to credit risk.

Second, insurance companies are also required to hold reserves to protect from future random claims whereas no such regulation is necessary for CDS protection sellers.

Another important difference between CDS and insurance policies comes from the nature of the insured risks. Typical risks covered in insurance policies are diversifiable whereas defaults (and in turn bankruptcies) are often tied to economic conditions which are much harder to diversify over a portfolio (see Chapter 1).

4.5 Commodity swaps

At inception, a **commodity swap** fixes the price paid for several *deliveries* of the underlying commodity, on a specified set of dates and forming the fixed leg. The floating leg of a commodity swap agreement consists of the unknown (at inception) values of this commodity at the predetermined dates. On each exchange date, the fixed-amount payer has the choice between receiving the commodity (physical settlement) or its cash equivalent (cash settlement).

A commodity swap can be viewed as a portfolio of commodity forwards. Indeed, to fix today the price we pay for delivery of a good at times $t_1, t_2, \ldots, t_n = T$, we can simply enter into

n commodity forwards with maturities at times $t_1, t_2, \ldots, t_n$, each at an initial cost of zero (as a forward contract costs nothing to enter). Because we can easily replicate the cash flows of a commodity swap with a portfolio of forward contracts, the initial no-arbitrage price of a commodity swap should also be equal to zero.

Valuation of a commodity swap is therefore based upon forward commodity prices. Recall that F_t^T is the delivery price determined at time t and to be paid at time T (for delivery at time T). Let us define by η_i the number of units of the commodity the swap promises to deliver (or its cash equivalent) at time t_i in exchange for a fixed payment of c. The value of the fixed-leg side of the swap is

$$B_t^{\times} = \sum_{i:t_i>t}^{n} \frac{c}{\left(1 + r_t^{t_i-t}\right)^{t_i-t}}$$

i.e. the present value of future fixed payments whereas the value of the floating-leg is

$$B_t^{\ell} = \sum_{i:t_i>t}^{n} \frac{\eta_i \times F_t^{t_i-t}}{\left(1 + r_t^{t_i-t}\right)^{t_i-t}}$$

i.e. the value of the remaining forwards in the swap. Because the delivery price is paid at future dates t_i (with respect to time t), we need to discount these prices back to time t. Finally, at inception, the fixed premium c is the unique one such that

$$B_0^{\ell} = B_0^{\times}.$$

Example 4.5.1 *Crude oil swap*

An airline company would like to manage the risk tied to the price of jet fuel. To do so, it enters a commodity swap on the crude oil barrel. Although the company does not have the expertise to transform crude oil into jet fuel, it enters a swap on crude oil as its price is highly correlated with the price of jet fuel. Of course, the swap will be cash settled. Such a crude oil swap allows the airline to receive the cash value of 1000 barrels of crude oil every month, for the next 3 months, in exchange for a fixed periodic payment. Let us find the fixed periodic premium given the following scenario for the term structure of crude oil forward prices and the term structure of Treasury rates:

Maturity	Forward price	Treasury rate
1 month	30	1%
2 months	31	1.2%
3 months	32	1.5%

At inception, we must have $B_0^{\ell} = B_0^{\times}$. In this scenario, on one side, we have

$$B_0^{\times} = c\left(\frac{1}{1.01^{1/12}} + \frac{1}{1.012^{2/12}} + \frac{1}{1.015^{3/12}}\right),$$

and, on the other side, we have

$$B_0^{\ell} = \frac{30}{1.01^{1/12}} + \frac{31}{1.012^{2/12}} + \frac{32}{1.015^{3/12}}.$$

Combining the two equations, we get $2.993469795c = 92.79467725$ and thus the periodic premium is $c = 30.999$. ■

4.6 Summary

Swaps

- Periodic exchange of (fixed or variable) cash flows related to the value of a given asset, at predetermined dates. Usually, a fixed payment is exchanged for a variable (random) payment.
- Interest rate swap: exchange a variable interest rate, applied to a notional, for a fixed interest rate, applied on the same notional.
- No cash flows (no premium) exchanged at inception (at time 0).
- Currency swap: exchange a principal and its interests in one currency for the same principal and its interests in another currency.
- Credit default swap: compensation in case a bond defaults in exchange for a fixed periodic premium.
- Commodity swap: periodic delivery of a commodity (or its cash equivalent) for a fixed periodic payment.

Terminology

- Maturity, swap term or swap tenor, is the date of the last cash flow.
- Fixed leg: stream of the fixed payments.
- Floating/variable leg: stream of the floating/variable/random payments.

Interest rate swaps

- Equivalent to exchanging a fixed-rate loan/bond for a floating-rate loan/bond.
- Fixed cash flows: periodic coupon and principal paid back at maturity.
- Variable cash flows: interests paid at the end of each period, and computed with a variable rate, and principal paid back at maturiy.

Valuation of a fixed-rate loan/bond

- Principal: F.
- Payment dates: $t_1, t_2, \dots, t_n = T$.
- Fixed payment (coupon): c.
- Fixed-rate bond's time-t value for $t \in [0, t_n)$:

$$B_t^{\mathrm{x}} = \sum_{i:t_i>t} \frac{c}{\left(1 + r_t^{t_i-t}\right)^{t_i-t}} + \frac{F}{\left(1 + r_t^{t_n-t}\right)^{t_n-t}}.$$

Valuation of a floating-rate loan/bond

- Value at inception: $B_0^{\ell} = F$.
- Value immediately after a coupon payment: $B_k^{\ell} = F$, for all $k = 1, 2, \dots, mT$.
- Value between two coupon payments: for $t \in [k-1, k)$,

$$B_t^{\ell} = \frac{F\left(1 + r_{k-1}^{1}\right)}{\left(1 + r_t^{k-t}\right)^{k-t}}.$$

Valuation of interest rate swaps

- Pays fixed, receives floating: $V_t^{\ell} = B_t^{\ell} - B_t^{\mathrm{x}}$.
- Pays variable, receives fixed: $V_t^{\mathrm{x}} = B_t^{\mathrm{x}} - B_t^{\ell}$.
- Swap rate: fixed (coupon) rate such that $B_0^{\mathrm{x}} = F$ (bond is priced at par).

Currency swaps

- Fixed-for-fixed: fixed interest in one currency against fixed interest in another currency.
- Fixed-for-floating: fixed interest in one currency against floating interest in another currency.
- Floating-for-floating: floating interest in one currency against floating interest in another currency.
- Time-t value of the domestic bond (in the domestic currency): B_t^d.
- Domestic currency: U.S. dollar.
- Time-t value of the foreign bond (in the foreign currency): B_t^f.
- Time-t price (in USD) to purchase one unit of the foreign currency: S_t (exchange rate at time t).
- Swap value for the party receiving the domestic currency: $V_t^d = B_t^d - S_t B_t^f$.
- Swap value for the party receiving the foreign currency: $V_t^f = S_t B_t^f - B_t^d$.

Credit default swaps

- Issuer of a bond may be unable to meet its obligations: default.
- Credit default swap (CDS): offers protection to the bondholder in case of default.
- Fixed periodic premium paid in exchange for a unknown/random compensation in case of default.
- Protection buyer: pays fixed premium until maturity or until default.
- Protection seller: pays compensation in case of default.
- Valuation principle: defaultable bond + CDS = risk-free (Treasury) bond.

Commodity swaps

- Equivalent to a portfolio of commodity forwards but with a constant periodic premium.
- Fixed leg:

$$B_t^{\mathrm{x}} = \sum_{i:t_i>t}^{n} \frac{c}{\left(1 + r_t^{t_i - t}\right)^{t_i - t}}.$$

- Number of units of the commodity delivered at time t_i: η_i.
- T-forward commodity price at time t: F_t^T.
- Floating leg:

$$B_t^{\ell} = \sum_{i:t_i>t}^{n} \frac{\eta_i \times F_t^{t_i - t}}{\left(1 + r_t^{t_i - t}\right)^{t_i - t}}.$$

- Periodic cash flow c determined such that $B_0^{\mathrm{x}} = B_0^{\ell}$.

4.7 Exercises

4.1 You are given the following zero-coupon bond prices.

Maturity (term)	Price
1 year	99.01
2 years	97.07
3 years	93.54
4 years	88.85
5 years	82.19

A 5-year interest rate swap exchanging fixed for floating is initiated today. Cash flows are exchanged annually.

(a) Find the swap rate.

(b) Suppose the swap is prepaid. As a fixed-rate payer, how much do you need to pay at inception to enter such a swap?

(c) Now suppose instead the swap is deferred by 1 year, meaning the first payment occurs at time 2. If a 6-year zero-coupon bond currently trades at \$78.13, find the swap rate of this 1-year deferred swap.

4.2 Over the years 2011 to 2015, you have recorded the following interest rates (spot):

Date/Maturity	Jan 1st, 2011	Jan 1st, 2012	Jan 1st, 2013	Jan 1st, 2014	Jan 1st, 2015
1 year	1%	1.10%	1.54%	1.39%	1.94%
2 years	1.46%	1.61%	2.25%	2.02%	2.83%
3 years	2.12%	2.33%	3.26%	2.94%	4.11%
4 years	2.79%	3.07%	4.30%	3.87%	5.41%
5 years	3.47%	3.82%	5.34%	4.81%	6.73%

A 5-year interest rate swap (with annual payments) has been issued on January 1st, 2011.

(a) What is the swap rate?

(b) Calculate the cash flows of the fixed and floating legs over the entire duration of the swap.

(c) What is the value of the swap on January 1st 2012 for the side receiving floating and paying fixed?

4.3 Two loans of \$10,000 have been issued 3 months ago. Interest is paid annually and principal is paid at maturity. You have the following data regarding interest rates:

- The one-year rate at inception of the loans was 4.2%.
- The current term structure of interest rates is given by the function $r_{0.25}^{T} = 0.04 + 0.004T$.

(a) Calculate the value as of today of a 5-year floating-rate loan for the issuing bank.

(b) Calculate the value as of today of a 5-year fixed-rate loan with an annual coupon of 4.5% for the issuing bank.

4.4 A 4-year fixed-for-floating interest rate swap with semi-annual payments has been issued today. The 6-month LIBOR rate is exchanged for some fixed rate. All rates are compounded semi-annually.

(a) The current LIBOR curve is $r_0^T = 0.04 + 0.004T$. Calculate the (semi-annual) swap rate.

(b) Thirteen months later, the LIBOR curve evolves to $r_{13/12}^{T} = 0.038 + 0.0041T$. Calculate the value of the swap for the fixed payer. Note that $r_1^{0.5} = 0.041$.

4.5 A fixed-for-floating interest rate swap is initiated between two parties by an intermediary. The fixed-rate payer agrees to pay 4% in exchange for the 6-month LIBOR. The intermediary charges a spread of 0.1% to both sides of the swap. The current 6-month LIBOR rate is 3.5%. Assume all interest rates are semi-annually compounded and that the 6-month LIBOR rate 6 months from now is 3%. Describe the cash flows to the three parties in 6 months and in 1 year for a notional of 10,000.

4.6 A 3-year fixed-for-floating currency swap is initiated today. The fixed side agrees to pay $100c\%$ annually in USD in exchange for the 1-year EUR LIBOR rate.

- The current exchange rate is 1.10 USD/EUR.
- The USD LIBOR curve is: $r_0^T = 0.01T$.
- The EUR LIBOR curve is: $r_0^T = 0.009T$.

Find c such that the swap costs nothing at inception.

4.7 A fixed-for-fixed currency swap was initiated some time ago. Fixed interest of 3.5% in USD is exchanged for 3.0% in CAD. You also know that:

- the exchange rate at inception was 1.33 CAD/USD;
- the current exchange rate is 0.85 USD/CAD;
- notional at inception was 100 USD;
- there are 4 years remaining in the swap;
- the USD LIBOR curve is: $r_0^T = 0.01T$;
- the CAD LIBOR curve is: $r_0^T = 0.012T$.

Find the current value of the swap in USD for the party that pays in USD.

4.8 ABC inc. some time ago issued a 10-year corporate bond with semi-annual coupons of 4% which is currently traded at \$91.25. If ABC inc. defaults on such bond, creditors will recover 40% of par upon default. Describe the cash flows of a 5-year credit default swap with semi-annual payments of 0.25% of par in two situations:

(a) ABC inc. does not default on the bond.

(b) ABC inc. defaults on the bond 4 years from now and the compensation upon default is 60% of a Treasury bond whose price is \$95.25.

4.9 You require delivery of 500 pounds of copper in 1 month, 600 pounds in 2 months and 650 pounds in 3 months. The term structure of interest rates and forward copper price (per pound) are shown in the next table.

Maturity (term)	Spot rate	Copper price
1 month	1.0%	2.10
2 months	1.1%	2.25
3 months	1.3%	2.45

(a) Describe the strategy necessary to fix today the price of future copper delivery using forwards. Calculate the related cash flows at each month (including inception).

(b) Compute the swap rate of a 3-month copper swap initiated today to receive 500, 600 and 650 pounds in 1, 2 and 3 months respectively.

(c) Compare the cash flows of the swap in (b) with the cash flows in (a). Why are they different?

5

Options

If you hold a long position in a forward contract, you have the *obligation* to buy the underlying asset at the maturity date, no matter the discrepancy between the delivery price and the price of the asset, if any. In a forward contract, the investor with the short position has the opposite *obligation*.

In an option contract, the obligation is replaced by the possibility to *run away* whenever the situation is not profitable. Indeed, if you have the long position in an *option to buy*, you have the *right* to buy the underlying asset at the maturity date. Of course, this will be the case only when the *delivery price* is less than the price of the asset, i.e. only when it is rational to do so. On the other hand, the investor with the short position still has the *obligation* to sell you this asset. However, this *optionality* in the option contract comes at a price: a premium must be paid (at inception).

Example 5.0.1 *Comparing a forward and an option to buy*
Suppose that you can find the following two securities:

- a forward maturing in 3 months and written on a stock with a 3-month forward price of \$48;
- an option to buy a stock maturing in 3 months, with a delivery price of \$48 and actually trading for \$2 (premium).

Let us describe the cash flows in the following two scenarios: in 3 months, the price of the stock will be \$45 or it will be \$53.

At inception, buying this option requires the payment of a premium of \$2 while the forward can be entered at no cost.

In the first scenario, the stock trades for \$45. If you have a long position in the forward contract, you are obliged to pay \$48 for an asset worth \$45, so you lose \$3. If you have a long position in this option to buy, you simply use your right not to buy the stock and therefore avoid any cash outflow.

In the second scenario, the stock trades for \$53. If you have a long position in the forward contract, you have to pay \$48 for an asset worth \$53, so you gain \$5. If you have a long position in the option, you use your right to buy the stock and also gain \$5.

Actuarial Finance: Derivatives, Quantitative Models and Risk Management, First Edition.
Mathieu Boudreault and Jean-François Renaud.

The cash flows are summarized in the next table.

		Long forward		Option to buy	
Scenario	Stock at maturity	CF at inception	CF at maturity	CF at inception	CF at maturity
1	45	0	−3	−2	0
2	53	0	5	−2	5

■

Example 5.0.1 contrasts the cash flows of a long forward contract with those of an option to buy.

Options and option-like contracts are very common and can be found outside financial markets. For example, insurance companies sell option-like riders in some of their policies (an important application in life insurance is presented in Chapter 8). You might also have encountered options if you leased a car and have been offered the possibility of buying the car back from the dealer at the end of the contract.

Example 5.0.2 *Car lease and buy-back option*
You have just signed a contract with a car dealer to lease a brand new vehicle for the next four years (at a given monthly payment). The contract offers the opportunity to buy the car back (from the dealer) at a price of $5,000 once the contract is over. This end-of-lease option is clearly not free: its value is already included in the monthly payments.

At the end of the contract, you analyze the option to buy the car. You go online, look at the ads and feel you could sell the car for $6,000. Therefore, you decide to use the option, i.e. buy the car from the dealer, and then you sell it, for an immediate benefit of $1,000. Suppose instead the market value of the car at the end of the contract is about $4,500. In this case, you will not buy the car from the dealer, as it is an asset worth only $4,500 on the market. In this scenario, you will return the car to the dealer. ■

The objective of this chapter is to introduce the reader to standard options. The specific objectives are to:

- understand options to buy (call) and options to sell (put) an asset;
- understand the difference between options and forward contracts;
- determine whether an option is used for hedging/risk management or speculating purposes;
- describe various investment strategies using options.

5.1 Framework

After forward contracts, options are the simplest examples of derivatives or contingent claims. A **call option** (or simply a *call*) is a financial contract, written on a given asset, that gives its owner the right, but not the obligation, to *buy* this asset for a given price at a future date (or set of dates). Similarly, a **put option** (or simply a *put*) is a financial contract, written on a given asset, that gives its owner the right, but not the obligation, to *sell* this asset for a given price at a future date (or set of dates). To summarize, a call is an option to buy whereas a put is an option to sell.

The asset on which the option is written is known as the **underlying asset** and it is specified in the option contract. It could be a security such as a stock or a bond, a commodity, or even a non-financial asset. We say an option is *written on* an underlying asset because the option gets its value from that asset.

The price at which the underlying asset can be bought or sold upon exercise is also specified in the contract. It is called the **strike price**, the **exercise price** or even the **delivery price** (as in forward contracts). This is similar to the delivery price in forward/futures contracts but with options, *delivery* might not take place. Buying or selling the underlying asset is known as **exercising** the option.

Finally, the duration, or life, of the option is also specified in the option contract. The date when the contract is initiated, i.e. the issuance date, is known as *inception* whereas the last day when the option can be exercised is called **maturity**, the maturity date or expiry date.

As discussed in Chapters 2 and 3, we will use the words *owner* and *buyer* (of the option) interchangeably to represent the investor with the long position in the option (long option holder), and the words *writer* and *seller* (of the option) for the investor with the short position (short option holder). One has to be careful with the terminology: it is possible to buy options to sell (put options) and sell options to buy (call options).

As seen above, not being required to exercise an option has value because the owner can avoid losing money at maturity. Therefore a **premium** must be paid at inception, or a price must be paid at later times, to acquire an options contract. Options commonly traded on financial markets require an up-front premium but many options-like riders found in insurance policies are paid during the life of the contract.

Therefore, options generate cash flows at two specific moments:

- At inception, the option buyer pays a premium to the option seller.
- At maturity, if the option is *exercised*, the asset and an amount of cash will be exchanged by the two parties involved. The net value is known as the option **payoff**.

The **profit** (or **loss** when negative) from buying or selling an option is the amount received, net of the amount paid. For example, when buying an option, the profit is calculated as the payoff net of the premium and when selling an option, the profit is the premium that we receive net of the payoff that we owe. Note that we do not take into account the time value of money; for most options, maturities are relatively short.

An option contract also requires to specify when, during the life of the option, exercise may happen. There are three categories:

- when an option can be exercised only at maturity, the option is said to be **European**;
- when an option can be exercised at any date between inception and maturity, the option is said to be **American**;
- when an option can be exercised on given dates between inception and maturity, the option is said to be **Bermudan**.

For example, an American put option gives its holder the right to sell the underlying asset for the strike price anytime before maturity.

Options' names

One may also wonder why some options' names are linked to a continent, country or city. Modern financial history generally dates back to the early 1970s with the creation of options markets in the U.S. However, option-like contracts were created long before the 20th century.

Typical option-like contracts traded in Europe in the 17th century could be exercised at maturity only whereas American option-like contracts traded in the U.S. in the late 18th century could be exercised anytime before maturity [17]. The names European and American options were popularized by the economist Paul Samuelson.

The name Bermudan option was coined because Bermuda is geographically located between Europe and the U.S. but closer to the U.S. (a Bermudan option is indeed closer to an American option than to a European option). There is also the Asian option, which is a type of exotic option whose payoff is determined by the average price (see Chapter 7). The name comes from the traders who were in Tokyo in the late 1980s when they developed the methodology to price such options.

The relationship between the asset price and the strike price determines the option's current *intrinsic value* and its *moneyness*. The **intrinsic value** of an option is the amount that would be received today if the option were exercised, regardless of whether such an exercise is allowed. The **moneyness** of an option is given by the relative positions of the underlying asset's price and the strike price. More precisely, an option is said to be:

- **in the money** if the intrinsic value is positive;
- **out of the money** if the intrinsic value is negative;
- **at the money** if the intrinsic value is zero.

Therefore, when an option is exercised at maturity, the intrinsic value and the payoff are equivalent.

In what follows, we will analyze standard call and put options only, also known as **vanilla options**. Unless stated otherwise, we will consider only European calls and puts whereas American options will be investigated in Chapter 6.

5.2 Basic options

As before, one should think of time 0 as *today*, while any other time t, for $t > 0$, is set *in the future*. The evolution of the price of a given asset (stock, index, commodity, etc.) will be represented by $S = \{S_t, t \geq 0\}$, where S_t stands for its (future) time-t price. In particular, S_0 is today's spot price, the price at which the asset is currently traded.

We will use the following notation for the main features of an option written on S:

- K is the strike price;
- T is the maturity date.

In other words, this option is *struck at K* and *matures at time T*.

The underlying asset price at the option's expiry is given by S_T. As of today, this value is unknown, because it is set *in the future*. Therefore, S_T is a random variable, and so is S_t, for any other $t > 0$.

5.2.1 Call options

A call option gives the right to buy the underlying asset S for a price of K at time T. At maturity, there are two possibilities:

- If S_T is less than K, then the option is out of the money. Buying the asset for K is not reasonable and the option is left unexercised.
- If S_T is greater than K, then the option is in the money. Buying the asset for K is rational and the option is exercised.

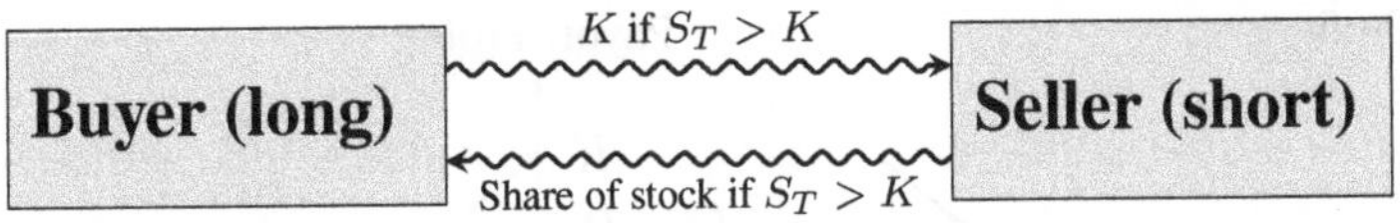

Figure 5.1 Exchanges between the buyer and the seller of a call option at maturity for a physical settlement

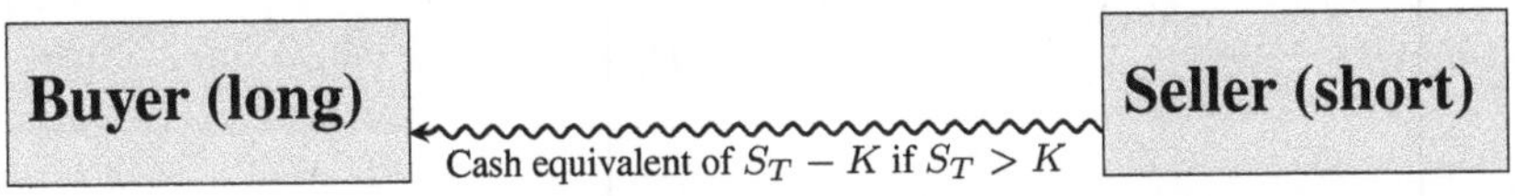

Figure 5.2 Exchanges between the buyer and the seller of a call option at maturity for a cash settlement

Figure 5.1 illustrates the cash flows between the buyer and the seller of the call option. If $S_T > K$, then the option's buyer pays K to the option's seller in exchange for an asset worth S_T. Overall, the buyer's wealth has increased by an amount of $S_T - K$.

When the asset is delivered to the call option buyer, we say the call is *physically* settled (**physical settlement**). Instead of delivering the asset, the call option seller could simply pay the cash equivalent of $S_T - K$ upon exercise. This is known as **cash settlement**. Cash settlement is illustrated in Figure 5.2.

For a liquid asset, physical and cash settlements are equivalent. Indeed, upon delivery of the asset, the call option owner could immediately sell the asset in the market and receive S_T in cash. Recall that we assume all assets are liquid.

Regardless of the type of settlement, the payoff of a **long call option** is given mathematically by

$$(S_T - K)_+ = \max(S_T - K, 0) = \begin{cases} 0, & \text{if } S_T \leq K, \\ S_T - K, & \text{if } S_T > K. \end{cases}$$

Here, we used the following mathematical function: $(x)_+ = \max(x, 0)$. In other words, for any given real number x, this function returns x if it is a positive number, otherwise it returns 0.

Recall that entering the long position in a call requires the payment of an up-front premium $C_0 > 0$. The following table summarizes the cash flows occurring at time 0 (inception) and at time T (maturity) for a long call option.

Time/Scenario	$S_T < K$	$S_T > K$
time 0 (inception)	$-C_0$	$-C_0$
time T (maturity)	0	$S_T - K$

We observe that no matter the scenario, the option buyer has to pay the premium, while the (random) payoff of the option depends on the terminal value S_T of the underlying asset. The upper-left plot of Figure 5.3 also shows the payoff and profit diagrams for a long call option.

A long call option with strike price K has all the upside benefits of a long *forward contract with delivery price K* without the downside risk. However, an up-front premium needs to be paid by the long call option holder.

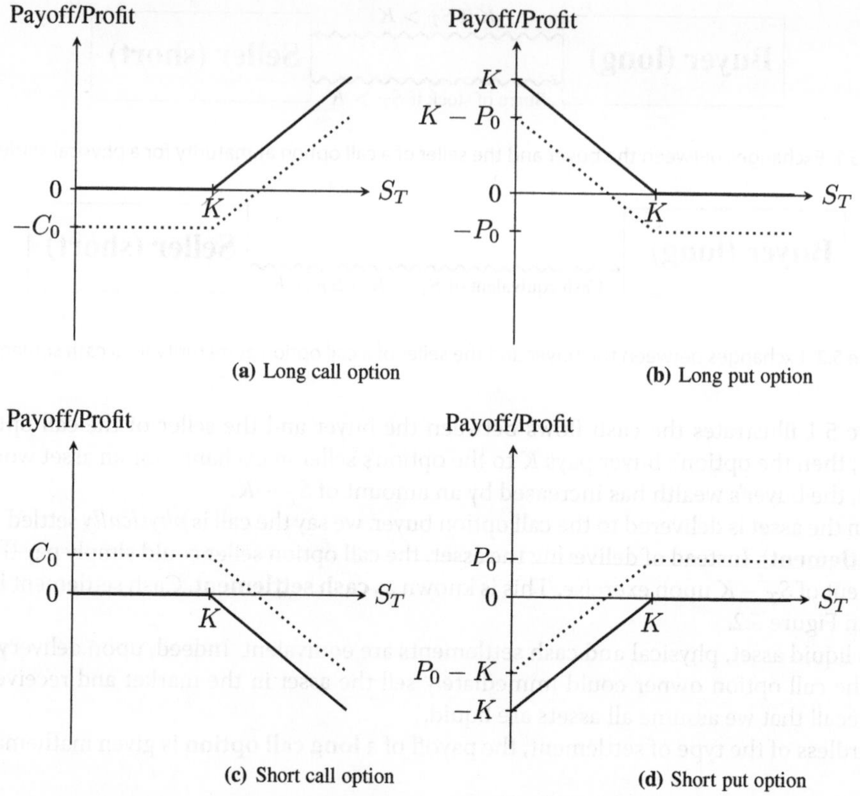

Figure 5.3 Payoff (continuous line) and profit (dashed line) of long/short call and put options

Example 5.2.1 *Long call option*
A 6-month European call option on a stock has just been issued. The call has a strike price of \$55 and currently trades at \$2.50. Let us describe the cash flows of a long position in this option for the following two scenarios: if the stock trades at \$60 or if it trades at \$50, 6 months from now.

At inception, the long call option holder needs to pay the premium of \$2.50.

At maturity, if the stock price is at \$60, then the option will provide the possibility to buy the stock at \$55 rather than at \$60. This payoff is equal to \$5 and the corresponding profit is equal to \$2.50.

At maturity, if the stock price is at \$50, the option will not be exercised. The payoff is thus equal to \$0 and the corresponding profit in this case is a loss of \$2.50.

The cash flows are summarized in the table.

Time/Scenario	$S_T = 50$	$S_T = 60$
At inception	−2.50	−2.50
At maturity	0	60 − 55 = 5

■

We now take the point of view of the option writer, i.e. the investor with the short position in the call option. For this investor, all cash flows are reversed, as shown in the following table.

Time/Scenario	$S_T < K$	$S_T > K$
time 0 (inception)	$+C_0$	$+C_0$
time T (maturity)	0	$-(S_T - K)$

Indeed, the option seller always receives a premium at inception but if the call option is exercised by the buyer, then the writer needs to deliver a unit of the asset and will receive K in exchange. The interactions between the option buyer and seller are summarized in Figures 5.1 and 5.2. Moreover, the lower left plot of Figure 5.3 shows the payoff and profit diagrams for a short call option.

Mathematically, the payoff of a **short call option** can be written as

$$-(S_T - K)_+ = -\max(S_T - K, 0) = \begin{cases} 0 & \text{if } S_T \le K, \\ -(S_T - K) & \text{if } S_T > K. \end{cases}$$

Example 5.2.2 *Short call option*
Using the values and scenarios provided in example 5.2.1, let us describe the cash flows of a call option writer.

At inception, the option seller receives a premium of \$2.50.

If the stock price goes down to \$50, then the option buyer will not exercise the option. In this scenario, the payoff will be \$0 and the profit for the option seller will be \$2.50.

If the stock price goes up to \$60, then the option buyer will exercise the option. This means the option seller will have to deliver a share of stock worth \$60 in exchange for \$55. In this scenario, the payoff will be −\$5 and the corresponding loss will be \$2.50.

The cash flows are summarized in the table.

Time/Scenario	$S_T = 50$	$S_T = 60$
At inception	+2.50	+2.50
At maturity	0	$-(60 - 55) = -5$

■

5.2.2 Put options

A put option gives the right to sell the underlying asset S for a price of K at time T. At maturity, there are two possibilities:

- If S_T is less than K, then the option is in the money. Selling the asset for K is rational and the option is exercised.
- If S_T is greater than K, then the option is out of the money. Selling the asset for K is not reasonable and the option is left unexercised.

Figures 5.4 and 5.5 illustrate the cash flows between the buyer and the seller of the put option. If $S_T < K$, then the option's buyer receives K from the option's seller in exchange for an asset worth S_T. Overall, the buyer's wealth has increased by an amount of $K - S_T$.

Mathematically, the payoff of a **long put option** is given by

$$(K - S_T)_+ = \max(K - S_T, 0) = \begin{cases} K - S_T & \text{if } S_T < K, \\ 0 & \text{if } S_T \ge K. \end{cases}$$

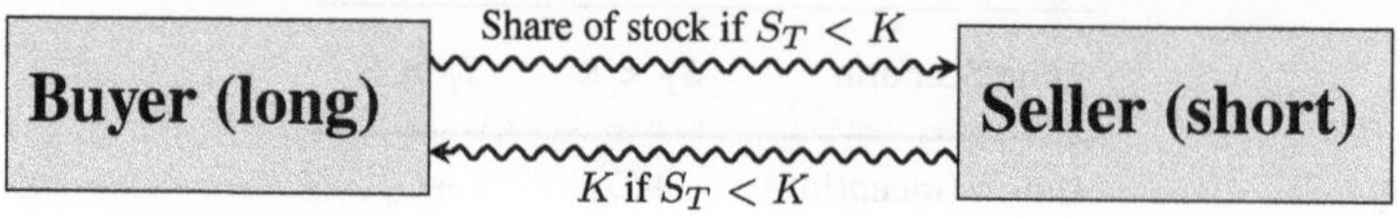

Figure 5.4 Exchanges between the buyer and the seller of a put option at maturity for a physical settlement

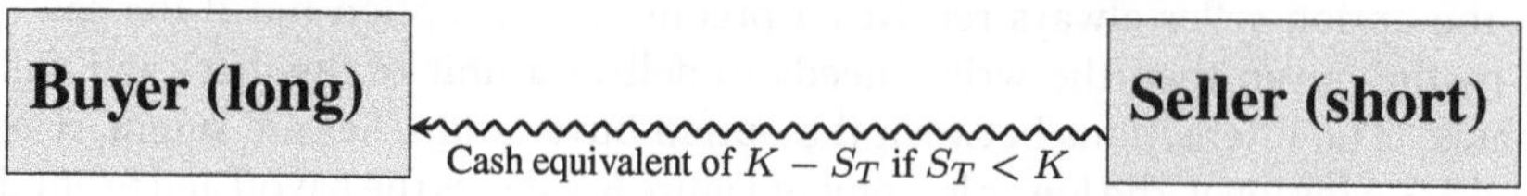

Figure 5.5 Exchanges between the buyer and the seller of a put option at maturity for a cash settlement

Here are the cash flows of a put option buyer:

Time/Scenario	$S_T < K$	$S_T > K$
time 0 (inception)	$-P_0$	$-P_0$
time T (maturity)	$K - S_T$	0

Also, the upper-right plot in Figure 5.3 shows the payoff and profit diagrams for a long put option.

A long put option with strike price K has all the benefits (if the underlying asset price goes down) of a short *forward contract with delivery price* K without the disadvantages (if the underlying asset price goes up). However, an up-front premium needs to be paid by the long put option holder.

Example 5.2.3 *Long put option*

A 4-month European put option on a stock has just been issued. The put has a strike price of \$60 and currently trades at \$3. Let us describe the cash flows of a long position in this option for the following two scenarios: if the stock trades at \$55 or if it trades at \$65, 4 months from now.

At inception, the long put option holder needs to pay the premium of \$3.

At maturity, if the stock price is at \$65, then the option will provide the possibility to sell the stock at \$60 rather than at \$65. The option will not be exercised, so the payoff is equal to \$0 and the corresponding loss is equal to \$3.

At maturity, if the stock price is at \$55, then the investor with the long position is allowed to sell for \$60 a stock worth \$55. The payoff is \$5 and the profit is then \$2.

The cash flows are summarized in the next table.

Time/Scenario	$S_T = 55$	$S_T = 65$
At inception	-3	-3
At maturity	$60 - 55 = 5$	0

■

Taking the point of view of the put option seller, then again all cash flows are reversed as shown in the following table.

Time/Scenario	$S_T < K$	$S_T > K$
time 0 (inception)	$+P_0$	$+P_0$
time T (maturity)	$-(K - S_T)$	0

Based upon Figure 5.4, we see that the investor with the long position in the put option will deliver/sell a share of the underlying asset for K (to the investor with the short position in the put option) only if $K > S_T$. Therefore, the investor with the short position in the put option will effectively be buying a unit of the asset for K. This is done in exchange for an up-front premium that the seller has received at inception. The bottom-right plot of Figure 5.3 shows the payoff and profit diagrams for a short put option.

Mathematically, the payoff of a **short put option** can be written as

$$-(K - S_T)_+ = -\max(K - S_T, 0) = \begin{cases} -(K - S_T) & \text{if } S_T < K, \\ 0 & \text{if } S_T \geq K. \end{cases}$$

Example 5.2.4 *Short put option*
Using the values and scenarios provided in example 5.2.3, let us describe the cash flows of a put option writer.

At inception, the investor issuing the put option receives a premium of \$3.

At maturity, if the stock price ends up at \$65, then the put option would provide its buyer with the possibility to sell the stock for \$60 (rather than \$65). Of course, she will abandon this possibility. The payoff from the seller's perspective is zero and the profit in this scenario would be \$3, i.e. the value of the initial premium.

At maturity, if the stock price ends up at \$55, then the option buyer will exercise its right of selling a share of stock for \$60 (rather than \$55). Therefore, the put option buyer delivers one share of stock to the option seller in exchange for \$60. The payoff is −5 and the loss is \$2.

The cash flows are summarized in the next table.

Time/Scenario	$S_T = 55$	$S_T = 65$
At inception	+3	+3
At maturity	−(60 − 55) = −5	0

■

5.3 Main uses of options

In Chapter 2, we mentioned the two main uses of derivatives – including of course options – which are hedging/risk management and speculation. The main difference between the two is in how the profits from the option are used: to offset a potential loss (hedging) or to magnify a potential gain (speculation).

5.3.1 Hedging and risk management

Options can be used for hedging, i.e. to attenuate risks as part of a risk management strategy. Insurance companies and pension plan sponsors have assets and liabilities that are exposed to variations in stock prices, interest rates, exchange rates, etc. They can use options to help mitigate those risks. As mentioned already in Chapter 2, when insurance companies use derivatives (including options), they mostly use them for hedging purposes.

Example 5.3.1 *Protective put*
Suppose the assets of a pension plan will decrease (resp. increase) significantly if the price of a stock goes down (resp. up). This is similar to being long that stock. As the investment actuary of that pension plan, you are interested in keeping the upside potential of your investment but you are seeking protection against the downside risk.

A long put option can play this role. In exchange for an up-front premium P_0, the put option gains whenever the stock price goes down whereas the payoff is zero otherwise. The next table shows how we can hedge a long stock position by simultaneously being long a put option.

Security/Scenario	$S_T \leq K$	$S_T > K$
Long stock	S_T	S_T
Long put	$K - S_T$	0
Total	K	S_T

We see from this table that buying a put option essentially limits the losses due to a drop in the stock price. It sets a lower limit on the portfolio value and this is why it is also known as a *floor*. The put option acts as an *insurance on the long stock position* and an up-front premium is required for this *insurance*. This strategy is also known as a **protective put.** ■

Example 5.3.2 *Call option for hedging purposes*
Suppose now that an investor is exposed to variations in the stock price such that she loses if the price goes up. This can be the case whenever an investor has shorted a stock she needs to buy back later. Indeed, she benefits whenever the stock price goes down but she is severely hit if the stock price goes up. Obviously, this investor would want to continue to profit from stock price decreases but limit losses if the stock price goes up. A long call option can play this role. In exchange for an up-front premium of C_0, the call option profits whenever the stock price goes up whereas the payoff is zero otherwise. The next table shows how we can hedge a short stock position with a long call option.

Security/Scenario	$S_T \leq K$	$S_T > K$
Short stock	$-S_T$	$-S_T$
Long call	0	$S_T - K$
Total	$-S_T$	$-K$

It is important to insist that, at inception, the investor will receive $S_0 - C_0 > 0$, which is the result of the short sale net of the cost of the call option. At maturity, the cash outflow is either S_T or K, whichever is the smallest. Note that in the scenario where $S_T > K$, the final payoff is $-K$ which in this case is greater than $-S_T$.

The short stock position is costly if the stock price increases so the call option caps at K the price at which the stock will have to be bought back. The call option acts as an *insurance on the short stock position* by limiting losses and this is why it is called a *cap*. Therefore, an up-front premium is required for this *insurance*. ■

Covered and naked options

For an option writer, a **covered option** is simply a short option combined with an offsetting position in the underlying asset, as opposed to a **naked option** which is just the short option.

We know that a call option writer should be prepared to deliver a share of stock if the option is exercised at maturity. Therefore, a **naked call** is simply a short call whereas a **covered call** is a strategy combining a short call and a long stock. In this case, if exercise occurs at maturity, the call option writer can deliver the share of stock he already owns. The payoff of a covered call is summarized in the following table.

Security/Scenario	$S_T \leq K$	$S_T > K$
Long stock	S_T	S_T
Short call	0	$-(S_T - K)$
Total	S_T	K

Similarly, a put option writer should be prepared to buy a share of stock if the option is exercised. Therefore, a **naked put** is simply a short put option whereas a **covered put** is a strategy combining a short put and a short stock. In this case, if exercise occurs at maturity, the put option writer will cover its short position in the stock. The payoff of this strategy is summarized in the following table.

Security/Scenario	$S_T \leq K$	$S_T > K$
Short stock	$-S_T$	$-S_T$
Short put	$-(K - S_T)$	0
Total	$-K$	$-S_T$

Although the final payoffs are negative, the investor receives a substantial income of $S_0 + P_0$ at inception.

5.3.2 Speculation

To speculate means taking investment positions to profit from anticipated asset price movements. It can also be interpreted as a *bet* because even though we believe the stock price will go up or down, it might not happen.[1] Therefore, an investor can use calls (resp. puts) to speculate that the stock price will go up (resp. down) in the future.

Example 5.3.3 *Call option for speculation purposes*
Suppose you have \$1000 to invest. A share of a stock trades at \$100 and a 3-month at-the-money call trades at \$5. Assume for simplicity that the stock price can only increase and end up at \$125 or decrease and end up at \$75 in 3 months.

You have analyzed the risk-return profile of this stock and truly believe its price will increase to \$125. You want to compare two investment opportunities:

1) \$1000 worth of stocks;
2) \$1000 worth of call options on that stock.

1 The words *speculation* or *bet* shall not be interpreted in a pejorative manner. Before entering into such a trade, investors will thoroughly analyze the risk-return profile and all available information.

Buying 10 shares of stock costs \$1000. If your prediction is realized, then it will yield a profit of \$250 in 3 months (\$25 per share). On the other hand, \$1000 also buys 200 call options. For this other financial position, if your prediction is realized, then the payoff would be \$25 per option and the profit would be of \$20 per option or \$4000 in total. The profit of this second investment opportunity is much larger than the first one, if your prediction is realized.

Call options provide what is known as *leverage* because you would need to buy 20 shares of stock to have the same gain as a single call option. It is important to note that if your prediction is not realized and the stock price goes down to \$75, the loss incurred by buying the stocks is limited to \$250, but with 200 call options, you would lose your entire investment, i.e. \$1000.

■

Example 5.3.4 *Put option for speculation purposes*
Whenever an investor believes the stock price of a company will go down, she can speculate by:

1) shorting the stock; or
2) buying a put option.

Suppose the current stock price is \$50 and it can only increase to \$60 or decrease to \$40 6 months from now.

Let us begin by looking at what happens if we short the stock. We receive \$50 by selling the stock today, knowing we will have to buy it back in 6 months. If the stock price were to decrease to \$40, we would make a profit of \$10 but, if the stock price were instead to increase to \$60, we would suffer a loss of \$10.

Suppose there is an at-the-money put option available (i.e. with a strike price of \$50) for a premium of \$4. We buy one unit of this option. If the stock price decreases to \$40, we will exercise the option, i.e. sell at \$50 and immediately buy one stock at \$40, for a payoff of \$10. The profit is \$6 and we will have a return on investment of 150%. But if the stock price goes up, the option will be left unexercised and we will lose the entire investment of \$4, for a return of −100%.

■

In conclusion, buying call and put options instead of buying or short-selling the underlying asset amplifies the exposure to changes in the asset price, which is known as the **leverage effect**. We need to remember that it exaggerates both gains and losses.

5.4 Investment strategies with basic options

The previous section showed that a put (or a call) option can be used to profit from a price drop (increase) in the underlying asset. We now look at how to use combinations of options to benefit from, or protect our position against, different types of movements of the underlying asset price. In what follows, unless stated otherwise, all options are written on the same underlying asset S and have the same maturity date T.

Bull spreads

A *bull spread* is a strategy that benefits when the price of the underlying asset goes up. Instead of buying a single call option, the idea of a bull spread is to reduce the overall cost by reducing the upside potential. Therefore, a **bull spread** consists in buying a call with strike price K_1 and

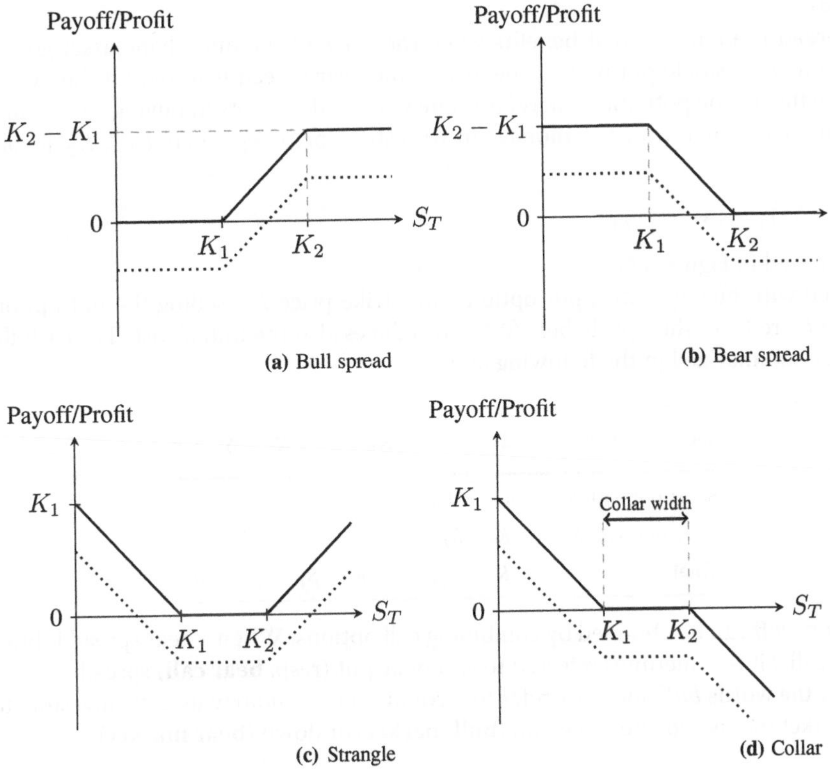

Figure 5.6 Payoff (continuous line) and profit (dashed line) of various strategies involving options

selling an out-of-the-money call with strike price K_2, where $K_1 < K_2$. The payoff of the bull spread is given by

$$(S_T - K_1)_+ - (S_T - K_2)_+$$

and is illustrated in Figure 5.6.

Compared with buying only a call option with strike price K_1, selling the call option with strike price K_2 reduces the upside benefit but also reduces the initial cost. Indeed, the lower the strike price, the higher the value of the call. The cash flows at maturity are summarized in the following table.

Security/Scenario	$S_T \leq K_1$	$K_1 < S_T < K_2$	$S_T \geq K_2$
Long call with K_1	0	$S_T - K_1$	$S_T - K_1$
Short call with K_2	0	0	$-(S_T - K_2)$
Total	0	$S_T - K_1$	$K_2 - K_1$

Note that it is also possible to obtain a similar payoff by combining put options. When the bull spread is constructed with call (resp. put) options, it is referred to as a **bull call** (resp. **bull put**) spread.

Bear spreads

A **bear spread** is a strategy that benefits when the price of the underlying asset goes down. Instead of buying a single put option, the idea of the bear spread is to reduce the overall cost by reducing the upside potential. Therefore, a **bear spread** consists in buying a put with strike price K_2 and selling an out-of-the-money put with strike price K_1, where $K_1 < K_2$. Its payoff is given by

$$(K_2 - S_T)_+ - (K_1 - S_T)_+$$

and is illustrated in Figure 5.6.

Compared with buying only a put option with strike price K_2, selling the put option with strike price K_1 reduces the upside benefit but it reduces also the initial cost. The cash flows at maturity are summarized in the following table.

Security/Scenario	$S_T \leq K_1$	$K_1 < S_T < K_2$	$S_T \geq K_2$
Short put with K_1	$-(K_1 - S_T)$	0	0
Long put with K_2	$K_2 - S_T$	$K_2 - S_T$	0
Total	$K_2 - K_1$	$K_2 - S_T$	0

A similar payoff can be obtained by combining call options. When a bear spread is built with puts (resp. calls) it is sometimes referred to as a **bear put** (resp. **bear call**) spread.

Note that the words *bull* and *bear* refer to a common terminology used by investors to designate a market that is expected to go up (**bull market**) or down (**bear market**).

Ratio spreads

We can generalize bull and bear spreads by buying n options with strike price K_1 and selling m options with strike price K_2. This strategy is known as a **ratio spread** and can be built with $m + n$ calls or $m + n$ puts.

Ratio spreads are usually quoted using the notation $n : m$ designating the ratio of the number of options bought and sold. Since m and n are flexible, it is possible to build a *zero-cost ratio spread.*

Variations on the ratio spread

There are other spread strategies that are similar in spirit to a ratio spread:

- A vertical spread is a ratio spread with $m = n$.
- A box spread can be obtained by combining a bull spread and a bear spread. It results in a risk-free position of $K_2 - K_1$.
- A butterfly spread is a strategy that involves buying two calls, one with strike K_1 and another with strike K_2, in addition to selling two calls with strike K, where $K_1 < K < K_2$.

Straddle

A **straddle** is a strategy used when (large) movements, upward or downward, in the price of the underlying asset are expected. It consists in buying a call and a put with the same strike price. The payoff of the strategy is then given by

$$(S_T - K)_+ + (K - S_T)_+ = |S_T - K|,$$

as illustrated in Figure 5.7.

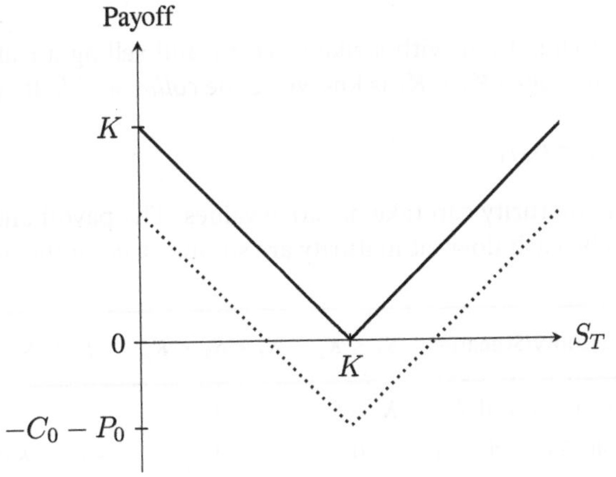

Figure 5.7 Payoff (continuous line) and profit (dashed line) of a straddle

With such a strategy, we benefit when S_T is far from K, but we have to pay $C_0 + P_0$. The cash flows at maturity are summarized in the following table.

Security/Scenario	$S_T \leq K$	$S_T > K$
Long call	0	$S_T - K$
Long put	$K - S_T$	0
Total	$K - S_T$	$S_T - K$

Strangle

We can generalize a straddle: it is called a **strangle**. This strategy can be used when (large) movements, upward or downward, in the price of the underlying asset are expected, but with a reduced cost.

A **strangle** consists in buying a call and a put option with different strike prices. More precisely, the strangle involves buying a call with a strike price $K_2 > K$ and a put with strike price $K_1 < K$. Both options are cheaper than their equivalent with a strike of K.

The payoff of the strategy is then given by

$$(S_T - K_2)_+ + (K_1 - S_T)_+$$

and is illustrated in Figure 5.6.

With such a strategy, we benefit when S_T is outside the interval (K_1, K_2). Again, we have to buy two options, but it is cheaper than a straddle with strike price K. The cash flows at maturity are summarized in the following table.

Security/Scenario	$S_T \leq K_1$	$K_1 < S_T < K_2$	$S_T > K_2$
Long put with K_1	$K_1 - S_T$	0	0
Long call with K_2	0	0	$S_T - K_2$
Total	$K_1 - S_T$	0	$S_T - K_2$

Clearly, if $K_1 = K_2$, then we recover a straddle.

Collar

A **collar** consists in buying a put with strike price K_1 and selling a call with strike price K_2, where $K_1 < K_2$. The difference $K_2 - K_1$ is known as the *collar width*. Its payoff is given by

$$(K_1 - S_T)_+ - (S_T - K_2)_+.$$

Note that the payoff at maturity can take negative values. The payoff and profit of a collar are shown in Figure 5.6. The cash flows at maturity are summarized in the following table.

Security/Scenario	$S_T \leq K_1$	$K_1 < S_T < K_2$	$S_T > K_2$
Long put with K_1	$K_1 - S_T$	0	0
Short call with K_2	0	0	$-(S_T - K_2)$
Total	$K_1 - S_T$	0	$-(S_T - K_2)$

A *zero-cost collar* is a collar such that strike prices are adjusted to have a zero initial cost. One possible case is when both K_1 and K_2 are equal to the forward price of the underlying asset. More details are given in the next chapter.

5.5 Summary

Options

- Underlying asset: S.
- Strike/exercise price: K.
- Maturity/expiry date: T.
- Provide the right, rather than the obligation, to buy/sell the asset S, for the strike price K, at maturity T.
- Difference from forwards/futures:
  - right to buy/sell the asset, not an obligation.
  - a premium must be paid.

Types of exercises

- European option: exercise is allowed only at maturity.
- American option: exercise is allowed anytime before or at maturity.
- Bermudan option: exercise is allowed only on pre-specified dates before or at maturity.

Payoff and profit

- Payoff: net value at maturity.
- Profit: payoff minus the premium.

Intrinsic value and moneyness

- Intrinsic value: amount that would be received today if the option were exercised, regardless of whether such an exercise is allowed.
- Moneyness: at any given time, an option is
  - in the money: if the intrinsic value is positive;
  - out of the money: if the intrinsic value is negative;
  - at the money: if the intrinsic value is zero.

Settlement

- Physical settlement: asset and strike price are physically exchanged between the option buyer and the option seller.
- Cash settlement: payoff is paid in cash.

Call options

- Provide the right to buy the underlying asset S for the strike price of K at maturity time T.
- Premium: C_0.
- Payoff: $(S_T - K)_+ = \max(S_T - K, 0)$.
- Intrinsic value at time t: $(S_t - K)_+ = \max(S_t - K, 0)$.

Put options

- Provide the right to sell the underlying asset S for the strike price of K at maturity time T.
- Premium: P_0.
- Payoff: $(K - S_T)_+ = \max(K - S_T, 0)$.
- Intrinsic value at time t: $(K - S_t)_+ = \max(K - S_t, 0)$.

Calls, puts and forward contracts

Position/Scenario	$S_T < K$	$S_T > K$
Long call	0	$S_T - K > 0$
Short call	0	$-(S_T - K) < 0$
Long put	$K - S_T > 0$	0
Short put	$-(K - S_T) < 0$	0
Long forward	$S_T - K < 0$	$S_T - K > 0$
Short forward	$K - S_T > 0$	$K - S_T < 0$

Main uses

- Hedging: attenuate risks as part of a risk management strategy; the gain is used to offset a potential loss.
  - Protective put, i.e. long stock + long put option: puts a lower limit on the portfolio value (*floor*).
  - Short stock + long call: puts an upper limit on the amount required to cover the short stock position (*cap*).
- Speculation: taking investment positions to profit from anticipated asset price movements.

Investment strategies with basic options

- Assumption: $K_1 < K < K_2$.
- Bull spread, i.e. long call K_1 + short call K_2:

$$(S_T - K_1)_+ - (S_T - K_2)_+.$$

- Bear spread, i.e. long put K_2 + short put K_1:

$$(K_2 - S_T)_+ - (K_1 - S_T)_+.$$

- Ratio spread, i.e. long n options K_1 + short m options K_2 (options should be of the same type):

$$n \times (S_T - K_1)_+ - m \times (S_T - K_2)_+$$

or

$$n \times (K_1 - S_T)_+ - m \times (K_2 - S_T)_+.$$

- Straddle, i.e. long call K + long put K:

$$(S_T - K)_+ + (K - S_T)_+ = |S_T - K|.$$

- Strangle, i.e. long call K_2 + long put K_1:

$$(S_T - K_2)_+ + (K_1 - S_T)_+.$$

- Collar, i.e. long put K_1 + short call K_2:

$$(K_1 - S_T)_+ - (S_T - K_2)_+.$$

5.6 Exercises

5.1 Butterfly spread
A butterfly spread is a strategy that involves buying one call each with strikes K_1 and K_2 and selling two calls with strike K, where $K_1 < K < K_2$. Note also that we should have $K_2 - K = K - K_1$.
(a) Detail in a table the payoff of each call option and the total payoff of the strategy.
(b) Plot the payoff diagram for this strategy.
(c) Describe the scenarios where the butterfly spread would be valuable.

5.2 Strips and straps
A strip combines one call option with two put options, whereas a strap combines two call options with one put. In all cases, the strike price is K. For the strip and the strap:
(a) Detail in a table the payoff of each option and the total payoff of the strategy.
(b) Plot the payoff diagram for this strategy.
(c) Describe the scenarios where strips and straps would be valuable.

5.3 Covered and naked options
(a) A stock currently trades for \$50 whereas a 1-month at-the-money call option on that stock trades for \$2. Assume that over the next month the stock price can only increase to \$55 or decrease to \$45. Describe the payoff and profit at maturity for a naked call and a covered call.
(b) Suppose a 1-month at-the-money put option trades at \$2 as well (for simplicity). Describe the payoff and profit at maturity for a naked put and a covered put option.

5.4 Options and forwards
(a) Suppose that in the context of example 5.2.1, a forward contract is also available for a delivery price of \$55. Compare the cash flows from a short forward and a short call option position.
(b) In the context of example 5.2.3, compare the profits from a long forward contract or a short put option.

5.5 Comparing a long call option and a long forward contract
Suppose that in the context of example 5.2.1, a forward contract is also available for a delivery/forward price of \$55 with delivery in 6 months. Compare the cash flows of a long forward and a long call along with the payoff and profit from these positions.

5.6 Comparing a long put option and a short forward contract
In the context of example 5.2.3, suppose a forward contract is also available such that the 4-month forward price is \$60. Compare the cash flows of a short forward contract with those of a long put option along with the payoff and profit from these positions.

5.7 Investment strategies
What securities, or combinations thereof, would you use to gain from each of the following situations?

(a) ABC inc. is being sued for patent infringement on an important product it sells. The court will make its decision public within 3 months. Its decision, positive or negative, is likely to affect future profitability of the company.

(b) Rumors are persistent on the bankruptcy of ABC inc.

(c) The government intends to tax sugar. Suppose ABC inc. refines sugar.

5.8 A box spread combines bull and bear spreads.

(a) Show that a box spread results in a risk-free position of $K_2 - K_1$.

(b) What should be the no-arbitrage price of this box spread?

(c) When is the box spread costless at inception?

5.9 Collars

(a) What is the no-arbitrage price of a collar such that $K = K_2 = K_1$?

(b) Show that when K corresponds to the T-forward price of the underlying asset, then the collar has zero cost.

5.10 Comparing options with different strike prices
Suppose we analyze two call options with strike prices $K_1 < K_2$. Let C_0^1 and C_0^2 be the up-front premiums for these two call options.

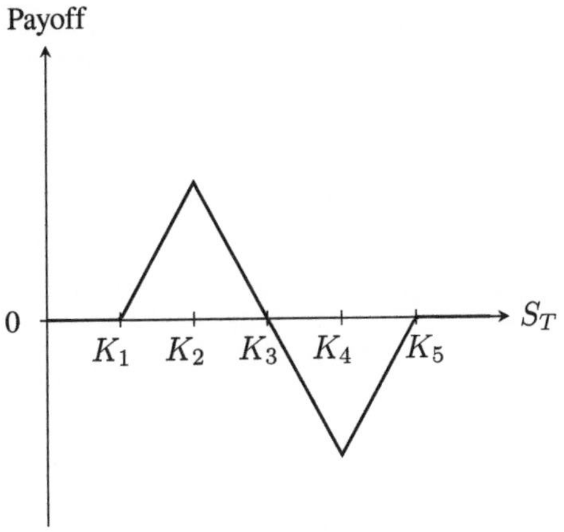

Figure 5.8 Payoff of a *heartbeat* option

(a) Explain intuitively why C_0^1 should be higher than C_0^2.
(b) Using no-arbitrage arguments, show that $C_0^1 > C_0^2$.

5.11 Heartbeat option
A heartbeat option is a totally fake option whose payoff diagram is given in Figure 5.8. Reverse engineer this option to find a strategy that exactly replicates the latter payoff diagram. Hint: butterfly.

5.12 Zero-cost ratio spread
A 40-strike call option currently sells for \$4.25 whereas a 42-strike call options trades for \$3.75. A ratio spread can be built by buying n 40-strike call options and selling m 42-strike call options. Find the range of values of m and n such that the ratio spread is costless at inception.

6

Engineering basic options

The word *engineering* usually involves the application of mathematics to study the structure and design of machines, tools, etc. As a result, the term *financial engineering* became increasingly popular, with the growth of derivatives markets in the 1990s, to designate the study of the structure and design of financial derivatives.

In Chapter 5, we introduced basic (vanilla) options such as call and put options and determined how they can be used for various investment purposes. The focus of this chapter is more on the mathematical structure of payoffs with the goal of designing and replicating simple financial products, and also obtaining parity relationships. *Engineering* also applies to common life insurance policies (see Chapter 8) where death and maturity benefits are designed using the mechanics of options.

The main objective of this chapter is to understand the basic financial engineering tools, i.e. to understand how to build and relate simple payoffs and then use no-arbitrage arguments to derive parity relationships. The specific objectives are to:

- use simple mathematical functions to design simple payoffs and relate basic options;
- create synthetic versions of basic options;
- obtain parity relationships between stocks, bonds and simple derivatives;
- understand the payoff structures of binary options and gap options;
- derive relationships between the prices of binary options, gap options and vanilla options;
- understand when American options should be (early-)exercised.

6.1 Simple mathematical functions for financial engineering

In this section, we analyze the properties of the *maximum function* and two of its relatives known as the *positive part* and the *stop-loss functions*. We will also look at indicator functions. All these functions determine the mechanics of call and put options payoffs. Some of them have strong roots in actuarial science, e.g. when losses are subject to deductibles and limits in (re)insurance policies. The reader familiar with these simple mathematical functions can easily skip this section.

6.1.1 Positive part function

We define the **positive part function**[1] by

$$(x)_+ = \max\{x, 0\} = \begin{cases} 0 & \text{if } x \leq 0, \\ x & \text{if } x > 0. \end{cases}$$

1 Note that in the literature it is also represented by $(x)^+$ instead of $(x)_+$.

Actuarial Finance: Derivatives, Quantitative Models and Risk Management, First Edition.
Mathieu Boudreault and Jean-François Renaud.

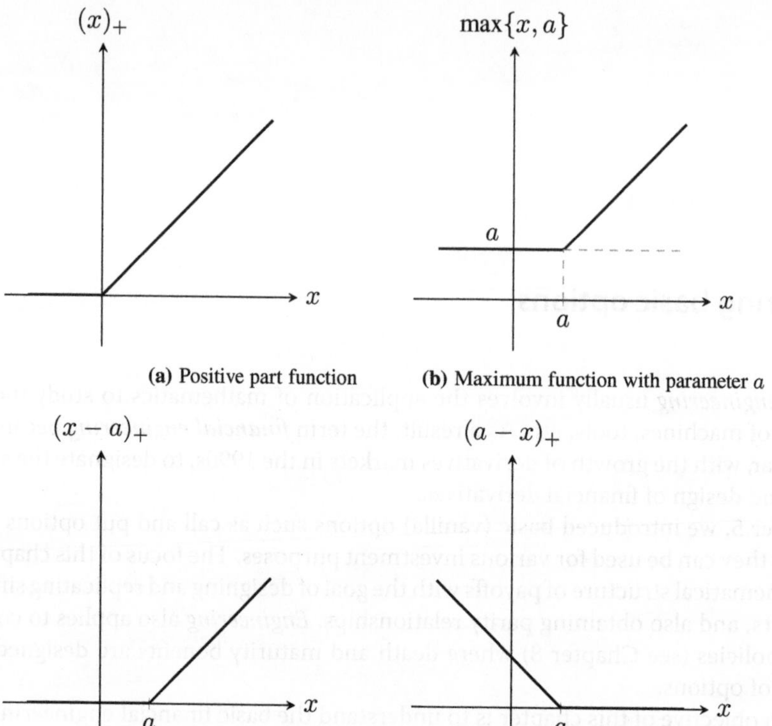

(a) Positive part function (b) Maximum function with parameter a

(c) Stop-loss function at a (d) Reverse stop-loss function at a

Figure 6.1 Maximum function and related functions

The plot of the function is shown in the top-left part of Figure 6.1. One may recognize the notation $(\cdot)_+$ as it was introduced in Chapter 5 to shorten the writing of call and put options' payoffs.

6.1.2 Maximum function

We can generalize the positive part function as follows. For a fixed real number a, we consider the **maximum function** $x \mapsto \max\{x, a\}$, defined by

$$\max\{x, a\} = \begin{cases} a & \text{if } x \leq a, \\ x & \text{if } x > a. \end{cases}$$

With this notation, a is a parameter of the function while x is the variable. The maximum function is illustrated in the top-right plot of Figure 6.1. Note that if we take $a = 0$, then we recover the positive part function.

When working with call and put options, one very important property of the maximum function is its behavior with respect to translations. For any real number b, we have

$$\max\{x, a\} + b = \max\{x + b, a + b\}$$

or, equivalently,

$$\max\{x, a\} - b = \max\{x - b, a - b\}.$$

These properties are easily deduced by looking at the plot of $\max\{x, a\}$, as given in Figure 6.1, and shifting it by $\pm b$.

6.1.3 Stop-loss function

We can also generalize the positive part function in another direction. For a fixed real number a, we will call the function $x \mapsto (x - a)_+$, the **stop-loss function** at a. From the definition of the positive part function, we can write

$$(x - a)_+ = \max\{x - a, 0\} = \begin{cases} 0 & \text{if } x \le a, \\ x - a & \text{if } x > a. \end{cases}$$

Again, with this notation, a is a parameter of the function while x is the variable.

One should easily recognize from the stop-loss function the structure of a call option payoff. Moreover, the stop-loss function is well known in actuarial science as $(x - a)_+$ determines the amount paid by the insurer when a claim x is subject to a deductible a.

It is also possible to look at the *reversed version* of the stop-loss function. For a fixed real number a, we have

$$(a - x)_+ = \max\{a - x, 0\} = \begin{cases} a - x & \text{if } x \le a, \\ 0 & \text{if } x > a. \end{cases}$$

Again, with this notation, a is a parameter while x is the variable. Both the stop-loss and reversed stop-loss functions are illustrated in the bottom panels of Figure 6.1. One should recognize from the reversed stop-loss function the payoff structure of a put option.

Using the above properties of the maximum function, we get that, for any other real number b,

$$(x - a)_+ + b = \max\{x - a + b, b\} \tag{6.1.1}$$

and, in particular,

$$(x - a)_+ + a = \max\{x, a\}. \tag{6.1.2}$$

Finally, if we take the difference of a stop-loss function with its reserved version, we get the following simple relationship:

$$(x - a)_+ - (a - x)_+ = x - a. \tag{6.1.3}$$

Indeed, since the functions $(x - a)_+$ and $(a - x)_+$ are never different from zero at the same time (for the same value of the variable x), we have

$$(x - a)_+ - (a - x)_+ = \begin{cases} 0 - (a - x) & \text{if } x \le a, \\ (x - a) - 0 & \text{if } x > a. \end{cases}$$

The identities in equations (6.1.2) and (6.1.3) are at the core of the upcoming parity relationships between vanilla options and the underlying asset.

6.1.4 Indicator function

For a fixed real number a, we define the **indicator function** of $\{x > a\}$ by

$$x \mapsto \mathbb{1}_{\{x>a\}} = \begin{cases} 0 & \text{if } x \le a, \\ 1 & \text{if } x > a. \end{cases}$$

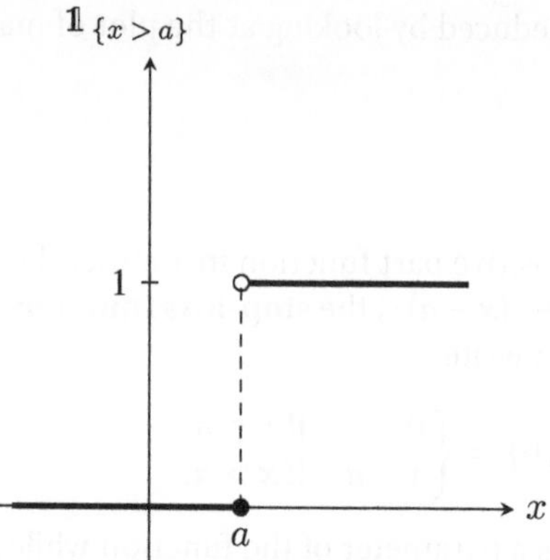

Figure 6.2 Indicator function of the event $\{x > a\}$

With this notation, a is a parameter of the function while x is the variable. The function takes the value 1 if and only if the variable x is greater than a; otherwise it takes the value 0. The same function can also be written as

$$x \mapsto \mathbb{1}_{(a,\infty)}(x).$$

In fact, in general, for any interval (a, b), we can define

$$x \mapsto \mathbb{1}_{(a,b)}(x) = \begin{cases} 0 & \text{if } x \notin (a,b), \\ 1 & \text{if } x \in (a,b). \end{cases}$$

In other words, if the variable x is in the interval (a, b) then the indicator function is equal to 1, otherwise it is equal to 0. For example, the graph of $\mathbb{1}_{\{x>a\}}$ is given in Figure 6.2. Indicator functions will be used mainly in Section 6.3.

6.2 Parity relationships

In this section, we make use of the mathematical properties of simple payoffs to deduce how we can build other financial products.

But first, recall from Section 2.5.2 that two products (or portfolios) having the same cash flows between inception and maturity must have the same (no-arbitrage) price at any time. For most derivatives, it is sufficient to compare the payoffs and premiums, as there are no other cash flows.

When manipulating payoffs/prices, we must recognize that a positive (negative) sign in a price equation means that we are taking a long (short) position. For example, if A, B and C are financial assets with time-t prices given by A_t, B_t and C_t, then

$$A_t + B_t = C_t$$

means that, at time t, being simultaneously long asset A and asset B is equivalent to being long asset C.

Therefore, being long asset A can be mimicked by being long asset C and being short asset B.

6.2.1 Simple payoff design

Let us consider the following four assets:

- a risky asset not generating any income (for example, a non-dividend paying stock) and whose final payout at time T is S_T;
- a call and a put option issued on the above (underlying) asset, both with maturity T and strike price K;
- a zero-coupon bond with maturity time T and face value K or, equivalently, a risk-free bank account whose accumulated balance with capital and interest is K at time T.

Using the maximum function, we will illustrate how to design two simple payoffs:

1) $S_T - K$, which is a payoff similar to that of a forward contract with maturity time T and delivery price K;
2) $\max(S_T, K)$, i.e. the payoff of an **investment guarantee** with maturity time T.

First, using equation (6.1.3) with $a = K$ and $x = S_T$, we can write

$$(S_T - K)_+ - (K - S_T)_+ = S_T - K. \tag{6.2.1}$$

Therefore, the payoff of a strategy which consists in being long a call option and short a put option is equal to the payoff of a strategy consisting in being long one unit of the underlying asset and having a loan worth K, principal and interests included. No matter the scenario, i.e. the value of the random variable S_T, the two positions will have the same payoff.

Reorganizing equation (6.2.1), we can write

$$(S_T - K)_+ = (K - S_T)_+ + S_T - K,$$

from which we deduce that a long call option can be replicated by a long put option together with a long position in a contract with payoff $S_T - K$ (or, said differently, one unit of the underlying asset and a loan worth K, principal and interests included). Again, no matter the value of S_T, the two positions will have the same payoff. This is a **synthetic call**. We use the word *synthetic* because legally we do not own a call. Of course, we can also reorganize equation (6.2.1) to create a **synthetic put**.

Second, an investment guarantee with payoff $\max(S_T, K)$, i.e. a product providing the largest value between the asset price S_T and the guarantee K at maturity, also plays an important role in life insurance as it forms the basis of many equity-linked insurance and annuity policies (see Chapter 8).

Using the translation property of the maximum function as given in equation (6.1.1), we get

$$\max(S_T, K) = (S_T - K)_+ + K \tag{6.2.2}$$

or, equivalently,

$$\max(S_T, K) = S_T + (K - S_T)_+. \tag{6.2.3}$$

Therefore, being long an investment guarantee can be mimicked by one of the two following strategies:

- a long call option together with a risk-free investment;
- a long protective put, i.e. a long put option together with one unit of the underlying asset.

Those strategies are summarized in the next table.

Position/Scenario	$S_T \leq K$	$S_T > K$
Long call	0	$S_T - K$
Risk-free investment	K	K
TOTAL	K	S_T
Long put	$K - S_T$	0
Long stock	S_T	S_T
TOTAL	K	S_T

Again, reorganizing equation (6.2.3), we can write

$$(K - S_T)_+ = \max(S_T, K) - S_T,$$

which says that a long put option can be replicated by buying an investment guarantee and shorting the underlying asset. This is another synthetic put. Of course, we can also reorganize equation (6.2.2) to create another synthetic call.

6.2.2 Put-call parity

Using some of the results from the previous section, we will now derive the so-called *put-call parity*.

Using equation (6.2.1) (or combining equations (6.2.2) and (6.2.3)), we can write

$$(S_T - K)_+ + K = (K - S_T)_+ + S_T.$$

In other words, the payoff of a call option and a risk-free investment (with final value K) is equal to the payoff of a put and one unit of the underlying asset.

By the no-arbitrage principle, since none of these securities generates any cash flows between inception time 0 and maturity time T, their prices at any previous times must also satisfy this relationship: for any $0 \leq t \leq T$,

$$C_t + Ke^{-r(T-t)} = P_t + S_t, \tag{6.2.4}$$

where C_t and P_t are the call and put prices at time t. In particular, at inception, we have

$$C_0 + Ke^{-rT} = P_0 + S_0. \tag{6.2.5}$$

This is the classical **put-call parity** relationship.

In other words, at time 0, buying a call and investing Ke^{-rT} at the risk-free rate r will generate the same payoff as buying a put and buying the underlying asset.

Example 6.2.1 *Call and put prices*

A stock currently trades at \$25 and a 3-month at-the-money European call option on that stock sells for \$2. If the annual interest rate is 2.7% (continuously compounded), what is the current (no-arbitrage) price of an otherwise equivalent put option?

We have $C_0 = 2$, $S_0 = K = 25$, $r = 0.027$ and $T = 0.25$. To avoid arbitrage opportunities, the put-call parity in (6.2.5) must hold. Therefore, we must have

$$2 + 25e^{-0.027\times 0.25} = P_0 + 25,$$

from which we deduce that $P_0 = 1.8318$. ■

Example 6.2.2 *Risk-free rates*
A 2-month European call currently sells for \$2 whereas a European put option sells for \$3. Their common strike price is \$53 and the stock trades for \$51. Let us determine the risk-free rate you can lock in by trading in the stock, the call and the put.

Using the put-call parity of equation (6.2.5), we can replicate the payoff of a long bond with capital K (i.e. a risk-free investment) using a long put, a long stock and a short call:

$$Ke^{-rT} = P_0 + S_0 - C_0 = 3 + 51 - 2 = 52,$$

where $K = 53$ and $T = 1/6$. Therefore, the annual risk-free rate (continuously compounded) that we can lock in by trading in the other assets is

$$r = -6 \times \ln(52/53) = 0.1143.$$

■

Example 6.2.3 *Arbitrage opportunities arising from the put-call parity*
With a strike price of \$50, a 1-month European call currently trades for \$3 whereas the corresponding European put option sells for \$2.50. The stock currently trades for \$51. Assume the 1-month rate is 0%. Let us determine whether there are arbitrage opportunities between the four assets and, if so, let us describe how to exploit this opportunity.

First of all, we verify whether the put-call parity of equation (6.2.5) holds or not. Since we have

$$C_0 + Ke^{-rT} = 3 + 50 = 53$$

and

$$P_0 + S_0 = 2.50 + 51 = 53.50,$$

the put-call parity is violated. Therefore, there are arbitrage opportunities. Indeed, in the current situation, a long put with a long stock is overpriced compared with a long call and an investment at the risk-free rate. In order to benefit from this mismatch, we can buy the cheapest portfolio (call and bond) and sell the costliest one (put and stock). Hence, we short-sell/write a put and short-sell the stock, and we buy the call and invest at the risk-free rate. This strategy generates an immediate cash amount of 0.50 and, as we saw above, at maturity all positions will offset each other. We have thus identified one arbitrage opportunity arising from the (violated) put-call parity. ■

In conclusion, the put-call parity in equation (6.2.4) links four financial assets through a no-arbitrage relationship: a call, a put, a stock and a zero-coupon bond (risk-free investment). Therefore, it is always possible to replicate one of these assets using the other three: to replicate a

1) long call: we need a long put, a long stock and a short bond (i.e. a loan);
2) long put: we need a long call, a long bond and a short stock;
3) long stock: we need a long call, a long bond and a short put;
4) long bond: we need a long put, a long stock and a short call.

6.3 Additional payoff design with calls and puts

We saw in Section 6.2.1 how to build simple financial payoffs/products with calls and puts using the mathematical properties of their payoff functions. It allowed us:

- to design a contract known as an investment guarantee;
- to synthetically replicate a forward contract;
- to derive no-arbitrage relationships between a risk-free bond, a risky asset, a call and a put issued on that asset.

We will now illustrate how to engineer two additional payoffs, namely those of binary (or digital) options and gap options.

6.3.1 Decomposing call and put options

Using indicator functions we can rewrite the payoff of a call as

$$(S_T - K)_+ = (S_T - K)\mathbb{1}_{\{S_T>K\}} = S_T\mathbb{1}_{\{S_T>K\}} - K\mathbb{1}_{\{S_T>K\}} \tag{6.3.1}$$

and the payoff of a put as

$$(K - S_T)_+ = (K - S_T)\mathbb{1}_{\{S_T<K\}} = K\mathbb{1}_{\{S_T<K\}} - S_T\mathbb{1}_{\{S_T<K\}}. \tag{6.3.2}$$

Therefore, a call option is the combination of two simple products (a similar decomposition can also be obtained for the put option):

- a long position in a derivative that pays S_T, if $S_T > K$, and 0 otherwise;
- a short position in a derivative that pays K, if $S_T > K$, and 0 otherwise.

It turns out that these two derivatives exist and are collectively known as *binary* (or *digital*) *options*.

6.3.2 Binary or digital options

A **binary or digital option** is a basic derivative in which, at maturity, the holder receives an amount of money or the underlying asset, if the underlying asset price at maturity is small/large enough, or nothing otherwise. Two elements define binary options: the condition under which money is received and what is paid in that case (cash or asset).

A **binary call option** is based upon the condition that $S_T > K$, i.e. the option is *activated* if and only if $S_T > K$ (at maturity), while a **binary put option** is based on the condition $S_T < K$, i.e. the option is activated if and only if $S_T < K$.

Then, for both binary calls and binary puts, the amount received at maturity is either fixed or calculated with the underlying asset price. For an **asset-or-nothing option**, the buyer receives the *cash equivalent* of the stock price at maturity when the condition is activated and for a **cash-or-nothing option**, the buyer receives \$1. Therefore, most binary options are settled in cash, but it is possible to opt for a physical delivery of the asset in an asset-for-nothing binary option.

To summarize, the payoffs of binary options are given by

- asset-or-nothing call: $S_T\mathbb{1}_{\{S_T>K\}}$;
- asset-or-nothing put: $S_T\mathbb{1}_{\{S_T<K\}}$;
- cash-or-nothing call: $\mathbb{1}_{\{S_T>K\}}$;
- cash-or-nothing put: $\mathbb{1}_{\{S_T<K\}}$.

They are illustrated in Figure 6.3.

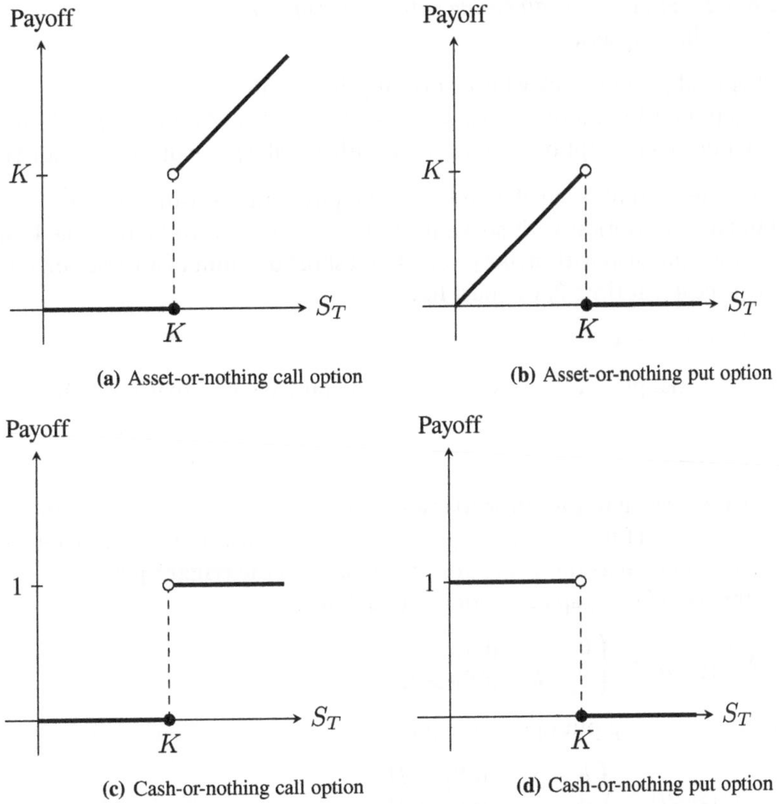

Figure 6.3 Asset-or-nothing and cash-or-nothing binary options

Example 6.3.1 *Payoff of binary options*
Three-month binary options are issued on the stock of ABC inc. For a strike price of \$50, compute the payoffs of the four binary options in the scenario where the final stock price is equal to \$52 (after 3 months).

In the scenario where $S_T = 52$, we get:

- asset-or-nothing call: the condition $S_T > K$ is met because $52 > 50$. Therefore, the holder receives \$52;
- asset-or-nothing put: the condition $S_T < K$ is not met because $52 > 50$. Therefore, the holder receives nothing;
- cash-or-nothing call: the condition $S_T > K$ is met because $52 > 50$. Therefore, the holder receives \$1;
- cash-or-nothing put: the condition $S_T < K$ is not met because $52 > 50$. Therefore, the holder receives nothing. ■

Finally, from equation (6.3.1), the payoff of a standard call option is equal to being long one unit of an asset-or-nothing call and short K units of a cash-or-nothing call, all options having the same strike price K. Similarly, from equation (6.3.2), the payoff of a standard put option is equal to being long K units of a cash-or-nothing put and short one unit of an asset-or-nothing call, all options having the same strike price K.

Example 6.3.2 *Financial engineering with binary options*
Consider the following assets:

- a non-dividend paying stock whose current price is \$45;
- a standard put option on that stock with a strike price of \$41 currently sells for \$2;
- a cash-or-nothing put option on that stock with a strike price of \$41 sells for \$0.27.

Let us find the current price of an otherwise-equivalent asset-or-nothing put.

As a standard put option with strike price 41 is equivalent to being long 41 units of a cash-or-nothing put also with strike price 41 and short one unit of an asset-or-nothing call with strike price 41, at time 0, we must have

$$2 = 41 \times 0.27 - x,$$

where x is the initial price for an asset-or-nothing put. We find that $x = 9.07$. ■

6.3.3 Gap options

A **gap call option** (resp. **gap put option**) is an option to buy (resp. to sell) the underlying asset for K if the stock price at maturity S_T is greater than (resp. less than) a predetermined trigger level H. Here, K is known as the strike price whereas H is the **trigger price**.

Therefore, the payoff of a gap call option is equal to

$$(S_T - K)\mathbb{1}_{\{S_T > H\}} = \begin{cases} 0 & \text{if } S_T \leq H, \\ S_T - K & \text{if } S_T > H, \end{cases}$$

whereas the payoff of a gap put option is equal to

$$(K - S_T)\mathbb{1}_{\{S_T < H\}} = \begin{cases} K - S_T & \text{if } S_T < H, \\ 0 & \text{if } S_T \geq H. \end{cases}$$

Figure 6.4 illustrates the payoff of both the gap call and put options. Clearly, if $H = K$, then a gap option is equivalent to its otherwise-equivalent standard vanilla option.

! It is important to note that, depending on the relationship between K and H, it is possible that the payoff of a gap option takes on *negative* values. Indeed, for a gap call, if $H < K$ and if at maturity the final underlying price is such that $H < S_T < K$, then in this scenario $(S_T - K)\mathbb{1}_{\{S_T > H\}} = S_T - K < 0$. There is a similar possibility for a gap put.

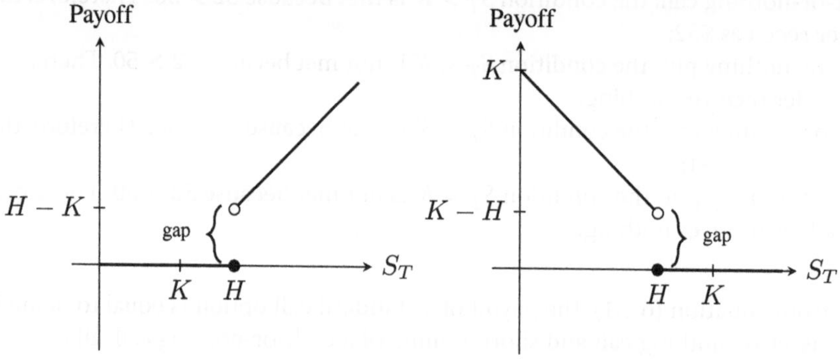

(a) Gap call option (b) Gap put option

Figure 6.4 Gap call and put options

Note that we can engineer the payoffs of gap options as follows:

$$(S_T - K)\mathbb{1}_{\{S_T>H\}} = S_T\mathbb{1}_{\{S_T>H\}} - K\mathbb{1}_{\{S_T>H\}}$$
$$(K - S_T)\mathbb{1}_{\{S_T<H\}} = K\mathbb{1}_{\{S_T<H\}} - S_T\mathbb{1}_{\{S_T<H\}}.$$

Therefore, we deduce that a gap call is equivalent to being long one unit of an asset-or-nothing call (with strike price H) and being short K units of a cash-or-nothing call (also with strike price H). Similarly, a gap put is equivalent to being long K units of a cash-or-nothing put (with strike price H) and being short one unit of an asset-or-nothing put (also with strike price H).

Example 6.3.3 *Payoff of a gap put option*
The strike price of a gap put option is \$74 whereas its trigger price is \$70. Compute the payoff of this gap put option in the scenario that the stock price is equal to \$72 at maturity.

The gap put option's payoff is activated if the stock price at maturity is below the trigger price, otherwise it is worthless. In this scenario, since $S_T = 72 > 70 = H$, the payoff is equal to 0. ■

Example 6.3.4 *Price of a gap call option*
The price of an asset-or-nothing call option with a strike price of \$30 is \$9 and the price of a cash-or-nothing call option also with a strike price of \$30 is \$0.25. Find the price of a gap call option with a trigger price of \$30 and a strike price of \$25.

As we saw before, the payoff of this gap call can be written as

$$(S_T - 25)\mathbb{1}_{\{S_T>30\}} = S_T\mathbb{1}_{\{S_T>30\}} - 25\mathbb{1}_{\{S_T>30\}}.$$

This payoff is the same as being long one unit of the asset-or-nothing call and being short 25 units of the cash-or-nothing call.

Therefore, the actual price of the gap call option is

$$9 - 25 \times 0.25 = 2.75.$$ ■

6.4 More on the put-call parity

Now let us have a second look at the put-call parity. First, we will obtain bounds on option prices with the help of the put-call parity obtained earlier, and second, we will extend the put-call parity to dividend-paying assets.

6.4.1 Bounds on European options prices

Let us establish bounds on the price of call and put options. Simply said, price bounds are values that a no-arbitrage price cannot exceed. The main focus will be on the financial arguments needed to obtain those bounds rather than their *sharpness*, i.e. how close they are from the true value. In fact, as we will see in the examples below, most of those bounds are not very informative. The bounds are perhaps more useful to better understand the design of call and put options and the behavior of American options, as we will see in Section 6.5.

Again, unless stated otherwise, the calls and puts are written on the same underlying asset S not generating any cash flows/dividends, have the same strike price K and maturity date T.

Let us start by looking for lower bounds. As the payoff of a call and the payoff of a put option are always (in all scenarios) non-negative, we must have $C_0 \geq 0$ and $P_0 \geq 0$, or clearly there is an arbitrage opportunity.

We can say more. From the put-call parity in equation (6.2.5), we have $P_0 = C_0 + Ke^{-rT} - S_0$ and, since $P_0 \geq 0$, we then have

$$C_0 + Ke^{-rT} - S_0 \geq 0,$$

from which we deduce that

$$C_0 \geq S_0 - Ke^{-rT}.$$

Finally, since we must also have that $C_0 \geq 0$, we conclude with the following:

$$C_0 \geq \max\{S_0 - Ke^{-rT}, 0\}. \tag{6.4.1}$$

Using similar arguments, we can also obtain the following lower bound for a put option's initial price:

$$P_0 \geq \max\{Ke^{-rT} - S_0, 0\}. \tag{6.4.2}$$

Example 6.4.1 *Lower bounds on call and put prices*
The stock of ABC inc. currently trades at \$34. If the 3-month rate is 1% (continuously compounded), identify lower bounds on at-the-money call and put options prices both maturing in 3 months.

From equation (6.4.1), we have

$$C_0 \geq \max\{S_0 - Ke^{-rT}, 0\} = \max\{34 - 34e^{-0.01\times 0.25}, 0\} = 0.0849$$

and from equation (6.4.2), we have

$$P_0 \geq \max\{Ke^{-rT} - S_0, 0\} = \max\{34e^{-0.01\times 0.25} - 34, 0\} = \max(-0.0849, 0) = 0.$$

Therefore, for a strike price of \$34, if we find a call trading for less than 8.49 cents, there is an arbitrage opportunity. On the other hand, the bound on the put option price is not informative. ■

Let us now look for upper bounds. First, using arbitrage arguments, we will show that $C_0 < S_0$. To do so, let us assume the opposite, i.e. let us assume that $S_0 \leq C_0$. We consider the following portfolio:

- sell the call;
- buy the underlying asset;
- lend $C_0 - S_0$ at the risk-free rate.

This portfolio does not require any investment at time 0. Therefore, if it yields a non-negative payoff at maturity, it is an arbitrage opportunity and the inequality $S_0 \leq C_0$ cannot be true.

At maturity, the payoffs are depicted in the following table.

Position/Scenario	$S_T \leq K$	$S_T > K$
Short call	0	$-(S_T - K)$
Long stock	S_T	S_T
Total	S_T	K

Selling a call and buying the underlying asset yields the strictly positive final payoff $\min\{S_T, K\}$. Moreover, our investment at the risk-free rate yields $(C_0 - S_0)e^{rT} \geq 0$ at maturity. Overall, the total value of the portfolio at maturity is given by the strictly positive payoff

$$\min\{S_T, K\} + (C_0 - S_0)e^{rT}.$$

This is an arbitrage opportunity. Consequently, we must have $C_0 < S_0$.

For the put, using the put-call parity together with the newly obtained inequality $C_0 < S_0$, we get

$$C_0 = P_0 - Ke^{-rT} + S_0 < S_0,$$

yielding $P_0 < Ke^{-rT}$.

In conclusion, we have obtained the following lower and upper bounds on the initial prices of European call and put options (on an underlying that is not generating cash flows, i.e. a non-dividend-paying asset):

$$\max\{S_0 - Ke^{-rT}, 0\} \leq C_0 < S_0$$
$$\max\{Ke^{-rT} - S_0, 0\} \leq P_0 < Ke^{-rT}.$$

Example 6.4.2 *Bounds on call and put options*
Recall example 6.4.1. We now compute the upper bounds on those call and put prices. The present value of the strike price is $Ke^{-rT} = 34e^{-0.01\times0.25} = 33.915$. Therefore, the bounds on the call option premium are

$$0.0849 \leq C_0 < 34$$

whereas the bounds on the put option premium are

$$0 \leq P_0 < 33.915.$$

Any call or put option (with these features) traded for a price outside these bounds will generate an arbitrage opportunity. ■

Since these relationships will hold at any other time t in between inception and maturity, we have: for all $0 \leq t \leq T$,

$$\max\{S_t - Ke^{-r(T-t)}, 0\} \leq C_t < S_t \tag{6.4.3}$$
$$\max\{Ke^{-r(T-t)} - S_t, 0\} \leq P_t < Ke^{-r(T-t)}. \tag{6.4.4}$$

6.4.2 Put-call parity with dividend-paying assets

We already know that holding a stock or having a long position in a forward contract is quite different when dividends are paid by the stock. This is also the case for options. When holding a dividend-paying stock, the shareholder receives the dividends (cash flows between time 0 and time T). However, when holding a call option on this stock, no dividends are received during the life of the option.

Suppose now that S_t is the ex-dividend time-t price of a stock, i.e. after payment of dividends. We know that the payoff of a call (or a put) is calculated on the ex-dividend price. From equation (6.2.1), which says that

$$(S_T - K)_+ - (K - S_T)_+ = S_T - K,$$

we know that being long a call option and short a put option is equal to the payoff of a long forward contract with delivery price K (not the forward price, as it is the convention on the market). We obtained, in Chapter 3, the value of such a contract.

By the no-arbitrage principle, this relationship also holds at any prior time t, even for dividend-paying assets, because neither the call, the put nor the forward generates any cash flows between inception and maturity. Consequently, using the same notation as in Chapter 3, we have that, for any time t such that $0 \leq t \leq T$:

- If dividends are paid discretely (cash dividends) and reinvested at the risk-free rate for a total accumulated amount of D_T, then

$$\underbrace{C_t - P_t}_{\text{Long call and short put}} = \underbrace{S_t - Ke^{-r(T-t)} - D_t}_{\text{Long forward}}, \tag{6.4.5}$$

  where we used the fact that the time-t value of a (long) forward contract on a discrete-dividend paying asset is given by $S_t - Ke^{-r(T-t)} - D_t$ (see Chapter 3). In particular, at time 0, we have

$$C_0 - P_0 = S_0 - Ke^{-rT} - D_0.$$

  This is the put-call parity relationship when the underlying asset is paying discrete dividends.
- If dividends are paid continuously at rate γ (dividend yield) and reinvested in the asset, then

$$\underbrace{C_t - P_t}_{\text{Long call and short put}} = \underbrace{S_t e^{-\gamma(T-t)} - Ke^{-r(T-t)}}_{\text{Long forward}}, \tag{6.4.6}$$

  where we used the fact that the time-t value of a (long) forward contract on an asset paying continuous dividends is given by $S_t e^{-\gamma(T-t)} - Ke^{-r(T-t)}$. In particular, at time 0, we have

$$C_0 - P_0 = S_0 e^{-\gamma T} - Ke^{-rT}.$$

  This is the put-call parity relationship when the underlying asset is paying continuous dividends.

Again, if one of these relationships is violated, then there is an arbitrage opportunity.

Example 6.4.3 *Put option on a stock paying discrete dividends*
A dividend-paying stock currently trades at \$35 and a \$1 dividend is expected in a month. An at-the-money 2-month call option on that stock trades at \$4 and the risk-free rate is 3% (continuously compounded). To avoid arbitrage opportunities, what should be the price of a 2-month at-the-money put option?

Using the put-call parity of equation (6.4.5), we get

$$4 - P_0 = 35 - 35e^{-0.03\times 2/12} - e^{-0.03\times 1/12}$$

and deduce that $P_0 = 4.8229$. ■

Example 6.4.4 *Put-call parity on a stock index*
The S&P 500 index is currently at 1000. A 6-month option to sell the index for 990 currently trades for 25 whereas a similar option to buy the index trades for 50. You also know that the (average) dividend yield generated by the stocks in the index is 3% (continuously compounded). What is the price of a 6-month Treasury zero-coupon bond?

Let B_0 be the initial price of a 6-month zero-coupon bond (with face value 1). Using the put-call parity of equation (6.4.6), we get

$$50 - 25 = 1000e^{-0.03\times 0.5} - 990 \times B_0.$$

We get $B_0 = 0.9698$. ■

6.5 American options

As discussed in Chapter 5, European and American options differ on *when* the investor is allowed to exercise its option, i.e. when she can buy or sell the underlying asset. In an American option, the holder has the right to *exercise* at any time before or at maturity time T whereas an European option can be exercised only at the expiration date.

In practice most traded stock options are American. Moreover, many equity-linked insurance policies have American-like riders, i.e. insureds have additional *rights* that can be exercised at any convenient date for the policyholder.

A very important question that arises is *when an option buyer should exercise.* Unfortunately this is not an easy question to answer; we will come back to it later in the book when assuming a dynamic for the stock price. The focus of this section is rather to determine whether or not an American option should be exercised before maturity, and how early exercise affects an option price.

6.5.1 Lower bounds on American options prices

We will first compare European and American options that are otherwise identical, i.e. they are written on the same underlying asset, have the same strike price and maturity. To simplify the comparison, let us denote by P_t^A and P_t^E the time-t value of an American put and a European put, respectively. Also, let us denote by C_t^A and C_t^E the time-t value of an American call and a European call, respectively. We will use this notation temporarily, until the end of this section.[2]

Comparing European and American option prices

Intuitively, an American option should be worth at least as much as its European counterpart because an American option holder is able to exercise any time prior to or at maturity. As it gives more flexibility, it should have more value; otherwise there would be an arbitrage opportunity because a rational investor should not lose from having the possibility to exercise early. This is the case for both call and put options.

Indeed, assume instead that the European call is worth more than its American counterpart, i.e. assume $C_0^A < C_0^E$. In this case, we buy the American option, sell the European option and invest the (positive) difference $C_0^E - C_0^A$ at the risk-free rate. We hold on to the American call until maturity, i.e. we do not exercise early, so both options will offset each other at maturity.

2 In this book, and unless stated otherwise, the option is European.

We will thus end up with the profits generated by the risk-free investment. The same strategy works for put options and for any other time t between inception and maturity.

In conclusion, we have obtained the following relationships:

$$P_t^{\text{A}} \geq P_t^{\text{E}} \quad \text{and} \quad C_t^{\text{A}} \geq C_t^{\text{E}}, \tag{6.5.1}$$

for all $0 \leq t \leq T$.

Comparing with the exercise value

The **(early-)exercise value** or **intrinsic value** of an American option is the amount received upon exercise. Take, for example, an American put struck at K with expiration date T. The holder can decide at any time $0 \leq t \leq T$ to receive K in exchange for delivery of an asset worth S_t: the (early-)exercise value of this put is then $K - S_t$.

From equation (6.4.3) and equation (6.5.1), we have that

$$C_t^{\text{A}} \geq C_t^{\text{E}} \geq \max\{S_t - Ke^{-r(T-t)}, 0\} \geq S_t - K, \tag{6.5.2}$$

for any $0 \leq t \leq T$. In particular, we have $C_t^{\text{A}} \geq S_t - K$. In other words, at any time t during its life, the price of the American call option is larger than its exercise value $S_t - K$.

This is also true for American put options: for any $0 \leq t \leq T$, we have $P_t^{\text{A}} \geq K - S_t$. If, at some time t, we observe $P_t^{\text{A}} < K - S_t$, then there is an arbitrage opportunity: we buy the American put and exercise it right away, making an immediate risk-less profit of $K - S_t - P_t^{\text{A}} > 0$.

Note that we have not said that American option prices are *strictly* greater than their early-exercise values, as they could be equal. Those (random) times when the American option price equals its early-exercise value play an important role in the timing of early exercise. As the latter question will only be answered later, we now focus our attention on whether a rational investor should early-exercise an American call or put.

6.5.2 Early exercise of American calls

In equation (6.5.1), we have verified that an American call is always worth at least as much as its European counterpart, i.e. $C_t^{\text{A}} \geq C_t^{\text{E}}$. It would be natural to think/believe that this last inequality is in fact a strict inequality. The idea behind this belief is that the underlying stock S may reach its maximum value at some point in time during the life of the option, so that early exercise would be optimal at that time. In fact, this is never the case for American calls written on non-dividend-paying stocks:

It is never optimal to early-exercise an American call written on a non-dividend-paying stock.

Indeed, even if an investor believes the stock price has reached its peak value at some time t, then the American call price C_t^{A} still is larger than its early-exercise value, as obtained above. Mathematically, for all t, we have

$$C_t^{\text{A}} \geq S_t - K,$$

(see also equation (6.5.2)) no matter how large S_t is. This means that if the underlying stock S reaches a lifetime maximum, then the American call option price C^{A} will also contain this information and be worth more. Consequently, early-exercising an American call option at any time t is sub-optimal in the sense that we replace a financial position worth C_t^{A} by another position worth less, that is $S_t - K$.

To summarize, getting rid of an American call (on a non-dividend-paying asset) by exercising it before maturity is never a good idea. This would be equivalent to exchanging an asset that is worth more (the American call) with something of a lesser value (the exercise value). As shown by the bounds, selling the American call yields a greater income than exercising it.

Finally, note that if dividends are paid by the underlying asset S, then it could be optimal to early-exercise an American call option. For example, if dividends are discrete, then early exercise would occur right before the dividend payment.

6.5.3 Early exercise of American puts

Whether we should exercise an American put option early or not is more complicated. Using the put-call parity in equation (6.2.4), we get easily that

$$P_t^{\mathrm{E}} = (K - S_t) + C_t^{\mathrm{E}} - K(1 - \mathrm{e}^{-r(T-t)}),$$

for all $0 \leq t \leq T$. The objective is to compare P_t^{E} with the exercise value $K - S_t$.

In this case, since $C_t^{\mathrm{E}} \geq 0$ and $K(1 - \mathrm{e}^{-r(T-t)}) \geq 0$, we cannot deduce that $C_t^{\mathrm{E}} - K(1 - \mathrm{e}^{-r(T-t)})$ is always positive or always negative. In fact, it is sometimes positive, sometimes negative. Therefore, we cannot say if the European put is always worth more than its exercise value as we did for the European call.

If we were to exercise the American put option at time t, we would receive $K - S_t$. However, the European put option price P_t^{E} might be worth more or less than the exercise value. The same can be said about the American put option price P_t^{A}. In conclusion: it is not always *optimal* to early-exercise (or not) an American put option on a non-dividend-paying stock.

Early exercise of an American put on a non-dividend-paying stock

With the above analysis, we see that it will be *optimal* to early-exercise an American put at (random) time τ as soon as

$$C_\tau^{\mathrm{E}} < K(1 - \mathrm{e}^{-r(T-\tau)})$$

because, in this scenario, we will have

$$P_\tau^{\mathrm{E}} < K - S_\tau = P_\tau^{\mathrm{A}}$$

which is a strict inequality. Holding on to the contract (which is here given by the European put value) is worth less than the exercise value $K - S_\tau$, which at that time is equal to the American put value P_τ^{A}. Note that time τ is a random time, as we do not know if and when such a scenario will occur.

If the stock price drops dramatically (approaches 0), then the call option price will also get closer to zero. If this is the case at another (random) time τ, then the European put option is worth

$$P_\tau^{\mathrm{E}} \approx K\mathrm{e}^{-r(T-\tau)},$$

since then $C_\tau^{\mathrm{E}} \approx 0$ whereas the exercise value of the corresponding American put is approximately K.

6.6 Summary

Simple mathematical functions

- Positive part function: $(x)_+ = \max\{x, 0\}$.
- Stop-loss function: $(x - a)_+$.
- Reversed stop-loss function: $(a - x)_+$.
- Indicator function: $\mathbb{1}_{(a,b)}(x) = 1$, if $x \in (a, b)$, and $\mathbb{1}_{(a,b)}(x) = 0$, otherwise.

Relationships between simple payoffs

- Investment guarantee: $\max(S_T, K)$.
- Synthetic contracts:
  - $(S_T - K)_+ - (K - S_T)_+ = S_T - K$;
  - $\max(S_T, K) = (S_T - K)_+ + K$;
  - $\max(S_T, K) = S_T + (K - S_T)_+$.

Binary (or digital) and gap options

- Binary/digital options:
  - asset-or-nothing call: $S_T \mathbb{1}_{\{S_T>K\}}$;
  - asset-or-nothing put: $S_T \mathbb{1}_{\{S_T<K\}}$;
  - cash-or-nothing call: $\mathbb{1}_{\{S_T>K\}}$;
  - cash-or-nothing put: $\mathbb{1}_{\{S_T<K\}}$.
- Long one call = long one asset-or-nothing call + short K cash-or-nothing calls:
$$(S_T - K)_+ = S_T \mathbb{1}_{\{S_T>K\}} - K\mathbb{1}_{\{S_T>K\}}.$$
- Long one put = long K cash-or-nothing puts + short one asset-or-nothing put:
$$(K - S_T)_+ = K\mathbb{1}_{\{S_T<K\}} - S_T \mathbb{1}_{\{S_T<K\}}.$$
- Gap call option:
$$(S_T - K)\mathbb{1}_{\{S_T>H\}} = S_T \mathbb{1}_{\{S_T>H\}} - K\mathbb{1}_{\{S_T>H\}}.$$
- Long one gap call = long one asset-or-nothing call with strike price H + short K cash-or-nothing calls with strike price H.
- Gap put option:
$$(K - S_T)\mathbb{1}_{\{S_T<H\}} = K\mathbb{1}_{\{S_T<H\}} - S_T \mathbb{1}_{\{S_T<H\}}.$$
- Long one gap put = long K cash-or-nothing puts with strike price H + short one asset-or-nothing put with strike price H.
- When $H = K$, gap call/put options are equivalent to regular call/put options.

Parity relationships

- C_t is the time-t price of a European call.
- P_t is the time-t price of a European put.
- If S does not pay dividends, then the put-call parity is
$$C_t + K\mathrm{e}^{-r(T-t)} = P_t + S_t.$$
- If S pays discrete dividends (reinvested at the risk-free rate), then the put-call parity is
$$C_t + K\mathrm{e}^{-r(T-t)} = P_t + S_t - D_t.$$
- If S pays dividends continuously at rate γ (reinvested in the stock), then the put-call parity is
$$C_t + K\mathrm{e}^{-r(T-t)} = P_t + S_t\mathrm{e}^{-\gamma(T-t)}.$$

- To replicate a
  - long call: we need a long put, a long stock and a short bond (i.e. a loan);
  - long put: we need a long call, a long bond and a short stock;
  - long stock: we need a long call, a long bond and a short put;
  - long bond: we need a long put, a long stock and a short call.

Bounds on European options prices

- Bounds always hold because otherwise there would be arbitrage opportunities.
- Call options:

$$\max\{S_t - Ke^{-r(T-t)}, 0\} \leq C_t < S_t.$$

- Put options:

$$\max\{Ke^{-r(T-t)} - S_t, 0\} \leq P_t < Ke^{-r(T-t)}.$$

American options

- Options that can be exercised anytime before or at maturity.
- Exercise/intrinsic value: amount received upon exercise.
- It is never optimal to early exercise an American call written on a non-dividend-paying stock.
- It could be optimal to early exercise an American put written on a non-dividend-paying stock.

6.7 Exercises

6.1 Find the value of the strike price K for which a call and a put, written on the same underlying, with the same maturity date and strike price K, will have the same initial price, i.e. $C_0 = P_0$.

6.2 A financial market is composed of the following four assets:
- A share of stock paying quarterly dividends sells for \$47. The next dividend is due in 1 month.
- A 45-strike 6-month call option on that stock sells for \$3.25.
- A put option having similar characteristics trades for \$2.79.
- The term structure of interest rates is flat at 3% (continuously compounded).

Find the value of the quarterly dividend.

6.3 You hold a long forward position on a stock. A call option currently trades for \$4.50 whereas a similar put option sells for \$4. The call, the put and the forward have been issued at the same time and share the same delivery/strike price. How much will you receive or pay today to clear your long forward position in the markets?

6.4 The 3-month forward price for a barrel of crude oil is currently \$60. An option to buy a barrel for a price of \$62 in 3 months currently sells for \$4. A 3-month Treasury zero-coupon bond trades at 98 (face value of 100). Find the price of the corresponding put option to avoid arbitrage opportunities.

6.5 The risk-free rate is continuously compounded at a rate of 3%. A share of a non-dividend paying stock sells for \$54. Compute upper and lower bounds on prices of at-the-money call and put options maturing in 6 months.

6.6 A share of a non-dividend-paying stock sells for $18 and a 6-month zero-coupon bond trades for $95 (face value of $100). Two options are issued in this market: an at-the-money 6-month call option sells for $3 and a put option with similar characteristics trades for $2. Is there an arbitrage opportunity in this market and if this is the case, show how to exploit it.

6.7 A share of stock currently sells for $10 and can take two possible values in a year: $12 or $9. An at-the-money put option on that stock sells for $0.50.
(a) Calculate the payoff (at maturity) of an investment guarantee and the profit from the latter in each possible scenario.
(b) Calculate the possible returns in each scenario from the investment in (a) (profit over initial investment).
(c) Calculate the possible returns from investing in the stock only.
(d) Compare and comment on the latter two investment products: (i) investing in a stock, (ii) buying an investment guarantee.

6.8 Early exercise of American options
(a) The stock price of ABC inc. has been plunging for the last couple of days. It now trades at 50 cents. European call and put options issued 1 month ago with a strike price of $10 currently trade for $0.02 and $9.25. Options expire in 2 months. Determine whether it is optimal to exercise an American put option on the latter stock.
(b) The stock price of ABC inc. has been surging for the last couple of days and you believe it can only fall over the next weeks. Stock trades for $125 and you hold a 90-strike American call option. Explain whether you should exercise or not your American call option.

6.9 Bull and bear spreads
In Chapter 5, we built bull spreads using calls and bear spreads using puts. Using the put-call parity, show how to build:
(a) a bull spread using put options and other assets available in the market;
(b) a bear spread using call options and other assets available in the market;
In all cases, write the payoff table and compare with Chapter 5.

6.10 Using binary options, describe how to build the following strategies:
(a) bull spread;
(b) bear spread;
(c) straddle;
(d) strangle;
(e) collar.

6.11 As of today, you can find the current price of the following binary options:

	$K = 40$	$K = 42$
Cash-or-nothing call option	0.648	0.5176
Asset-or-nothing call option	29.87	24.53
Cash-or-nothing put option	0.332	0.4624
Asset-or-nothing put option	12.13	17.47

Compute the price of the following assets or strategies:
(a) a share of stock;
(b) zero-coupon bond having the same maturity as the binary options;
(c) gap call option with strike price of 40 and trigger price of 42;
(d) a bull spread with strikes 40 and 42;
(e) a bear spread with strikes 40 and 42;
(f) a strangle with strikes 40 and 42;
(g) a collar with strikes 40 and 42.

6.12 Bounds on options prices
In this chapter, we have determined bounds on call and put option prices when the underlying asset does not pay dividends. Using the appropriate put-call parity relationship, derive lower and upper bounds on options prices using similar steps than when the asset does not pay dividends.
(a) dividends are discrete and reinvested at the risk-free rate;
(b) dividends are paid continuously and also continuously reinvested in the stock.

6.13 Early exercise of an American call option on dividend-paying stocks
A share of stock trades for \$40 and a dividend of \$5 will be paid 1 week prior to maturity. Using the bounds derived in exercise 6.12 and the put-call parity, explain whether the following American options should be exercised and when such exercise should occur.
(a) American call option;
(b) American put option.

7

Engineering advanced derivatives

In Chapter 6, we engineered *basic derivatives*, including an investment guarantee, binary options and gap options. European call and put options, in addition to the abovementioned derivatives, fall into the category of **simple options** because their payoffs depend only on the underlying asset price S_T at the maturity of the derivative.

However, on financial markets, institutional investors such as banks, insurance companies and pension plans design and trade more sophisticated contracts tailored for their specific needs. Those derivatives/options are said to be *exotic* or *path-dependent*, while others are called *event-triggered derivatives*.

A derivative is said to be **exotic** or **path-dependent** if its payoff depends on the underlying asset price at more than one date during the life of the derivative. **Event-triggered derivatives** are contracts whose payoff depends on the occurrence and/or severity of events such as natural catastrophes. Just like forwards, futures and simple options, exotic options and event-triggered derivatives are used by investors for speculation or hedging purposes.

The main objective of this chapter is to familiarize the reader with exotic/path-dependent options and event-triggered derivatives. The specific objectives are to:

- understand and compute the payoffs of barrier, Asian, lookback and exchange options;
- understand and compute the payoffs of weather, catastrophe and longevity derivatives;
- understand why complex derivatives exist and how they can be used;
- use no-arbitrage arguments to identify relationships between the prices of some of these derivatives.

7.1 Exotic options

Introduced originally as fully-customized agreements between two investors, most exotic options are now commonly traded *over the counter*. We will now look at the following four categories of exotic options:

- barrier options;
- Asian options;
- lookback options;
- exchange options.

7.1.1 Barrier options

A **barrier option** is an option to buy (call) or an option to sell (put) an asset S for a strike price K at maturity time T if the underlying asset price crosses (or not) a predetermined barrier level H during the life of the option.

Actuarial Finance: Derivatives, Quantitative Models and Risk Management, First Edition.
Mathieu Boudreault and Jean-François Renaud.

There are two main types of barrier options: *knock-in* and *knock-out*. A **knock-in** barrier option is said to be activated (or triggered) if the asset price reaches the barrier level H before maturity. A **knock-out** option is activated if the asset price never reaches the barrier level H before maturity. Said differently, a knock-out option is deactivated if the barrier level is crossed.

Depending on whether S_0 is below or above H, knock-in and knock-out barrier options can be further divided into two sub-groups:

- a knock-in option can be either an **up-and-in (barrier) option** or a **down-and-in (barrier) option**;
- a knock-out option can be either an **up-and-out (barrier) option** or a **down-and-out (barrier) option**.

To be more precise, first assume $S_0 < H$ and then denote by τ_H^+ the first (random) time the stock price crosses the barrier level H (coming from below). Mathematically,

$$\tau_H^+ = \min\{t \in [0, T] : S_t \geq H\}. \tag{7.1.1}$$

Similarly, if we assume $S_0 > H$, then we can define τ_H^-, the first (random) time the stock price attains the barrier level H (coming from above), by

$$\tau_H^- = \min\{t \in [0, T] : S_t \leq H\}. \tag{7.1.2}$$

Figure 7.1 gives possible realizations of the random variables τ_H^+ and τ_H^-, i.e. when barrier options are activated/deactivated.

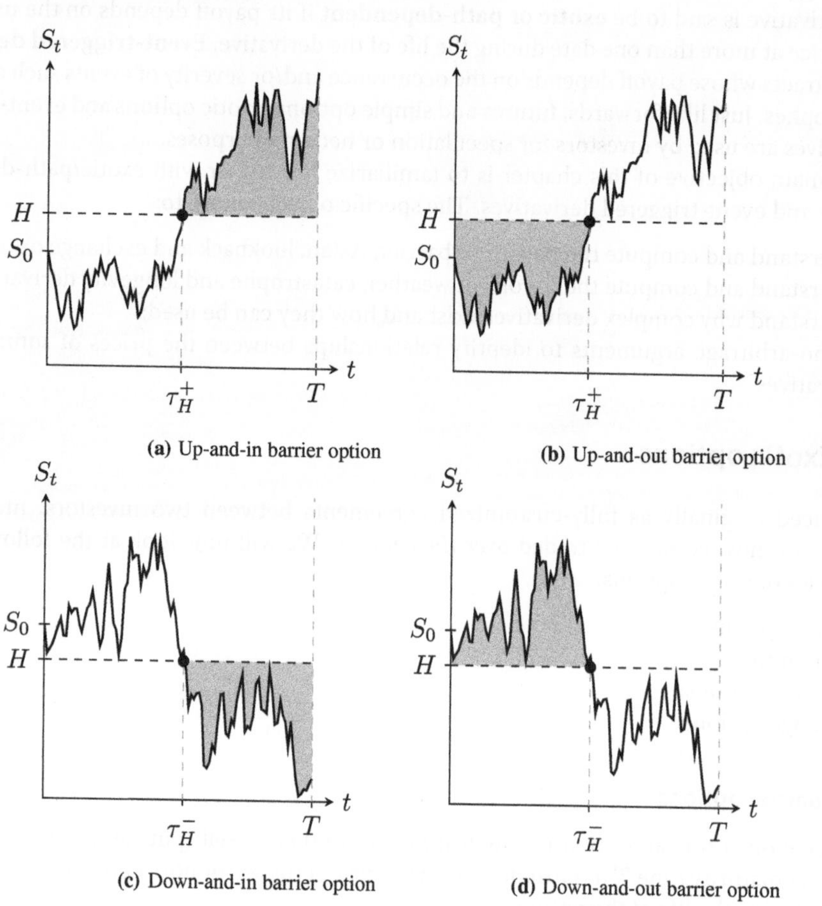

Figure 7.1 Knock-in and knock-out barrier options. Shaded areas imply the option is activated

Therefore, the event $\{\tau_H^+ \leq T\}$ is realized whenever the barrier is crossed during the life of the option, i.e. there is a time $0 \leq t \leq T$ such that $S_t \geq H$. Similarly, the event $\{\tau_H^- \leq T\}$ occurs when the stock price attains the barrier during the life of the option, i.e. there is a time $0 \leq t \leq T$ such that $S_t \leq H$.

These events can be used to define indicator random variables determining whether a barrier has been attained or not. For example, $\mathbb{1}_{\{\tau_H^+ \leq T\}}$ is equal to 1 if the barrier is crossed (coming from below) during the life of the option and it is equal to 0 otherwise. This notation allows us to summarize the eight different barrier options whose payoffs are described below:

- up-and-in options: in these cases, we have $S_0 < H$ and the payoffs are
  - up-and-in call (UIC): $(S_T - K)_+ \mathbb{1}_{\{\tau_H^+ \leq T\}}$
  - up-and-in put (UIP): $(K - S_T)_+ \mathbb{1}_{\{\tau_H^+ \leq T\}}$
- up-and-out options: in these cases, we have $S_0 < H$ and the payoffs are
  - up-and-out call (UOC): $(S_T - K)_+ \mathbb{1}_{\{\tau_H^+ > T\}}$
  - up-and-out put (UOP): $(K - S_T)_+ \mathbb{1}_{\{\tau_H^+ > T\}}$
- down-and-in options: in these cases, we have $S_0 > H$ and the payoffs are
  - down-and-in call (DIC): $(S_T - K)_+ \mathbb{1}_{\{\tau_H^- \leq T\}}$
  - down-and-in put (DIP): $(K - S_T)_+ \mathbb{1}_{\{\tau_H^- \leq T\}}$
- down-and-out options: in these cases, we have $S_0 > H$ and the payoffs are
  - down-and-out call (DOC): $(S_T - K)_+ \mathbb{1}_{\{\tau_H^- > T\}}$
  - down-and-out put (DOP): $(K - S_T)_+ \mathbb{1}_{\{\tau_H^- > T\}}$

For example, an up-and-in call option provides the opportunity to buy a share of stock for K only if the stock price has attained H (and $S_0 < H$) before time T. When a barrier option has been activated, it does not mean the option will necessarily be exercised. In the case of the up-and-in call option, the option to buy is exercised only if $S_T > K$.

Example 7.1.1 *Payoffs of barrier options*
Consider down-and-in and down-and-out call and put options with maturity in 3 months, initial stock price \$50, strike price of \$50 and barrier level at \$47. Assume the following scenario has occurred on the market: the stock price (after 3 months) is \$45 and, in between the option's inception and maturity, the minimum stock price reached \$41 (after 20 days). Let us compute the payoffs in this scenario.

First of all, the stock price started at $S_0 = 50 > 47 = H$, reached 41 after 20 days and ended up at $S_{0.25} = 45$ at maturity time $T = 0.25$. Therefore, the barrier $H = 47$ has been crossed (coming from above).

Therefore, we have

- DIC: the barrier is crossed, the option is activated but the call option is out of the money at maturity and consequently the payoff is

$$(S_T - K)_+ \mathbb{1}_{\{\tau_H^- \leq T\}} = (45 - 50)_+ \times 1 = 0.$$

- DIP: the barrier is crossed, the option is activated and the put option is in the money at maturity and consequently the payoff is

$$(K - S_T)_+ \mathbb{1}_{\{\tau_H^- \leq T\}} = (50 - 45)_+ \times 1 = 5.$$

- DOC or DOP: the barrier is crossed so both knock-out options are deactivated and consequently their payoffs are 0. ■

It should be emphasized, despite the similarities from a notational point of view, that barrier options are significantly different from gap options studied in Chapter 6. Even if both types of options have final payments depending on the crossing of a barrier, those events differ a lot. To illustrate the situation, let us consider:

1) an up-and-in call, i.e. an option with payoff $(S_T - K)_+ \mathbb{1}_{\{\tau_H^+ \leq T\}}$; and
2) a gap call option, i.e. an option with payoff $(S_T - K)\mathbb{1}_{\{S_T > H\}}$.

For the up-and-in call, the payoff is strictly positive if the underlying stock price crosses level H before or at maturity and if the final stock price is greater than K. Its payoff is always non-negative. For the gap call, the payoff is strictly positive if the final stock price is greater than both H and K, it is equal to zero if the final stock price is less than H, and it can be negative in some situations (if $H < S_T < K$).

Financial engineering with barrier options

Barrier options can be combined to create vanilla options. Indeed, as the barrier is either crossed or not crossed, we have

$$\mathbb{1}_{\{\tau_H^+ \leq T\}} + \mathbb{1}_{\{\tau_H^+ > T\}} = 1$$
$$\mathbb{1}_{\{\tau_H^- \leq T\}} + \mathbb{1}_{\{\tau_H^- > T\}} = 1.$$

The financial interpretation of the latter is that

$$\text{knock-in option} + \text{knock-out option} = \text{vanilla option}.$$

Consequently, the payoff of a vanilla call option $(S_T - K)_+$ can be further decomposed as the sum of two barrier options. First, we can write

$$(S_T - K)_+ = (S_T - K)_+ \mathbb{1}_{\{\tau_H^+ \leq T\}} + (S_T - K)_+ \mathbb{1}_{\{\tau_H^+ > T\}}.$$

In other words, a vanilla call option can be replicated by a long up-and-in call and a long up-and-out call. Second, we can write

$$(S_T - K)_+ = (S_T - K)_+ \mathbb{1}_{\{\tau_H^- \leq T\}} + (S_T - K)_+ \mathbb{1}_{\{\tau_H^- > T\}}.$$

In other words, a vanilla call option can be replicated by a long down-and-in call and a long down-and-out call. Of course, similar relationships hold for barrier put options.

Therefore, by the no-arbitrage assumption, we must also observe, at any time $0 \leq t \leq T$:

$$C_t = \text{DIC}_t + \text{DOC}_t = \text{UIC}_t + \text{UOC}_t$$

and

$$P_t = \text{DIP}_t + \text{DOP}_t = \text{UIP}_t + \text{UOP}_t,$$

where for example DIC_t is the time-t value of a down-and-in call option.

Example 7.1.2 *Financial engineering with barrier options*
You are given the following:

- a share of stock trades for \$75;
- a 72-strike call option trades for \$7.50;
- an up-an-out put option with strike \$72 and barrier \$80 sells for \$3;
- a 3-month Treasury zero-coupon bond trades for \$98.50 (face value of \$100);
- all options mature in 3 months.

Let us find the price of an up-and-in put option.

Using the put-call parity of equation (6.2.4), we have

$$\begin{aligned} P_0 &= C_0 + e^{-rT}K - S_0 \\ &= 7.50 + 0.985 \times 72 - 75 \\ &= 3.42. \end{aligned}$$

Moreover,

$$\mathrm{UIP}_0 = P_0 - \mathrm{UOP}_0 = 3.42 - 3$$

and therefore, to avoid arbitrage opportunities, the up-and-in put should sell for \$0.42. ■

Alternative representation with max and min

There are important relationships between the first-passage-time random variables τ_H^+ and τ_H^- and the minimum/maximum value of the underlying asset price. Let us define the *continuously monitored* maximum and minimum values taken by S over the time interval $[0, T]$ by

$$M_T^S = \max_{0 \le t \le T} S_t \quad \text{and} \quad m_T^S = \min_{0 \le t \le T} S_t,$$

where

$$\begin{aligned} \max_{0 \le t \le T} S_t &= \max\{S_t : 0 \le t \le T\}, \\ \min_{0 \le t \le T} S_t &= \min\{S_t : 0 \le t \le T\}. \end{aligned}$$

We deduce the following equalities of events:

$$\{\tau_H^+ \le T\} = \{M_T^S \ge H\}$$

and

$$\{\tau_H^- > T\} = \{m_T^S > H\}.$$

If $S_0 < H$, then the stock price has to increase to reach H at the random time τ_H^+. This also means the maximum attained by the stock price between time 0 and time T has to be larger than H for the stock price to cross H. A similar reasoning applies if $S_0 > H$. Therefore, we can rewrite each of the above eight payoffs using the maximum or the minimum value of the underlying asset during the life of the option.

In practice, the minimum/maximum is monitored daily, weekly or monthly, i.e. at the beginning or end of each day, week or month. First, let us set the monitoring dates $0 \le t_1 < t_2 < \cdots < t_n \le T$. Then, we can define the corresponding *discretely monitored* maximum and minimum values taken by S over the time interval $[0, T]$ by

$$M_T^S = \max\{S_{t_1}, S_{t_2}, \ldots, S_{t_n}\} \quad \text{and} \quad m_T^S = \min\{S_{t_1}, S_{t_2}, \ldots, S_{t_n}\}.$$

Of course, the monitoring frequency affects the value of the maximum/minimum: the continuously monitored maximum is always greater than or equal to the discretely monitored maximum, while the continuously monitored minimum is always less than or equal to the discretely monitored minimum.

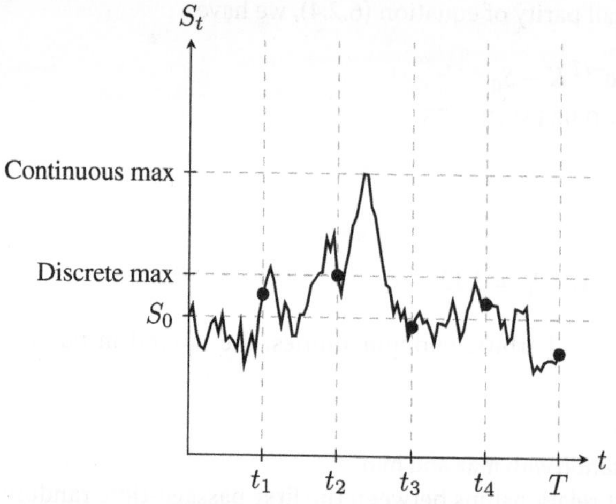

Figure 7.2 A path of the stock price along with the continuously and discretely monitored maxima

Figure 7.2 shows a sample path (or times series) for the price of a stock. In this example, the continuously monitored maximum is observed between time t_2 and time t_3. If the maximum is discretely monitored at times t_1, t_2, t_3, t_4 and T, then the maximum is observed at time t_2. As illustrated by this figure, the monitoring frequency affects the value of the maximum/minimum.

Consequently, we could have a second look at the eight different barrier options and define *discretely monitored barrier options.*

In general, *discretely monitored exotic options* are difficult to price so continuously monitored exotic options are often used as an approximation.

7.1.2 Lookback options

Lookback options are options whose payoff is based upon the minimum or maximum asset price observed during the life of the option. The contract usually specifies the frequency at which the minimum or maximum is monitored: continuously or discretely.

There are two types of lookback options: **floating-strike** and **fixed-strike**. In floating-strike lookback call and put options, we replace the strike price of a vanilla option by the minimum and maximum price of the underlying, respectively. The payoffs are then given by

$$S_T - m_T^S \quad \text{and} \quad M_T^S - S_T,$$

for a floating-strike lookback call and a floating-strike lookback put, respectively. Because these payoffs are already positive, there is no need to use the positive part function.

In a fixed-strike lookback call or put option, we replace the asset price in a vanilla option by the minimum or maximum price of the underlying. There are four possibilities for the payoff:

- calls: $(m_T - K)_+$ or $(M_T - K)_+$;
- puts: $(K - m_T)_+$ or $(K - M_T)_+$.

Example 7.1.3 *Payoffs of lookback options*

One-month lookback options are issued on the stock of ABC inc. The fixed strike on the minimum is \$12 and the fixed strike on the maximum is \$20. Assume you observe the following scenario over the next month:

- stock price after a month: \$16;
- minimum stock price, during that month, monitored at the end of each business day: \$10.25;
- maximum stock price, during that month, monitored at the end of each business day: \$22.84.

Let us compute the payoffs of lookback call and put options of either floating-strike or fixed-strike types.

We get:

- floating-strike call option: the payoff is $S_T - m_T = 16 - 10.25 = 5.75$;
- floating-strike put option: the payoff is $M_T - S_T = 22.84 - 16 = 6.84$;
- fixed-strike call option on the maximum: the payoff is $(M_T - K)_+ = (22.84 - 20)_+ = 2.84$;
- fixed-strike call option on the minimum: the payoff is $(m_T - K)_+ = (10.25 - 12)_+ = 0$;
- fixed-strike put option on the maximum: the payoff is $(K - M_T)_+ = (20 - 22.84)_+ = 0$;
- fixed-strike put option on the minimum: the payoff is $(K - m_T)_+ = (12 - 10.25)_+ = 1.75$. ■

Equity-linked insurance and annuities (see Chapter 8) offer benefits that are similar to discretely monitored lookback options. For example, the high watermark indexing scheme offered in equity-indexed annuities is based on the maximum value of a reference index observed on anniversary or tri-anniversary dates. Variable annuities may also offer automatic resets, a feature similar to a protective put whose strike price is updated periodically (once every year or two years) to reflect the best market conditions.

7.1.3 Asian options

An **Asian option** is an option whose payoff is based upon the average price of the underlying asset during the life of the option. Asian options limit the effects, on the value of the option, of speculative transactions on the underlying stock. They are generally cheaper than vanilla call and put options.

Let us denote by $\overline{S}_T$ the *average* asset price between inception and maturity. In general, the calculation of the average is specified in the contract with respect to:

1) monitoring, i.e. the frequency at which prices are computed in the average;
2) type of average, i.e. arithmetic or geometric.

Using the fixed set of dates $0 \leq t_1 < t_2 < \cdots < t_n \leq T$ for discrete monitoring, we have the following four possible ways of computing the average stock price $\overline{S}_T$:

1) Discrete monitoring, arithmetic average: $\overline{S}_T = \frac{1}{n} \sum_{i=1}^{n} S_{t_i}$.
2) Discrete monitoring, geometric average: $\overline{S}_T = \sqrt[n]{\prod_{i=1}^{n} S_{t_i}}$.
3) Continuous monitoring, arithmetic average: $\overline{S}_T = \frac{1}{T} \int_0^T S_t \mathrm{d}t$.
4) Continuous monitoring, geometric average: $\overline{S}_T = \exp\left(\frac{1}{T} \int_0^T \ln(S_t)\mathrm{d}t\right)$.

Note that the continuous versions are obtained as limits of the discrete versions when the fixed set of dates $0 \leq t_1 < t_2 < \cdots < t_n \leq T$ gets *larger* in order to cover the whole time-interval $[0, T]$.

Asian options can be of two types: **average price** or **average strike.** For an average price Asian option, the payoff is based on the difference between the average price and a fixed strike price K. For an average strike Asian option, the price at which we can buy/sell the underlying asset is random and given by the average stock price. More precisely, the payoff of an *average price Asian call option* (resp. *average price Asian put option*) with strike price K is defined as

$$(\overline{S}_T - K)_+ \qquad \text{(resp. } (K - \overline{S}_T)_+\text{)}$$

and the payoff of an *average strike Asian call option* (resp. *average strike Asian put option*) by

$$(S_T - \overline{S}_T)_+ \qquad \text{(resp. } (\overline{S}_T - S_T)_+\text{)}.$$

In practice, the most popular type of average is the arithmetic average computed over a discrete set of dates (discrete monitoring). For example, the average stock price could be calculated as the arithmetic mean of each end-of-day price. However, in many financial models, calculating the exact no-arbitrage price of Asian options in that popular case is very difficult and the geometric average or the continuous monitoring are used as a way to approximate the price of such options.

Example 7.1.4 *Payoff of Asian options*
In an Asian option contract having a 6-month maturity, the average is said to be computed on the observed stock prices at the end of each week (every Friday). Average price Asian call/put options both have a strike of \$70.

Here is a possible scenario for the next 6 months: the average stock price, during those 6 months, is \$69.41 and the final stock price is \$72. Let us calculate the payoff of average strike/price Asian call/put options in this scenario.

We have $T = 0.5$ and $K = 70$. In this scenario, we also have $\overline{S}_{0.5} = 69.41$ and $S_{0.5} = 72$. Consequently,

- average strike Asian call: $(S_T - \overline{S}_T)_+ = (72 - 69.41)_+ = 2.59$;
- average strike Asian put: $(\overline{S}_T - S_T)_+ = (69.41 - 72)_+ = 0$;
- average price Asian call: $(\overline{S}_T - K)_+ = (69.41 - 70)_+ = 0$;
- average price Asian put: $(K - \overline{S}_T)_+ = (70 - 69.41)_+ = 0.59$. ■

7.1.4 Exchange options

An **exchange option** gives its holder the right to exchange an asset for another asset. It is a generalization of vanilla call and put options, which are options to exchange the underlying asset for a cash amount K.

More generally, let us denote by $S_t^{(1)}$ and $S_t^{(2)}$ the time-t prices of two risky assets. The payoff of an exchange option to obtain a share of $S^{(1)}$ in exchange for a share of $S^{(2)}$ at maturity time T is given by

$$\left(S_T^{(1)} - S_T^{(2)}\right)_+ = \max\left(S_T^{(1)} - S_T^{(2)}, 0\right)$$

while the payoff to obtain a share of $S^{(2)}$ in exchange for a share of $S^{(1)}$ is given by

$$\left(S_T^{(2)} - S_T^{(1)}\right)_+ = \max\left(S_T^{(2)} - S_T^{(1)}, 0\right).$$

Example 7.1.5 *Payoff of an exchange option*
You enter into an exchange option to buy one share of ABC inc. in exchange for one share of XYZ inc. Both shares currently trade at 100. Find the payoff of that exchange option if, at maturity, the stock of ABC inc. is worth 105 whereas the stock of XYZ is worth 112.

The option gives you the right, not the obligation, to buy a share of ABC inc. in exchange for a share of XYZ. Therefore, you can buy for 112 an asset worth 105 and you decide not to exercise this option. The payoff is 0. ■

Of course, exchange options make sense when the two assets being exchanged have comparable prices. In general, an exchange option gives its holder the right to exchange a given quantity of $S^{(1)}$ for another given quantity of $S^{(2)}$.

Using the translation property of the maximum function, we can write

$$\max\left(S_T^{(1)} - S_T^{(2)}, 0\right) + S_T^{(2)} = \max\left(S_T^{(2)} - S_T^{(1)}, 0\right) + S_T^{(1)} = \max\left(S_T^{(1)}, S_T^{(2)}\right).$$

Denoting by C_t the value at time t of the option to buy $S^{(1)}$ for $S^{(2)}$ and P_t the price at time t of the option to buy $S^{(2)}$ for $S^{(1)}$, we can deduce the following parity relationship between exchange options:

$$C_t + S_t^{(2)} = P_t + S_t^{(1)},$$

for all $0 \leq t \leq T$.

This is a generalization of the classical put-call parity of Chapter 6. Indeed, suppose that $S^{(1)}$ is a stock such that $S_t^{(1)} = S_t$ whereas $S^{(2)}$ is a zero-coupon bond with face value K and time-t value $S_t^{(2)} = Ke^{-r(T-t)}$. As previously discussed, an option to buy $S^{(1)}$ ($S^{(2)}$) for $S^{(2)}$ ($S^{(1)}$) is a regular call (put) option. Therefore, and as expected, the above parity relationship becomes $C_t + Ke^{-r(T-t)} = P_t + S_t$, which is the put-call parity of equation (6.2.4).

Other exotic options

It is generally recognized that Mark Rubinstein[1] coined the term "exotic options" in the early 1990s to name a set of derivatives whose payoff structure is unconventional. The expression has remained in contrast to *vanilla* options that comprise European and American call/put options. Vanilla options are usually exchange-traded whereas most exotic options are traded OTC.

In addition to barrier, Asian, lookback and exchange options, the following designs have appeared regularly in the markets (shown alphabetically):

- Chooser option: option that allows the holder to choose whether the contract is a European call or put prior to maturity.
- Cliquet (or ratchet) option: portfolio of forward start options (see below) such that the k-th option applies only during the k-th period.
- Compound option: option written on another option (e.g. call on a call, call on a put, etc.), i.e. in which the underlying asset is another option.
- Forward start option: option issued today, *starting* at time $T_1 > 0$ and expiring at time $T_2 > T_1$. More precisely, the strike price is given by S_{T_1}, i.e. the option is *started* at-the-money and so it is unknown at inception.
- Parisian option and step option: option whose payoff depends on the time spent by the underlying asset below, above or between barrier(s).
- Package: portfolio of standard European/American call/put options. These typically involve the strategies of Section 5.4.
- Reset option: option allowing the strike price to be reset before maturity.

1 Professor Emeritus of Finance, Haas School of Business, University of California, Berkeley

7.2 Event-triggered derivatives

An **event-triggered derivative** is a financial contract whose payoff is tied to the occurrence and/or severity of one or several events or variables, such as temperature, rainfall, natural catastrophes, mortality or longevity of a pool of individuals. These derivatives are useful to transfer the risk tied to these events to other investors in exchange for a premium. For example, pension plan sponsors use longevity derivatives as a risk transfer mechanism to offset the adverse effects of longevity risk (see Chapter 1).

Why would someone else accept to bear weather, natural catastrophe or longevity risks? Simply because investors can use these instruments to diversify their typical investment portfolios (stock, bonds, commodities, retail, etc.). Indeed, it is widely recognized that weather, natural catastrophe or longevity risks are relatively independent from common financial variables such as stock prices and interest rates. Those derivatives can also be used for speculation.

Catastrophe and longevity derivatives are the most common examples of **insurance-linked securities** (ILS). It is a class of instruments whose payoff is linked to insurance losses. They are used by insurance companies for risk transfer.

In this section, we will look at weather, catastrophe and longevity derivatives and we will see how pension plans, life and property and casualty insurers can use these derivatives to risk manage their liabilities.

7.2.1 Weather derivatives

A **weather derivative** is a financial contract whose payoff is contingent on the temperature or amount of precipitation (rain, snow) observed at a given location during a predetermined period. For industries whose profitability is largely determined by the weather, it is easy to imagine how weather derivatives can be used for risk management:

- An industrial farmer loses whenever there is too much rain (flooded lands) or not enough (droughts). He could enter into a weather derivative on rainfall to receive a payment if one of these adverse events occurs.
- A ski station suffers a lot if snowfall is low during winter. It could enter into a weather derivative on snowfall or temperature so that it receives a payment if snowfall is well below average or if temperature during winter is too high.
- A city pays for snow plowing of its streets and if snowfall is well above average, it will incur major costs. It could enter into a weather derivative to receive a payment if snowfall is well above average.

The first weather derivative was created between two energy companies in 1996. The contract offered a rebate to one party if temperature was cooler than expected. Weather derivatives have been trading OTC since 1997 and on the Chicago Mercentile Exchange (CME) since 1999. For example, the CME offers weather derivatives on the temperature for major cities in the U.S. and in Canada, and rain/snowfall derivatives for some cities in the U.S.

Example 7.2.1 *Weather derivative for a P&C insurer*
A property and casualty insurance company covers flood risk in its homeowner's insurance policies. The company would like to offset losses resulting from a flood occurring on a specific river. The actuary knows that whenever the accumulated rainfall during a week is more than 10 inches near that river, a flood is very likely to occur.

A rainfall derivative is available: it pays \$1 per inch of accumulated rain in excess of 10 inches, measured between May 1st and May 7th, at a specific weather station. If the cumulative rainfall is below 10 inches, the derivative pays 0. If the cumulative rainfall is 13 inches, then the derivative will pay $3 per unit.

Hopefully, this will be enough to offset the losses the company will incur. Remember that, in all scenarios, the insurance company is required to pay an initial premium on these derivatives. ■

The most common measures to determine the payoff of temperature-based weather derivatives are **heating degree days (HDD)** and **cooling degree days (CDD)**. The idea is that for a temperature below 65°F (18°C), people start heating buildings whereas for temperatures above 65°F, people need cooling (air conditioning). The HDD (resp. CDD) is a measure of the extent by which heating (resp. cooling) is required. On a given day, the HDD and the CDD are computed as follows:

$$HDD = \max\{65 - X, 0\} = (65 - X)_+$$

and

$$CDD = \max\{X - 65, 0\} = (X - 65)_+,$$

where X is the average between the minimum and maximum temperature on that day. For a day whose average temperature X is 80°F, the corresponding HDD is 0 (no heating required) and the CDD is 15 (cooling required). Finally, we define the **cumulative HDD (resp. CDD)** as the sum of the HDD (resp. CDD) over a given number of days.

Example 7.2.2 *Weather derivative for a life insurer*
A life insurance company wants to protect itself against excess mortality due to heat waves for the first week of July. It decides to enter into a temperature-based weather derivative. This derivative will pay $1 per cumulative CDD in excess of 105. Let us calculate the payoff of the weather derivative in the scenario where temperatures are as follows during that week:

Temp/Day	S	M	T	W	T	F	S
Min	55	57	59	65	64	63	59
Max	75	80	85	90	88	89	78

We calculate the daily mean between the minimum and maximum temperature, in addition to the daily CDD. The results for this scenario are shown in the next table.

Temp/Day	S	M	T	W	T	F	S
Mean	65	68.5	72	77.5	76	76	68.5
CDD	0	3.5	7	12.5	11	11	3.5

For example, for Monday, 68.5 is calculated as $(80 + 57)/2$ and the CDD on Monday is $68.5 - 65$. Summing the CDD over the seven days of the week, we get that the cumulative CDD is 48.5. Hence, in this scenario, the weather derivative will not pay since the cumulative CDD is well below 105. ■

7.2.2 Catastrophe derivatives

For an insurance company covering losses resulting from natural catastrophes, a common risk-transfer mechanism is reinsurance: in exchange for a premium, the reinsurer covers losses in excess of some predetermined amount. Catastrophe derivatives are an alternative to reinsurance: instead of transferring the risk to one of many reinsurers, other investors can assume a part of the natural catastrophe risk in exchange for a premium.

A **catastrophe (CAT) derivative** is a financial contract whose payment is affected by the occurrence and/or severity of natural catastrophes in a given region during a given period of time. According to the OECD, they have "[...] appeared in the aftermath of Hurricane Andrew in 1992 in the belief that the capacity offered by the traditional reinsurance market and the retrocession market would shrink." The most important CAT derivatives are CAT bonds, options and futures.

The first CAT derivative that appeared on the market was the CAT bond. A **CAT bond** is like a regular bond but principal (and maybe coupons as well) is reduced whenever a catastrophic event occurs. The triggering events have to be well defined and they are usually based on either the actual losses of the issuer or based upon a *catastrophe loss index*. It was created in 1994 as a customized agreement between insurers and reinsurers.

CAT options and **CAT futures** are issued on catastrophe loss indices and provide the opportunity to buy or sell the index for a predetermined price. Such derivatives were introduced by the Chicago Board of Trade in 1995. Nowadays, the main providers of catastrophe loss indices are the Property Claims Services (PCS) and the CME. They are meant as proxies for industry losses due to hurricanes, earthquakes or other natural catastrophes.

We illustrate two popular CAT derivatives in the examples below.

Example 7.2.3 *Earthquake bond*
In January, a 5-year earthquake bond with annual coupons of 8% and principal of $100 is issued by an insurance company. Principal is reduced by 50% if the Earthquake Loss Index (computed in a given region) is over 500 at maturity. Let us describe the cash flows of this bond if it is issued at par, i.e. for a price of $100.

At inception, the insurance company receives $100 and pays annual coupons of $8. In the meantime, if earthquakes are rare and the index ends up below 500 after 5 years, then the insurance company will repay the full principal, i.e. $100. But if earthquake losses are high during this 5-year period, it may have to repay only 50% of the principal (as specified in the contract), i.e. $50. Investors in the market lose $50 in this scenario but they are usually compensated by a high coupon rate that provides a large return whenever earthquake losses are small. ■

Example 7.2.4 *Hurricane options*
A call option on the Hurricane Loss Index is issued on May 31st just before hurricane season in the North Atlantic. The index is at 100. The option provides the right *to buy* the index for 130 on November 30th, at the end of hurricane season. The call option currently sells for $15. Let us describe the cash flows of the call option if we assume there are only two possible scenarios: the Hurricane Loss Index is either 160 or 110, at maturity.

For a company exposed to hurricane losses, this call option acts as an insurance. The company pays $15 at inception. If there are many severe hurricanes, the index will increase a lot and the option will mature in the money. If the index is 160 at maturity, the payment will be $160 - 130 = 30$ per unit of option, which should help offset losses an insurer might suffer for a severe hurricane season. Otherwise, for an index at 110 on November 30th, the call option will be out of the money (payoff equal to zero) at maturity. ■

7.2.3 Longevity derivatives

As defined in Chapter 1, *longevity risk* is the uncertainty related to the overall improvement of life expectancy. It is difficult to predict how much longer people will live in the future. Longevity risk is an important systematic risk faced by pension plans, putting a significant pressure on social security systems.

Life insurance companies are also exposed to longevity risk but to a lesser extent. Higher longevity means their annuity business is more expensive (additional annuity payments) whereas their insurance business is cheaper (lower present value). As a result, life insurers have a natural hedge against large variations in mortality patterns. Pension plan sponsors, as annuity providers/sellers, traditionally manage longevity risk by buying annuities from a life insurer (buy-in) or by selling the pension liability (along with its related assets) to a life insurer (buy-out).

Longevity derivatives were introduced in the early 2000s as an alternative scheme to transfer longevity risk. A **longevity derivative** is a financial contract whose payoff is contingent on how many individuals, from a given group, survive during a given period of time. For example, the group could be the retirees of a pension plan or it could be the 65-year old males in a given country. *Survivorship indices* are also common and are provided by Credit Suisse and the Life & Longevity Markets Association (LLMA).

Three types of longevity derivatives have appeared on the market: longevity bonds, forwards and swaps. A **longevity bond** is a bond whose coupons decrease with the number of survivors in the pool of individuals. As in Chapter 4, in a typical **longevity swap**, a fixed amount is exchanged periodically for a variable (floating) amount computed with fixed and floating mortality rates. A **longevity forward** is a single exchange of cash flows based upon a fixed and a floating mortality rate. These derivatives are illustrated in the following examples.

Example 7.2.5 *Longevity bond*

A bank issues a longevity bond with a principal of \$100. The annual coupons decrease according to the number of survivors in a population of 100,000 65-year-old individuals. The first coupon is \$5 while the subsequent coupons are equal to the proportion of survivors times \$5. Let us describe the first few cash flows of the bond in the following scenario: after 1 year, the number of survivors is 98,500, and after 2 years it is 97,000.

First of all, the principal is received at inception and it is repaid at maturity. The first coupon paid, at the end of the first year, is 5. In this scenario, the second coupon, paid after 2 years, will be $0.985 \times 5 = 4.925$, whereas the third coupon, paid after 3 years, will be $0.97 \times 5 = 4.85$.

If the mortality experience of a pension plan is similar to this population (65-year-old individuals) and if each retiree had identical benefits, then the pension plan sponsor could invest in this longevity bond to (partially) hedge out the longevity risk, as the decreasing coupon pattern could (partially) offset the annuity payments made by the plan. Indeed, in this case, it would not matter (for the sponsor) how much longer retirees might live. ■

Example 7.2.6 *Longevity forward*

You need to manage longevity risk tied to 70-year-old females. You enter into a longevity forward known as a **q-forward** whose payoff design is simple: for a zero initial cost, you pay 100 times the mortality rate experienced by 70-year-old females next year and you receive a fixed amount of 1.56 in exchange. Let us describe the cash flows in the following two scenarios: the mortality experience of 70-year-old females is either 1500 or 1600 deaths per 100,000.

In a year, you will receive 1.56 for a notional of 100, which translates into a fixed mortality rate of 1560 per 100,000. In the case where 1500 people die, you would pay 1.50 and receive 1.56 at maturity, for a gain of 6 cents per 100 of notional. Hence, to protect against improvements in mortality, you should be the fixed-rate receiver. If 1600 people die, you would pay 1.60 and still receive 1.56 for a loss of 4 cents per 100 of notional.

Depending on whether you are the fixed-rate receiver or payer, a longevity forward can protect against decreases/increases in mortality (longevity). ■

7.3 Summary

Barrier option

- Option to buy/sell an asset only if a barrier has been crossed (or never been crossed) prior to or at maturity.
- Knock-in: activated if the barrier has been crossed.
- Knock-out: deactivated if the barrier has been crossed.
- Types:
  - up-and-in call (UIC): $(S_T - K)_+ \mathbb{1}_{\{\tau_H^+ \leq T\}}$
  - up-and-in put (UIP): $(K - S_T)_+ \mathbb{1}_{\{\tau_H^+ \leq T\}}$
  - up-and-out call (UOC): $(S_T - K)_+ \mathbb{1}_{\{\tau_H^+ > T\}}$
  - up-and-out put (UOP): $(K - S_T)_+ \mathbb{1}_{\{\tau_H^+ > T\}}$
  - down-and-in call (DIC): $(S_T - K)_+ \mathbb{1}_{\{\tau_H^- \leq T\}}$
  - down-and-in put (DIP): $(K - S_T)_+ \mathbb{1}_{\{\tau_H^- \leq T\}}$
  - down-and-out call (DOC): $(S_T - K)_+ \mathbb{1}_{\{\tau_H^- > T\}}$
  - down-and-out put (DOP): $(K - S_T)_+ \mathbb{1}_{\{\tau_H^- > T\}}$
- Parity relationships:

$$C_t = \text{DIC}_t + \text{DOC}_t = \text{UIC}_t + \text{UOC}_t$$

$$P_t = \text{DIP}_t + \text{DOP}_t = \text{UIP}_t + \text{UOP}_t$$

Lookback options

- Options based on the minimum/maximum price of the underlying asset.
- Minimum price of the underlying asset: m_T^S.
- Maximum price of the underlying asset: M_T^S.
- Types:
  - Floating-strike lookback call: $S_T - m_T^S$.
  - Floating-strike lookback put: $M_T^S - S_T$.
  - Fixed-strike lookback call: $(m_T - K)_+$ and $(M_T - K)_+$.
  - Fixed-strike lookback put: $(K - m_T)_+$ and $(K - M_T)_+$.
- Types of monitoring:
  - Continuous:

$$M_T^S = \max_{0 \leq t \leq T} S_t \quad \text{and} \quad m_T^S = \min_{0 \leq t \leq T} S_t.$$

  - Discrete (every week, month, etc.):

$$M_T^S = \max\{S_{t_1}, S_{t_2}, \ldots, S_{t_n}\} \quad \text{and} \quad m_T^S = \min\{S_{t_1}, S_{t_2}, \ldots, S_{t_n}\}.$$

Asian options

- Option based on the average price of the underlying asset.
- Average price of the underlying asset: $\overline{S}_T$.
- Types:
  - Average price Asian call: $(\overline{S}_T - K)_+$.
  - Average price Asian put: $(K - \overline{S}_T)_+$.
  - Average strike Asian call: $(S_T - \overline{S}_T)_+$.
  - Average strike Asian put: $(\overline{S}_T - S_T)_+$.
- Types of monitoring and averages:
  - Discrete monitoring, arithmetic average: $\overline{S}_T = \frac{1}{n}\sum_{i=1}^{n} S_{t_i}$.
  - Discrete monitoring, geometric average: $\overline{S}_T = \sqrt[n]{\prod_{i=1}^{n} S_{t_i}}$.
  - Continuous monitoring, arithmetic average: $\overline{S}_T = \frac{1}{T}\int_0^T S_t \mathrm{d}t$.
  - Continuous monitoring, geometric average: $\overline{S}_T = \exp\left(\frac{1}{T}\int_0^T \ln(S_t)\mathrm{d}t\right)$.

Exchange options

- Option to exchange one risky asset for another.
- Call option: receive $S_T^{(1)}$ in exchange for $S_T^{(2)}$.
- Put option: receive $S_T^{(2)}$ in exchange for $S_T^{(1)}$.
- Regular call and put options: exchange cash (strike price) for a stock (or vice versa).
- Similar parity relationship:

$$C_t + S_t^{(2)} = P_t + S_t^{(1)}.$$

Weather derivative

- Derivative whose payoff is contingent on the temperature or amount of precipitation (rain, snow) observed at a given location during a predetermined period.
- Temperature-based weather derivatives are generally built with cumulative HDD and CDD:

$$HDD = (65 - X)_+$$
$$CDD = (X - 65)_+$$

where X is the average between the minimum and maximum temperature observed on a given day.

Catastrophe derivative

- Derivative whose payoff is based upon the occurrence and/or severity of specified natural catastrophes during a given period of time and in a given region.
- Types: bonds, futures, options.
- Payoff based on a catastrophe loss index.

Longevity derivative

- Derivative whose payoff is contingent on how many individuals, from a given group, survive during a given period of time.
- Types: bonds, swaps and forwards.
- Payoff based on a survivorship index.

7.4 Exercises

7.1 In Figure 7.3 and in Table 7.1, you will find the (fictional) evolution of the stock price of ABC inc. Calculate the payoff of the following exotic options (or investment strategies) given that they are purchased (or initiated) at the beginning of the year and mature at the end.

Note that in this example a year is made up of 240 business days or 48 business weeks of 5 days or 12 months of 20 business days.

(a) Call option with strike price of \$50.
(b) Covered call with strike price of \$49.
(c) Straddle at \$50.
(d) Gap put option with strike price of \$52 and trigger price of \$50.
(e) Up-and-in call option with barrier of \$56 and strike price of \$51.
(f) Average strike Asian call option whose arithmetic average is computed at the end of each month (on $S_{20}, S_{40}, \dots$).
(g) Floating-strike lookback put option where the maximum/minimum is monitored daily.
(h) Cash-or-nothing put option with strike price of \$50.
(i) Fixed-strike lookback call option on the minimum where the maximum/minimum is monitored every week on Fridays (i.e. $k = 5, 10, 15, \dots$) with a strike of \$40.
(j) Bear spread with strike prices of \$47 and \$52.
(k) Investment guarantee with $K = 1.01S_0$ (minimum return of 1% on the initial investment).

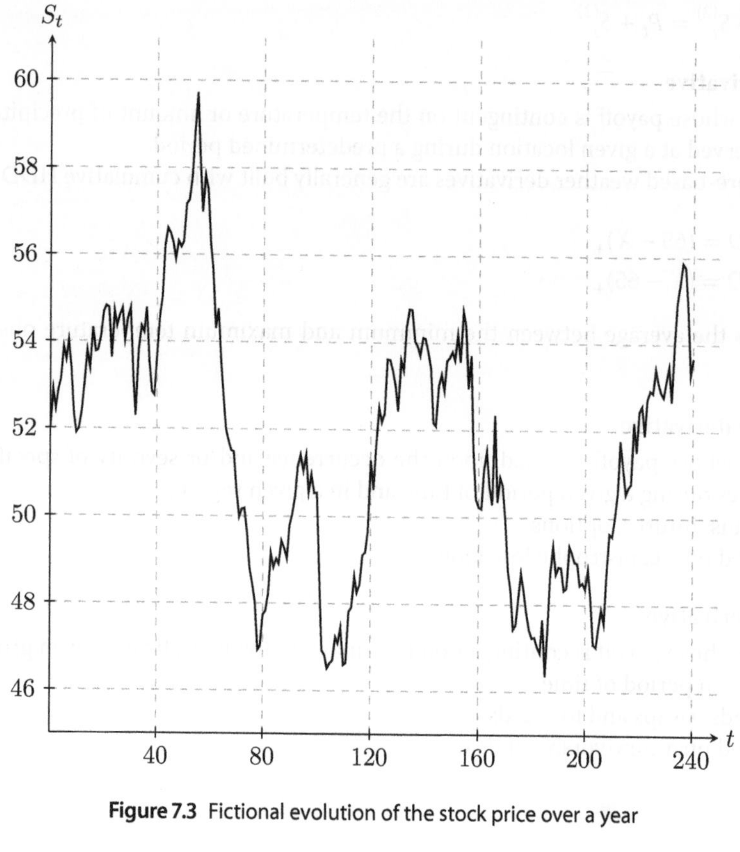

Figure 7.3 Fictional evolution of the stock price over a year

Table 7.1 Fictional evolution of the stock price over a year. Note that k labels the days within the year, i.e. $k = 0, 1, 2, \dots, 240$

k	S_k	k	S_k	k	S_k	k	S_k	k	S_k	k	S_k
0	52.00										
1	52.90	41	54.57	81	47.82	121	50.63	161	50.20	201	48.81
2	52.45	42	55.05	82	48.24	122	51.61	162	51.08	202	48.18
3	52.86	43	56.21	83	49.16	123	52.51	163	51.51	203	47.09
4	53.14	44	56.61	84	48.83	124	52.18	164	50.70	204	47.03
5	53.88	45	56.48	85	49.02	125	52.39	165	50.30	205	47.47
6	53.57	46	56.29	86	48.91	126	53.62	166	52.35	206	47.89
7	54.01	47	55.90	87	49.44	127	53.60	167	50.36	207	47.57
8	53.37	48	56.30	88	49.27	128	53.35	168	50.92	208	48.81
9	52.32	49	56.18	89	49.03	129	52.99	169	50.74	209	49.63
10	51.90	50	56.30	90	49.79	130	52.44	170	49.92	210	49.57
11	52.01	51	57.16	91	50.06	131	53.60	171	48.79	211	50.01
12	52.40	52	57.26	92	50.47	132	53.37	172	48.33	212	50.58
13	53.10	53	57.78	93	51.15	133	54.35	173	47.43	213	51.77
14	53.78	54	58.79	94	51.41	134	54.77	174	47.54	214	51.15
15	53.39	55	59.71	95	51.04	135	54.75	175	48.17	215	50.54
16	54.31	56	58.15	96	51.41	136	54.04	176	48.73	216	50.70
17	53.77	57	56.98	97	50.55	137	53.93	177	48.33	217	50.78
18	53.80	58	57.82	98	50.96	138	53.72	178	47.80	218	52.20
19	54.13	59	57.49	99	50.54	139	54.12	179	47.41	219	51.52
20	54.68	60	56.23	100	50.54	140	54.10	180	47.23	220	52.45
21	54.82	61	55.66	101	48.82	141	53.80	181	47.25	221	52.57
22	54.78	62	54.30	102	46.94	142	53.43	182	47.03	222	51.90
23	53.63	63	54.75	103	46.81	143	52.16	183	46.89	223	52.74
24	54.71	64	53.54	104	46.52	144	52.09	184	47.65	224	52.93
25	54.35	65	52.79	105	46.63	145	53.02	185	46.36	225	52.88
26	54.60	66	52.00	106	46.70	146	53.28	186	47.49	226	53.43
27	54.77	67	51.73	107	47.23	147	52.91	187	48.72	227	53.14
28	54.03	68	51.52	108	47.09	148	53.60	188	49.34	228	52.90
29	54.53	69	51.47	109	47.44	149	53.66	189	48.71	229	52.79
30	54.77	70	50.58	110	46.60	150	53.79	190	48.87	230	53.17
31	53.22	71	49.98	111	46.64	151	53.60	191	48.86	231	53.48
32	52.31	72	50.16	112	47.81	152	54.19	192	48.33	232	52.68
33	53.79	73	50.19	113	47.92	153	53.57	193	48.39	233	54.03
34	54.35	74	49.59	114	48.63	154	54.72	194	49.35	234	54.96
35	54.21	75	48.81	115	48.24	155	54.40	195	49.20	235	55.38
36	54.78	76	48.47	116	48.43	156	52.97	196	48.98	236	55.87
37	53.56	77	48.03	117	49.12	157	53.71	197	48.95	237	55.77
38	52.84	78	46.93	118	49.19	158	51.57	198	48.44	238	54.84
39	52.77	79	47.02	119	49.67	159	50.42	199	48.51	239	53.29
40	53.69	80	47.73	120	51.29	160	50.23	200	48.35	240	53.68

7.2 You are given the following prices for barrier options with strike price $50 all maturing at the same time:

	$H = 40$		$H = 60$	
	Call	Put	Call	Put
DO	5.18	0.81	0	0
DI	0.05	1.98	5.23	2.79
UO	0	0	0.59	2.68
UI	5.23	2.79	4.64	0.11

The current stock price is $50.

(a) What is the price of regular call and put options?

(b) Calculate the current price of an at-the-money call option that is deactivated if the stock price ever attains $40 or $60 during the life of the option. Repeat for a similar put option.

7.3 An average strike Asian call option was issued 5 months ago and matures in a month. The average stock price over the last 5 months is $14 and the payoff of the option is $2. If the average is computed arithmetically on prices observed at the end of each month, determine the stock price at maturity.

7.4 Suppose that you can buy today for $5.75 a contract that pays off $\overline{S}_T$ at maturity. Given that a stock currently trades for $6.14 and an average strike Asian call option sells for $1.78, find the current no-arbitrage price of an average strike Asian put option.

7.5 Four floating-strike lookback put options maturing in 6 months are available in the market. They differ only in the monitoring frequency of the maximum:

- Lookback # 1: maximum is monitored continuously.
- Lookback # 2: maximum is monitored monthly.
- Lookback # 3: maximum is monitored weekly.
- Lookback # 4: maximum is monitored daily.

To avoid arbitrage opportunities, order the price of these lookback options, starting from the lowest to the highest price.

7.6 Suppose the temperature (°F) in Montpelier, VT, in the first week of February is as follows:

	S	M	T	W	T	F	S
Min	6	8	14	18	2	12	0
Max	37	20	37	22	39	38	20

Calculate the cumulative HDD over the 7 days.

8

Equity-linked insurance and annuities

Equity-linked insurance or annuity (ELIA) is the generic term to designate a life insurance or annuity that includes benefits tied to the return of a reference portfolio of assets. For example, instead of adjusting the death benefit with inflation (or some predetermined rate of return), the benefit could increase with the returns of a financial index, e.g. the S&P 500. An ELIA is a hybrid between a life insurance (or annuity) policy and a pure investment product as it gives the policyholder an opportunity to benefit from the upside potential while being protected against the downside risk. It competes with mutual funds[1] with the important distinction that ELIAs include various guarantees at the maturity of the contract and/or on the death of the policyholder.

During their active years, policyholders invest their savings in preparation for retirement. Then, at retirement, they annuitize their savings, i.e. the accumulated savings are used to buy an annuity-type product that provides a regular stream of income until death. These two steps are known as the *accumulation* and the *annuitization* phases. ELIAs, depending on their features, can be used for one or both phases.

It is important at this point to mention that in actuarial and financial mathematics, there is a clear distinction between a life *insurance contract* and an *annuity*. The former provides a benefit at death whereas the latter provides a regular stream of income until death. But in the investment business, the term *annuity* refers to the entire investment package for both the accumulation and annuitization phases. Consequently, in this chapter, an **annuity** will be known as a savings product that has an insurance component in case of death.

Example 8.0.1 *Equity-linked insurance tied to the S&P 500*
A 60-year-old policyholder buys an equity-linked insurance, maturing 8 years from now, with a benefit of \$100,000 indexed to the returns of the S&P 500. The contract will pay, at maturity, the greatest value between: (1) \$100,000 credited with returns of the S&P 500, or (2) the guaranteed benefit of \$100,000. Assume that the S&P 500 index is currently at 1500.

Consider the following scenario: in 8 years, the index will be at 2000. In this case, the accumulation factor is $\frac{2000}{1500} = 1.333$. The insured being entitled to the greatest value between \$133,333 and \$100,000 (the guaranteed benefit), the benefit paid in this scenario would be \$133,333. ■

1 A mutual fund is a diversified portfolio of assets (stocks, bonds, etc.) sold (in pieces) to small investors by banks and other investment firms. The name comes from the fact that small investments are pooled and managed by professionals. It allows small investors to hold a diversified portfolio, e.g. of stocks and bonds, without having to buy the actual shares.

Actuarial Finance: Derivatives, Quantitative Models and Risk Management, First Edition.
Mathieu Boudreault and Jean-François Renaud.

Because an ELIA's value is contingent on the value of a reference index or portfolio, one can interpret an ELIA as a derivative sold by life insurance companies. Just like *optionality* for vanilla options, guarantees embedded in ELIAs come at a price. Therefore, we should expect contributions, *premiums* or other form of payments to be made by the investor, in order to benefit from these embedded guarantees.

However, there are four important differences between typical financial derivatives (such as call and put options) and ELIAs:

1) Life contingent cash flows: life insurance benefits are paid upon death or at maturity of the contract and annuities are paid periodically until the death of the policyholder. It is the single most important difference with common derivatives.
2) Very long maturities: whereas financial derivatives have fixed maturities below 1 year (generally a few months), ELIAs have *effective maturities* between 10 and 50 years. We use the term *effective maturity* because ELIAs are contingent upon death.
3) A premium is paid up-front to own a derivative but guaranteed benefits in ELIAs are generally paid for by giving up some of the upside potential on the reference portfolio.
4) ELIAs are held by small investors (savings for/during retirement) whereas financial derivatives are held mainly by institutional investors (such as banks, insurance companies, investment banks).

This chapter is not meant to be a thorough review of ELIA contracts and their practical aspects. The main objective of this chapter is to introduce the reader to a large class of insurance products known as ELIAs and to link their cash flow structure to financial derivatives and options. To classify ELIAs, we will use the most popular names in the United States, i.e. *equity-indexed annuities* and *variable annuities*. This simple classification has been chosen to ease the presentation and it is not necessarily the one prevailing on the market.

More specific objectives are to:

- understand the relationships and differences between ELIAs and other derivatives;
- understand the following three indexing methods: point-to-point, ratchet and high water-mark;
- compute the benefit(s) of typical guarantees included in ELIAs;
- recognize how equity-indexed annuities and variable annuities are funded;
- analyze the loss tied to equity-indexed annuities and variable annuities;
- explain how mortality is accounted for when risk managing ELIAs.

Because maturity benefits are important *building blocks* for a full analysis of ELIA contracts, we first focus on this type of benefit in Sections 8.1-8.4. Mortality risk will be incorporated only in Section 8.5 with a formal treatment of death benefits.

Other applications of ELIAs will be spread out throughout the book once different models for the underlying asset's price (the binomial model, Black-Scholes-Merton, etc.) are presented.

8.1 Definitions and notations

Before we proceed to the analysis of specific policies, let us introduce some notation and define important quantities to better understand how benefits are computed.

First, a policyholder investing in an ELIA has to select an index or a reference investment portfolio as the underlying asset of the contract. As before, let us use S_t to denote the time-t value of the chosen index or reference portfolio, where once again time 0 stands for the contract's inception. Recall that for a fixed time t, the price S_t is a random variable. As for options, let T be the maturity time of a given ELIA contract. The underlying asset's value is the main *source of risk* in such contracts.

Second, ELIAs include a guarantee at maturity and/or provide a guaranteed income. Let G_t be the time-t value of this guaranteed amount (at maturity). This guaranteed amount is usually tied to the initial investment I made by the policyholder and, at maturity time T, is given by

$$G_T = I \times \phi(1 + \gamma)^T \tag{8.1.1}$$

where $0 < \phi < 1$ is the fraction/percentage of the initial investment I that is guaranteed and γ is the guaranteed minimum return, also known as *roll-up rate*. Of course, if T is expressed in years, then γ is an *annual* rate. In other words, for each dollar invested, an amount of $\phi(1 + \gamma)^T$ is guaranteed at maturity. Usually, the guarantee at maturity is a deterministic (non-random quantity), i.e. its value is already known at inception. We will come back to guaranteed income later in this chapter.

Example 8.1.1 *Equity-linked insurance tied to the S&P 500 (continued)*
Consider again the contract of example 8.0.1 and let us identify the parameters of the maturity guarantee as defined in (8.1.1). For this contract, the maturity time is $T = 8$ and we have

$$I = G_8 = 100000.$$

This means that $\phi = 1$ and $\gamma = 0$. The maturity benefit was implicitly given by

$$\max\left(100000 \times \frac{S_8}{S_0}, G_8\right),$$

where S_8 is the (random) value of the S&P 500, 8 years from now. ■

As mentioned above, guarantees in ELIAs are *funded* or *paid for* by single or multiple contributions/payments made by the policyholder during the life of the contract (between time 0 and time T) and/or by a limited access to the upside gains of the underlying asset. These are expected to cover the value of the guarantee, in addition to management fees, operating expenses, taxes, etc. When we analyze more specific products, we will be more precise as to how these guarantees are paid for.

In the next two sections, we will study two types of ELIAs. We will use the names *equity-indexed annuities* and *variable annuities* to classify an ELIA according to the method used to pay for the embedded guarantee. As mentioned earlier, these names may differ from the ones used in the industry and in some countries.

8.2 Equity-indexed annuities

In what follows, we will use the name **equity-indexed annuity** (EIA) for an investment product sold by insurance companies providing:

1) a participation in the growth of an index or a reference portfolio;
2) a financial guarantee at maturity.

EIAs form a popular sub-class of ELIAs mostly used by investors in the accumulation phase, as they usually provide a single benefit at maturity time T (the maturity benefit). Of course, that amount can then be annuitized with another financial product (issued at time T or later).

EIAs are classified according to their indexing methods, i.e. how the maturity benefit is tied to the underlying asset. There are three popular methods of indexation:

- point-to-point;
- ratchet;
- high watermark.

8.2.1 Additional notation

Let R_t be the (periodic) cumulative returns of the underlying asset S between time 0 and time t. Mathematically,

$$R_t = \frac{S_t}{S_0} - 1.$$

For time points $0 = t_0 < t_1 < t_2 < \cdots < t_n = T$, the return earned by S during the k-th period, i.e. between time t_{k-1} and time t_k, is simply

$$y_k = \frac{S_{t_k}}{S_{t_{k-1}}} - 1, \tag{8.2.1}$$

which can also be written as

$$y_k = \frac{1 + R_{t_k}}{1 + R_{t_{k-1}}} - 1.$$

For example, if we are interested in annual returns, we only need to choose $t_k = k$, for each k. Note that for a fixed time t, since the price S_t is a random variable, then the returns R_t and y_t are also random variables.

8.2.2 Indexing methods

As we will see, equity-indexed annuities have a similar payoff structure to investment guarantees studied in Chapter 6. Recall that the payoff of an investment guarantee is given by $\max(S_T, K)$. Consequently, it provides a *full participation* in the asset, if its final price S_T is larger than K, or a guaranteed cash amount of K, otherwise. In exchange, a premium must be paid.

Equity-indexed annuities are very similar to investment guarantees: the *full participation* in the asset will be replaced by another way of participating in the returns of the asset, while the guaranteed amount will be denoted by G_T, as defined in equation (8.1.1), instead of K. We will look at three ways of participating in the returns, generally known as *indexing methods*.

The **point-to-point (PTP) indexing method** yields a maturity benefit given by

$$\max(I \times (1 + \beta R_T), G_T) \tag{8.2.2}$$

where β is known as the **participation rate** and is typically such that $0 < \beta < 1$, and where G_T is given in equation (8.1.1). In other words, for each dollar invested, an EIA with a PTP indexing method provides a maturity benefit of

$$\max(1 + \beta R_T, \phi(1 + \gamma)^T).$$

The policyholder thus gives up part of the upside potential in exchange for a protection against the downside risk. Indeed, contrarily to financial derivatives such as options, no upfront premium is required to be entitled to the guarantee embedded in an EIA, but the participation rate being lower than 100% ($\beta < 1$) serves that purpose. In this case, finding the *fair value* of the participation rate β, i.e. its no-arbitrage value for a given maturity guarantee G_T, is similar to finding the *fair price* of an option. We will come back to this issue in Chapter 18.

Example 8.2.1 *Point-to-point indexing method*
Suppose that in example 8.0.1, the policyholder is entitled to 80% of the accumulated returns on the S&P 500 with a guarantee applying to 100% of the initial investment. Using the same scenario (values of the index), let us compute the maturity benefit.

For this PTP EIA, the maturity benefit of equation (8.2.2) is such that we have $\beta = 0.8$ and $I = G_8 = 100000$. Using the same scenario as in example 8.0.1, in 8 years we would have $R_8 = 0.333$ and therefore the maturity benefit would be

$$\begin{aligned}\max(I(1+\beta R_8), G_8) &= \max(100000 \times (1 + 0.8 \times 0.333), 100000)\\ &= \max(126640, 100000) = 126640.\end{aligned}$$

■

Another popular type of indexing is the (compound periodic) **ratchet indexing method**. In a T-year EIA contract with a ratchet indexing scheme applied at time points $0 = t_0 < t_1 < t_2 < \dots < t_n = T$, the maturity benefit is given by

$$\max\left(I \times \prod_{k=1}^{n}(1 + \beta \max(y_k, 0)), G_T\right), \tag{8.2.3}$$

where the periodic returns $y_k, k = 1, 2, \dots, n$, were defined in (8.2.1). Again, β is known as the participation rate and is typically such that $0 < \beta < 1$. The ratchet indexing method mostly differs from the PTP indexing method as a minimal return applies periodically, usually annually, rather than at maturity only.

Example 8.2.2 *Ratchet indexing method*
Let us consider a 2-year EIA whose ratchet indexing scheme applies annually. Moreover, the participation rate is 90% and the initial investment is fully guaranteed, i.e. $G_2 = I$. Therefore, the maturity benefit of (8.2.3) is here given by

$$\max(I \times (1 + 0.9 \max(y_1, 0)) \times (1 + 0.9 \max(y_2, 0)),\ G_2).$$

Recall that y_1 and y_2 are random variables; therefore, at inception, their values are unknown.

Today, say we invest $I = 250000$ in this ratchet EIA. Consider the following scenario: in the first year, the return is 10% and during the second year, it is negative. Then, in this scenario

$$(1 + 0.9 \max(y_1, 0)) = (1 + 0.9 \max(0.1, 0)) = 1.09$$

and

$$(1 + 0.9 \max(y_2, 0)) = 1 + 0 = 1.$$

Therefore, in this scenario, the maturity benefit would be equal to

$$\max(250000 \times 1.09 \times 1, 250000) = 272500,$$

2 years from now, as $G_2 = I = 250000$.

■

Finally, the **high watermark indexing method** applies the *maximum cumulative return*, observed over the life of the EIA, to the whole life of the contract. Borrowing notation from lookback options, as seen in Chapter 7, we have that the maturity benefit of a *high watermark EIA* is given by

$$\max\left(I \times \left(1 + \beta R_T^{\max}\right), G_T\right) \tag{8.2.4}$$

where $R_T^{\max} = M_T^S / S_0 - 1$ is the maximum cumulative return, where $M_T^S = \max\{S_{t_1}, S_{t_2}, \ldots, S_{t_n}\}$ is the discretely monitored maximum value of the underlying asset/portfolio (as already defined for lookback options in Chapter 7). The maximum is generally monitored at some pre-specified dates $0 < t_1 < \cdots < t_n = T$, usually yearly or bi-annually. Again, β is known as the participation rate and is usually such that $0 < \beta < 1$. Note the similarity with the maturity benefit of a PTP EIA as given in (8.2.2).

Since G_T is usually smaller than the maximal cumulative return, in that case the maturity benefit will simplify to

$$I \times \left(1 + \beta R_T^{\max}\right).$$

Example 8.2.3 *Comparison of indexing methods*
Consider a 10-year EIA with a participation rate of 80% (for each of the indexing methods) and an initial investment guaranteed at 100% at maturity. For simplicity, assume this initial investment is 100. Therefore, $T = 10$, $\beta = 0.8$ and $I = G_{10} = 100$.

Suppose that over the course of the next 10 years the reference portfolio evolves as follows: 1000, 1022.16, 1042.98, ..., 1317.11, 1237.97 (details are provided in the next table). In this scenario, let us compute the maturity benefit for the three indexing methods.

k	S_k	R_k	y_k	$\max(y_k, 0)$
0	1000			
1	1022.16	0.02216	2.22%	2.22%
2	1042.98	0.04298	2.04%	2.04%
3	1008.88	0.00888	−3.27%	0.00%
4	1010.83	0.01083	0.19%	0.19%
5	1067.89	0.06789	5.64%	5.64%
6	1102.42	0.10242	3.23%	3.23%
7	1051.51	0.05151	−4.62%	0.00%
8	1125.86	0.12586	7.07%	7.07%
9	1317.11	0.31711	16.99%	16.99%
10	1237.97	0.23797	−6.00%	0.00%

For the PTP scheme, we have

$$\begin{aligned}\max(I \times (1 + \beta R_T), G_T) &= \max(100(1 + 0.8 \times 0.23797), 100)\\ &= \max(119.0376, 100) = 119.0376.\end{aligned}$$

For the compound annual ratchet, the computation is tedious, but we have

$$\max\left(I \times \prod_{k=1}^{10}(1+\beta\max(y_k,0)), G_T\right) = (1+0.8\times 0.0222)\times(1+0.8\times 0.0204)\times\ldots$$
$$\times(1+0.8\times 0) = 1.332979.$$

Thus, the maturity benefit is

$$\max(100\times 1.332979, 100) = 133.30.$$

Finally, the maximum value of the reference index is observed at time 9 (at the end of the 9-th year), i.e. $M_{10}^S = 1317.11$ and hence the high watermark benefit is

$$I\times\left(1+\beta R_T^{\max}\right) = 100(1+0.8\times 0.31771) = 125.3688.$$ ∎

8.3 Variable annuities

In what follows, we will use the name **variable annuity (VA)** for another popular sub-class of ELIA contracts. They are constructed with a *separate account*, also known as the *sub-account* of the contract, where the initial investment is deposited and credited with returns from a reference portfolio (underlying asset). The policyholder is then allowed to withdraw from this sub-account either for liquidity purposes during the accumulation phase or simply to purchase an annuity at retirement (annuitization phase). This sub-account is subject to a set of guarantees that apply at maturity or death, making VAs typical hybrids between insurance and investment.

One of the most important difference between EIAs and VAs is how guarantees are financed, i.e. how they are paid for by the policyholder. As mentioned before, in an EIA contract, the cost of the guarantee is paid implicitly through the participation rate β that lowers the upside potential. In a VA contract, premiums are withdrawn periodically from the sub-account and are usually set as a fixed percentage of the sub-account balance, acting again like a *penalty* on the credited returns. As before, these premiums also include management fees, operating expenses, taxes, etc. The sum of these expenses is known as the **management and expense ratio** (MER).[2]

The generic name for this type of ELIA contract is **separate account policy**. In the U.S., it is known as a VA, in Canada, as a *segregated fund*, and in the U.K., as a *unit-linked contract*. They borrow their name from typical regulations that prevent insurance companies from mixing assets backing variable annuities with other investments.

8.3.1 Sub-account dynamics

The insured's initial investment $A_0 = I$ is deposited in a sub-account. This sub-account is then credited with the returns of the reference portfolio. Also, it is from this sub-account that fees are deducted and that *policyholder withdrawals* will be made.

2 In the industry, the initial investment is often known as the *premium* whereas periodic contributions made by the policyholder to pay for the guarantee and other costs (MER) is known as a *fee* instead of a premium. We will keep using the terms *initial investment* and *periodic premium* so that the vocabulary is consistent with actuarial and financial mathematics.

Let the sub-account value at time k (after k months or years) be denoted by A_k, where $k = 1, 2, \ldots, n$. The sub-account value is adjusted periodically and it is not allowed to become negative so that deductions (withdrawals, fees) must be such that $A_k \geq 0$ for each k. With this notation, the index n corresponds to the maturity time T. More specifically, the sub-account dynamic is given by

$$A_k = A_{k-1} \times \frac{S_k}{S_{k-1}} \times (1 - \alpha) - \omega_k \quad \text{as long as } A_k \geq 0, \tag{8.3.1}$$

where α is the fee/premium rate deducted from the sub-account at the end of the period and where ω_k is the amount withdrawn by the policyholder, at the end of the k-th period (as long as it is less than the sub-account value prior to the withdrawal).

We can describe equation (8.3.1) as follows: the sub-account balance A_k is obtained by crediting returns $\frac{S_k}{S_{k-1}}$ to the previous sub-account balance A_{k-1}, then deducting the proportional periodic premium (done by multiplying by $(1 - \alpha)$) and finally deducting the policyholder's withdrawal ω_k. Again, unless stated otherwise, all the parameters and quantities are given and computed on an annual basis. Finding the *fair value* of α is a typical problem for insurance companies and is a concept related to finding the no-arbitrage price of an option: more details in Chapter 18.

If we set $A_{k-} = A_{k-1} \times \frac{S_k}{S_{k-1}} \times (1 - \alpha)$, i.e. if A_{k-} is the value of the sub-account just after deducing the fee but just before the k-th withdrawal, then we can rewrite (8.3.1) as follows:

$$A_k = A_{k-} - \omega_k \quad \text{if } \omega_k < A_{k-}.$$

Note that depending on the VA policy, there might be restrictions on withdrawals and guarantees might be adjusted based on the amounts withdrawn.

Example 8.3.1 *Sub-account balance after 1 year*
An investor puts \$100 in the sub-account tied to a VA policy. The periodic premium is set to 1% per annum. The investor plans on withdrawing \$5 at the end of the year, if the funds are available.

Assume the following scenario: the reference portfolio will grow by 4% over the next year. Let us compute the value of the sub-account balance at the end of the year, i.e. after the withdrawal, in this scenario.

Here, we have $A_0 = 100$, $\omega_1 = 5$, $\alpha = 0.01$. In this scenario, we have $S_1/S_0 = 1.04$ and therefore the sub-account balance at year-end would be

$$A_1 = A_0 \times \frac{S_1}{S_0} \times (1 - \alpha) - \omega_1 = 100 \times 1.04 \times 0.99 - 5 = 102.96 - 5 = 97.96.$$

Note that we were allowed to withdraw $\omega_1 = 5$ from the sub-account since $\omega_1 = 5 < A_{1-} = 102.96$.

In this scenario, in the second year, the return of the reference portfolio would be applied to a sub-account balance of 97.96. ■

Example 8.3.2 *Sub-account balance during a year*
A retiree puts \$120 in the sub-account tied to a VA policy. If the premium paid at the end of each month is 0.1% of the (monthly) balance and if the contract allows for withdrawals of

$10 at the end of each month (or the remaining balance), let us determine the sub-account balance at the end of the year in the scenario depicted in the following table:

Month k	S_k	A_k
0	1000	120.00
1	966	105.80
2	1015	101.06
3	999	89.37
4	950	74.90
5	992	68.13
6	953	55.39
7	865	40.22
8	837	28.88
9	808	17.85
10	898	9.82
11	981	0.72
12	1012	0

Here, $A_0 = 120$, $\alpha = 0.001$ (monthly) and $\omega_k = \min(10, A_{k-})$ for all $k = 1, 2, \ldots, 10$. For example, here is how A_1 was computed:

$$A_1 = A_0 \times \frac{S_1}{S_0} \times (1-\alpha) - \omega_1 = 120 \times \frac{966}{1000} \times 0.999 - \min(10, 115.80) = 105.80.$$

Note that in the above scenario (given by the table), $\omega_k = 10$, for each $k = 1, 2, \ldots, 11$, while $\omega_{12} = A_{12-} = 0.74$, as this is the sub-account balance just after crediting the returns of the 12th month (i.e. $0.74 = 0.72 \times \frac{1012}{981}$). In other words, in this scenario, there are not enough funds left at time 12 to withdraw $10, so the remaining balance of $0.74 is withdrawn instead. ■

8.3.2 Typical guarantees

In this section, we will look at two popular guarantees included in VAs. The first one is a guarantee that applies on the sub-account balance at maturity and is known as a **guaranteed minimum maturity benefit (GMMB)** rider. The second guarantee assures the policyholder can withdraw predetermined amounts from the sub-account and is known as a **guaranteed minimum withdrawal benefits (GMWB)**[3] rider. The first guarantee is widely used in the accumulation phase, while the second guarantee is designed for the annuitization phase.

8.3.2.1 GMMB

Let us begin by assuming the policyholder cannot withdraw from the sub-account, so that $\omega_k = 0$, for each k. Therefore, the sub-account balance at time k is given by

$$A_k = A_0 \frac{S_k}{S_0}(1-\alpha)^k.$$

3 Each company selling VAs can brand these guarantees as it wants. As a result, there is no consistent and uniform name for these guarantees in the industry nor in the scientific literature. We will therefore follow [15].

In a GMMB, the maturity benefit is given by

$$\max(A_T, G_T) \tag{8.3.2}$$

where as in (8.1.1) the maturity guarantee is given by $G_T = I \times \phi(1+\gamma)^T$ and where $0 < \phi \leq 1$ and $\gamma > 0$. However, in practice, we have typically $0.75 < \phi \leq 1$ and $0 < \gamma < 0.02$, i.e. only 75–100% of the initial investment is subject to a minimum guaranteed return of 0–2%.

As shown in equation (8.3.2), at maturity, the guarantee G_T is compared to the final sub-account balance A_T, and not to the underlying asset value S_T. In a scenario where the final balance A_T ends up below G_T, the guarantee will apply and the policyholder will receive G_T.

Example 8.3.3 *Guaranteed minimum maturity benefit*
Suppose that a 10-year GMMB is issued and the reference portfolio is tracking the Dow Jones 30 index. A premium of 1.5% per year is paid to fund the maturity guarantee, which is 105% of the initial investment. The value of this index is 10,000 at issuance of the contract.

In the scenario that the index reaches a value of 12,000 at maturity, let us determine the payoff of the GMMB for an initial investment of \$500.

The guaranteed amount at maturity is given by:

$$G_{10} = 500 \times 1.05 = 525.$$

In the given scenario, the final sub-account balance is easy to compute:

$$A_{10} = A_0 \frac{S_{10}}{S_0}(1-\alpha)^{10} = 500 \frac{12000}{10000}(1-0.015)^{10} = 515.8383.$$

Consequently, the maturity benefit of this GMMB would be

$$\max(A_T, G_T) = \max(515.8383, 525) = 525.$$ ■

8.3.2.2 GMWB

A GMWB rider guarantees a regular stream of income to the (retired) investor. If returns on the reference portfolio are favorable, then the investor may also receive the final balance of the sub-account.

The nature of a GMWB is to allow the policyholder to withdraw from the sub-account *when needed*. However, for modeling purposes, we assume that withdrawals are constant and pre-determined. More specifically, the initial investment $I = A_0$ can be withdrawn uniformly, i.e. $\omega_k = \omega = A_0/n$ at the end of each period, for each $k = 1, 2, \ldots, n$, if there are n periodic withdrawals allowed. These withdrawals are guaranteed, no matter the value of the sub-account on a given year-end.

Let us consider a T-year GMWB that allows for annual withdrawals (it is easy to generalize to monthly or periodic withdrawals). Then, after the k-th withdrawal, the sub-account balance is

$$A_k = A_{k-} - \omega \quad \text{as long as } A_k \geq 0,$$

or, written differently,

$$A_k = \max(0, A_{k-} - \omega),$$

for each $k = 1, 2, \ldots, n$.

If returns are favorable, i.e. if $A_T > 0$, then the policyholder will recover this final sub-account balance A_T, in addition to the guaranteed withdrawals already collected.

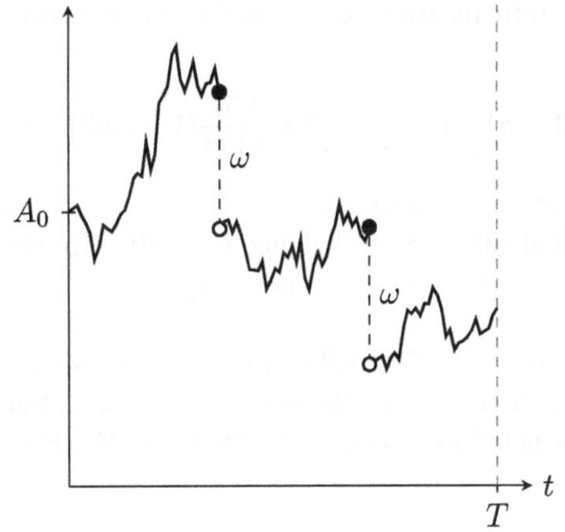

Figure 8.1 Evolution of a sub-account balance for a GMWB with periodic withdrawal ω

Figure 8.1 shows the sample path of the sub-account balance after periodic withdrawals. In the case of this scenario, the sub-account balance being positive at maturity, this amount would be returned to the policyholder.

It is important to note that even if the sub-account balance were to reach zero before the maturity of the contract (due to unfavorable returns), the policyholder would still be entitled to the remaining guaranteed withdrawals. In that case, the insurer would suffer a loss. This will be discussed at length in the next section.

The next two examples illustrate the cash flows of a GMWB in two market scenarios.

Example 8.3.4 *Sub-account of a GMWB – favorable returns*
A policyholder invests \$500 in a 5-year GMWB with annual fee payments and withdrawals. If the annual fee rate is 2% and if the underlying index evolves as follows during the next five years, i.e. 1000, 1174, 957, 1048, 1220, 1296, then let us determine the cash flows of this GMWB's sub-account.

The following table illustrates the cash flows in this scenario:

year k	S_k	Withdrawal ω	A_k
0	1000		500
1	1174	100	475.26
2	957	100	279.67
3	1048	100	200.13
4	1220	100	128.32
5	1296	100	33.59

Recall that with a GMWB, no matter what happens, the policyholder is guaranteed to receive \$100 at the end of each year.

After one year, the sub-account balance is

$$A_1 = A_0 \times \frac{S_1}{S_0} \times (1 - \alpha) - \omega = 500 \times \frac{1174}{1000} \times (1 - 0.02) - 100 = 475.26.$$

For the second year, the returns are credited on the previous account balance of 475.26, so that

$$A_2 = A_1 \times \frac{S_2}{S_1} \times (1-\alpha) - \omega = 475.26 \times \frac{957}{1174}(1 - 0.02) - 100 = 279.67.$$

The rest of the table is computed similarly.

Finally, because the final sub-account balance is positive (in this scenario), the policyholder is entitled to an *extra* $A_5 = \$33.59$ at maturity. ■

Example 8.3.5 *Sub-account of a GMWB – unfavorable returns*
Consider the same GMWB rider as in the previous example, but suppose now that the index will evolve instead as follows: 1000, 977, 978, 1396, 954, 879.

year k	S_k	Withdrawal ω	A_k
0	1000		500
1	977	100	378.73
2	978	100	271.54
3	1396	100	279.84
4	954	100	87.41
5	879	100	0

Again, the policyholder is entitled to \$100 at the end of each year, as this is guaranteed by the contract design and is not affected by the scenario.

However, the returns credited to the sub-account are not enough to cover the withdrawals. The policyholder will receive all five withdrawals of \$100, but she will not receive any additional amount at maturity since

$$A_{5-} = 87.41 \times \frac{879}{954} \times (1 - 0.02) = 78.93 < 100 = \omega.$$

In this scenario, the insurance company would suffer a loss of $100 - 78.93 = 21.07$. ■

Should I stay or should I go: the incentive to lapse

To lapse or to surrender a VA policy is a right given to the policyholder to cease premium payments and forego the underlying protection. We have ignored this feature in the above discussion.

We know that derivatives and variable annuities differ in various aspects, but one very important element is that an option is paid by an upfront premium whereas periodic premiums are paid throughout the life of a VA policy. This may raise an interesting question when the sub-account balance is well over the guarantee: should a policyholder continue to pay premiums for a guarantee that is unlikely to be triggered? In this case, there is an *incentive to lapse*.

Suppose 2 years before maturity, the sub-account balance is 200 and the minimum guaranteed maturity benefit is 100. If the investor believes it is nearly impossible for the sub-account balance to go below 100 during the next 2 years, then it might be profitable to stop paying premiums for such a guarantee.

In conclusion, depending on the market conditions and the structure of the contract, it might be optimal for a policyholder to lapse a VA. This is an American-like feature of a VA in the sense that an informed investor may make a decision that optimizes her wealth just as for American put options.

Insurance companies try to prevent lapses by having a periodic premium structure that discourages early surrenders (to recoup initial costs) or by including riders that reset the guaranteed amount whenever market returns are favorable (automatic and voluntary resets).

8.4 Insurer's loss

Before we analyze the insurer's loss tied to an ELIA contract, let us look again at an *investment guarantee*, a product that we presented first in Chapter 6. Recall that its payoff can be decomposed as

$$\max(S_T, K) = S_T + \max(K - S_T, 0)$$

or

$$\max(S_T, K) = K + \max(S_T - K, 0).$$

In the first case, we see that writing an investment guarantee is equivalent to selling a share of stock in addition to writing a put option. If the company already owns or buys a share of the stock, then it is *liable* to provide, at maturity, the following put payoff:

$$\max(S_T, K) - S_T = \max(K - S_T, 0).$$

In the second case, we see that writing an investment guarantee is equivalent to selling risk-free bonds with a total principal value of K in addition to writing a call option. If the company already owns or buys the risk-free bonds, then it is *liable* to provide, at maturity, the following call payoff:

$$\max(S_T, K) - K = \max(S_T - K, 0).$$

In both cases, the remaining risk arising from these long-term option payoffs needs to be actively managed using adequate investment strategies, i.e. with hedging portfolios. We will come back to this issue later in the book.

8.4.1 Equity-indexed annuities

EIAs are typically branded as a product offering participation in the upside potential of an index or investment portfolio. This is because, in practice, insurance companies typically invest the policyholder's initial investment I in risk-free bonds and can then easily provide a payoff of G_T, i.e. the initial investment accumulated at some fixed rate, usually below the risk-free rate, if needed.

Therefore, the insurer's loss on an EIA is the maturity benefit paid in excess of G_T. Mathematically speaking, the loss on a PTP EIA is defined as

$$\max(I(1 + \beta R_T), G_T) - G_T.$$

Using the properties of the maximum operator, the loss can be written as follows:

$$\max(I(1 + \beta R_T), G_T) - G_T = \max(I(1 + \beta R_T) - G_T, 0)$$

This loss can be further rewritten (with some work) as

$$\frac{\beta I}{S_0} \max\left(S_T - \frac{S_0}{\beta}\left(\frac{G_T}{I} - (1-\beta)\right), 0\right).$$

In conclusion, the insurer's loss arising from a PTP EIA has the same value as a certain number of call options. More precisely, the loss is equivalent to $\frac{\beta I}{S_0}$ units of a call option (written on the same underlying asset) with a strike price of $\frac{S_0}{\beta}\left(\frac{G_T}{I} - (1-\beta)\right)$.

Calculating the insurer's loss from an EIA with a ratchet or high watermark indexing scheme is similar.

Example 8.4.1 *Loss on several EIAs*
Recall the policies considered in example 8.2.3 and the given scenario. For the PTP indexing method, the maturity benefit was 119.0376 (in that scenario) whereas the guaranteed amount was 100. Therefore, the loss for this policy would be 19.0376 (in that scenario). Similarly, the loss on the compound annual ratchet is 33.30 and 25.3688 on the high watermark benefit. ■

8.4.2 Variable annuities

A VA is a separate account policy and, as such, the sub-account belongs to the policyholder. It is held in her name by the insurance company. At maturity, the policyholder will cash the final sub-account balance A_T if it is positive. For the insurer, this is a riskless component of the contract. Therefore, it is natural to define the loss on a VA policy as the benefit net of the sub-account balance.

Mathematically, the insurer's loss on a GMMB is defined as

$$\max(A_T, G_T) - A_T.$$

Again, it can be rewritten as follows:

$$\max(A_T, G_T) - A_T = \max(G_T - A_T, 0) = (G_T - A_T)_+.$$

The loss has the same payoff structure as that of a put option written on the sub-account value and with a strike price of G_T, even though such a derivative is not traded.

Example 8.4.2 *Loss on a GMMB*
We continue example 8.3.3 by computing the loss of this specific contract with the observed trajectory of the reference portfolio. We know that the maturity benefit in that scenario would be equal to

$$\max(A_T, G_T) = \max(516.425, 525) = 525.$$

Since the final sub-account value would be 516.425, the loss (for this specific trajectory) would then be

$$(G_T - A_T)_+ = (525 - 516.425)_+ = 8.575.$$

■

The loss of a GMWB is more complex than that of a GMMB because the guarantee allows the policyholder to withdraw the specified periodic amounts no matter what occurs on the financial markets. Let us contrast two scenarios, depending on whether or not the sub-account is sufficient to fund the withdrawals:

- Sub-account is *sufficient* to fund all withdrawals: the returns credited on the sub-account were sufficient to cover both the withdrawals and the premiums and hence the insurer suffers no loss. The sub-account remaining balance is paid back in full to the policyholder, providing the policyholder with upside potential.

- Sub-account is *not sufficient* to fund the withdrawals: the sub-account balance reaches zero before maturity and the policyholder is still entitled to the remaining withdrawals. The insurer suffers a loss equivalent to the sum of all withdrawals the sub-account could not provide. In this scenario, the guarantee has *kicked in* and provided the policyholder with downside protection on the withdrawals.

To summarize, the loss on a GMWB is the sum of withdrawals owed in excess of the sub-account balance. It is possibly best illustrated with a numerical example.

Example 8.4.3 *Loss of a GMWB*
We compute the loss of the GMWB in both the favorable and unfavorable market scenarios of example 8.3.4 and example 8.3.5.

In the favorable market scenario, the sub-account balance at the end of the contract is positive and therefore the sub-account balance $A_5 = 33.59$ is returned in full to the policyholder. Since all withdrawals were funded by the sub-account, the insurance company has suffered no loss.

In the unfavorable market scenario, the returns were insufficient to fund the final withdrawal. In that scenario, the loss would be equal to 21.07, i.e. the portion of the final withdrawal that would not be covered by the sub-account. ■

8.5 Mortality risk

Although we have only treated maturity benefits so far, most ELIAs offer a protection until maturity or death, whichever comes first. The typical death benefit is very similar to the maturity benefit in the sense that a similar guarantee may apply at the time of death, if death occurs during the life of the contract. We will begin by analyzing general investment guarantees with death benefits and apply the same principles to EIAs and VAs.

Let τ be the (random) time of death of some given individual. The lifetime distribution of τ depends on the age of the policyholder and other characteristics such as gender, smoking habits, etc. The *effective maturity* time of an ELIA contract is given by the following random variable:

$$T^\star = \min(T, \tau) = \begin{cases} T & \text{if death does not occur before time } T, \text{i.e. if } \tau > T, \\ \tau & \text{if death occurs before time } T, \text{i.e. if } \tau \leq T. \end{cases}$$

In other words, the contract lasts until one of the following two possible events occurs: death of the policyholder at the random time τ, which is less than the maturity time T, or survival until time T.

Example 8.5.1 *Investment guarantee applying at death or at maturity*
A policyholder enters into an investment guarantee that protects 100% of the initial capital in case of death and 105% of the initial investment at maturity (in 10 years). Today, the reference index is worth 1000. Suppose the insured invests \$500 in this product.

Let us consider the following two scenarios:

(a) the reference index is worth 980 after 3 years and 1350 after 10 years, and the policyholder dies at the end of the third year;

(b) the reference index is worth 980 after 3 years and 1350 after 10 years, and the policyholder is alive after 10 years.

Let us compute the value of the payoff of this contract in each scenario.

In scenario (a), the effective maturity is at time $\min(T, \tau) = \min(10, 3) = 3$, i.e. the guarantee applies at time of death (at time 3), and the corresponding death benefit is $\max(500, 500 \times \frac{980}{1000}) = \max(500, 490) = 500$. Note that in this scenario, the fact that the reference index will reach level 1350 at time 10 does not have any impact on the benefit.

In scenario (b), the effective maturity is at time 10 and, since the accumulated investment is larger than the guaranteed 5% return, the maturity benefit is $\max(500 \times 1.05, 500 \times \frac{1350}{1000}) = \max(525, 675) = 675$. ■

For the rest of this section, we will discuss how death benefits are priced and managed at the level of the insurance company. We will assume the insurer manages a *large* portfolio of ELIA contracts sold to (independent) individuals having the same age, living habits, etc., and thus having the same *mortality distribution*. The insurance company is exposed to two risks: mortality and fluctuations of the (common) reference portfolio. As seen in Chapter 1, the risk associated with the reference portfolio is a *systematic risk* for the insurance company because it is common to all contracts, whereas the *mortality risk* is assumed to be fully diversifiable. As usual, we assume that those two risks are independent.

Even if an insurer cannot exactly identify the time of death of each individual, diversification of the mortality risk in a very large portfolio allows the insurer to *approximate* the number of policyholders who will die every year. As seen in Chapter 1, this is a direct consequence of the law of large numbers. Therefore, even if at the individual level the effective maturity of the policy is random (it is the random variable denoted above by $T^\star$), at the portfolio level the insurer can estimate the number of policies maturing after 1 month, 2 months, …, after 1 year, 2 years, etc., based upon the expected number of people who will die after 1 month, 2 months, etc.

Therefore, the insurer can approximate its portfolio by considering it as a pool of ELIA contracts all having various *deterministic* maturities

$$T = \frac{1}{12}, \frac{2}{12}, \ldots, 1, \ldots, 2, \ldots,$$

as if each of these contracts did not include the *mortality component*. This way, we can use all the previous results of this chapter. The following example will illustrate this very important idea.

Example 8.5.2 *Risk management of investment guarantees with a death benefit*
An insurance company sells an investment guarantee contract to 100,000 insureds assumed to have the same age and living habits. The actuary has determined that approximately 1% of the 100,000 individuals will die in each year over the next 10 years.

Let us explain how to manage this portfolio of investment guarantees protecting 100% of the capital in case of death and 105% of the capital until maturity if the policy matures in 10 years.

Out of the 100,000 policies issued, the actuary expects[4] that 1,000 will terminate in each year due to death (over 10 years), and therefore cashing in the death benefit, leaving 90,000

4 It is a mathematical expectation, in the probabilistic sense.

policies to terminate with the maturity benefit. In other words, 1,000 policyholders will die during the first year, another 1,000 during the second year, …, and finally a last 1,000 during the last year; all those policies will receive the death benefit equal to 100% of the initial investment. The other policies, i.e. 90% of the 100,000 in this portfolio, will receive the maturity benefit whose guarantee is 105% of the initial capital.

Therefore, the insurance company can approximate its actual portfolio by considering that it holds the following portfolio of 11 different investment guarantees with deterministic maturities:

- 1,000 contracts maturing in 1 year with a maturity benefit corresponding to 100% of the investment;
- 1,000 contracts maturing in 2 years with a maturity benefit corresponding to 100% of the investment;
- …
- 1,000 contracts maturing in 10 years with a maturity benefit corresponding to 100% of the investment;
- 90,000 contracts maturing in 10 years with a maturity benefit corresponding to 105% of the investment.

As we did before, we can decompose each investment guarantee's maturity benefit $\max(S_T, K)$ as the sum of a unit of the underlying S_T and a put option $\max(K - S_T)_+$, where $T = 1, 2, \dots, 10$. Then, the approximate portfolio is equivalent to holding

- 1,000 put options maturing in k year(s) with a strike of $100\% \times S_0$, for each $k = 1, 2, \dots, 10$;
- 90,000 put options maturing in 10 years with a strike of $105\% \times S_0$;
- 100,000 shares of the underlying stock. ∎

EIA and VA contracts also include death benefits that are very similar to the maturity benefits described previously. Therefore, the point-to-point, the compound annual ratchet and the high watermark indexing schemes may also apply at the time of death, if it occurs before maturity.

Similar to a GMMB, a guarantee that applies at death is known as a **guaranteed minimum death benefit (GMDB)**. Therefore, to receive a benefit at death or maturity, whichever comes first, we need to combine a GMMB and a GMDB. Much like a GMWB, a **guaranteed lifetime withdrawal benefit (GLWB)** allows the retiree to withdraw guaranteed amounts until death whereas the remaining sub-account balance (if any) is paid upon death to the beneficiaries.

Example 8.5.3 *Illustration of a GLWB*
A 65-year-old retiree has accumulated savings of $20,000 which she invests in a policy allowing withdrawals of $1,000 per year until death and subject to an annual premium of 2% of the sub-account balance.

Consider the following scenario: the policyholder will die after 8 years, while the evolution of the index will be 1, 1.1047, 1.0559, 1.3563, 1.0952, 0.9798, 1.0774, 1.1667, 0.9904 at the end of those 8 years. Let us compute the death benefit of this contract.

We have computed the sub-account balance for the next 8 years, according to equation (8.3.1), in this scenario. The results are presented in the next table.

k	S_k	Withdrawals	A_k
0	1		20000
1	1.1047	1000	20652.12
2	1.0559	1000	18345.0186
3	1.3563	1000	22092.8324
4	1.0952	1000	16482.9674
5	0.9798	1000	13451.252
6	1.0774	1000	13495.3372
7	1.1667	1000	13321.6165
8	0.9904	1000	10082.4157

In this scenario, the policyholder will die in the 8th year and the sub-account balance at the end of that year will be 10082.4157. Therefore the policyholder's heirs will receive \$10082.42 at that time, while the retiree will have received \$1000 per year until death. ■

Clearly, a GLWB provides annuity payments in addition to participation in the upside potential of the reference portolio. The insurance company will thus provide protection if the insured dies very old and/or if returns are unfavorable. This contrasts with basic annuities that cease at death and do not provide any right to favorable market returns (i.e. the sub-account balance at time of death).

Similarly, as in example 8.5.2, the loss tied to an ELIA is a portfolio of call/put options having different maturities whose quantities are determined by the expected number of people who will die each year.

Example 8.5.4 *Loss of two ELIAs with death benefits*
Suppose we have a pool of 100,000 insureds having the same age and living habits. The actuary has determined that, on average, 1.5% of those insureds will die in each of the following 6 years (1500 per year), whereas the rest will survive the next 6 years. Describe the loss of a 6-year PTP EIA and the loss of a 6-year policy combining a GMDB and GMMB.

We found earlier that the liability tied to a PTP EIA with maturity T (and no death benefit) is equivalent to $\frac{\beta I}{S_0}$ units of a call option on the stock with a strike price of $\frac{S_0}{\beta}\left(\frac{G_T}{I} - (1-\beta)\right)$. However, we can approximate the annual number of deaths, so that the total loss is

- $100000 \times 0.015 \times \frac{\beta I}{S_0}$ units of call options maturing in T years, with strike $\frac{S_0}{\beta}\left(\frac{G_T}{I} - (1-\beta)\right)$, for each $T = 1, 2, \dots, 6$;
- $100000 \times 0.91 \times \frac{\beta I}{S_0}$ units of call options maturing in 6 years, with strike $\frac{S_0}{\beta}\left(\frac{G_6}{I} - (1-\beta)\right)$.

Recall that the loss tied to a GMMB with maturity T (and no death benefit) is a put option on the sub-account balance with strike price equivalent to the guaranteed amount G_T. Similarly, the total loss of the GMMB combined with a GMDB both issued to 100,000 policyholders is

- 100000×0.015 put options maturing in T years with strike G_T, for each $T = 1, 2, \dots, 6$;

- 100000 × 0.91 put options maturing in 6 years with strike G_6.

Depending on the clauses of the policies, note that G_6 could be different whether the insured died in the 6th year or survived the 6 years. ■

In conclusion, it is important to remember that risk management of death benefits in ELIAs is based upon diversification, i.e. by approximating the expected number of people dying/surviving every year. Then, the insurance portfolio can be treated as a pool of contracts each having fixed maturities (i.e. no death benefit).

8.6 Summary

Equity-linked insurance or annuity (ELIA): generic name to designate a life insurance/annuity including living and/or death benefits tied to the returns of a reference portfolio of assets.

Differences between an ELIA and other derivatives

- Life contingent cash flows.
- Very long maturities.
- No premium paid up-front.
- Held by small investors (savings for/during retirement).

General notation and terminology

- Time-t price of the underlying asset/portfolio: S_t.
- Maturity date: T.
- Guaranteed amount at maturity: G_T.
- Time-t value of the guaranteed amount: G_t.
- Initial investment: I
- Percentage of I that is guaranteed: $\phi \in (0, 1)$.
- Roll-up rate (guaranteed minimum return): γ
- Guaranteed amount at maturity:

$$G_T = I \times \phi(1+\gamma)^T.$$

- Cumulative returns of S between time 0 and time t:

$$R_t = \frac{S_t}{S_0} - 1.$$

- Periodic monitoring: $0 = t_0 < t_1 < t_2 < \cdots < t_n = T$.
- Return earned by S during the k-th period:

$$y_k = \frac{S_{t_k}}{S_{t_{k-1}}} - 1.$$

- Maximum price of S: $M_T^S = \max\{S_{t_1}, S_{t_2}, \dots, S_{t_n}\}$.
- Maximum cumulative return: $R_T^{\max} = M_T^S / S_0 - 1$.

Equity-indexed annuities (EIAs)

- Specific features:
  - Participation rate: β.
  - Guarantee is paid for (by the policyholder) because β is less than 1.

- Maturity benefits for different indexing methods/schemes:
  - point-to-point:
$$\max(I \times (1 + \beta R_T), G_T).$$
  - ratchet:
$$\max\left(I \times \prod_{k=1}^{n}(1 + \beta \max(y_k, 0)), G_T\right).$$
  - high watermark:
$$\max\left(I \times \left(1 + \beta R_T^{\max}\right), G_T\right).$$

Variable annuities (VAs)

- Sub-account increases: initial investment, credited returns.
- Sub-account decreases: fees/premium payments, withdrawals.
- Sub-account value at time k: A_k.
- Initial investment: $I = A_0$.
- Fee/premium rate: α.
- Withdrawal (if allowed) at the end of the k-th period: ω_k.
- Value of the sub-account after the fee but before k-th withdrawal:
$$A_{k-} = A_{k-1} \times \frac{S_k}{S_{k-1}} \times (1 - \alpha).$$
- Sub-account dynamics:
$$A_k = A_{k-} - \omega_k \quad \text{if } \omega_k < A_{k-}.$$
- Guaranteed minimum maturity benefit (GMMB):
  - No withdrawal, i.e. $\omega_k = 0$, for all k.
  - Sub-account balance at time k:
$$A_k = A_0 \frac{S_k}{S_0}(1 - \alpha)^k.$$
  - Maturity benefit: $\max(A_T, G_T)$.
- Guaranteed minimum withdrawal benefit (GMWB):
  - Guaranteed withdrawals.
  - Assume ω_k is constant and equal to $\omega = A_0/n$.
  - After the k-th withdrawal:
$$A_k = A_{k-} - \omega \quad \text{as long as } A_k \geq 0,$$
or, written differently,
$$A_k = \max(0, A_{k-} - \omega).$$
  - If $A_T > 0$, then policyholder receives A_T.

Insurer's loss

- For an EIA, the loss is the maturity benefit paid in excess of G_T:
$$\text{Loss} = \text{maturity benefit} - G_T.$$
- Loss on a PTP EIA: same value as a number of call options.
- For a VA, the loss is the benefit net of the sub-account balance:
  - Loss for a GMMB: $\max(A_T, G_T) - A_T = (G_T - A_T)_+$ (payoff of a put option).
  - Loss for a GMWB: sum of the withdrawals in excess of the sub-account balance.

Mortality risk and death benefits

- Time of death (random) of an individual: τ.
- Effective maturity: $\min(T, \tau)$.
- Guaranteed minimum death benefit (GMDB): like a GMMB, but guarantee applies at time τ.
- GMMB + GMDB: benefit at death or at maturity.
- Guaranteed lifetime withdrawal benefits (GLWB): like a GMWB, but ends at time τ.
- Mortality risk can be diversified: approximate the expected number of people dying/surviving every year.
- Risk management: treat the portfolio as a pool of contracts with fixed maturities and no death benefit.

8.7 Exercises

8.1 *Benefit of equity-indexed annuities*
Suppose a 10-year EIA is issued on a stock index whose fictional evolution is given in Table 8.1. For a participation rate of 75%, an initial investment of 1000 and a guaranteed amount of 1050, compute the maturity benefit of an EIA under each of the following indexing schemes:
(a) point-to-point;
(b) annual ratchet;
(c) high watermark.

8.2 *Benefit of variable annuities*
Suppose a policyholder enters into a 10-year VA whose main reference index evolved as in Table 8.1 over the next 10 years. The initial investment is 1000.
(a) Compute the benefit of a GMMB if the annual premium (paid as a fraction of the sub-account balance) is 1.5% per year and 75% of the initial investment is subject to a minimum annual return of 1%.

Table 8.1 Fictional path of the reference index (over 10 years)

k	S_k
0	15000
1	19189
2	20537
3	25652
4	32206
5	37799
6	38022
7	32732
8	31119
9	30857
10	26407

(b) Compute the maturity benefit of a GMWB if the annual premium (paid as a fraction of the sub-account balance) is 2% per year and the annual withdrawal (withdrawn after return is credited) is 100.

8.3 *Total amount of fees paid*
Consider the sub-account and the scenario given in example 8.3.2. Compute the total amount of fees paid by the policyholder between inception and maturity.

8.4 *Loss of EIAs*
Give the mathematical details leading to the insurer's losses given in example 8.4.1.

8.5 *Cash flows and loss of a GMWB*
A policyholder invests 1000 in a 10-year GMWB with an annual premium of 2% per year and the annual withdrawal (withdrawn after return is credited) is 100.

As the actuary managing this contract, you analyze the cash flows and liability of this GMWB over two possible scenarios for the reference index. Those two scenarios are given in the following table:

	S_k	
k	Scenario # 1	Scenario # 2
0	15000	15000
1	20150	14608
2	14957	17345
3	11260	20132
4	11701	19511
5	12067	22273
6	11749	22079
7	15549	23285
8	14624	20689
9	17733	20588
10	19263	19946

For each scenario:

(a) Calculate the cash flow for the policyholder at times $k = 0, 1, \dots, 9, 10$.
(b) Calculate the cash flow for the insurer at times $k = 0, 1, \dots, 9, 10$.
(c) Calculate the loss to the insurance company.

8.6 Suppose that a high watermark EIA is issued such that $G_T = I$. Write down the loss of this policy and relate it to a lookback option and a risk-free bond.

8.7 Show that the loss of a PTP indexing scheme is

$$\frac{\beta I}{S_0} \max\left(S_T - \frac{S_0}{\beta}\left(\frac{G_T}{I} - (1-\beta)\right), 0\right).$$

8.8 A 65-year-old policyholder wishes to invest in an ELIA for 10 years but seeks protection for his heirs in case of death. Two products are presented: a PTP EIA and a combination of a GMMB + GMDB. In both cases, the guaranteed amount is 100% of the initial investment which is 1000.

Suppose the reference index evolves as follows and the death benefit is paid at the end of the year:

k	S_k
0	15000
1	20150
2	14957
3	11260
4	11701
5	12067
6	11749
7	15549
8	14624
9	17733
10	19263

For each of the following products, calculate the death benefit if the individual dies in the third year. What is the benefit if the individual survives 10 years?

(a) PTP EIA with an 80% participation rate;

(b) protection that combines a GMMB and GMDB with an annual premium of 1.9%.

8.9 A 65-year-old retiree enters into a GLWB. The initial investment is \$100,000 and the guaranteed minimum withdrawal is \$7000 annually. Suppose that the annual premium is 2.4% and the reference index evolves as in the next table.

k	Stock index
0	14000
1	14545
2	16450
3	18105
4	10073
5	9075
6	10065
7	12561
8	12729
9	16781
10	13047
11	20004
12	16575
13	15025
14	13993
15	14656

Compute the sub-account balance, the death benefit and the loss if the policyholder dies:

(a) in the fifth year;

(b) in the 10-th year;

(c) in the 15-th year.

You can assume that upon death, cash flows are paid at year-end.

8.10 A policyholder invests \$100,000 in an investment portfolio equally-weighted between a stock and bond index. The investment is protected by a 5-year GMMB with a guarantee corresponding to 95% of the initial investment. For a hypothetical scenario depicted in the next table, calculate the sub-account balance at the end of each year and the maturity benefit of the GMMB if the annual premium is 1.75%.

k	Stock index	Bond index
0	14000	1200
1	17158	1466
2	14984	1357
3	11423	1589
4	15336	1406
5	17557	1136

8.11 Your company sold 100,000 identical policies to insureds all aged 65. Each policy has the following characteristics:

- maturity benefit is at least 90% of the initial investment (GMMB);
- death benefit is at least 105% of the initial investment (GMDB);
- annual premium is 2.2%;
- initial investment is 100.

You have determined that for the latter portfolio:

- 0.5% of the initial pool will die every year;
- 95% of them will survive 10 years.

Explain the behavior of the insurer's loss.

Part II

Binomial and trinomial tree models

9

One-period binomial tree model

In the first part of this book, whenever we discussed options and other derivatives, we always considered the initial premium as given. We briefly discussed that to find the price of an option, a mathematical model for the evolution of the underlying asset price would be needed.

In the next chapters, we will introduce our very first class of market models for the price of the underlying asset, namely the *binomial tree model*. It is probably the most famous discrete-time financial model to represent the evolution of a stock price, an index value, etc. We will mostly focus on pricing and hedging financial and actuarial derivatives written on a risky asset whose price follows such a model.

Before introducing the general multi-period binomial tree model, we will have a first (but deep) look at the very important one-period and two-period versions of this model. These two special cases will allow us to introduce all the main ideas and concepts in a technically (and financially) easy fashion.

The main objective of this chapter is to investigate the one-period binomial tree model to price options and other derivatives. The specific objectives are to:

- describe the basic assets available in a one-period binomial model;
- identify the assumptions on which a one-period binomial model is based;
- understand how to build a one-period binomial tree;
- understand how to price a derivative by replicating its payoff;
- rewrite the price of a derivative using risk-neutral probabilities;
- recognize that risk-neutral probabilities are only a mathematical convenience.

9.1 Model

In a one-period model there are two time points: the beginning of the period is time 0 and the end of the period is time 1. In most cases, time 0 will represent inception or issuance of a derivative contract while time 1 will stand for its maturity. One should think of time 0 as *today* while time 1 is in the *future*.

A one-period binomial model is a *frictionless* financial market (see Chapter 2) composed of only two assets:

- a risk-free asset (a bank account or a bond), which evolves according to the risk-free interest rate r;
- a risky asset (a stock or an index) with a known initial value and an unknown future value.

9.1.1 Risk-free asset

A risk-free asset is an asset for which capital and interest are repaid with certainty at maturity. It is said to be risk-free in the sense that there is no default: the borrower always meets its

Actuarial Finance: Derivatives, Quantitative Models and Risk Management, First Edition.
Mathieu Boudreault and Jean-François Renaud.

obligations. We denote by $B = \{B_0, B_1\}$ the evolution of the price of this asset with $B_1 \geq B_0$, which is the risk-free condition. In other words, B_0 is today's price while B_1 is its value at the end of the period.

In the one-step binomial tree model, we assume there exists a constant interest rate $r \geq 0$ for which we can either write

$$B_1 = \begin{cases} B_0 e^r & \text{if the rate is continuously compounded,} \\ B_0(1+r) & \text{if the rate is periodically compounded.} \end{cases}$$

Therefore, in this one-period model, r is *not* an annual rate.

Depending on whether we fix B_0 or B_1, B can be interpreted as the evolution of a bank account (in which case, we deposit B_0 that accumulates with capital and interest to B_1) or a zero-coupon bond (in which case, B_1 is equal to the face value of the bond and B_0 is the present value of B_1).

For simplicity and without loss of generality, in what follows we assume that $B_0 = 1$. In summary, we have

$$B_1 = \begin{cases} e^r & \text{if the rate is continuously compounded,} \\ (1+r) & \text{if the rate is periodically compounded.} \end{cases}$$

9.1.2 Risky asset

The evolution of the risky asset price is represented by $S = \{S_0, S_1\}$ where S_0 is today's price and S_1 is the asset price at the end of the period. We have that S_0 is a known (observed) constant while S_1 is a random variable. We assume that this random variable S_1 takes values in $\{s^u, s^d\}$ with probability p and $1-p$ respectively, where $0 < p < 1$ and $s^d < s^u$. The probability distribution of S_1 (with respect to $\mathbb{P}$) is simply:

$$\mathbb{P}(S_1 = s^u) = p = 1 - \mathbb{P}(S_1 = s^d),$$

which is similar to a (biased) coin toss experiment.

One could think of the scenario/event $\{S_1 = s^u\}$ as a *bull market* scenario (u for up) and $\{S_1 = s^d\}$ as a *bear market* scenario (d for down). Therefore, the one-period binomial tree model is a 5-parameter model: those parameters are B_1 (or equivalently r), S_0, s^d, s^u and p.

The one-period model and its randomness are often represented graphically by a tree:

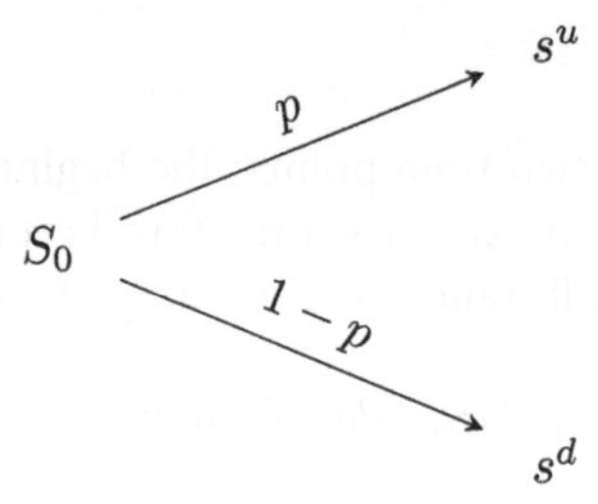

Example 9.1.1 *Construction of a binomial tree*
We represent the possible prices of a stock in 3 months by a one-step binomial tree. The current stock price is \$52 and a 3-month Treasury zero-coupon bond currently trades at \$99 (for a face value of \$100). According to financial analysts, the stock price in 3 months will either be \$55 or \$50 with a 60% chance of going up. Let us build the corresponding one-step binomial tree.

We have $B_0 = 1, B_1 = \frac{1}{0.99} = 1.0101$ and the binomial tree for the stock is

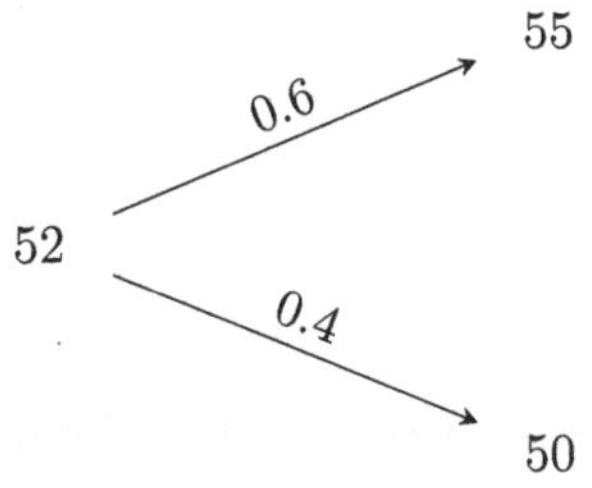

■

As illustrated by the previous example, a period in our model can represent either a day, a month, 6 months, a year, etc. As a result, the time length of the period has an impact on the interpretation of the parameters. For example, the interest rate r is not annualized even though it is the case in practice. One can also work directly with B_0 and B_1 to remove any ambiguity.

Finally, in order to make sure that the model is arbitrage-free, we must have

$$s^d < S_0 B_1 < s^u, \tag{9.1.1}$$

which means that no asset is *doing better* than the other one in both scenarios. Say an investor has $S_0\$$ in cash and needs to determine whether to invest $S_0\$$ in the risk-free asset B or purchase a share of S. Then the arbitrage-free condition states that neither of the two choices will perform better than the other in both scenarios.

If (9.1.1) is not verified, say if $s^d < s^u < S_0 B_1$, then we can construct an arbitrage by going long the risk-free asset and short-selling the risky asset. The inequalities in (9.1.1) will be referred to as the **no-arbitrage condition**. Note that the probability p does not appear in this condition.

Example 9.1.2 *Absence of arbitrage between the stock and the bond*
The current stock price is \$10 and it will either increase by \$1 or \$2 over the next month. Finally, if we deposit \$1 in a bank account, it will be worth 1.01 in a month. Determine whether there is an arbitrage opportunity between the risky and risk-free assets and how to exploit it.

Here, we have $S_0 = 10$, $s^d = 11$, $s^u = 12$ and $B_1 = 1.01$. In this case, the no-arbitrage condition of equation (9.1.1) is not verified: we have

$$s^d = 11 \nless S_0 B_1 = 10 \times 1.01 = 10.10 < s^u = 12.$$

In other words, investing in the stock is always (in the two scenarios) a better investment than investing in the risk-free asset.

To exploit this opportunity, we can borrow \$10 at the risk-free rate and immediately buy the stock for \$10. At time 0, the net initial cash flow is 0. At time 1, we have a value in the stock that is either 11 or 12: in both cases, this is enough to pay back the loan with capital and interest (a total of 10.10). Therefore, there is a certain (at no risk) profit available, which is either 0.90 or 1.90. ■

9.1.2.1 Simplified tree

We can use an alternative representation[1] for the stock price at maturity using an *up factor u* and a *down factor d* as follows: set

$$u = \frac{s^u}{S_0} \qquad \text{and} \qquad d = \frac{s^d}{S_0}.$$

We can now write

$$s^u = uS_0 \qquad \text{and} \qquad s^d = dS_0.$$

Since it has been assumed that $s^d < s^u$, we deduce that $d < u$.

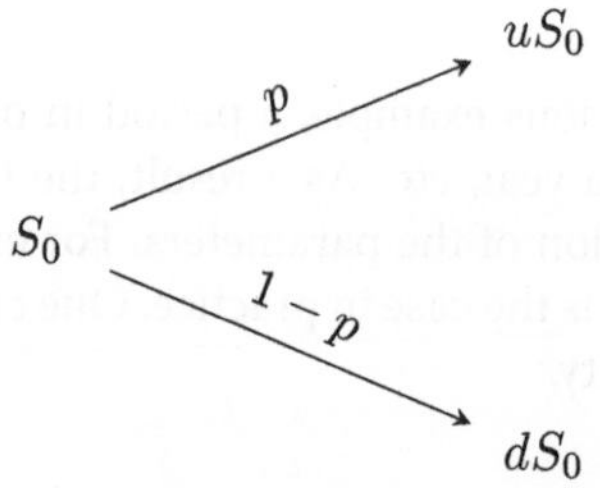

With this new notation, we can say that the one-period binomial tree model is parameterized by B_1(or r), S_0, d, u, p and the no-arbitrage condition (9.1.1) is now equivalent to

$$d < B_1 < u.$$

Note that this alternative representation will play a crucial role for the multi-period binomial tree, to be studied in Chapter 11.

Example 9.1.3 *Binomial tree using up and down factors*
Suppose a stock can either increase by 10% or decrease by 5% by the end of the period. Therefore $u = 1.1$ and $d = 0.95$. In the context of example 9.1.1, we have $u = \frac{55}{52} = 1.05769$ and $d = \frac{50}{52} = 0.961538$. ■

9.1.3 Derivatives

We now introduce a third asset in the market: it is an option or a derivative written on the risky asset S. The price of the derivative over the period is denoted by $V = \{V_0, V_1\}$ with V_0 being today's price while V_1 is the derivative's value at the end of the period, i.e. V_1 is the payoff. As for the risky asset, V_0 is a (still unknown) constant but V_1 is a random variable. The derivative is said to be written on the risky asset S because its payoff is such that $V_1 = v^u$ when $S_1 = s^u$, whereas $V_1 = v^d$ when $S_1 = s^d$. As a result, V_1 is a random variable taking values in the set $\{v^u, v^d\}$. Consequently, V_1 has the following probability distribution (with respect to $\mathbb{P}$):

$$\mathbb{P}(V_1 = v^u) = p = 1 - \mathbb{P}(V_1 = v^d).$$

For example, we can say that the probability that an option with payoff V_1 expires in-the-money is given by $\mathbb{P}(V_1 > 0)$.

1 Indeed, for the one-period tree, this is an alternative representation rather than a simplified version of the tree. However, for two or more periods, this will make a difference.

In the next section, we will determine the value of V_0, such that there are no arbitrage opportunities.

Underlying probability model and the real-world probability measure

It is sometimes useful to describe the full probability model on which the one-period binomial tree is constructed. As for coin tossing, one can choose its sample space Ω_1 to be

$$\Omega_1 = \{\omega : \omega = \mathrm{u} \text{ or } \mathrm{d}\}.$$

It is then clear that each state of nature $\omega \in \Omega_1$, that is each scenario, corresponds to a branch in the tree. Clearly, the possible prices at maturity are

$$S_1(\mathrm{u}) = s^u \equiv uS_0 \qquad \text{and} \qquad S_1(\mathrm{d}) = s^d \equiv dS_0.$$

Similarly, the random variable V_1 is

$$V_1(\mathrm{u}) = v^u \qquad \text{and} \qquad V_1(\mathrm{d}) = v^d.$$

In other words, the random variable V_1 takes the value v^u in the up scenario ($\omega = \mathrm{u}$) which is when S_1 takes the value s^u. Moreover, V_1 takes the value v^d in the down scenario ($\omega = \mathrm{d}$) which is when S_1 takes the value s^d.

The objective probability measure $\mathbb{P}$: events $\to [0,1]$ contains the information of how investors view the market and consequently p is the likelihood (w.r.t. $\mathbb{P}$) of the event $\{S_1 = s^u\}$ which characterizes the distribution of S_1. The value of p could be estimated using data on the price of the risky asset being modeled.

We can now superimpose the evolution of the derivative price V (in boxes) and the evolution of the stock price S.

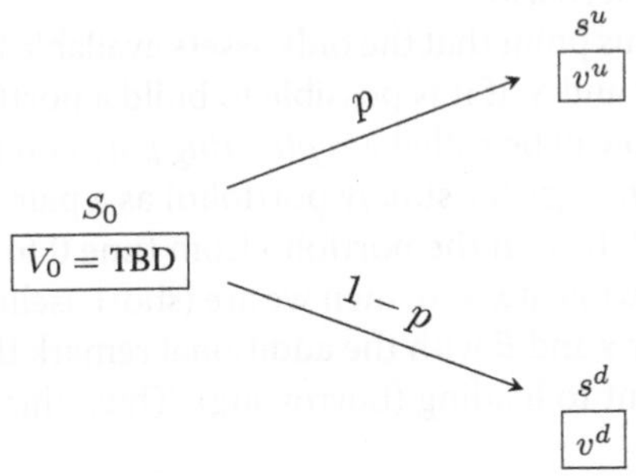

Because they are very important cases to analyze, instead of the notation $V = \{V_0, V_1\}$, we will use the notation $C = \{C_0, C_1\}$ and $P = \{P_0, P_1\}$ for a call option and a put option, respectively. Therefore, we have

$$C_1 = (S_1 - K)_+$$
$$P_1 = (K - S_1)_+,$$

if the options are both struck at K and if they both mature at time 1.

Example 9.1.4 *Payoff of call and put options*

A stock currently trades for \$100 and will either increase or decrease by 10% over the next 6 months. It was determined by financial analysts that the stock price has a probability of 55% of increasing. Suppose an at-the-money call option is issued today. Determine the payoff of the call option at maturity and the probability that it expires in the money. Repeat with an at-the-money put option.

We have $C_1 = (S_1 - 100)_+$ and therefore $C_1 = 10$ when $S_1 = 110$ and $C_1 = 0$ when $S_1 = 90$. As for the put option, $P_1 = (100 - S_1)_+$ and therefore $P_1 = 0$ when $S_1 = 110$ and $P_1 = 10$ when $S_1 = 90$. Graphically, we have

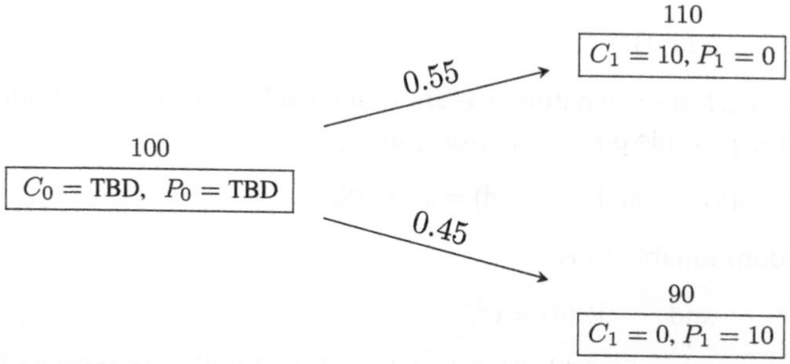

Then, the probability that the call option expires in the money is 0.55 while there is a 45% chance for the put option to expire in the money. ■

9.2 Pricing by replication

We are now ready to price derivatives in the one-period binomial tree model. In other words, we seek to find the price (premium) V_0 that a buyer should pay to be entitled to the random payment V_1 at time 1. In a market where we should avoid arbitrage opportunities, a derivative V should have the same value as an investment strategy/portfolio replicating its payoff (its only cash flow) in each and every scenario.

It must be emphasized at this point that the only assets available to invest in, and thus to replicate V, are the basic assets B and S. If it is possible to build a portfolio matching the derivative payoff, the resulting portfolio will be called a *replicating portfolio* for V_1.

First, we define a *trading strategy* (or simply portfolio) as a pair (x, y), where x (resp. y) is the number of units of S (resp. B) held in the portfolio from time 0 to time 1. If $x > 0$, then we are buying x units of S at time 0, while if $x < 0$, then we are (short-)selling x units of S at time 0. The same interpretation holds for y and B with the additional remark that a long (short) position in the bank account is equivalent to lending (borrowing). Thus, this strategy's value over time is given by $\Pi = \{\Pi_0, \Pi_1\}$ with

$$\Pi_0 = xS_0 + yB_0$$

for today's value whereas

$$\Pi_1 = xS_1 + yB_1$$

is the (random) value of the strategy at the end of the period.

If our goal is to replicate the payoff V_1, then we should choose the pair (x, y) to make sure that

$$\Pi_1 = V_1$$

in *each possible scenario*. Equivalently, we need to find (x, y) such that

$$\begin{cases} xs^u + yB_1 = v^u, \\ xs^d + yB_1 = v^d. \end{cases} \tag{9.2.1}$$

In other words, the number of units of the basic assets to hold should be such that the portfolio's value mimics the values v^u and v^d, i.e. the payoff in the up and down scenarios, respectively.

Therefore, we need to solve (9.2.1), a simple system of two equations with two unknowns, where B_1, s^d, s^u are given model parameters and v^d, v^u are given by the derivative payoff at maturity. The values of x and y are the only two unknowns.

The cost for setting up this replicating strategy at time 0 is simply

$$\Pi_0 = xS_0 + yB_0.$$

And since the binomial model is arbitrage-free, the time-0 value (cost) of this replicating portfolio must also correspond to the time-0 price of the derivative because otherwise, there would be an arbitrage opportunity. In other words, if (x, y) are chosen as in (9.2.1), then the initial no-arbitrage price of the derivative must be

$$V_0 = \Pi_0. \tag{9.2.2}$$

Example 9.2.1 *Price of a call option*
Financial analysts have determined that ABC inc.'s stock, which is currently traded at \$100, will either win or lose 10% of its value over the next year. Unfortunately, they have not been able to specify the likelihood of each scenario. If the interest rate is 6% (compounded annually), what is the replicating portfolio and the (no-arbitrage) price of a call option on ABC inc.'s stock with strike price \$105 and expiring a year from now?

We can model the stock price using a one-period binomial model with parameters $S_0 = 100, d = 0.9$ and $u = 1.1$ (or equivalently $s^u = 110$ and $s^d = 90$), and $B_1 = 1.06$. Moreover, the call option payoff is given by $C_1 = (S_1 - 105)_+$. Therefore, we have the following tree:

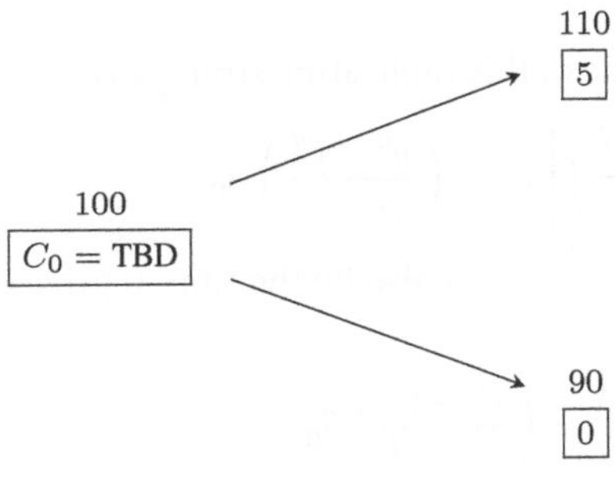

To replicate $C_1 = (S_1 - 105)_+$, we need to choose (x, y) such that

$$\Pi_1 = C_1,$$

in each possible scenario, or equivalently solve

$$\begin{cases} 110x + 1.06y = 5, \\ 90x + 1.06y = 0. \end{cases}$$

Subtracting one equation from the other, we easily get

$$x = \frac{5 - 0}{110 - 90} = 0.25$$

and inserting $x = 0.25$ in any of these two equations, we get

$$y = \frac{1}{1.06}(5 - 110 \times 0.25) = -21.226.$$

So, in order to replicate this call option, at time 0 we must buy 0.25 unit of ABC inc.'s stock and sell 21.226 units of the risk-free asset (borrow $21.226 \times B_0 = \$21.226$). The cost for setting up this strategy is

$$\Pi_0 = 0.25 \times 100 - 21.226 = 3.774.$$

To avoid arbitrage opportunities, the initial price of the call option must be equal to the initial value of this replicating portfolio, that is we must have $C_0 = \$3.774$. ■

If $V_0 \neq \Pi_0$, then it is possible to construct an arbitrage by buying the cheapest (the derivative or the replicating portfolio) and selling the costliest (the replicating portfolio or the derivative). This is illustrated in detail in example 9.2.2.

In a one-period binomial tree model, for any derivative with payoff V_1, there exists a replicating strategy. Said differently, the linear system in (9.2.1) can always be solved algebraically (and its solution is unique). It suffices to subtract one equation from the other in the system of (9.2.1) in order to get

$$x = \frac{v^u - v^d}{s^u - s^d} \tag{9.2.3}$$

and then

$$y = \frac{1}{B_1}(v^u - xs^u) = \frac{1}{B_1}(v^d - xs^d).$$

Using the expression for x given in (9.2.3), we get

$$y = \frac{1}{B_1}\left(\frac{v^d s^u - v^u s^d}{s^u - s^d}\right). \tag{9.2.4}$$

Consequently, the time-0 value of this replicating strategy is

$$\Pi_0 = \left[\frac{1}{B_1}\left(\frac{v^d s^u - v^u s^d}{s^u - s^d}\right)\right] B_0 + \left(\frac{v^u - v^d}{s^u - s^d}\right) S_0.$$

By the no-arbitrage principle, this must also be the time-0 price of this derivative and we conclude that

$$V_0 = \frac{1}{B_1}\left(\frac{v^d s^u - v^u s^d}{s^u - s^d}\right) + \left(\frac{v^u - v^d}{s^u - s^d}\right) S_0. \tag{9.2.5}$$

Overall, in the one-period binomial tree model, all derivatives can be replicated and have a unique no-arbitrage price. Such a market model is also said to be *complete* (market (in)completeness and its impact in risk management will be discussed in detail in Chapter 13).

Example 9.2.2 *Exploiting arbitrage opportunities*
In the context of example 9.2.1, you find that a market maker is willing to buy or sell the call option for \$4. Determine whether there is an arbitrage opportunity and how to exploit it.

Setting up the replicating strategy, you know that the no-arbitrage price of the option should be \$3.774. In other words, at \$4, this option is *overpriced*. Therefore, you sell the \$4 call option to the market maker, set up the replicating strategy for \$3.774 (25 – 21.226) and get an instant profit of 22.6 cents.

We need to check that cash flows are covered at maturity in any given scenario. If we observe $S_1 = 110$, then the short call option is exercised and the position is worth –5.

However, the 0.25 share of stock you bought is worth 27.50 and you need to repay the loan with interest for 22.50. The net total value of the positions is 0. If we observe instead $S_1 = 90$, then the short call option is out of the money and the replicating portfolio is worth 0. Therefore, in both the up and down scenarios, everything is covered confirming the risk-free profit of 0.226.

In summary, we have:

Time	$t = 0$	$t = 1$	
		Up	Down
Short option	+4	−5	0
Long stock	−25	+27.5	+22.5
Short bond	+21.226	−22.5	−22.5
TOTAL	+0.226	0	0

■

It is very interesting and important to note that p, the probability of observing the up scenario, does not appear in the option price V_0. This should have been expected because the option payoff has been replicated with the strategy (x, y) *no matter how likely each scenario is*. The following example illustrates how important no-arbitrage pricing and replication are for actuaries.

Example 9.2.3 *Actuarial price*
A stock currently trades for \$30. You work for an insurance company selling a product that pays \$5 in one period if the stock price goes up (to \$35) and \$10 if the stock price goes down (to \$25). Based on findings from financial analysts, the probability that the stock price goes down is 20%. Moreover, interest rates are very low so you believe that assuming $r = 0$ is a conservative assumption.

Based on your experience as an actuary, you think that the premium for this insurance product should be

$$\mathbb{E}[V_1] = 5 \times 0.8 + 10 \times 0.2 = 6.$$

Let us explain why this is not a *good* price for this product.

First of all, this is a product that is easily replicated. Using (9.2.3) and (9.2.4), we find

$$x = \frac{5 - 10}{35 - 25} = -0.5$$

and

$$y = \frac{10 \times 35 - 5 \times 25}{35 - 25} = 22.50.$$

Therefore, short selling 0.5 share of the stock and investing 22.5 at the risk-free rate will yield the same payoff as the insurance product described above. The initial cost of this replicating strategy is

$$\Pi_0 = -0.5 \times S_0 + 22.5 \times B_0 = -0.5 \times 30 + 22.5 \times 1 = \$7.50.$$

Second, in a frictionless market where rational agents exploit arbitrage opportunities, the premium of this product should be \$7.50, i.e. we should have $V_0 = \$7.50$. Therefore, the product is underpriced at \$6.

To exploit this mispricing, it suffices to buy the product from the insurance company (for \$6) and replicate a short position in the above portfolio (buy 0.5 share of the stock and borrow 22.50) to lock in a risk-free profit of \$1.50, with offsetting cash flows at maturity. ■

The last example shows that trading a derivative at any price different from its replicating portfolio's cost yields an arbitrage opportunity.

Example 9.2.4 *Forward contracts*
The payoff of a one-period forward contract with delivery price K is $S_1 - K$. We want to find the replicating strategy and the no-arbitrage price of this derivative in a one-period binomial tree model.

In the up scenario, the payment will be $s^u - K$ whereas the payment will be $s^d - K$ in the down scenario. To replicate this payoff, we use equations (9.2.3) and (9.2.4) and deduce that, at time 0, the number of shares of stock we need to buy is

$$x = \frac{v^u - v^d}{s^u - s^d} = \frac{(s^u - K) - (s^d - K)}{s^u - s^d} = 1$$

and the position in the risk-free asset is

$$y = \frac{1}{B_1}(v^u - xs^u) = \frac{1}{B_1}(s^u - K - s^u) = \frac{-K}{B_1}.$$

Therefore, to replicate the payoff $S_1 - K$, which occurs at maturity, we need to hold one share of stock and borrow $\frac{K}{B_1}$ at inception. This is the same strategy as the one obtained in Chapter 3, which was a model-free replicating strategy.

In conclusion, the initial cost of replicating this portfolio and consequently the no-arbitrage time-0 value of a forward contract with delivery price K are both equal to

$$S_0 - \frac{K}{B_1}.$$

Recall that a forward contract is usually set up such that its initial value is zero. Hence, the corresponding delivery price, known as the forward price and denoted here by F_0^1, is

$$F_0^1 = S_0 B_1,$$

as expected. It is the same expression as the one obtained in Chapter 3. ■

We conclude this section with an example that illustrates how to manage a more realistic product sold by insurance companies, namely an investment guarantee, as discussed in Chapters 6 and 8.

Example 9.2.5 *Investment guarantees and risk management*
A life insurer sells a very simple variable annuity written on a risky asset S. The policy promises to credit the return on S after one period subject to a minimum of 2%. Using the context of example 9.2.1, let us calculate how much a policyholder should pay for this product.

We seek to price a derivative whose final payoff is $V_1 = \max(S_1, S_0 \times 1.02) = \max(S_1, 102)$. This product is known as an investment guarantee.

The corresponding binomial tree is depicted as follows:

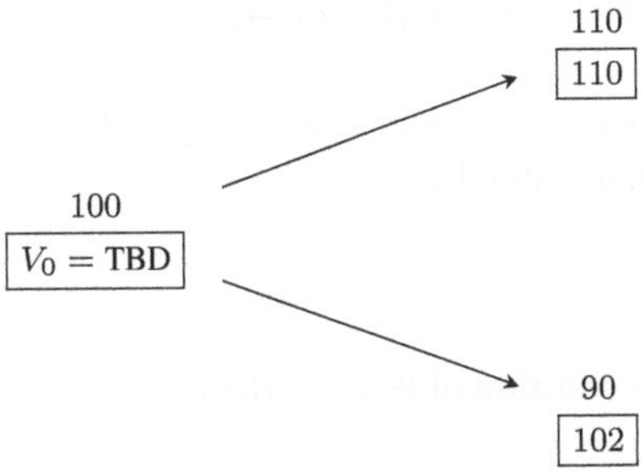

Using equations (9.2.3) and (9.2.4), we find that to replicate this investment guarantee, we need to buy $x = 0.4$ shares of the stock and invest $y = 62.26415094$ at the risk-free rate. Therefore, the initial no-arbitrage price of this derivative is $V_0 = 0.4 \times 100 + 62.26415094 = 102.26415094$.

Note that, as we did in Section 6.2.1 of Chapter 6, using basic financial engineering arguments, we can decompose this payoff as

$$V_1 = \max(S_1, 102) = S_1 + (102 - S_1)_+ = S_1 + P_1.$$

In other words, the insurance company has sold a package *containing* a stock and a put option on that stock. When an insured buys this product, it is therefore long a stock and long a put. Consequently, we have

$$V_0 = S_0 + P_0.$$

Pricing the investment guarantee is somewhat equivalent to pricing this put option. ■

9.3 Pricing with risk-neutral probabilities

We will now appeal to elementary algebraic manipulations to reorganize our pricing formula of equation (9.2.5) and somehow simplify its expression. In fact, we will be able to write the initial price of any derivative in a probabilistic form.

First, notice that the formula in (9.2.5) can be rewritten as follows:

$$\begin{aligned}
V_0 &= \frac{1}{B_1}\left(\frac{v^d s^u - v^u s^d}{s^u - s^d}\right) + \left(\frac{v^u - v^d}{s^u - s^d}\right) S_0 \\
&= \frac{1}{B_1}\left(\frac{v^d s^u - v^u s^d}{s^u - s^d} + B_1 \frac{v^u - v^d}{s^u - s^d} S_0\right) \\
&= \frac{1}{B_1}\left(\frac{v^d s^u - v^u s^d + v^u S_0 B_1 - v^d S_0 B_1}{s^u - s^d}\right) \\
&= \frac{1}{B_1}\left(v^d \frac{s^u - S_0 B_1}{s^u - s^d} + v^u \frac{S_0 B_1 - s^d}{s^u - s^d}\right).
\end{aligned}$$

Consequently, if we define

$$q = \frac{S_0 B_1 - s^d}{s^u - s^d}, \tag{9.3.1}$$

then we can write

$$V_0 = \frac{1}{B_1}(qv^u + (1-q)v^d) = q\frac{v^u}{B_1} + (1-q)\frac{v^d}{B_1}. \tag{9.3.2}$$

Alternatively, if the binomial tree is defined using up and down factors, i.e. if $s^u = uS_0$ and $s^d = dS_0$, then we can rewrite q defined in (9.3.1) as

$$q = \frac{B_1 - d}{u - d}.$$

Recalling the no-arbitrage condition of (9.1.1), which is

$$s^d < S_0 B_1 < s^u,$$

then it should be clear that $0 < q < 1$.

Obviously, the pricing formula obtained in (9.3.2) is equal to the one previously obtained in (9.2.5). However, it has a familiar look: it is an expectation. Indeed, it can be considered as the *expected value of the discounted payoff*, if we use the *new probability weights* q and $1-q$, where the discounting is done at the risk-free rate. Consequently, we will say that q is the *risk-neutral probability* of an up-move and that $1-q$ is the *risk-neutral probability* of a down-move.

In conclusion, we have obtained our first **risk-neutral pricing formula:** in a one-period binomial tree, for any given payoff V_1, we have

$$V_0 = \frac{1}{B_1}\mathbb{E}^{\mathbb{Q}}[V_1], \tag{9.3.3}$$

where the superscript $\mathbb{Q}$ in the expectation emphasizes the use of the risk-neutral probability weights q and $1-q$.

Example 9.3.1 *Pricing with risk-neutral probabilities*
In the context of example 9.2.1, the risk-neutral probability q is

$$q = \frac{S_0 B_1 - s^d}{s^u - s^d} = \frac{100 \times 1.06 - 90}{110 - 90} = 0.80$$

or alternatively

$$q = \frac{B_1 - d}{u - d} = \frac{1.06 - 0.9}{1.1 - 0.9} = 0.80.$$

Consequently, $1 - q = 0.20$.

Therefore, the price of the call option of example 9.2.1 can be computed with the risk-neutral pricing formula of equation (9.3.3):

$$C_0 = \frac{1}{1.06}(0.8 \times 5 + 0.2 \times 0) = 3.774.$$

Of course, this is the same price as in example 9.2.1.

Finally, the put option of example 9.2.5 can also be computed with the risk-neutral pricing formula of equation (9.3.3):

$$P_0 = \frac{1}{1.06}(0.8 \times 0 + 0.2 \times 12) = 2.264.$$

Again, this is the same price as in example 9.2.5. ■

The risk-neutral probability measure

To allow us to be more precise, let us define a new probability measure we will call the **risk-neutral probability measure**. We define $\mathbb{Q}$: events $\to [0,1]$ by

$$\mathbb{Q}(S_1 = s^u) = q, \qquad \mathbb{Q}(S_1 = s^d) = 1 - q,$$

where q is given by (9.3.1). It should be noted that the possible outcomes taken by the random variables S_1 and V_1 are not related to the choice of probability measure. However, their probability distributions and any expectation computed with these distributions will depend on whether they are identified with respect to $\mathbb{P}$ or $\mathbb{Q}$. Then, with respect to $\mathbb{Q}$, we have

$$\mathbb{Q}(V_1 = v^u) = q = 1 - \mathbb{Q}(V_1 = v^d).$$

In words, if we create a probability measure $\mathbb{Q}$ with weights q and $1 - q$ assigned to the up and down scenarios respectively, then we can write the price of a derivative

$$V_0 = \frac{1}{B_1}(qv^u + (1-q)v^d) = \frac{1}{B_1}\mathbb{E}^{\mathbb{Q}}[V_1]$$

as an expectation under the probability measure $\mathbb{Q}$.

Let us now illustrate what it means to compute expectations with the risk-neutral probabilities q and $1 - q$ rather than with the real-world or actuarial probabilities p and $1 - p$. First of all, the expected asset price at time 1 is simply given by

$$\mathbb{E}[S_1] = ps^u + (1-p)s^d.$$

This is the *real* or actuarial expected value of S_1, i.e. what is expected by actuaries and financial analysts (agreeing on this market model).

Second, if we use the risk-neutral probabilities q and $1 - q$, then

$$\mathbb{E}^{\mathbb{Q}}[S_1] = qs^u + (1-q)s^d.$$

From equation (9.3.1), we have that

$$q = \frac{S_0 B_1 - s^d}{s^u - s^d} \quad \text{and} \quad 1 - q = \frac{s^u - S_0 B_1}{s^u - s^d},$$

so we can further write

$$\mathbb{E}^{\mathbb{Q}}[S_1] = \frac{(B_1 S_0 - s^d)s^u + (s^u - S_0 B_1)s^d}{s^u - s^d} = \frac{B_1 S_0 (s^u - s^d)}{s^u - s^d} = S_0 B_1.$$

Therefore, in the *risk-neutral world*, on average, the stock earns the risk-free rate. This is clearly not equivalent to the actuarial expected value.

As it was made clear by the algebraic reorganization of the formula in (9.2.5) at the beginning of the section, the probabilities q and $1 - q$ make sense only whenever we want to find the no-arbitrage price of a derivative. The risk-neutral probabilities should not be used when assessing the likelihood of events.

Example 9.3.2 *Forward contracts (continued)*
We continue example 9.2.4. The risk-neutral pricing formula stipulates that

$$V_0 = \mathbb{E}^{\mathbb{Q}}\left[\frac{V_1}{B_1}\right] = \frac{1}{B_1}(\mathbb{E}^{\mathbb{Q}}[S_1] - K)$$

where

$$S_0 B_1 = \mathbb{E}^{\mathbb{Q}}[S_1].$$

Consequently,

$$V_0 = S_0 - \frac{K}{B_1}$$

and we find that the forward price is given by $F_0^1 = S_0 B_1$.

Moreover, as discussed above, it does not make sense for a risk-averse investor to earn the risk-free rate on average on a stock. Therefore,

$$\mathbb{E}[S_1] > F_0^1 = S_0 B_1 = \mathbb{E}^{\mathbb{Q}}[S_1]$$

and we conclude again that the expected stock price (in one period) should be larger than the one-period forward price. ■

At this point, we should not try to find more financial or actuarial meaning to these new probability weights q and $1 - q$. The fact that we can write the initial price V_0 as a risk-neutral expectation of the discounted payoff is a (very convenient) mathematical curiosity/coincidence.

The risk-neutral pricing formula is a consequence of the derivation of the replicating portfolio. This portfolio mimics the payoff in each and every scenario, no matter how likely each scenario actually is.

9.4 Summary

One-period binomial model

- Risk-free interest rate: r.
- Risk-free asset price: $B = \{B_0, B_1\}$, where $B_0 = 1$ and

$$B_1 = \begin{cases} B_0 e^r & \text{if the rate is continuously compounded,} \\ B_0(1+r) & \text{if the rate is periodically compounded.} \end{cases}$$

- Risky asset price: $S = \{S_0, S_1\}$, where S_1 is a random variable such that

$$S_1 = \begin{cases} s^u & \text{with probability } p, \\ s^d & \text{with probability } 1-p. \end{cases}$$

- No-arbitrage condition: $s^d < S_0 B_1 < s^u$.
- Derivative (written on S): $V = \{V_0, V_1\}$ where V_1 is a random variable such that

$$V_1 = \begin{cases} v^u & \text{if } S_1 = s^u, \\ v^d & \text{if } S_1 = s^d. \end{cases}$$

- Simplified tree: up and down factors $u > d$ such that

$$s^u = uS_0 \quad \text{and} \quad s^d = dS_0.$$

- Alternative no-arbitrage condition: $d < B_1 < u$.

Trading strategy/portfolio

- Portfolio: a pair (x, y).
- Number of units of S, the risky asset: x.
- Number of units of B, the risk-free asset: y.

- Portfolio's value: $\Pi = \{\Pi_0, \Pi_1\}$ where

$$\Pi_0 = xS_0 + yB_0 \quad \text{and} \quad \Pi_1 = xS_1 + yB_1.$$

Pricing by replication

- A strategy (x, y) replicates V_1 if $\Pi_1 = V_1$, i.e. if

$$\begin{cases} xs^u + yB_1 = v^u, \\ xs^d + yB_1 = v^d. \end{cases}$$

- Replicating portfolio for V_1:

$$x = \frac{v^u - v^d}{s^u - s^d},$$
$$y = \frac{1}{B_1}\left(\frac{v^d s^u - v^u s^d}{s^u - s^d}\right).$$

- By no-arbitrage assumption: $V_0 = \Pi_0 = xS_0 + yB_0$, i.e.

$$V_0 = \frac{1}{B_1}\left(\frac{v^d s^u - v^u s^d}{s^u - s^d}\right) + \left(\frac{v^u - v^d}{s^u - s^d}\right) S_0.$$

Pricing with risk-neutral probabilities

- Risk-neutral probabilities: $\{q, 1 - q\}$, where

$$q = \frac{S_0 B_1 - s^d}{s^u - s^d}.$$

- Risk-neutral pricing formula for a derivative with payoff V_1:

$$V_0 = \frac{1}{B_1}\mathbb{E}^{\mathbb{Q}}[V_1] = \frac{1}{B_1}(qv^u + (1 - q)v^d).$$

- Risk-neutral probabilities are only a mathematical convenience used when pricing under no arbitrage.

9.5 Exercises

9.1 Find the initial price of the following derivative: it provides one unit of the risky asset at time 1.

9.2 A stock currently trades for \$23 and a 3-month zero-coupon bond sells for \$99. In 3 months, a share of stock can take two possible values: \$22 and \$25. Is there an arbitrage opportunity between the stock and the bond? If it is the case, explain how to exploit such arbitrage.

9.3 The 6-month Treasury rate is 2% (compounded annually). We can find a share of stock of ABC inc. currently trading for \$37 that can take only two possible values in 6 months: \$38 or \$40. Is there an arbitrage opportunity between the stock and the bond? If it is the case, explain how to exploit such arbitrage.

9.4 Your company sold an at-the-money put option on a stock currently selling for \$75. The stock may sell for \$70 or \$85 in a year from now. Assume that the 1-year Treasury rate is 2.75% (compounded annually).

(a) What should be the no-arbitrage price of the put option?
(b) How many shares of the stock should you buy/sell at inception such that your company cancels out the liability tied to the put option?

9.5 A derivative pays \$3 when the stock price goes up to \$37 and the derivative pays \$6 when the stock price goes down to \$34. If the current stock price is \$35 and the 3-month Treasury rate is 1% (compounded annually), find the number of units of stock and risk-free bond to exactly replicate this 3-month derivative.

9.6 A share of stock currently sells for \$50 and can take up two values in a year: \$57 or \$48. It is believed by financial analysts that the probability of the stock going up is 85%.
(a) What is the expected return on the stock?
(b) If a 1-year Treasury zero-coupon bond currently sells for \$97, what is the no-arbitrage price of a 52-strike call option?

9.7 A share of stock sells for \$52 and can go up by 5% within 3 months or down by 2% in the same period. A 3-month zero-coupon Treasury bond trades for \$98.67 (face value of 100). Experts are unanimous on the outlook of the stock: they agree the probability of the stock going up is 75%. An at-the-money call option is issued.
(a) Compute the risk-neutral probabilities.
(b) What is the probability of the call option maturing out-of-the money?
(c) What is the expected payoff from the call option?
(d) What is the no-arbitrage price of the call option?

9.8 A 5-year at-the-money put option is sold as protection against long-term bear markets. The stock currently trades for 100 and may increase to 130 or decrease to 50 over the next 5 years. The 5-year Treasury rate is 4.23% (compounded annually). According to an expert, the probability of the stock going down is only 15%.
(a) What is the expected 5-year return on this stock?
(b) What is the expected payoff on this put option?
(c) Compute the present value (at the Treasury rate) of the expected payoff of this option.
(d) What is the no-arbitrage price of the put option?
(e) The issuer sells this put option for 1.25 times the amount in (c). Does it correspond to the no-arbitrage price?
(f) Suppose the market is frictionless. Explain how to exploit the arbitrage opportunity, if any.
(g) Suppose the market is frictionless with the exception that small investors cannot sell the put option. Does this affect your answer in (f)?

10

Two-period binomial tree model

For more flexibility and accuracy, we can add an intermediate time step, or equivalently another time period, to the one-period binomial tree model: the result is the two-period binomial tree model.

In the next chapter, we will focus on the *general multi-period binomial tree model*. In this chapter, we will take the time to provide a thorough treatment of its two-period version. Doing so will allow us to:

1) better understand the construction of the general binomial tree and the dynamics of the underlying asset over time;
2) focus on the replication procedure over more than one period, which is important for risk management purposes;
3) analyze the complexity of pricing path-dependent derivatives and variable annuities;
4) consider more complex situations, such as dollar dividends and stochastic interest rates.

Consequently, the main objective of this chapter is to replicate and price options and other derivatives in a two-period binomial tree model. The specific objectives are to:

- recognize how to build a two-period binomial tree using three one-period binomial trees;
- understand the difference between recombining and non-recombining trees;
- build the dynamic replicating strategy to price an option;
- apply the risk-neutral pricing principle in two periods;
- determine how to price options in more complex situations: path-dependent options, dollar dividends, variable annuities, stochastic interest rates.

10.1 Model

In a two-period model there are three time points: time 0, time 1 and time 2. In most cases, time 0 represents inception or issuance of a derivative contract, time 1 is an intermediate time, while time 2 stands for the contract's maturity. One should think of time 0 as *today* while time 1 and time 2 are in the *future*.

A two-period binomial model is a *frictionless* financial market (see Chapter 2) composed of only two assets:

- a risk-free asset (a bank account or a bond) which evolves according to the risk-free interest rate r;
- a risky asset (a stock or an index) with a known initial value and unknown future prices.

Actuarial Finance: Derivatives, Quantitative Models and Risk Management, First Edition.
Mathieu Boudreault and Jean-François Renaud.

10.1.1 Risk-free asset

Again, the risk-free asset is an asset for which capital and interest are repaid with certainty at maturity. We denote by $B = \{B_0, B_1, B_2\}$ the evolution of its price with B_0 being today's price while B_1 and B_2 are its values at time 1 and time 2. Moreover, because the asset is risk-free, we have $B_2 \geq B_1 \geq B_0$.

We assume that there exists an interest rate $r \geq 0$ for which we can either write, for $k = 1, 2$,

$$B_k = \begin{cases} B_{k-1}e^r & \text{if the rate is continuously compounded,} \\ B_{k-1}(1+r) & \text{if the rate is periodically compounded.} \end{cases}$$

As in the one-step binomial tree, we can set $B_0 = 1$, meaning that B represents the value of a dollar, or we can set $B_2 = 1$ and discount to obtain the price of a zero-coupon bond. In this chapter, unless stated otherwise, we will use the former approach.

10.1.2 Risky asset

The price of the risky asset evolves over time and is represented by $S = \{S_0, S_1, S_2\}$. Today's asset price is S_0 which is a known (observed) quantity. However, at times 1 and 2, the risky asset price is random and is represented by the random variables S_1 and S_2 respectively. Moreover, as in Chapter 9, the random variable S_1 takes values in $\{s^u, s^d\}$ whereas S_2 takes values in $\{s^{uu}, s^{ud}, s^{du}, s^{dd}\}$, with a notation that puts the emphasis on the four possible outcomes in this model. For simplicity, we assume $s^u > s^d$ and $s^{uu} > s^{ud}, s^{du} > s^{dd}$.

The dynamics of (the stochastic process) S is as follows. After one period (at time 1), the price of the risky asset admits the following probability distribution:

$$\mathbb{P}(S_1 = s^u) = p = 1 - \mathbb{P}(S_1 = s^d),$$

where $0 < p < 1$. After two periods (at time 2), the price of the risky asset admits the following *conditional* probability distribution:

$$\begin{aligned}
\mathbb{P}(S_2 = s^{uu} | S_1 = s^u) &= p^u, \\
\mathbb{P}(S_2 = s^{ud} | S_1 = s^u) &= 1 - p^u, \\
\mathbb{P}(S_2 = s^{du} | S_1 = s^d) &= p^d, \\
\mathbb{P}(S_2 = s^{dd} | S_1 = s^d) &= 1 - p^d,
\end{aligned}$$

where $0 < p^u, p^d < 1$. We deduce that the *marginal* probability distribution of S_2 is

$$\begin{aligned}
\mathbb{P}(S_2 = s^{uu}) &= p \times p^u, \\
\mathbb{P}(S_2 = s^{ud}) &= p \times (1 - p^u), \\
\mathbb{P}(S_2 = s^{du}) &= (1 - p) \times p^d, \\
\mathbb{P}(S_2 = s^{dd}) &= (1 - p) \times (1 - p^d).
\end{aligned}$$

More intuitively, the two-step binomial model is the concatenation of three one-step binomial trees:

- During the first period (from time 0 to time 1), the price of the risky asset evolves as in a one-step binomial model with root S_0: S_1 takes values in $\{s^u, s^d\}$ with probabilities p and $1 - p$.
- During the second period (from time 1 to time 2), *and given that after the first period* $S_1 = s^u$, the price of the risky asset evolves as in a one-step binomial model with root

$S_1 = s^u$: $S_2|S_1 = s^u$ (read: S_2 given that $S_1 = s^u$) takes values in $\{s^{uu}, s^{ud}\}$ with probabilities p^u and $1 - p^u$.

- During the second period (from time 1 to time 2), *and given that after the first period* $S_1 = s^d$, the price of the risky asset evolves as in a one-step binomial model with root $S_1 = s^d$: $S_2|S_1 = s^d$ (read: S_2 given that $S_1 = s^d$) takes values in $\{s^{du}, s^{dd}\}$ with probabilities p^d and $1 - p^d$.

It is illustrated by the following tree:

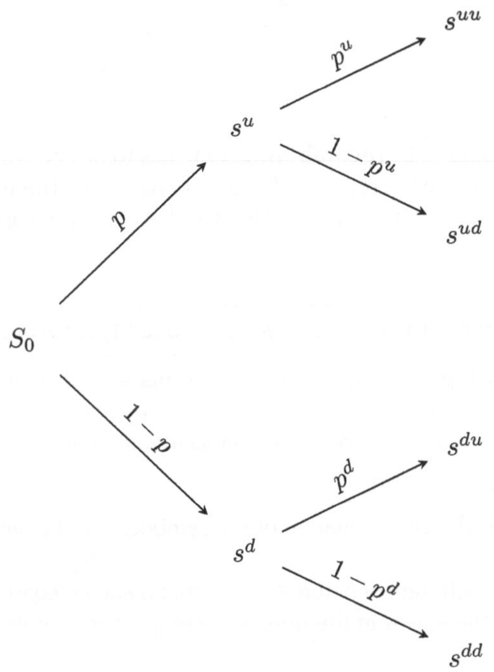

Clearly, the notation emphasizes that there are four possible outcomes in this model, or said differently, four possible paths in the tree:

1) up & up (s^u and s^{uu});
2) up & down (s^u and s^{ud});
3) down & up (s^d and s^{du});
4) down & down (s^d and s^{dd}).

In summary, the two-period binomial tree has multiple parameters:

- interest rate r or, equivalently, B_1 and B_2;
- stock prices at various *nodes*: $s^u, s^d, s^{uu}, s^{ud}, s^{du}, s^{dd}$;
- probabilities: p, p^u, p^d.

We can slightly simplify the two-step binomial tree by assuming that $s^{ud} = s^{du}$. As a result, there are still four possible outcomes in this model (paths in the tree) but the price of the risky asset can take only one of three values at maturity:

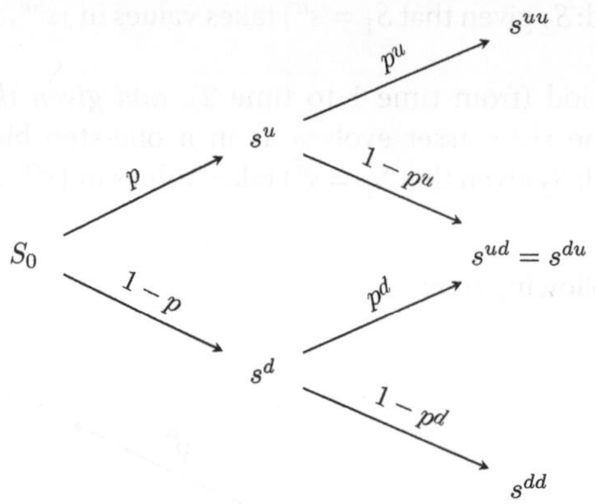

In this case, the tree is said to be **recombining**. One has to be careful when drawing a recombining two-period tree, as it may create confusion. It could give the impression that there are only three possible outcomes in the model. This will have an impact when considering path-dependent derivatives.

Underlying probability model and the real-world probability measure

The objective (or actuarial) probability measure $\mathbb{P}$: events $\to [0, 1]$ contains the information as to how investors view the market. Like the modelling of two consecutive coin tosses, one can choose the sample space Ω_2 of a two-period binomial model to be

$$\Omega_2 = \{\omega : \omega = \text{uu, ud, du, dd}\}$$

where, for example, ud is the concatenation of the symbols u and d, which have no numerical values.

It should be clear that each state of nature $\omega \in \Omega_2$, that is each scenario, corresponds to a path from the root to one of the leaves in the tree: for example, the sample path corresponding to $\omega = \text{ud}$ is

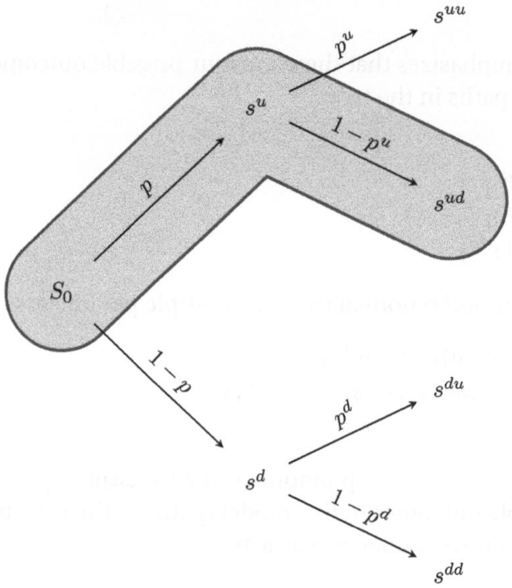

Example 10.1.1 *Construction of a two-period binomial tree*
In a frictionless market, a stock is traded whose initial price is 100 and the (annual) interest rate is constant at 4% (compounded four times a year). Financial analysts have determined that the price of the stock in 3 months will be either 107 (with a probability of 65%) or 97 (with a probability of 35%). Moreover, if the stock price rises to 107 after 3 months, then analysts believe that it will be at either 115 (with a probability of 75%) or 101 (with a probability of 25%) after 3 more months. Finally, if the stock price goes down to 97 after 3 months, then the price will be 103 or 95 (with a probability of 50% in each case) after 3 more months. Let us illustrate these possibilities in a two-step binomial tree.

As in Chapter 9, the choice of the time step is dictated by the context of the problem. Therefore, in this example, one step represents 3 months so that we have $B_0 = 1, B_1 = 1.01$ and $B_2 = 1.01^2$. The evolution of the stock price over 6 months along with the conditional probability distributions are depicted in the following tree:

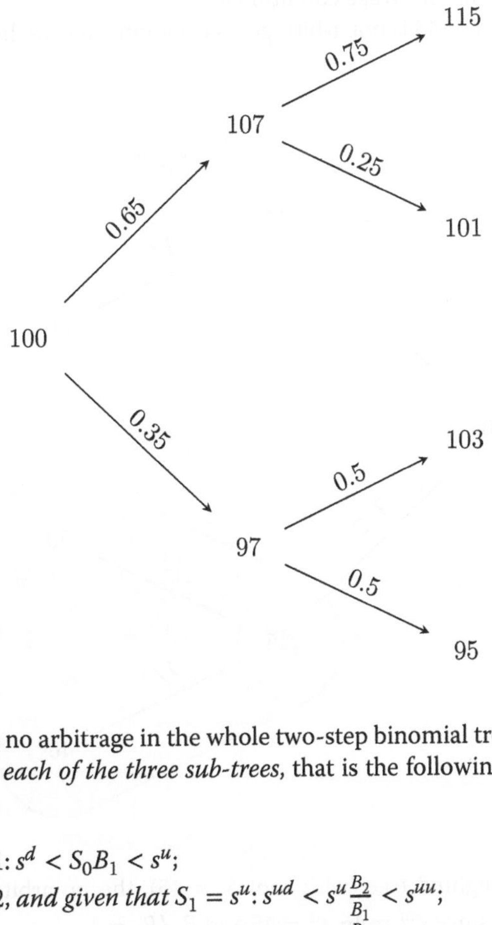

■

To make sure there is no arbitrage in the whole two-step binomial tree, one has to make sure there is *no arbitrage in each of the three sub-trees*, that is the following three conditions must hold:

- from time 0 to time 1: $s^d < S_0B_1 < s^u$;
- from time 1 to time 2, *and given that* $S_1 = s^u$: $s^{ud} < s^u\frac{B_2}{B_1} < s^{uu}$;
- from time 1 to time 2, *and given that* $S_1 = s^d$: $s^{dd} < s^d\frac{B_2}{B_1} < s^{du}$.

Therefore, no asset is *doing better* than the other one *in each of the three sub-trees*, i.e. no matter the scenario. If the condition is violated in only one of the sub-trees, then an arbitrage profit can be obtained.

Example 10.1.2 *Arbitrage in the two-step binomial tree*
Assume for simplicity the interest rate is zero. Moreover, the stock price evolves according to a two-period binomial tree with:

- initial price: 50;
- at time 1: either 45 or 55, with probability 1/2;
- at time 2, given that the stock price is 55 after one period: either 60 or 50, with probability 1/2;
- at time 2, given that the stock price is 45 after one period: either 46 or 50, with probability 1/2.

Determine whether there are arbitrage opportunities between the stock and the risk-free asset and how to exploit it if this is the case.

The zero interest rate greatly simplifies the comparison. Indeed, in this case $B_0 = B_1 = B_2 = 1$ and it suffices to determine whether the stock always wins or always loses in each of the three sub-trees (no-arbitrage conditions).

The tree with root $S_1 = 45$ is not arbitrage-free. Graphically, we have:

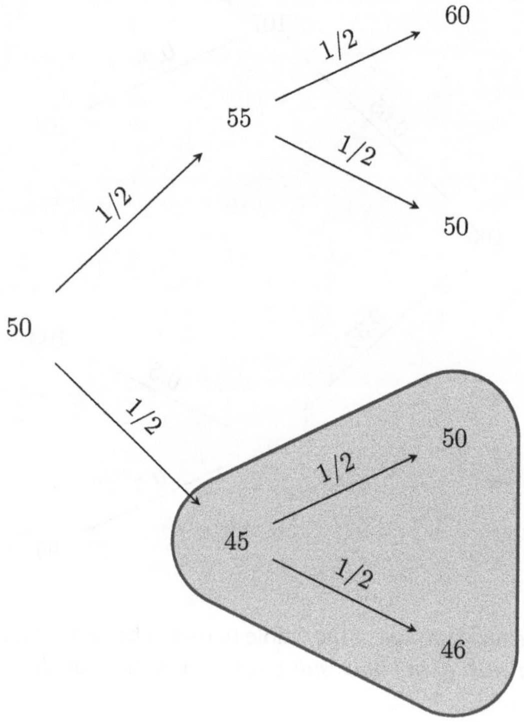

Indeed, in the highlighted tree (with root $S_1 = 45$), the no-arbitrage condition $s^{dd} < s^d \frac{B_2}{B_1} < s^{du}$ is violated, since $s^{dd} = 46$, $s^d = 45$ and $B_2/B_1 = 1$.

Here is how to exploit the arbitrage:

- At time 0, do nothing.
- At time 1, if the stock price is 55, then do nothing. At time 2, there is obviously no gain nor loss. This scenario occurs with probability 1/2.

- At time 1, if the stock price is 45, then buy the stock for 45 and borrow 45 at 0%, for a net cost of 0. At time 2, as the stock price increases to either 46 or 50, repay the loan, sell the stock and get a profit of 1 or 5 (each profit occurs with a probability 1/4).

Overall, following this strategy, the net cost is zero, there is a possibility of profit (if $S_1 = 45$), with a probability 1/2, and there is a possibility we should do nothing and break even (if $S_1 = 55$), with a probability 1/2. This is exactly the definition of an arbitrage. ■

10.1.2.1 Simplified tree

The construction of a two-step binomial tree can be cumbersome as it requires to specify the stock price at each node. We can also use instead up and down factors u and d at each time step, as it will be the case in the general n-period binomial tree. In other words, it is equivalent to assuming that whenever the stock price increases or decreases, it always does so by a factor of u or d, respectively. Therefore, for $d < u$, we can set

$$s^u = uS_0 \quad \text{and} \quad s^d = dS_0$$

and

$$s^{uu} = u^2 S_0, \quad s^{ud} = s^{du} = udS_0 \quad \text{and} \quad s^{dd} = d^2 S_0.$$

To further simplify the tree, we can also assume that the probability of a one-step price increase (resp. decrease) is constant and given by p (resp. $1 - p$). In other words, $p^u = p^d = p$.

Consequently, there is a *simplified version of the two-period binomial tree* which is constructed with only five parameters: r, S_0, d, u, p. The resulting tree is recombining and given by the following illustration:

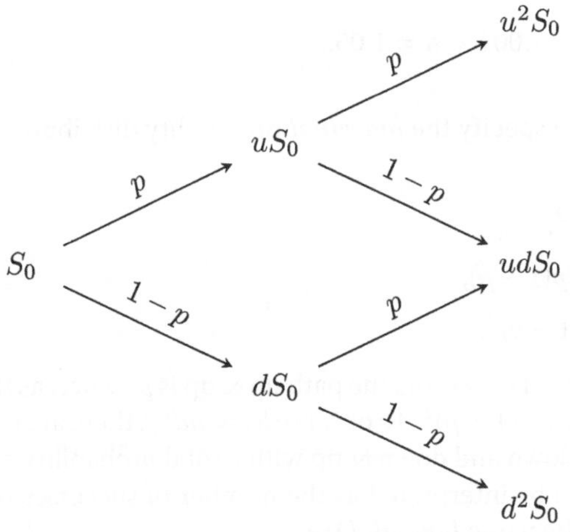

Moreover, in this case, the three no-arbitrage conditions collapse to a single condition as in the one-period model:

$$d < B_1 = \frac{B_2}{B_1} < u.$$

Recall that $B_1 = B_2/B_1$ is either equal to e^r or $1 + r$.

Example 10.1.3 *Construction of a simplified two-step binomial tree*
The annual interest rate is 1.2% (compounded monthly) and the current stock price is 100. It was determined by financial analysts that each month the stock can either increase by 5% or decrease by 3% and that the probability of an up-move is 60%. Build a two-step binomial tree assuming that one step is 1 month.

We have $B_0 = 1$, $B_1 = 1.001$ and $B_2 = 1.001^2$. We can easily verify that the tree is given by the following illustration:

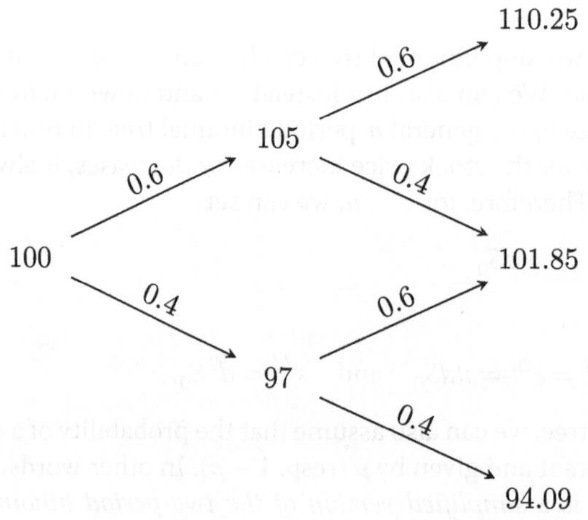

where, for example, $s^{ud} = s^{du} = 100 \times (1.05) \times (0.97) = 101.85$.

Finally, as supplementary information, note that there is no arbitrage in this tree because

$$d = 0.97 < B_1 = 1.001 < u = 1.05.$$

■

To end this section, we specify the *marginal* probability distribution for S_2 in the simplified tree. We get

$$\mathbb{P}(S_2 = u^2 S_0) = p^2,$$
$$\mathbb{P}(S_2 = udS_0) = 2p(1-p),$$
$$\mathbb{P}(S_2 = d^2 S_0) = (1-p)^2.$$

Therefore, the probability of observing the path up & up is p^2 whereas the probability of observing the path down & down is $(1-p)^2$. To observe $S_2 = udS_0$, there are two possible paths leading to this final value: up & down and down & up with a total probability of $2p(1-p)$. The number of up-moves $\{0, 1, 2\}$ can be interpreted as the number of successes out of two attempts (two time steps), and hence the name *binomial* tree.

10.1.3 Derivatives

As in the one-period model, we now introduce a third asset. It is an option or a derivative written on the risky asset S. The price of the derivative over time is denoted by $V = \{V_0, V_1, V_2\}$ where V_0 is today's price, which is a constant to be determined later, whereas V_1 and V_2 are the random derivative's prices at times 1 and 2.

The derivative is characterized by its payoff at maturity. As there are only four outcomes in the model, V_2 will take one of the four values $v^{uu}, v^{ud}, v^{du}, v^{dd}$. To make the notation consistent, the random variable V_2 takes, for example, the value v^{uu} in the up & up scenario or, said differently, along the path leading to the final value s^{uu}. Similarly, along the path leading to s^{ud}, then $V_2 = v^{ud}$, and the same goes for $V_2 = v^{du}$ when we observe $S_2 = s^{du}$ and $V_2 = v^{dd}$ if $S_2 = s^{dd}$.

In a manner consistent with Chapter 9, V_1 represents the time-1 value of the derivative with possible realizations v^u and v^d. The exact values of v^u and v^d are to be determined (TBD), just like V_0.

Superimposing the evolution of the derivative price V (in boxes) over the evolution of the stock price S, we can now represent our two-period model graphically as follows:

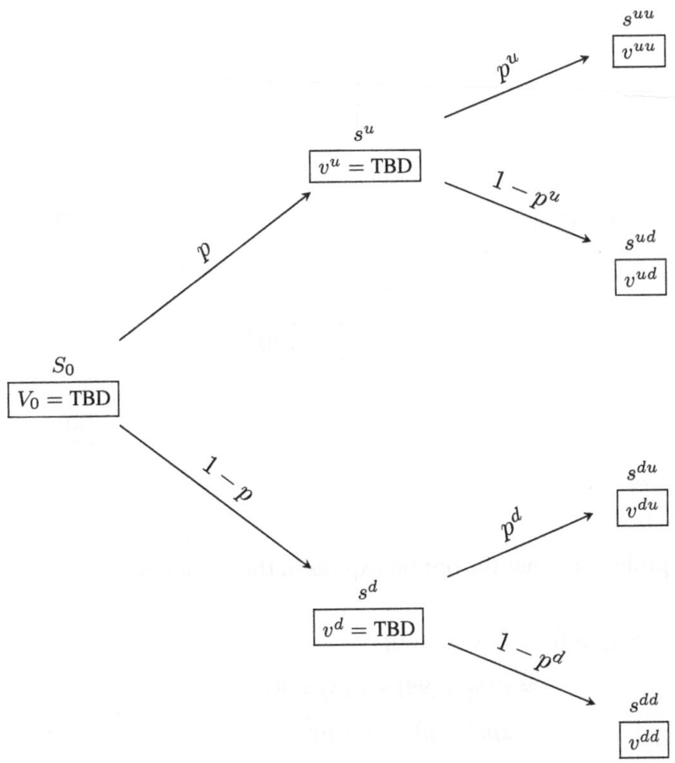

As in the one-period model, instead of the notation $V = \{V_0, V_1, V_2\}$, we will use the notation $C = \{C_0, C_1, C_2\}$ and $P = \{P_0, P_1, P_2\}$ for a call option and a put option, respectively. Therefore, we have

$$C_2 = (S_2 - K)_+$$
$$P_2 = (K - S_2)_+,$$

if the options are both struck at K and if they both mature at time 2.

Example 10.1.4 *Payoff of a put option in the two-step binomial tree*
A stock trades today at 100 and it can either increase/decrease by 10% every month. Using a two-step binomial tree, illustrate the evolution of the stock price and the payoff of an

at-the-money put option expiring in 2 months. Also, calculate the probability the option finishes in the money.

We have $S_0 = K = 100, u = 1.1, d = 0.9$ and the final payoff is

$$P_2 = (100 - S_2)_+.$$

Then the corresponding tree is

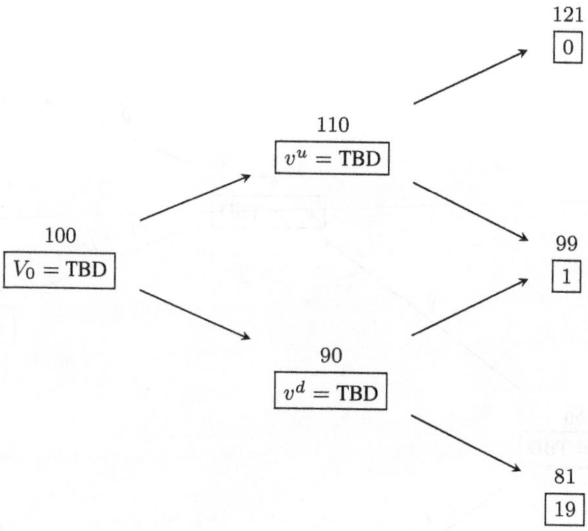

Finally, the probability that the option expires in the money is

$$\begin{aligned}\mathbb{P}((100 - S_2)_+ > 0) &= \mathbb{P}(S_2 < 100)\\ &= \mathbb{P}(S_2 = 99) + \mathbb{P}(S_2 = 81)\\ &= 2p(1-p) + (1-p)^2.\end{aligned}$$

■

Example 10.1.5 *Payoff of exotic options in the two-step binomial tree*
You are given the following three exotic options:

- up-and-in barrier call option with $H = 103$ and $K = 100$;
- average strike Asian call option;
- floating strike lookback put option.

Calculate the payoff of these three options, using the two-period tree of example 10.1.3, and compute the probability that each option expires in the money.

First of all, to avoid confusion, let us draw a non-recombining version of the tree of example 10.1.3:

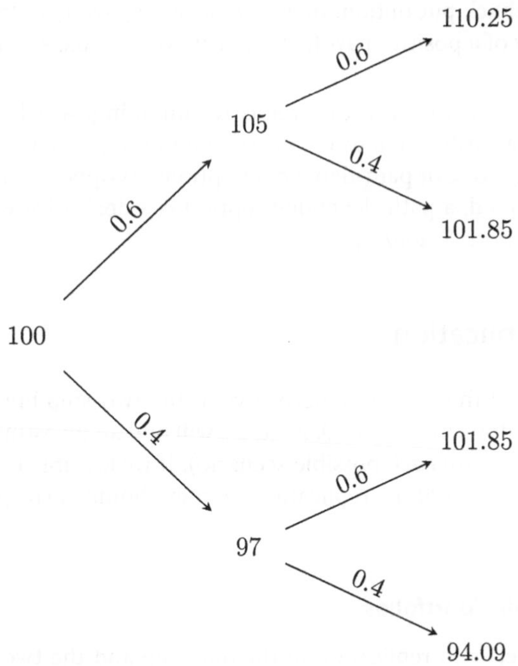

Each of these three exotic options will generate four different payoffs $\{v^{uu}, v^{ud}, v^{du}, v^{dd}\}$ corresponding to the four possible trajectories of the risky asset. The paths and some corresponding quantities are summarized in the following table:

Path	S_0	S_1	S_2	Barrier crossed?	Average	Maximum
Up & up	100	105	110.25	Yes	105.083	110.25
Up & down	100	105	101.85	Yes	102.283	105
Down & up	100	97	101.85	No	99.6166	101.85
Down & down	100	97	94.09	No	97.03	100

Therefore, we have:

- up-and-in barrier call option with $H = 103$ and $K = 100$: $v^{uu} = 10.25, v^{ud} = 1.85, v^{du} = v^{dd} = 0$;
- average strike Asian call option: $v^{uu} = 5.166, v^{ud} = 0, v^{du} = 2.23, v^{dd} = 0$;
- floating strike lookback put option: $v^{uu} = 0, v^{ud} = 3.15, v^{du} = 0, v^{dd} = 5.91$.

Recall that the probability that an option finishes in the money is $\mathbb{P}(V_2 > 0)$. We need to look at each possible path:

- up-and-in barrier call option with $H = 103$ and $K = 100$: it is in the money if there is an up-move after 1 period. Therefore, $\mathbb{P}(V_2 > 0) = \mathbb{P}(S_1 \geq 103, S_2 > 100) = \mathbb{P}(S_1 \geq 103) = 0.6$;
- average strike Asian call option: there are positive payoffs for the paths uu, du, and the total probability that the option finishes in the money is $0.6^2 + 0.4 \times 0.6 = 0.6$;

- floating strike lookback put option: there are positive payoffs in the paths ud, dd, and the probability of a positive payoff at maturity is $0.4 \times 0.6 + 0.4^2 = 0.4$. ■

It is important to note that even if the tree is recombining, which means here that $s^{ud} = s^{du}$, then v^{ud} is not necessarily equal to v^{du} (as shown in the previous example). This is a key difference when pricing exotic or path-dependent options, as opposed to simple options. In the notation previously defined, a path-dependent option is a derivative whose payoff V_2 can be written as a function of both S_1 *and* S_2.

10.2 Pricing by replication

We are now ready to find the price of a derivative in the two-step binomial tree. The idea is similar to what we have done in Chapter 9. First, we will find an investment strategy replicating the payoff of the derivative (*in each* possible scenario). If we are able to do so, then by the no-arbitrage assumption, the cost of the replicating strategy should correspond to the price of the derivative.

10.2.1 Trading strategies/portfolios

The major difference between replication in the one-step and the two-step binomial trees is that investment strategies in the latter tree can be **dynamic** as opposed to **static**. A dynamic investment strategy is a strategy which needs to be updated over time, when the underlying asset price changes, a procedure often called **rebalancing**.

Example 10.2.1 *Static and dynamic investment strategies*
Replicating a long position in a forward contract on a stock involves buying one share of stock and borrowing its cost at time 0. The replicating strategy of the forward contract is static because no matter how the stock price changes over time, the investor does not change its investment strategy: it holds onto its stock and loan. This is often called a buy-and-hold strategy.

An investment strategy that requires to hold 60% in stock and 40% in bonds has to be dynamic (60%/40% of the total value of investments). As the stock and bond prices evolve over time, the proportions in stocks and bonds will automatically change if nothing is done. To make sure the target weights are respected, the investor would need to sell stocks to buy bonds (or the opposite) to recover the 60–40 targeted weights (this is rebalancing the portfolio). Therefore, to keep weights the same over time, the investment strategy needs to be dynamically updated. ■

We now define a trading strategy (or portfolio), in a two-period tree, as a set of pairs

$$\{(x, y), (x^u, y^u), (x^d, y^d)\},$$

where:

- x (resp. y) is the number of units of S (resp. B) held in the portfolio from time 0 to time 1;
- x^u (resp. y^u) is the number of units of S (resp. B) held in the portfolio from time 1 to time 2 if $S_1 = s^u$ has been observed at time 1;
- x^d (resp. y^d) is the number of units of S (resp. B) held in the portfolio from time 1 to time 2 if $S_1 = s^d$ has been observed at time 1.

The values of (x, y), (x^u, y^u) and (x^d, y^d) can be chosen with specific investment-oriented goals in mind.

A strategy's value over time is denoted by $\Pi = \{\Pi_0, \Pi_1, \Pi_2\}$, where

$$\Pi_0 = xS_0 + yB_0$$

is the value today while Π_1 and Π_2 are the values at time 1 (after rebalancing) and at time 2. Consequently,

$$\Pi_1 = \begin{cases} x^u \times s^u + y^u \times B_1 & \text{if } S_1 = s^u, \\ x^d \times s^d + y^d \times B_1 & \text{if } S_1 = s^d, \end{cases}$$

and

$$\Pi_2 = \begin{cases} x^u \times S_2 + y^u \times B_2 & \text{if } S_1 = s^u, \\ x^d \times S_2 + y^d \times B_2 & \text{if } S_1 = s^d. \end{cases}$$

Note that no rebalancing is undertaken at time 2.

The following tree illustrates the randomness and the evolution of the trading strategy in a two-step binomial tree.

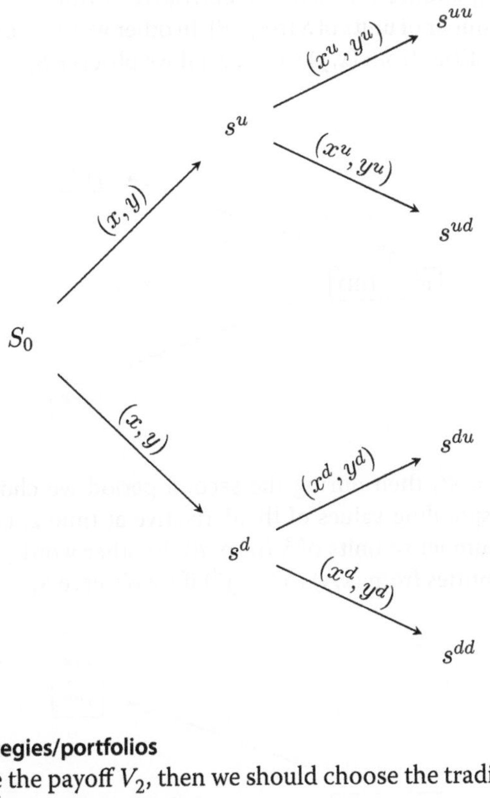

10.2.1.1 Replicating strategies/portfolios

If our goal is to replicate the payoff V_2, then we should choose the trading strategy, or the pairs (x, y), (x^u, y^u) and (x^d, y^d), to make sure that

$$\Pi_2 = V_2$$

in each possible scenario. We will again consider the fact that a two-period binomial tree is constructed using three one-period binomial trees. Hence, we will proceed as follows:

1) For the first period (from time 0 to time 1), we choose (x, y) to exactly replicate the values of the derivative at time 1, i.e. v^u and v^d. This is as in Chapter 9, except that v^u and v^d are still unknown quantities.

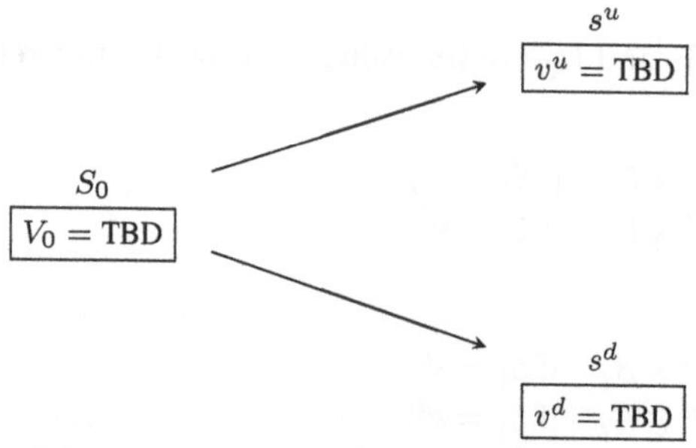

2) For the second period (from time 1 to time 2), depending on the observed stock price at time 1, we will rebalance the portfolio quantities as follows:
   (a) If we observe $S_1 = s^u$, then during the second period we choose (x^u, y^u) to replicate exactly the corresponding values of the derivative at time 2, i.e. v^{uu} and v^{ud}. Here, x^u (resp. y^u) is the number of units of S (resp. B). In other words, we will reallocate/rebalance the portfolio quantities from (x, y) to (x^u, y^u) if we observe $S_1 = s^u$.

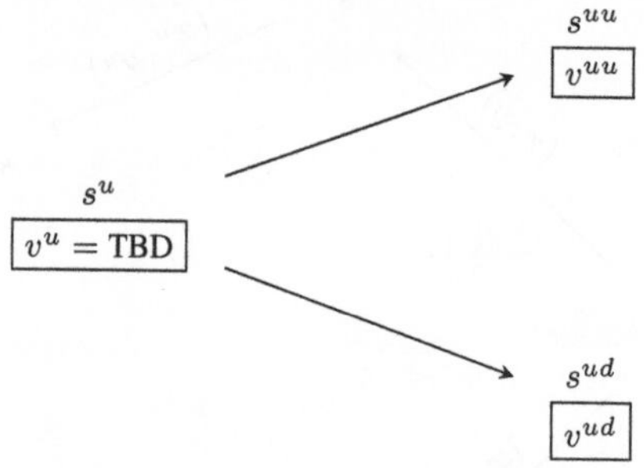

   (b) If we observe $S_1 = s^d$, then during the second period we choose (x^d, y^d) to replicate exactly the corresponding values of the derivative at time 2, i.e. v^{du} and v^{dd}. Here, x^d (resp. y^d) is the number of units of S (resp. B). In other words, we reallocate/rebalance the portfolio quantities from (x, y) to (x^d, y^d) if we observe $S_1 = s^d$.

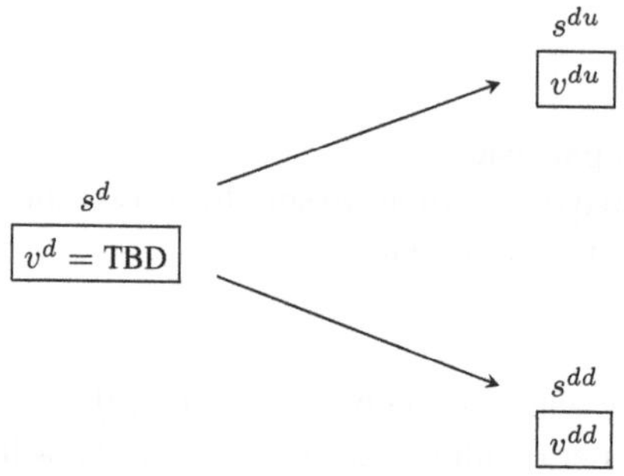

It should be noted that at time 0, an investor does not know which rebalancing procedure will take place, until the realization of S_1 is known. Therefore, it is a random procedure.

10.2.2 Backward recursive algorithm

As mentioned above, to replicate a derivative with payoff V_2, we must choose

$$\{(x, y), (x^u, y^u), (x^d, y^d)\}$$

such that $\Pi_2 = V_2$. To do so, we will work recursively in the tree from time 2 to time 1, and then from time 1 to time 0. The steps are as follows:

1) Depending on the underlying asset price at time 1, we will set up the portfolio quantities differently:
   (a) If we observe $S_1 = s^u$, at time 1 we will choose (x^u, y^u) such that

$$\begin{cases} y^u B_2 + x^u s^{uu} = v^{uu}, \\ y^u B_2 + x^u s^{ud} = v^{ud}. \end{cases}$$

   The only unknowns are (x^u, y^u) as the remainder of the variables are known. This is the first system of two equations with two unknowns to solve.

   We then need to determine the price of the derivative in this conditional (upon $S_1 = s^u$) one-period binomial tree. As we did in the one-period model, to avoid arbitrage opportunities the time-1 price realization v^u should reflect the cost to acquire the one-period portfolio (x^u, y^u) at time 1. Therefore,

$$v^u = x^u s^u + y^u B_1.$$

   (b) If we observe $S_1 = s^d$, at time 1 we will choose (x^d, y^d) such that

$$\begin{cases} y^d B_2 + x^d s^{du} = v^{du}, \\ y^d B_2 + x^d s^{dd} = v^{dd}. \end{cases}$$

   The only unknowns are (x^d, y^d) as the remainder of the variables are known. This is the second system of two equations with two unknowns to solve. We also need to determine the no-arbitrage price v^d of the derivative, conditional upon observing $S_1 = s^d$. Again, to avoid arbitrage opportunities,

$$v^d = x^d s^d + y^d B_1$$

   to reflect the cost of building the one-period replicating portfolio (x^d, y^d) at time 1.
2) For the first period, we need to choose (x, y) such that

$$\begin{cases} yB_1 + xs^u = v^u, \\ yB_1 + xs^d = v^d. \end{cases}$$

   In other words, we want to identify the one-period replicating strategy, set up at time 0, to make sure that we are able to create/replicate the portfolio's/derivative's values (v^u or v^d) one period later. Therefore, we choose (x, y) to make sure we have (a) v^u if $S_1 = s^u$ and (b) v^d if $S_1 = s^d$ one period later. In other words, it should provide the amounts required to set up the one-period strategies (x^u, y^u) or (x^d, y^d), depending on the movement of the underlying asset's price. The fact that a strategy is worth the same before and after rebalancing is known as a *self-financing strategy*; more details in Chapter 11.

The initial price of the derivative V_0 is equal to Π_0 the cost for setting up this portfolio:

$$V_0 = \Pi_0 = yB_0 + xS_0.$$

An example will help understand the whole process.

Example 10.2.2 *Pricing by replication in the two-step binomial tree*
Financial analysts have determined that a realistic model for ABC inc. stock's price is given by

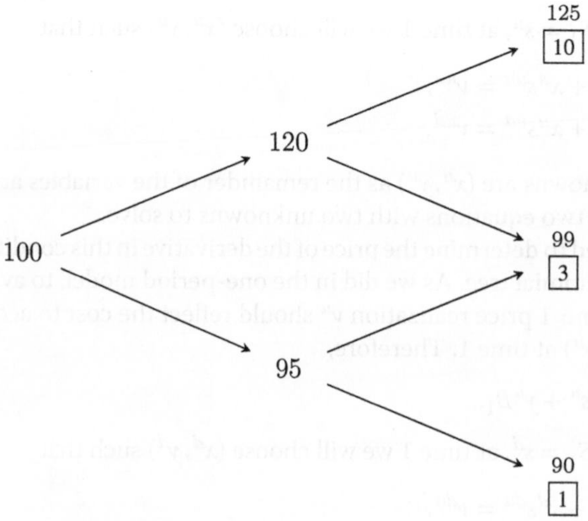

on which a derivative with payoff V_2 is written. Depending on the terminal asset price, the payoff of the derivative can be 10, 3 or 1, as indicated in the tree. The periodic interest rate is 2% and is compounded once every period. We are asked to find the strategy that will replicate the derivative in each node of the tree.

Clearly, we have $B_0 = 1$, $B_1 = 1.02$ and $B_2 = (1.02)^2$. It is also easy to verify that this model is arbitrage-free.

The first step is to determine what the replicating portfolio should be in the upper one-period sub-tree, i.e. the one with root $S_1 = 120$:

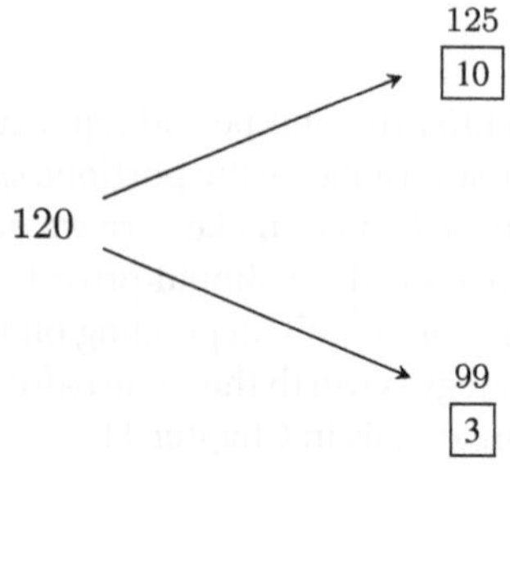

In other words, given that $S_1 = 120$, what should be the investment strategy that replicates the payments 10 or 3 no matter what happens in the last period (up & up, or up & down)? We must choose (x^u, y^u) such that

$$\begin{cases} 125x^u + (1.02)^2 y^u = 10, \\ 99x^u + (1.02)^2 y^u = 3. \end{cases}$$

As we did in Chapter 9, we easily find that $x^u = 0.26923077$ and $y^u = -22.735338$. Consequently, the derivative's price at this node must be equal to the portfolio's cost:

$$v^u = -22.735338(1.02) + 0.269230769(120) = 9.117647059.$$

If we reach the node where $S_1 = 120$ at time 1, then investing an amount of \$9.12 in the stock and bond according to the one-period strategy (x^u, y^u) will exactly replicate the derivative's payoff one period later (at time 2).

Similarly, for the lower one-period sub-tree, i.e. the one with root $S_1 = 95$,

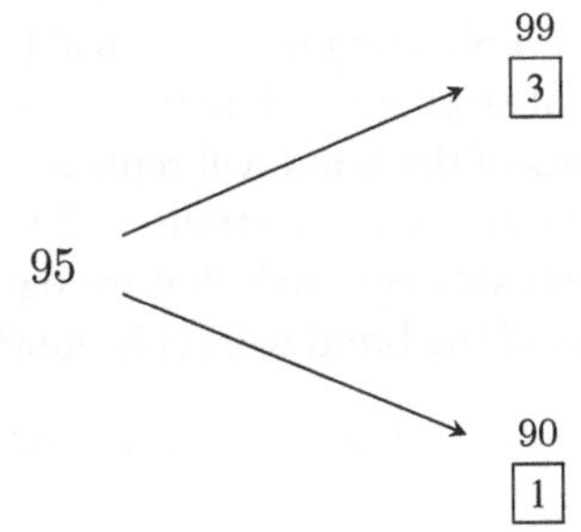

we must choose (x^d, y^d) such that

$$\begin{cases} 99x^d + (1.02)^2 y^d = 3, \\ 90x^d + (1.02)^2 y^d = 1. \end{cases}$$

Again, we easily find that $x^d = 0.2222222$ and $y^d = -18.262207$. Consequently, the derivative's price at this node is

$$v^d = -18.262207(1.02) + 0.2222222(95) = 2.483660131.$$

If we reach the node where $S_1 = 95$, then holding \$2.48 at time 1 and investing it in the stock and bond according to the one-period strategy (x^d, y^d) will exactly replicate the derivative's payoff.

At this point, we might ask: how can we make sure we will end up with either \$9.12 or \$2.48 at time 1? To do so, for the first one-period sub-tree

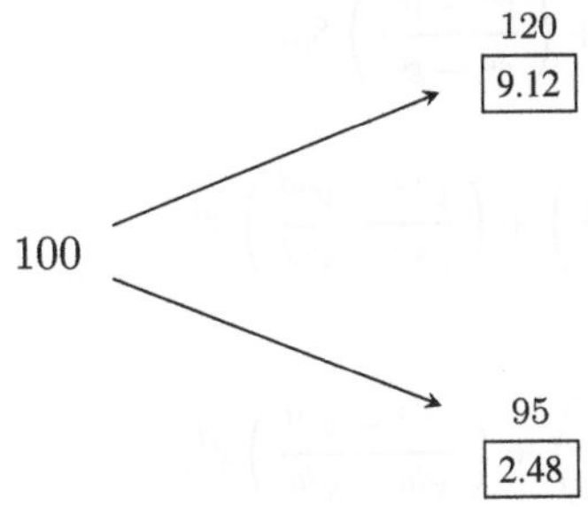

we must choose (x, y) such that

$$\begin{cases} 120x + 1.02y = 9.117647059, \\ 95x + 1.02y = 2.4836579. \end{cases}$$

Note that the values $v^u = 9.117647059$ and $v^d = 2.483660131$ have been obtained in the first two steps. Solving this last system, we get $x = 0.26535958$, $y = -22.279904$ and

$$V_0 = -22.279904(1) + 0.26535958(120) = 4.256055363.$$

To summarize, if we invest \$4.26 at time 0 and then apply the strategy

$$\{(x = 0.26535958, y = -22.279904), (x^u = 0.26923077, y^u = -22.735338),$$
$$(x^d = 0.2222222, y^d = -18.262207)\},$$

we can replicate the payoff of the derivative in all possible scenarios.

- At time 0, buy 0.2654 unit of a stock and borrow 22.28 units of the bond.
- At time 1 and if $S_1 = 120$, the replicating portfolio will be worth \$9.12. Reallocating/rebalancing the investment strategy such that we have a long position of 0.269 unit of stock and −22.735 units of the bond will replicate $v^{uu} = 10$ and $v^{ud} = 3$.
- At time 1 and if $S_1 = 95$, the replicating portfolio will be worth \$2.48. Reallocating/rebalancing the investment strategy such that we have a long position of 0.22 unit of stock and −18.26 units of the bond will replicate $v^{du} = 3$ and $v^{dd} = 1$.

Therefore, \$4.26 corresponds to the derivative price at time 0. ■

In general, we can solve explicitly the preceding systems of equations 1) (a), 1) (b) and 2) in terms of x, y, x^u, y^u, x^d and y^d. Consequently, we get

$$x = \frac{v^u - v^d}{s^u - s^d} \quad \text{and} \quad y = \frac{1}{B_1}\left(\frac{v^d s^u - v^u s^d}{s^u - s^d}\right), \tag{10.2.1}$$

$$x^u = \frac{v^{uu} - v^{ud}}{s^{uu} - s^{ud}} \quad \text{and} \quad y^u = \frac{1}{B_2}\left(\frac{v^{ud} s^{uu} - v^{uu} s^{ud}}{s^{uu} - s^{ud}}\right) \tag{10.2.2}$$

and

$$x^d = \frac{v^{du} - v^{dd}}{s^{du} - s^{dd}} \quad \text{and} \quad y^d = \frac{1}{B_2}\left(\frac{v^{dd} s^{du} - v^{du} s^{dd}}{s^{du} - s^{dd}}\right). \tag{10.2.3}$$

If we replace (x, y), (x^u, y^u) and (x^d, y^d) by the expressions obtained in (10.2.1), (10.2.2) and (10.2.3), we can write

$$V_0 = \frac{1}{B_1}\left(\frac{v^d s^u - v^u s^d}{s^u - s^d}\right) + \left(\frac{v^u - v^d}{s^u - s^d}\right) S_0, \tag{10.2.4}$$

$$v^u = \frac{B_1}{B_2}\left(\frac{v^{ud} s^{uu} - v^{uu} s^{ud}}{s^{uu} - s^{ud}}\right) + \left(\frac{v^{uu} - v^{ud}}{s^{uu} - s^{ud}}\right) s^u \tag{10.2.5}$$

and

$$v^d = \frac{B_1}{B_2}\left(\frac{v^{dd} s^{du} - v^{du} s^{dd}}{s^{du} - s^{dd}}\right) + \left(\frac{v^{du} - v^{dd}}{s^{du} - s^{dd}}\right) s^d. \tag{10.2.6}$$

In conclusion, investing V_0 (as given in (10.2.4)) to buy a derivative or holding $\Pi_0 = V_0$ and applying the latter replicating strategy will yield the same cash flows at any time (maturity, time 1 and inception) and in all scenarios.

Example 10.2.3 *Pricing a put option with a replicating portfolio*
Assume that your life insurance company has sold the put option from example 10.1.4 and suppose that $B_k = 1.03^k$, for $k = 0, 1, 2$. Determine the amount of money you should put aside at inception to hedge the cash flows of your liability and describe the dynamic investment strategy that you should implement in that case.

Your company's liability is a put option, i.e. you are short a put. Therefore, you need to synthetically replicate a long position in a put option through a dynamic replicating portfolio.

Let us work through the binomial tree from time 2 to time 0. First of all, at time 1, if we observe $S_1 = 110$, then the payoff at time 2, i.e. one period later, will be 0 or 1. Replicating v^{uu} and v^{ud} using the portfolio quantities (x^u, y^u), we find

$$x^u = \frac{v^{uu} - v^{ud}}{s^{uu} - s^{ud}} = \frac{0 - 1}{121 - 99} = \frac{-1}{22}$$

$$y^u = \frac{1}{B_2}\left(\frac{v^{ud}s^{uu} - v^{uu}s^{ud}}{s^{uu} - s^{ud}}\right) = \frac{1}{1.03^2}\left(\frac{1 \times 121 - 0 \times 99}{121 - 99}\right) = \frac{5.5}{1.03^2}.$$

Therefore, the value of the derivative at time 1 in this scenario (when $S_1 = 110$) is

$$v^u = s^u x^u + B_1 y^u = 110 \times \frac{-1}{22} + 1.03 \times \frac{5.5}{1.03^2} = 0.339805825.$$

Second, if we observe instead $S_1 = 90$, then the payoff at time 2 will be 1 or 19. Replicating v^{du} and v^{dd} using the portfolio quantities (x^d, y^d), we find

$$x^d = \frac{v^{du} - v^{dd}}{s^{du} - s^{dd}} = \frac{1 - 19}{99 - 81} = -1$$

$$y^d = \frac{1}{B_2}\left(\frac{v^{dd}s^{du} - v^{du}s^{dd}}{s^{du} - s^{dd}}\right) = \frac{1}{1.03^2}\left(\frac{19 \times 99 - 1 \times 81}{99 - 81}\right) = \frac{100}{1.03^2}.$$

Therefore, the value of the derivative at time 1 in this scenario (when $S_1 = 90$) is

$$v^d = s^d x^d + B_1 y^d = 90 \times -1 + 1.03 \times \frac{100}{1.03^2} = 7.087378641.$$

Finally, we need to determine the amount necessary at time 0 to set up the portfolio quantities (x, y) replicating the values $0.339805825 = v^u$ or $7.087378641 = v^d$. We find

$$x = \frac{v^u - v^d}{s^u - s^d} = \frac{0.339805825 - 7.087378641}{110 - 90} = -0.337378641$$

$$y = \frac{1}{B_1}\left(\frac{v^d s^u - v^u s^d}{s^u - s^d}\right)$$

$$= \frac{1}{1.03}\left(\frac{7.087378641 \times 110 - 0.339805825 \times 90}{110 - 90}\right) = 36.36063719$$

and therefore the time-0 value of this derivative is

$$V_0 = xS_0 + yB_0 = -0.337378641 \times 100 + 36.36063719 = 2.622773117.$$

In summary:

1) Your insurer has a liability because of its short position in the put option. You decide to hedge the payoff of that put with a replicating portfolio. You put aside 2.62 in a portfolio which is composed of an investment of 36.36 at the risk-free rate and a short sale of 0.3374 share of the stock.
2) At time 1, and if the stock price has risen to 110, the replicating portfolio will be worth 0.339805825. You reorganize the assets in your portfolio to make sure you have the following positions: a short position of 1/22 share of the stock and a long position of $\frac{5.5}{1.03^2}$ units of the risk-free asset. Therefore, you need to buy some additional stocks and withdraw from the risk-free investment (sell units of the bond).
3) At time 1, and if the stock price has dropped to 90, the replicating portfolio will then be worth 7.087378641. You notice that, in this scenario, the option will necessarily end up in-the-money at maturity (time 2). You now need to be short of an entire share of stock and hold $\frac{100}{1.03^2}$ units of the bond for the last period. Therefore, you will short-sell more stocks and invest the proceeds at the risk-free rate.
4) At time 2, the preceding dynamic strategy will exactly replicate the possible values in all possible scenarios. The dynamically updated portfolio exactly matches the only cash flow (the payoff) of this liability. ■

Finally, it is very important to note that the probabilities of up-moves p, p^u and p^d *do not appear* in the option price V. This was also the case in example 10.2.2 where such probabilities were not even specified. This should have been expected because in absence of arbitrage, we use replication arguments in each of the three one-period sub-trees, just as we did in Chapter 9.

10.3 Pricing with risk-neutral probabilities

In the one-step binomial tree, for any derivative, we were able to rewrite the cost of its replicating portfolio (hence the price of the derivative) as an expectation using risk-neutral probabilities. As a two-period binomial tree is the concatenation of three one-period sub-trees, we can hope to be able to rewrite V_0, v^u and v^d with similar expectations.

Let us look at the sub-tree with root $S_1 = s^u$. We know that v^u represents the corresponding price of the derivative at the root. We have obtained in equation (10.2.4) the expression for the cost of the replicating portfolio in the upper node s^u. If we define

$$q^u = \frac{s^u \times \frac{B_2}{B_1} - s^{ud}}{s^{uu} - s^{ud}},$$

then we can rewrite v^u, obtained in equation (10.2.5), as

$$v^u = \frac{B_1}{B_2}((1 - q^u)v^{ud} + q^u v^{uu}). \tag{10.3.1}$$

Applying a similar reasoning to v^d, obtained in equation (10.2.6), we get

$$v^d = \frac{B_1}{B_2}((1 - q^d)v^{dd} + q^d v^{du}), \tag{10.3.2}$$

if we define

$$q^d = \frac{s^d \times \frac{B_2}{B_1} - s^{dd}}{s^{du} - s^{dd}}.$$

Finally, in the first sub-tree, we have

$$V_0 = \frac{1}{B_1}(v^d(1-q) + v^u q) \tag{10.3.3}$$

with

$$q = \frac{S_0 \times \frac{B_1}{B_0} - s^d}{s^u - s^d},$$

as in Chapter 9.

Because no-arbitrage conditions have to apply in each sub-tree, we have that $0 < q < 1$, $0 < q^u < 1$ and $0 < q^d < 1$. Again, we can interpret the weights q, q^u and q^d as risk-neutral (conditional) probabilities, that is:

- q is the *risk-neutral* probability of an up-move from time 0 to time 1;
- q^u is the *risk-neutral* conditional probability of an up-move from time 1 to time 2, given that $S_1 = s^u$;
- q^d is the *risk-neutral* conditional probability of an up-move from time 1 to time 2, given that $S_1 = s^d$.

As in Chapter 9, risk-neutral probabilities are a mathematical convenience that simplifies computations of option prices in a no-arbitrage framework. It is not related to the true probabilities of up-moves given by p, p^u and p^d.

Finally, substituting the expressions for v^u and v^d of equations (10.3.1) and (10.3.2) into equation (10.3.3), we get the **risk-neutral pricing formula** in the two-period binomial tree:

$$V_0 = \frac{1}{B_2}(qq^u v^{uu} + q(1-q^u)v^{ud} + (1-q)q^d v^{du} + (1-q)(1-q^d)v^{dd}), \tag{10.3.4}$$

where qq^u is the risk-neutral probability of two consecutive up-moves, $q(1-q^u)$ is the risk-neutral probability of an up-move followed by a down move, etc.

The risk-neutral probability measure

As in the one-period model, to be more precise, we must define the risk-neutral probability measure as follows. We define $\mathbb{Q}$: events $\rightarrow [0,1]$ by

$$\mathbb{Q}(S_1 = s^u) = q, \qquad \mathbb{Q}(S_1 = s^d) = 1 - q$$

and

$$\mathbb{Q}(S_2 = s^{uu} | S_1 = s^u) = q^u,$$
$$\mathbb{Q}(S_2 = s^{du} | S_1 = s^d) = q^d,$$
$$\mathbb{Q}(S_2 = s^{ud} | S_1 = s^u) = 1 - q^u,$$
$$\mathbb{Q}(S_2 = s^{dd} | S_1 = s^d) = 1 - q^d.$$

It is then easy to deduce that

$$\mathbb{Q}(S_2 = s^{uu}) = qq^u, \quad \mathbb{Q}(S_2 = s^{ud}) = q(1-q^u),$$
$$\mathbb{Q}(S_2 = s^{du}) = (1-q)q^d \quad \text{and} \quad \mathbb{Q}(S_2 = s^{dd}) = (1-q)(1-q^d).$$

It should be recalled that the possible values taken by the random variables S_1, S_2, V_1 and V_2 are not related to the choice of probability measure. However, their *probability* distribution depends on whether it is identified with respect to $\mathbb{P}$ or $\mathbb{Q}$.

Finally,

$$\mathbb{P}(S_2 = s^{uu} | S_1 = s^d) = 0,$$

and

$$\mathbb{P}(S_2 = s^{dd} | S_1 = s^u) = 0$$

i.e. it is impossible to observe the price s^{uu} (s^{dd}) given that the stock price went down (up) after 1 period. Under $\mathbb{Q}$, we still have that

$$\mathbb{Q}(S_2 = s^{uu} | S_1 = s^d) = 0,$$

and

$$\mathbb{Q}(S_2 = s^{dd} | S_1 = s^u) = 0.$$

More generally, $\mathbb{P}$ and $\mathbb{Q}$ agree on events of probability 0 or 1. This is called *equivalence* of probability measures.

We can simplify the presentation of the risk-neutral pricing formula using the operator $\mathbb{E}^{\mathbb{Q}}$ to denote an expectation using the risk-neutral probabilities q, q^u and q^d. Therefore, the risk-neutral pricing formula of (10.3.4) can be simplified as follows:

$$V_0 = \frac{1}{B_2} \mathbb{E}^{\mathbb{Q}}[V_2],$$

where

$$\mathbb{E}^{\mathbb{Q}}[V_2] = qq^u v^{uu} + q(1 - q^u)v^{ud} + (1 - q)q^d v^{du} + (1 - q)(1 - q^d)v^{dd}.$$

This formula for the initial price can be interpreted as the discounted (from time 2 back to time 0) expected payoff using risk-neutral probabilities of observing each of the four possible scenarios/paths.

Similarly, the time-1 values of the derivative, as given in equations (10.3.1) and (10.3.2), can be simplified as follows:

$$v^u = \frac{B_1}{B_2} \mathbb{E}^{\mathbb{Q}}[V_2 | S_1 = s^u],$$

$$v^d = \frac{B_1}{B_2} \mathbb{E}^{\mathbb{Q}}[V_2 | S_1 = s^d],$$

where

$$\mathbb{E}^{\mathbb{Q}}[V_2 | S_1 = s^u] = (1 - q^u)v^{ud} + q^u v^{uu},$$
$$\mathbb{E}^{\mathbb{Q}}[V_2 | S_1 = s^d] = (1 - q^d)v^{dd} + q^d v^{du}.$$

Therefore, it allows us to write the time-1 value of the derivative as a (random) conditional expectation:

$$V_1 = \frac{B_1}{B_2} \mathbb{E}^{\mathbb{Q}}[V_2 | S_1]$$

or equivalently

$$\frac{V_1}{B_1} = \mathbb{E}^{\mathbb{Q}}\left[\frac{V_2}{B_2}\middle| S_1\right].$$

Hence, the time-1 value can be interpreted as the discounted (from time 2 back to time 1) conditional expected payoff using risk-neutral probabilities where $\frac{B_1}{B_2}$ discounts cash flows occurring from time 2 until time 1.

Finally, using the *iterated expectations formula*, we further get that

$$V_0 = \frac{1}{B_1}\mathbb{E}^{\mathbb{Q}}[V_1],$$

which is of the same form as the risk-neutral pricing formula of equation (9.3.3) in a one-period binomial tree.

Example 10.3.1 *Pricing with risk-neutral formulas in the two-step binomial tree*
Consider the two-step binomial tree of example 10.2.2. We will use risk-neutral pricing formulas to recompute the values of V_0 and V_1.

Recall that $s^u = 120$ and $s^d = 95$. When $S_1 = 120$, we compute the conditional risk-neutral probability of an up-move in the next period:

$$q^u = \frac{120 \times 1.02 - 99}{125 - 99} = 0.9.$$

Then,

$$\begin{aligned} v^u &= \frac{1}{1.02}\mathbb{E}^{\mathbb{Q}}[V_2|S_1 = 120] \\ &= \frac{1}{1.02}(q^u \times 10 + (1 - q^u) \times 3) \\ &= \frac{1}{1.02}(0.9 \times 10 + (1 - 0.9) \times 3) \\ &= 9.117647059 \end{aligned}$$

which is obviously the same value as in example 10.2.2.

When $S_1 = 95$, we compute the conditional risk-neutral probability of an up-move in the next period:

$$q^d = \frac{95 \times 1.02 - 90}{99 - 90} = 0.766666667.$$

Then,

$$\begin{aligned} v^d &= \frac{1}{1.02}\mathbb{E}^{\mathbb{Q}}[V_2|S_1 = 95] \\ &= \frac{1}{1.02}(q^d \times 3 + (1 - q^d) \times 1) \\ &= \frac{1}{1.02}(0.766 \times 3 + (1 - 0.766) \times 1) \\ &= 2.483660131 \end{aligned}$$

which is again the same value as in example 10.2.2.

Finally,

$$q = \frac{100 \times 1.02 - 95}{120 - 95} = 0.28$$

and, consequently,

$$\begin{aligned} V_0 &= \frac{1}{1.02}\mathbb{E}^{\mathbb{Q}}[V_1] \\ &= \frac{1}{1.02}(q \times 9.117647059 + (1-q) \times 2.483660131) \\ &= \frac{1}{1.02}(0.28 \times 9.117647059 + (1-0.28) \times 2.483660131) \\ &= 4.256055363. \end{aligned}$$

■

The last example showed that if we are only interested in the derivative's price, then it is much simpler to use the risk-neutral formulas than to compute the full replicating portfolio.

10.3.1 Simplified tree

We complete this section by looking at the simplified version of the two-period binomial tree with up and down factors u and d that are common to the whole tree, i.e. to each one-period sub-tree. In this case, we have the same risk-neutral probabilities at each time step:

$$q = q^u = q^d = \frac{B_1/B_0 - d}{u - d} = \frac{B_2/B_1 - d}{u - d}.$$

In particular, the risk-neutral pricing formula in (10.3.4) can be further simplified:

$$V_0 = \frac{1}{B_2}(q^2 v^{uu} + q(1-q)v^{ud} + (1-q)qv^{du} + (1-q)^2 v^{dd}).$$

Example 10.3.2 *Pricing a put option with risk-neutral formulas*
We continue example 10.2.3.

Because the up and down factors are constant across the tree, there is only one risk-neutral (conditional) probability q and it is equal to

$$q = q^u = q^d = \frac{1.03 - 0.9}{1.1 - 0.9} = 0.65.$$

Applying formulas (10.3.1) and (10.3.2), we get

$$\begin{aligned} v^u &= \frac{1}{1.03}(1 \times 0.35 + 0 \times 0.65) = 0.339805825, \\ v^d &= \frac{1}{1.03}(19 \times 0.35 + 1 \times 0.65) = 7.087378641 \end{aligned}$$

and finally

$$V_0 = \frac{1}{1.03}(0.35 \times 7.087378641 + 0.65 \times 0.339805825) = 2.622773117.$$

Therefore, the initial premium of the put option should be \$2.62 to avoid introducing an arbitrage opportunity in this market.

■

10.4 Advanced actuarial and financial examples

In this last section, we look at how to extend the two-period binomial tree model to consider:

1) stochastic interest rates;
2) stocks paying discrete dividends;
3) guaranteed minimal withdrawal benefits in a variable annuity contract.

In the first two cases we will see how to modify the dynamics of the *basic assets*, i.e. the risky asset S or the risk-free asset B, and in the third example we will compute the sub-account values assuming the reference index follows a two-period binomial tree. In each case, we will see how to adapt the pricing and replicating procedures.

10.4.1 Stochastic interest rates

So far, we have assumed that the risk-free asset was driven by a constant interest rate. It turns out that this assumption can be relaxed without changing our previous arguments. Therefore, let us now look at the following extension of the two-period binomial tree, where the possibility for stochastic interest rates is considered.

To be consistent with the notation in Chapter 2, assume that r_0 is the periodic interest rate (observed at time 0) that applies from time 0 to time 1. If, at time 1, the stock price is given by $S_1 = s^u$ (resp. $S_1 = s^d$), then the periodic interest rate, observed at time 1 and in effect from time 1 to time 2, is given by r_1^u (resp. r_1^d). In other words, there is a random variable R_1 taking values in $\{r_1^u, r_1^d\}$ and such that

$$R_1 = \begin{cases} r^u & \text{if } S_1 = s^u, \\ r^d & \text{if } S_1 = s^d. \end{cases}$$

Graphically, we have:

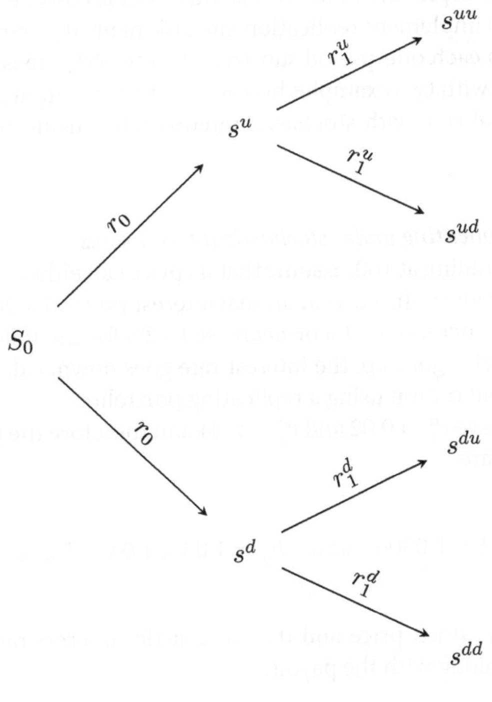

For simplicity, we assume interest rates are compounded periodically; the case of continuously compounded interest rates could be treated similarly. Note also that we do not necessarily assume that $r_1^u > r_1^d$ but this is obviously very convenient.

In summary, the risk-free asset price $B = \{B_0, B_1, B_2\}$ is now a *genuine* stochastic process, even if only B_2 is random. If, at time 0, an amount of $B_0 = 1$ is invested in the bank account, then it will accumulate to

$$B_1 = B_0(1 + r_0)$$

at time 1. This value is deterministic. Indeed, at time 0, we already know with certainty the value of B_1.

Then, one period later, B_1 will further accumulate to

$$B_2 = B_1(1 + R_1),$$

which is now a random quantity because R_1 is a random variable.

Let us denote the realizations of B_2 by $\{b_2^u, b_2^d\}$, where

$$b_2^u = B_0(1 + r_0)\left(1 + r_1^u\right) \quad \text{and} \quad b_2^d = B_0(1 + r_0)\left(1 + r_1^d\right).$$

Again, we cannot say that $b_2^u > b_2^d$.

In a way, B is *less random* than S because we know in advance the interest rate that will apply during the next period. Despite the randomness, this asset is considered a risk-free asset.

To understand how to implement replication and risk-neutral pricing as we did previously, it suffices to realize that in each one-period sub-tree, the interest rate is known.

We will now illustrate with two examples how we can find the no-arbitrage price of an option, in a two-period binomial tree with stochastic interest rates, using replicating portfolios and risk-neutral formulas.

Example 10.4.1 *Replicating under stochastic interest rates*

For a stock currently trading at 100, assume that its price can either go up by 7% or go down by 4% every year. Moreover, the current annual interest rate is 3% (for the upcoming year) and this rate can only increase to 4% or decrease to 2% for the following year. Assuming that when the stock price goes up, the interest rate goes down, calculate the no-arbitrage price of a 104-strike put option using a replicating portfolio.

First, note that we have $r_1^u = 0.02$ and $r_1^d = 0.04$ and therefore the possible time-2 values for the risk-free asset are

$$b_2^u = 1.03 \times 1.02 = 1.0506 \quad \text{and} \quad b_2^d = 1.03 \times 1.04 = 1.0712.$$

The evolution of the stock price and the (stochastic) interest rate are depicted in the following illustration along with the payoff:

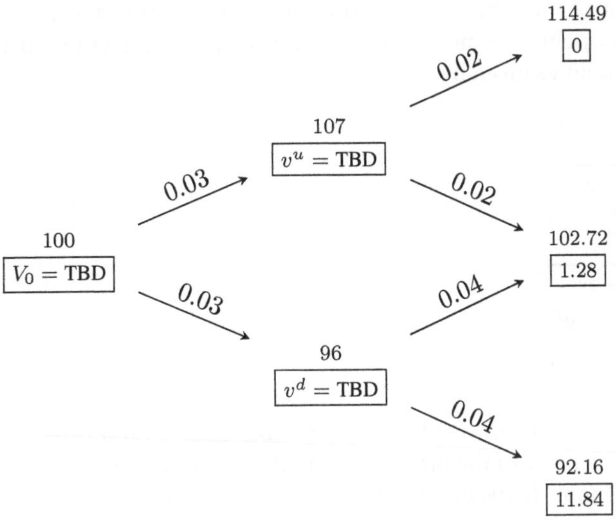

To replicate the option's payoff, we follow the same steps as in a *regular* two-step binomial tree, i.e. working backward one sub-tree at a time. We will set up the replicating portfolio depending on the values of the underlying asset and now also on the values of the risk-free asset:

- At time 1, if we observe $S_1 = 107$, then we will choose (x^u, y^u) such that

$$\begin{cases} 114.49x^u + 1.0506y^u = 0, \\ 102.72x^u + 1.0506y^u = 1.28. \end{cases}$$

  We find $x^u = -0.10875106202$ and $y^u = 11.851236523$. Therefore

$$v^u = 107x^u + 1.03y^u = 0.570409983.$$

- At time 1, if we observe $S_1 = 96$, then we will choose (x^d, y^d) such that

$$\begin{cases} 102.72x^d + 1.0712y^d = 1.28, \\ 92.16x^d + 1.0712y^d = 11.84. \end{cases}$$

  We find $x^d = -1$ and $y^d = 97.08737864$. Therefore,

$$v^d = 96x^d + 1.03y^d = 4.00.$$

Finally, at time 0, we choose (x, y) such that

$$\begin{cases} 107x + 1.03y = v^u = 0.570409983, \\ 96x + 1.03y = v^d = 4.00. \end{cases}$$

We find $x = -0.311781$ and $y = 32.9427$. So, the initial value of this replicating portfolio is given by

$$\Pi_0 = (-0.311781) \times 100 + 32.9427 = 1.7646.$$

The no-arbitrage price of the put option must be equal to the cost of this replicating portfolio, so we have

$$V_0 = 1.7646.$$

■

Deriving the cost of the replicating portfolio in each sub-tree and reorganizing the corresponding equations yields risk-neutral pricing formulas as in Section 10.3, with risk-neutral (conditional) probabilities given by

$$q^u = \frac{s^u \times \frac{b_2^u}{B_1} - s^{ud}}{s^{uu} - s^{ud}},$$

$$q^d = \frac{s^d \times \frac{b_2^d}{B_1} - s^{dd}}{s^{du} - s^{dd}},$$

$$q = \frac{S_0 \times \frac{B_1}{B_0} - s^d}{s^u - s^d}.$$

Example 10.4.2 *Risk-neutral valuation with stochastic interest rates*
We use the same context as the previous example. Let us find the conditional risk-neutral probabilities that apply between time 1 and time 2. We get

$$q^u = \frac{107 \times \frac{1.0506}{1.03} - 102.72}{114.49 - 102.72} = \frac{107 \times 1.02 - 102.72}{114.49 - 102.72} = 6/11,$$

$$q^d = \frac{96 \times \frac{1.0712}{1.03} - 92.16}{102.72 - 92.16} = \frac{96 \times 1.04 - 92.16}{102.72 - 92.16} = 8/11.$$

As expected, we obtain once more

$$v^u = \frac{B_1}{b_2^u}(0 \times q^u + 1.28 \times (1 - q^u)) = 0.570409982,$$

$$v^d = \frac{B_1}{b_2^d}(1.28 \times q^d + 11.84 \times (1 - q^d)) = 4.00.$$

Finally, the risk-neutral probability of an up-move from time 0 to time 1 being

$$q = \frac{100 \times 1.03 - 96}{107 - 96} = 7/11,$$

the no-arbitrage price of the option is

$$V_0 = \frac{1}{1.03} \times (v^u \times q + v^d \times (1 - q)) = \frac{1}{1.03} \times (0.570409982 \times q + 4.00 \times (1 - q)) = 1.7646,$$

as before. ■

10.4.2 Discrete dividends

This section analyzes how to account for discrete (dollar) dividends when valuing an option on a stock in a binomial tree. Recall from Chapter 2 how dividends affect the stock price and the value of derivatives on this stock:

- The dividend reduces the stock price by the amount of the dividend. From the company's point of view, it is a cash outflow.
- Holding a derivative on a stock is different from holding the stock itself. The derivative holder is *not* entitled to the dividend income.

An option's payoff is calculated on the ex-dividend stock price, but if we replicate the payoff with the underlying stock, we will need to account for the dividend income and adjust our replicating strategy accordingly.

Assume that a constant (non-random) dividend D is to be paid at time 1. More precisely, this dividend will be paid at time 1^-, i.e. *just before* time 1 when we are allowed to trade. There are no other dividend payments at time 2.

To ease the presentation, we will use the simplified tree with factors u and d. Therefore, at time 1,

- in the *up scenario*: before the payment of the dividend, the stock price will be S_0u and, after the payment of the dividend, the ex-dividend price will be $S_0u - D$;
- in the *down scenario*: before the payment of the dividend, the stock price will be S_0d and, after the payment of the dividend, the ex-dividend price will be $S_0d - D$.

During the second period, i.e. from time 1 to time 2, the ex-dividend stock price can increase by a factor u or decrease by a factor d. Since these factors apply on the ex-dividend price, the possible stock prices at time 2 are:

- in the *up & up scenario*: $(S_0u - D) \times u$;
- in the *up & down scenario*: $(S_0u - D) \times d$;
- in the *down & up scenario*: $(S_0d - D) \times u$;
- in the *down & down scenario*: $(S_0d - D) \times d$.

The resulting tree is illustrated below.

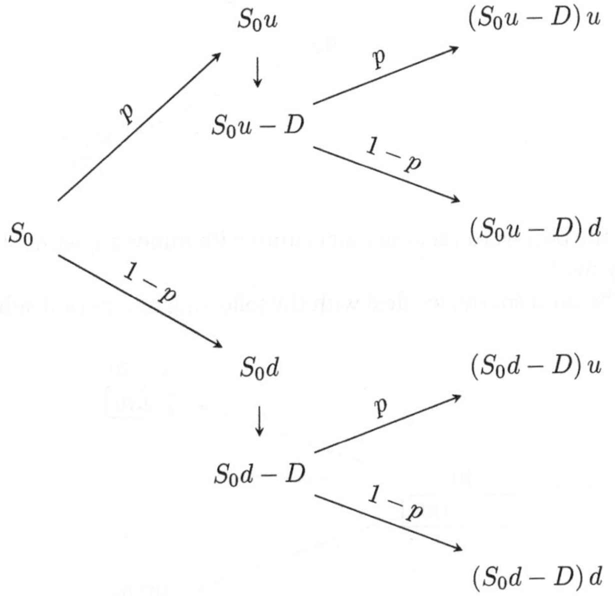

It should be clear that if the original tree (without dividends) is recombining, then adding a discrete (dollar) dividend at time 1 prevents the tree from recombining (at that specific time).

Example 10.4.3 illustrates how to use replicating portfolios to price an option on such a dividend-paying stock.

Example 10.4.3 *Replicating an option on a discrete dividend-paying stock*
Suppose that a stock currently trades at \$100 and can increase by 10% or decrease by 5% at each time step. A dividend of \$3 will be paid at time 1. The risk-free asset is given by $B_k = 1.02^k$, for $k = 0, 1, 2$. Using a replicating portfolio, let us find the no-arbitrage price of a two-period call option with a strike price of \$98.

At time 0, the stock price is 100 and, at time 1^-, i.e. prior to the dividend payment, the stock price will be 110 or 95. Consequently, after the dividend payment, the ex-dividend price will be 107 or 92. Finally, for the prices at time 2, we apply the up and down factors on 107 and 92. The tree is then:

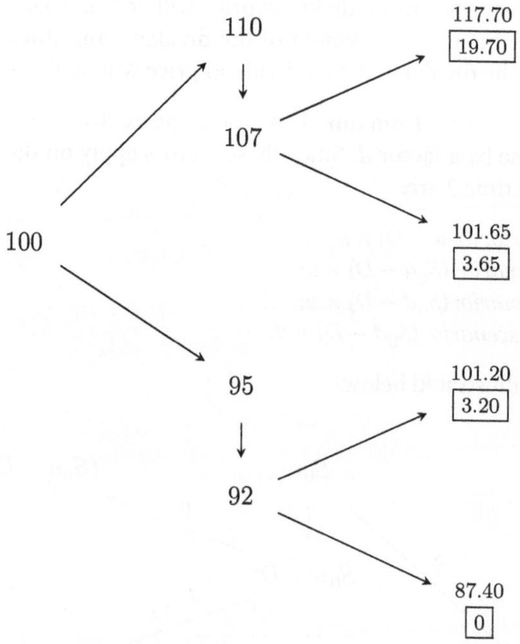

We can apply the backward recursive algorithm with minor adjustments to account for the dividend payment.

At time 1, in the up scenario, we deal with the following one-period sub-tree:

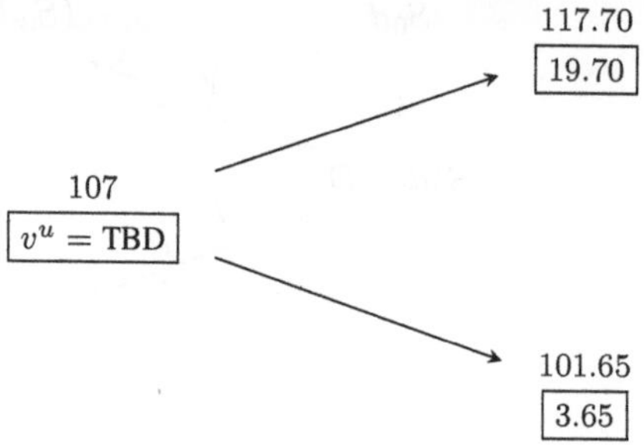

In other words, we need to solve the following system of equations:

$$\begin{cases} 117.70x^u + 1.02^2 y^u = 19.70, \\ 101.65x^u + 1.02^2 y^u = 3.65. \end{cases}$$

We easily find that $x^u = 1$ and $y^u = -94.19454056$. The rationale is simple: in that scenario, to replicate the derivative at time 2 (maturity), an investor would need to hold one share of stock, worth 107 just after the dividend payment, and be short 94.19 units of the risk-free asset. This will exactly replicate the payments of 19.70 and 3.65 one period later, no matter which one occurs. Therefore, by the no-arbitrage principle, the time-1 value of the option in the upper node is

$$\begin{aligned} v^u &= 107x^u + 1.02y^u \\ &= 107 \times 1 + 1.02 \times (-94.19454056) = 10.92156863. \end{aligned}$$

At time 1, in the down scenario, we have the following one-period sub-tree:

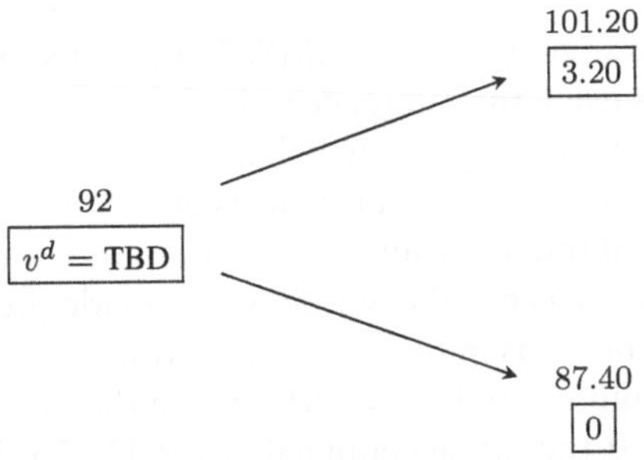

Similarly, we need to solve the following system of equations:

$$\begin{cases} 101.20x^d + 1.02^2y^d = 3.20, \\ 87.40x^d + 1.02^2y^d = 0. \end{cases}$$

We find that $x^d = 16/69$ and $y^d = -19.4796873$. Therefore, the time-1 value of the option in the lower node is

$$\begin{aligned} v^d &= 92x^d + 1.02y^d \\ &= 92 \times \frac{16}{69} + 1.02 \times (-19.4796873) = 1.464052288. \end{aligned}$$

We know that for the first period (from time 0 to time 1), we need to choose (x, y) to replicate the values of the derivative at time 1: $v^u = 10.92156863$ and $v^d = 1.464052288$. However, an investor buying shares of this stock at time 0 will be entitled to both the capital gain and dividends at time 1. Thus even if the stock price decreases by \$3 at time 1, this dividend is cashed by the stock holder, i.e. the effective stock price for that investor is either 110 or 95. Therefore, the last system of equations to solve is

$$\begin{cases} 110x + 1.02y = 10.92156863, \\ 95x + 1.02y = 1.464052288. \end{cases}$$

We find that $x = 0.630501089$ and $y = -57.28779529$. So the initial cost of the replicating strategy is

$$\Pi_0 = 100x + 1y = 100 \times 0.630501089 + 1 \times (-57.28779529) = 5.76231364.$$

This is also the time-0 price of the call option, i.e. $C_0 = 5.76231364$. ■

10.4.3 Guaranteed minimum withdrawal benefits

To conclude this chapter, we illustrate how to replicate the insurer's loss arising from a GMWB rider in a variable annuity contract in a *standard* two-period binomial tree model.

Example 10.4.4 *Risk management of a GMWB*
There are two years remaining in a variable annuity contract with a GMWB rider. The sub-account balance is currently 112.75 and two fixed withdrawals of 60 are still allowed in one and two years from now. A periodic premium of 1.5% per year is subtracted from the sub-account before the withdrawal. Finally, the returns credited on the sub-account are those of a reference index increasing by a factor of 8% or decreasing by 4% each year. We assume there is no mortality risk. Finally, the risk-free rate is at 2% and is compounded annually.

Let us determine the loss arising from the GMWB and let us see how an insurance company can replicate this loss using the reference index.

Recall from Chapter 8 that the sub-account balance in a GMWB evolves as follows: every year, it increases/decreases according to the returns of the underlying index while premiums and withdrawals are subtracted from this sub-account. Note that the return is credited on the sub-account balance after the withdrawal is made (just like u and d factors are applied on an ex-dividend price, as in the previous section).

In this context, the premium rate is $\alpha = 0.015$, while the factors are $u = 1.08$ and $d = 0.96$. Right now, i.e. at time 0, the sub-account balance is 112.75. Then, in one year (at time 1), the sub-account balance (before the withdrawal) will be

$$\begin{cases} 119.9435 = 112.75 \times 1.08 \times (1 - 0.015) & \text{in the up scenario,} \\ 106.6164 = 112.75 \times 0.96 \times (1 - 0.015) & \text{in the down scenario.} \end{cases}$$

Subtracting the withdrawal of 60, the two possible values at time 1 (of the sub-account balance) are 59.9435 and 46.6164. At time 2, returns and premiums are computed on 59.94345 and 46.6164. The evolution of the sub-account balance and the index is given in the next table.

Reference index value

Scenario	Time 0	Time 1	Time 2
Up & up	1	1.08	1.1664
Up & down	1	1.08	1.0368
Down & up	1	0.96	1.0368
Down & down	1	0.96	0.9216

Sub-account balance (after withdrawals)

Scenario	Time 0	Time 1	Time 2
Up & up	112.75	59.9435	3.76784211
Up & down	112.75	59.9435	−3.31747368
Down & up	112.75	46.6164	−10.4094737
Down & down	112.75	46.6164	−15.9195322

In the up & up scenario, the sub-account balance is positive at maturity: there is no loss for the insurance company. However, in the other three scenarios, the sub-account balance is insufficient to fund the last withdrawal and the insurer suffers a loss. Therefore, the values of the loss, in each scenario, are 0, 3.3175, 10.4095 and 15.9195, respectively. In other words, the loss is a random variable taking its values in $\{0, 3.3175, 10.4095, 15.9195\}$.

Next, we want to find a strategy that replicates this loss. The idea is that by setting up a portfolio that mimics the loss, the company will always be able to fund the withdrawals. We can treat this as if we were replicating a contract whose payoff takes its values in $\{0, 3.3175, 10.4095, 15.9195\}$.

To do this, the insurer will invest in the basic assets, i.e. in the underlying reference index (the risky asset) and in the risk-free asset. In the up scenario, the replicating strategy is found by solving the following system of equations:

$$\begin{cases} 1.1664x^u + 1.02^2y^u = 0, \\ 1.0368x^u + 1.02^2y^u = 3.31747368, \end{cases}$$

and we get $x^u = -25.59779074$ and $y^u = 28.6978692$. The (market) value of the loss is

$$v^u = 1.08x^u + 1.02y^u = 1.626212588.$$

In other words, at time 1 and given the reference index is $S_1 = 1.08$, shorting 25.59779074 units of the index and investing 28.6978692×1.02 at the risk-free rate will exactly replicate the loss generated by this GMWB.

Similarly, in the down scenario, the system we need to solve is

$$\begin{cases} 1.0368x^d + 1.02^2y^d = 10.40947368, \\ 0.9216x^d + 1.02^2y^d = 15.91953216, \end{cases}$$

and we find $x^d = -47.83036875$ and $y^d = 57.67012687$. The (market) value of the loss is

$$v^d = 0.96x^d + 1.02y^d = 12.90637541.$$

Finally, at time 0, the system of equations is

$$\begin{cases} 1.08x + 1.02y = 1.626212588, \\ 0.96x + 1.02y = 12.90637541 \end{cases}$$

and we obtain $x = -94.00135685$ and $y = 101.1251745$. The current (market) value of the loss is $V_0 = 1x + 1y = 7.123817646$.

To summarize, given that the periodic premium has been set at the contract's inception and there are two years remaining in the contract, the insurer can set aside \$7.12 (taken, for example, from previous premium inflows) and apply the above strategy to replicate the loss of this GMWB. ■

10.5 Summary

Two-period binomial tree model

- Risk-free interest rate: r.
- Risk-free asset price: $B = \{B_0, B_1, B_2\}$, where $B_0 = 1$ and, for $k = 1, 2$,

$$B_k = \begin{cases} B_{k-1}e^r & \text{if the rate is continuously compounded,} \\ B_{k-1}(1+r) & \text{if the rate is periodically compounded.} \end{cases}$$

- Risky asset price: $S = \{S_0, S_1, S_2\}$, where S_1 and S_2 are random variables such that

$$S_1 = \begin{cases} s^u & \text{with probability } p, \\ s^d & \text{with probability } 1-p \end{cases}$$

and

$$\begin{aligned}
&\mathbb{P}(S_2 = s^{uu} | S_1 = s^u) = p^u, \\
&\mathbb{P}(S_2 = s^{ud} | S_1 = s^u) = 1 - p^u, \\
&\mathbb{P}(S_2 = s^{du} | S_1 = s^d) = p^d, \\
&\mathbb{P}(S_2 = s^{dd} | S_1 = s^d) = 1 - p^d.
\end{aligned}$$

- No-arbitrage conditions:

$$s^d < S_0 B_1 < s^u, \quad s^{ud} < s^u \frac{B_2}{B_1} < s^{uu} \quad \text{and} \quad s^{dd} < s^d \frac{B_2}{B_1} < s^{du}.$$

- Derivative (written on S): $V = \{V_0, V_1, V_2\}$ where V_1 and V_2 are random variables such that

$$V_1 = \begin{cases} v^u & \text{if } S_1 = s^u, \\ v^d & \text{if } S_1 = s^d. \end{cases}$$

and

$$V_2 = \begin{cases} v^{uu} & \text{if } S_1 = s^{uu}, \\ v^{ud} & \text{if } S_1 = s^{ud}, \\ v^{du} & \text{if } S_1 = s^{du}, \\ v^{dd} & \text{if } S_1 = s^{dd}. \end{cases}$$

- Recombining tree: if $s^{ud} = s^{du}$.
- Simplified tree: up and down factors $u > d$ such that

$$s^u = uS_0 \quad \text{and} \quad s^d = dS_0$$

and

$$s^{uu} = u^2 S_0, \quad s^{ud} = s^{du} = udS_0 \quad \text{and} \quad s^{dd} = d^2 S_0.$$

- In a simplified tree, there is only one no-arbitrage condition: $d < B_1 < u$.

Trading strategy/portfolio

- Portfolio: three pairs $\{(x, y), (x^u, y^u), (x^d, y^d)\}$.
- Number of units of S held from time 0 to time 1: x.
- Number of units of S held from time 1 to time 2: x^u, if $S_1 = s^u$, or x^d, if $S_1 = s^d$.
- Number of units of B held from time 0 to time 1: y.
- Number of units of B held from time 1 to time 2: y^u, if $S_1 = s^u$, or y^d, if $S_1 = s^d$.
- Portfolio's value: $\Pi = \{\Pi_0, \Pi_1, \Pi_2\}$ where $\Pi_0 = xS_0 + yB_0$,

$$\Pi_1 = \begin{cases} x^u s^u + y^u B_1 & \text{if } S_1 = s^u, \\ x^d s^d + y^d B_1 & \text{if } S_1 = s^d, \end{cases}$$

and

$$\Pi_2 = \begin{cases} x^u S_2 + y^u B_2 & \text{if } S_1 = s^u, \\ x^d S_2 + y^d B_2 & \text{if } S_1 = s^d. \end{cases}$$

Pricing by replication

- A strategy $\{(x,y),(x^u,y^u),(x^d,y^d)\}$ replicates V_2 if $\Pi_2 = V_2$ and $\Pi_1 = V_1$.
- Replicating portfolio for V_2:

$$x = \frac{v^u - v^d}{s^u - s^d} \quad \text{and} \quad y = \frac{1}{B_1}\left(\frac{v^d s^u - v^u s^d}{s^u - s^d}\right),$$

$$x^u = \frac{v^{uu} - v^{ud}}{s^{uu} - s^{ud}} \quad \text{and} \quad y^u = \frac{1}{B_2}\left(\frac{v^{ud} s^{uu} - v^{uu} s^{ud}}{s^{uu} - s^{ud}}\right)$$

and

$$x^d = \frac{v^{du} - v^{dd}}{s^{du} - s^{dd}} \quad \text{and} \quad y^d = \frac{1}{B_2}\left(\frac{v^{dd} s^{du} - v^{du} s^{dd}}{s^{du} - s^{dd}}\right).$$

- By no-arbitrage assumption, we have

$$V_0 = \frac{1}{B_1}\left(\frac{v^d s^u - v^u s^d}{s^u - s^d}\right) + \left(\frac{v^u - v^d}{s^u - s^d}\right) S_0,$$

$$v^u = \frac{B_1}{B_2}\left(\frac{v^{ud} s^{uu} - v^{uu} s^{ud}}{s^{uu} - s^{ud}}\right) + \left(\frac{v^{uu} - v^{ud}}{s^{uu} - s^{ud}}\right) s^u$$

and

$$v^d = \frac{B_1}{B_2}\left(\frac{v^{dd} s^{du} - v^{du} s^{dd}}{s^{du} - s^{dd}}\right) + \left(\frac{v^{du} - v^{dd}}{s^{du} - s^{dd}}\right) s^d.$$

Pricing with risk-neutral probabilities

- Risk-neutral probabilities: $\{(q,1-q),(q^u,1-q^u),(q^d,1-q^d)\}$, where

$$q = \frac{S_0 \times \frac{B_1}{B_0} - s^d}{s^u - s^d},$$

$$q^u = \frac{s^u \times \frac{B_2}{B_1} - s^{ud}}{s^{uu} - s^{ud}},$$

$$q^d = \frac{s^d \times \frac{B_2}{B_1} - s^{dd}}{s^{du} - s^{dd}}.$$

- Risk-neutral pricing formula for a derivative with payoff V_2: for each $k = 0, 1$,

$$\frac{V_k}{B_k} = \mathbb{E}^{\mathbb{Q}}\left[\frac{V_{k+1}}{B_{k+1}}\middle| S_k\right]$$

or, written differently,

$$v^u = \frac{B_1}{B_2}\mathbb{E}^{\mathbb{Q}}[V_2|S_1 = s^u] = \frac{B_1}{B_2}((1-q^u)v^{ud} + q^u v^{uu}),$$

$$v^d = \frac{B_1}{B_2}\mathbb{E}^{\mathbb{Q}}[V_2|S_1 = s^d] = \frac{B_1}{B_2}((1-q^d)v^{dd} + q^d v^{du}),$$

$$V_0 = \frac{1}{B_1}\mathbb{E}^{\mathbb{Q}}[V_1] = \frac{1}{B_1}(v^d(1-q) + v^u q)$$

$$= \frac{1}{B_2}(qq^u v^{uu} + q(1-q^u)v^{ud} + (1-q)q^d v^{du} + (1-q)(1-q^d)v^{dd}).$$

- Risk-neutral probabilities are only a mathematical convenience used when pricing.

Advanced actuarial and financial examples

- Stochastic interest rates: different interest rate for each one-period sub-tree.
- Stocks paying discrete dividends: non-recombining trees with downward jumps in the risky asset values.
- Guaranteed minimal withdrawal benefits (GMWBs) in a variable annuity contract: must distinguish between the dynamics of the reference index and the sub-account balance.

10.6 Exercises

10.1 You are given that:
- $S_0 = 45$;
- $s^u = 47, s^d = 44$;
- $s^{uu} = 51, s^{ud} = s^{du} = 46, s^{dd} = 41$;
- $B_k = 1.02^k$.

Determine whether there are arbitrage opportunities in this binomial tree. If this is the case, determine how to exploit it.

10.2 A share of stock, whose current price is \$35, can take 5% per period or decrease by 3%. Analysts have determined that the probability of a stock price increase at each period is 68%.
- **(a)** What are the possible values taken by S_1 and S_2?
- **(b)** Given S_0, what are the unconditional probabilities of observing the values found in (a)?

10.3 A 45-strike call option maturing in two periods is issued on a stock whose current price is \$43. The stock price can increase by 6% or decrease by 4%. The Treasury rate is quoted as 1.5% per period. You believe the probability of stock price increases at each period is 75%.
- **(a)** According to your beliefs, what is the expected return from the stock?
- **(b)** According to your beliefs, what is the expected payoff from the call option?
- **(c)** How many shares of stock should you hold at time 1 to replicate the call option payoff at time 2 if the stock price increased from time 0 to time 1?
- **(d)** How much should you invest in the risk-free bond at time 0 to be able to replicate the option's payoff?
- **(e)** What is the no-arbitrage price of the call option at time 1 if the stock price went down from time 0 to time 1?

10.4 The price of a share of stock currently selling for \$67 can increase by 10% or decrease by 7%. An at-the-money call option trades for \$6.10. If $B_0 = 1$, find B_1 and B_2 such that there are no arbitrage opportunities in this market.

10.5 A forward start call option is a call option maturing at time 2 whose strike price is determined at time 1. In other words, the payoff of the option is $(S_2 - S_1)_+$. Using the binomial tree of question 10.1, determine the replicating strategy at each node of the tree and the no-arbitrage price of this option.

10.6 In the two-step binomial tree of question 10.1, calculate the no-arbitrage price of a fixed-strike Asian call option with strike 46. Note that the mean is computed over S_1 and S_2.

10.7 For a share of stock whose current price is \$38, analysts believe the probability that the stock price increases is 80%. The stock price is such that it can only increase by 10% per period or decrease by 3% otherwise. The periodic risk-free rate is 4.25%. Calculate the probability a 40-strike call option ends in-the-money in two periods.

10.8 You are given the following information about a financial market:

- Stock: current price is \$45. It can increase by \$3 in each period, or decrease by \$2.
- Bond: current two-period zero-coupon bond price is \$90 (face value of \$100).
- When the stock price increases to \$48, the bond price increases to \$96.
- When the stock price decreases to \$43, the bond price increases to \$94.

(a) Using trees, illustrate the evolution of the price of both assets.
(b) Describe in detail the replicating strategy necessary to replicate the payoff of a 47-strike put option.
(c) What should be the no-arbitrage price of this option?

10.9 In the context of Section 10.4.1, build the replicating portfolios and by reorganizing the equations, show that you can rewrite the option's value using an expectation and the following risk-neutral probabilities:

$$q^u = \frac{s^u \times \frac{b_2^u}{B_1} - s^{ud}}{s^{uu} - s^{ud}}$$

$$q^d = \frac{s^d \times \frac{b_2^d}{B_1} - s^{dd}}{s^{du} - s^{dd}}$$

$$q = \frac{S_0 \times \frac{B_1}{B_0} - s^d}{s^u - s^d}.$$

10.10 The price of a share of stock can increase by 7% or decrease by 3% per period. The stock currently trades for \$84 and a \$3 dividend will be paid at time 1. Find the no-arbitrage price of an at-the-money put option maturing after two periods if the periodic interest rate is 2%.

10.11 A two-period PTP EIA is issued with an initial investment of 100. The guaranteed amount is also set to 100. The contract is written on a stock whose price may increase by 10% or decrease by 4%. Find the participation rate β such that the no-arbitrage value of the benefit is equal to the initial investment.

10.12 A \$1000 two-year GMWB allows the policyholder to withdraw \$500 annually (at times 1 and 2). The GMWB is written on a stock whose price can increase by 8% per year or decrease by 4%. The annual premium is set to 2% and the risk-free is 3% (compounded continuously).

(a) For each scenario, calculate the loss of the insurer.
(b) If the risk-free rate is constant at 1% per period, calculate how much the insurer should set aside at time 0 to make sure it can replicate the loss of this contract.

11

Multi-period binomial tree model

The one-period and two-period binomial trees presented in Chapters 9 and 10 had the advantage of introducing important concepts and procedures, such as replication, portfolio dynamics and risk-neutral formulas, in fairly simple setups. Now, we seek to generalize these ideas to a model with more than two time steps.

The main objective of this chapter is to introduce the general multi-period binomial tree model and its own challenges. The algorithmic approach we will use throughout the chapter will allow a straightforward computer implementation of the model and it will lay the foundations for the limiting case known as the Black-Scholes-Merton model. The specific objectives are to:

- build a general binomial tree and relate the asset price observed at a given time step to the binomial distribution;
- understand the difference between simple options and path-dependent options;
- establish the dynamic replicating strategy to price an option;
- apply risk-neutral pricing in a multi-period model.

11.1 Model

Fix $n \geq 1$, the number of time steps or periods. In an n-period model there are $n+1$ time points: $0, 1, 2, \ldots, n$. One should think of time 0 as *today* while time 1, time 2, etc. are in the *future*. Often, time 0 will represent the issuance date of a derivative whose maturity occurs at time n.

When time is expressed without units, the timeline is as follows:

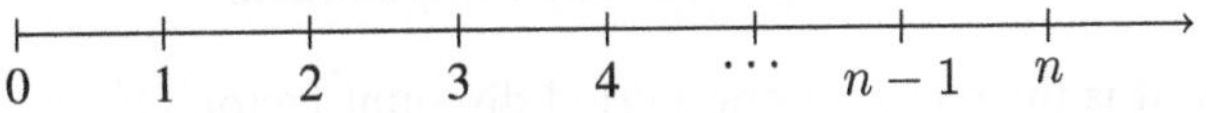

As is the case for most discrete-time financial models, it is often more convenient to express time points $0, 1, 2, \ldots, n$ in units of true time such as a year. Suppose we have a time horizon $[0, T]$ in mind, with 0 being the initial time and T the end of the investment period (in years). We divide the time interval $[0, T]$ in n periods given by $(0, T/n), (T/n, 2T/n), \ldots, ((n-1)T/n, T)$. The whole timeline is represented graphically by

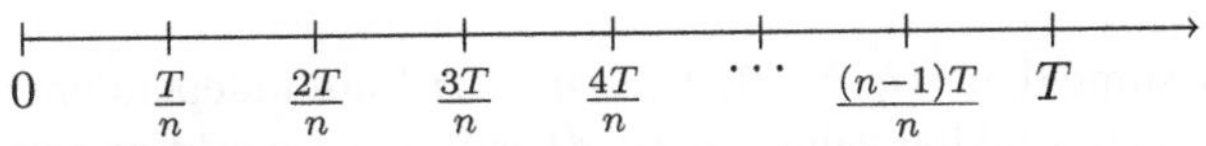

Actuarial Finance: Derivatives, Quantitative Models and Risk Management, First Edition.
Mathieu Boudreault and Jean-François Renaud.

Alternatively, defining $h = T/n$ as the length of a time step, then each time interval $(0, h), (h, 2h), \dots, ((n-1)h, nh)$ can be represented graphically by

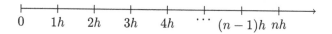

$$0 \quad 1h \quad 2h \quad 3h \quad 4h \quad \cdots \quad (n-1)h \quad nh$$

In this chapter, we will express time both ways.

Finally, an n-period binomial model is a *frictionless* financial market with only two basic assets:

- a risk-free asset (a bank account or a bond) which evolves according to the risk-free interest rate r;
- a risky asset (a stock or an index) with a known initial value and unknown future values.

Investors in this market act and make transactions only at those $n + 1$ time points and not in between.

11.1.1 Risk-free asset

The risk-free asset is modeled by a (deterministic) process $B = \{B_k, k = 0, 1, 2, n-1, n\}$ with $B_0 = 1$ and where $B_k \geq B_{k-1}$, for each $k = 1, 2, ..., n$. If r is the constant continuously compounded (annual) interest rate, then

$$B_k = B_{k-1}\mathrm{e}^{rh} = \mathrm{e}^{k(rh)},$$

and, if r is the periodically compounded (annual) interest rate, i.e. compounded n times per year, then

$$B_k = B_{k-1}(1 + rh) = (1 + rh)^k.$$

In other words,

$$\frac{B_k}{B_{k-1}} = \begin{cases} \mathrm{e}^r & \text{if the rate is continuously compounded,} \\ (1+r) & \text{if the rate is periodically compounded,} \end{cases}$$

is independent of k: it is the constant one-period discount factor. Of course, units of h and r must coincide for everything to make sense.

11.1.2 Risky asset

The risky asset is modeled by a stochastic process $S = \{S_k, k = 0, 1, 2, \dots, n-1, n\}$, where S_0 is a known (deterministic) quantity and

$$S_k = S_{k-1}U_k = S_0 U_1 U_2 \dots U_k.$$

For simplicity, we assume that $\{U_k, k = 1, 2, \dots, n-1, n\}$ are independent and identically distributed random variables taking values in $\{u, d\}$ with (real-world or actuarial) probabilities $\{p, 1-p\}$. Once again, we assume that $d < u$ and $0 < p < 1$.

As before, in order to verify the no-arbitrage assumption, we must make sure that

$$d < e^{rh} < u, \tag{11.1.1}$$

when interest is compounded continuously, or

$$d < 1 + rh < u,$$

when interest is compounded at each period. In plain words, one dollar accumulated at the risk-free rate over each period does not earn systematically more (or less) than an equivalent investment in the risky asset.

It should be clear that in this multi-period setup, we restrict ourselves to the *simplified version* of the binomial tree model as discussed in Chapter 10. Consequently, the n-period binomial tree model is based upon five parameters (r, S_0, u, d, p) and it is recombining. Because the tree is recombining, the 2^n possible trajectories lead to only $n+1$ possible prices at time n.

The following illustration shows all possible trajectories of the risky asset price (time series) in a three-step binomial tree:

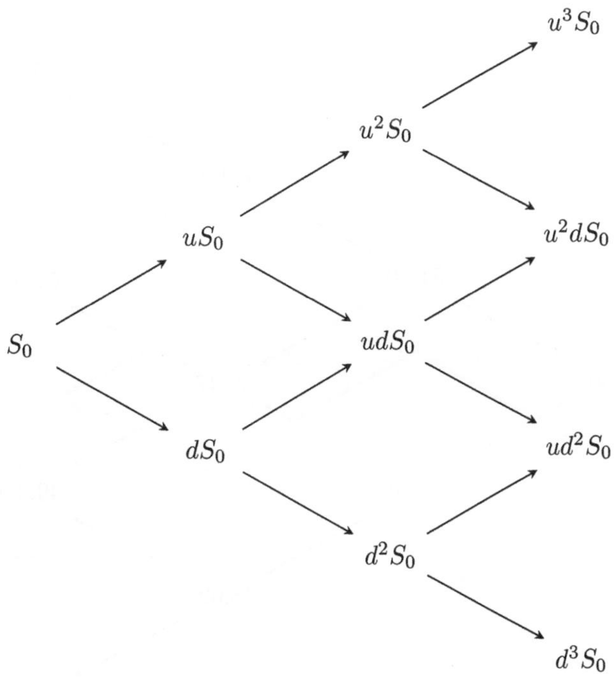

The sample space and the price of the risky asset

It is sometimes useful to describe the full probability model on which the binomial tree is constructed. For an n-period binomial model, one can choose its sample space Ω_n as follows:

$$\Omega_n = \{\omega = \omega_1\omega_2\ldots\omega_n : \omega_i = \text{u or d},\ i = 1, 2, \ldots, n\}.$$

It is then clear that each state of nature $\omega \in \Omega_n$, that is each scenario, corresponds to a full trajectory of the risky asset S (a path in the tree). For example, if $n = 3$ as in the previous illustration, then the possible paths and prices at maturity are

$$S_3(\text{uuu}) = u^3 S_0,$$
$$S_3(\text{uud}) = S_3(\text{udu}) = S_3(\text{duu}) = u^2 d S_0,$$
$$S_3(\text{udd}) = S_3(\text{dud}) = S_3(\text{ddu}) = u d^2 S_0,$$
$$S_3(\text{ddd}) = d^3 S_0.$$

Example 11.1.1 *Construction of a three-step binomial tree*
Assume that, each week, a stock price can increase by 3% or decrease by 2%. If the initial stock price is 50, build the corresponding three-step binomial tree.

For each time point $k = 1, 2, 3$, the corresponding random variable S_k, modeling the stock price in k weeks from now, can take one of the following values:

$$S_k = 50 \times 1.03^i \times 0.98^{k-i},$$

where $i = 0, 1, \ldots, k$. The corresponding tree is

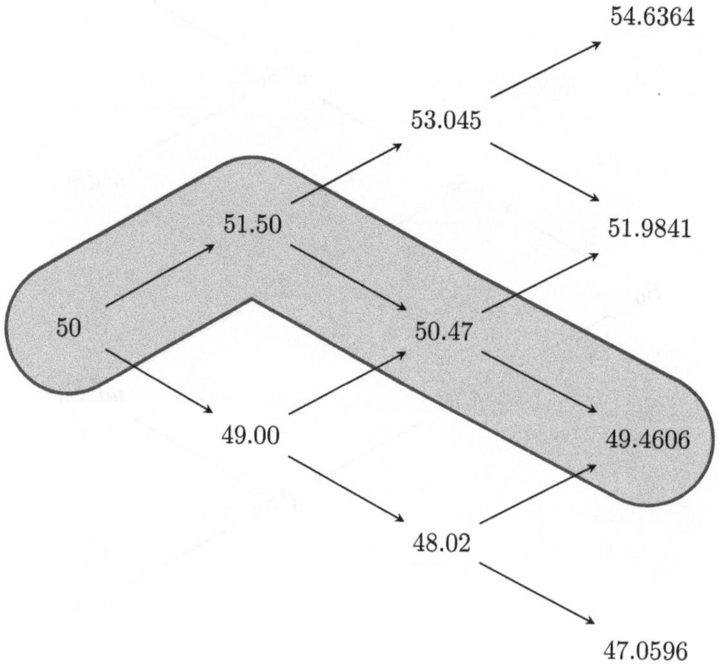

where the highlighted path $\{50, 51.50, 50.47, 49.46\}$ is a possible realization (time series) for the weekly stock price process S. ■

As illustrated by the previous trees (with $n = 3$), after k periods/steps there are $k + 1$ (different) nodes or, in other words, $k + 1$ different time-k prices. But overall, after k periods/steps, there are 2^k different price trajectories for the process $S = \{S_k, k = 0, 1, 2, \ldots, n-1, n\}$. One of these particular paths was illustrated in the preceding and the next trees: one upward movement followed by two downward movements.

At maturity (after n periods), the rationale is similar: there are $n+1$ different prices for the risky asset but a total of 2^n paths leading to those values. Such a distinction is fundamental when pricing path-dependent derivatives as opposed to simple vanilla derivatives (see Section 11.1.3).

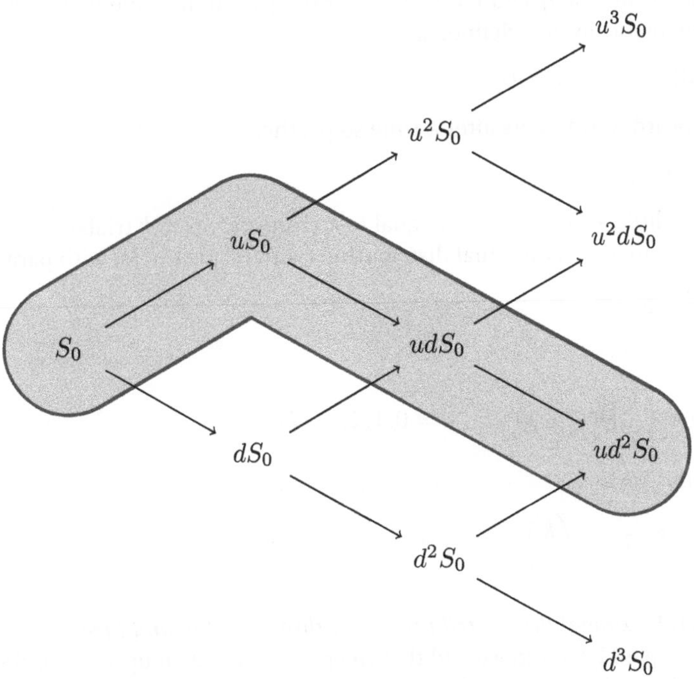

Two popular choices for u and d

There are two popular specifications for u and d in the finance literature:

- Cox-Ross-Rubinstein (CRR) model:

$$u = \exp\left(\sigma\sqrt{h}\right),$$
$$d = \exp\left(-\sigma\sqrt{h}\right) = \frac{1}{u},$$

- Jarrow-Rudd (JR) model:

$$u = \exp\left((r - \sigma^2/2)h + \sigma\sqrt{h}\right),$$
$$d = \exp\left((r - \sigma^2/2)h - \sigma\sqrt{h}\right),$$

where σ stands for the (annualized) *volatility* of the risky asset return. In practice, σ is in the range of $[0.1, 0.4]$. These formulas for u and d are practical because:

1) they automatically adjust with the length of the time step h;
2) they will become useful later to relate the binomial tree to the lognormal distribution (and hence the Black-Scholes formula).

Therefore, once n (or h) is fixed, the CRR or JR trees have four parameters: S_0, r, σ and p.

11.1.2.1 Probability distribution

We now look at the probability distribution of the risky asset time-k price, where $k = 1, 2, \dots, n$. We can write

$$S_k = S_0 u^{I_k} d^{k-I_k},$$

where the random variable I_k counts the number of upward movements in the trajectory after k periods. Mathematically, it is defined as

$$I_k = \text{card}\{i = 1, \dots, k \ : \ U_i = u\}.$$

If there are i upward movements after k time steps, then

$$S_k = S_0 u^i d^{k-i},$$

where i can be as little as 0 but at most equal to k (k up-moves in k trials).

It is clear that I_k follows a binomial distribution (with respect to $\mathbb{P}$) with parameters (k, p):

$$I_k \overset{\mathbb{P}}{\sim} \text{Bin}(k, p),$$

which means that

$$\mathbb{P}(I_k = i) = \binom{k}{i} p^i (1-p)^{k-i}, \quad i = 0, 1, 2, \dots, k,$$

and consequently

$$\mathbb{P}(S_k = u^i d^{k-i} S_0) = \binom{k}{i} p^i (1-p)^{k-i}, \quad i = 0, 1, 2, \dots, k.$$

Example 11.1.2 *Probability distribution in a three-step binomial tree*
We follow example 11.1.1 and we add that the probability of an up-move is 58%. We want to determine the probability distribution of the random variables S_1, S_2 and S_3.

The probability distribution of S_1 is simply

$$\mathbb{P}(S_1 = 51.50) = 0.58 = 1 - \mathbb{P}(S_1 = 49.00).$$

Also, we find that

$$\mathbb{P}(S_2 = 53.045) = 0.58^2, \qquad \mathbb{P}(S_2 = 50.47) = 2 \times 0.58 \times 0.42$$

because there are two possible paths leading to a price of 50.47 (up & down, down & up). Moreover, $\mathbb{P}(S_2 = 48.02) = 0.42^2$. Finally,

$$\mathbb{P}(S_3 = 54.6364) = 0.58^3 \times 0.42^0,$$
$$\mathbb{P}(S_3 = 51.9841) = 3 \times 0.58^2 \times 0.42^1,$$
$$\mathbb{P}(S_3 = 49.4606) = 3 \times 0.58^1 \times 0.42^2,$$
$$\mathbb{P}(S_3 = 47.0596) = 0.58^0 \times 0.42^3,$$

where again the constant 3 comes from the three possible paths leading to the price 51.9841 (same reasoning applies to 49.4606). ■

11.1.3 Derivatives

We now introduce a derivative (a third asset) in our model which is an asset whose value V is derived from the risky asset S. Therefore, the derivative's price is modeled by a stochastic process $V = \{V_k, k = 0, 1, \dots, n\}$, where once again V_0 is today's value and $V_1, V_2, \dots, V_n$ are (random) future values. It is assumed that the derivative matures at time n which means that V_n is a random variable representing the payoff.

We will consider two types of derivatives:

- simple (vanilla) options, whose final payoff V_n is some function g of the *final asset price*, i.e. $V_n = g(S_n)$. Examples include standard call and put options, binary and gap options, etc.;
- exotic or path-dependent options, whose final payoff V_n is some function g of the *entire trajectory*, i.e. $V_n = g(S_1, S_2, \dots S_n)$. Examples include Asian, lookback and barrier options.

Again, when dealing with call (resp. put) options, we will use C (resp. P) instead of V to represent the option's price.

11.1.4 Labelling the nodes

The notation of the form v^{uu}, x^d or s^{ud} (with superscripts) was appropriate for a one-period or a two-period binomial tree to emphasize on the path taken by the price of the risky asset. Unfortunately, for n large, this notation is not convenient.

Therefore, we will use a more efficient notation relying on the fact that the underlying tree is recombining. The idea is that a node in the tree is uniquely identified by:

1) time;
2) number of upward movements needed to reach this node.

Therefore, for $k = 1, 2, \dots, n$ and $j = 0, 1, \dots, k$, the pair (k, j) uniquely identifies the node at time k reached by a total of j upward movements (and therefore $k - j$ downward movements) after k steps. This notation will be useful to simplify and formalize some, but not all, computations and eventually for coding the binomial tree model with a programming language.

The following tree illustrates how each node is identified in a three-step binomial tree. We see that, for each time point k, nodes are numbered $j = 0, 1, \dots, k$, from *bottom to top*.

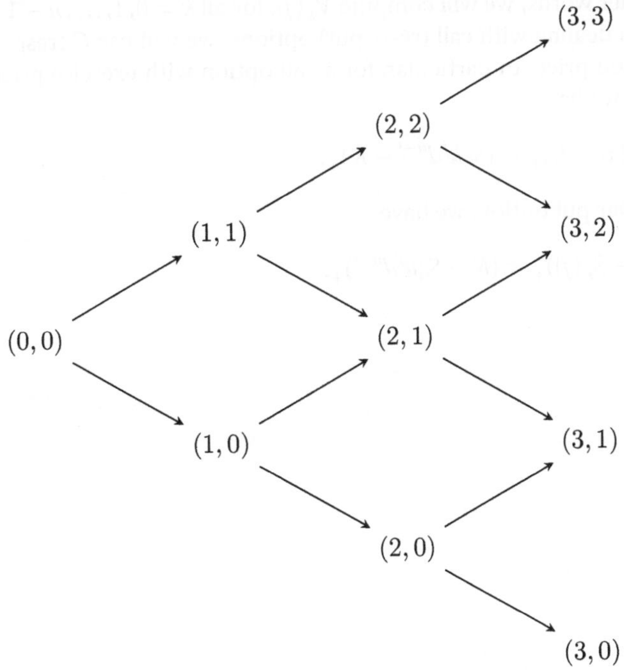

To code the tree in a spreadsheet, or in an array-oriented programming language, we can also organize the tree as a table/matrix as illustrated next, where j corresponds to rows and k to columns. Note that nodes are numbered $j = 0, 1, \dots, k$ from *top to bottom* which is much more convenient for referencing array cells.

j/k	0	1	2	3
0	(0, 0)	(1, 0)	(2, 0)	(3, 0)
1	–	(1, 1)	(2, 1)	(3, 1)
2	–	–	(2, 2)	(3, 2)
3	–	–	–	(3, 3)

As a first application of this notation, we can define $S_k(j)$ as the realization/value of the random variable S_k corresponding to the j-th node. For a given path, in which we observe j upward movements from time 0 to time k, we define: for each $j = 0, 1, \dots, k$,

$$S_k(j) = u^j d^{k-j} S_0.$$

Simple vanilla options are easier to handle than path-dependent options as there is no need to analyze all possible paths: only the final asset price matters. Therefore, we define $V_k(j)$ as the realization/value of the random variable V_k corresponding to the j-th node.

We have the following relationship at maturity: for each $j = 0, 1, \dots, n$,

$$V_n(j) = g(S_n(j)) = g(u^j d^{n-j} S_0).$$

Our goal, in the next sections, is to determine the derivative's price for all times and nodes prior to maturity. In other words, we will compute $V_k(j)$, for all $k = 0, 1, \dots, n-1$ and $j = 0, 1, \dots, k$.

Moreover, when dealing with call (resp. put) options, we will use C (resp. P) instead of V to represent the option price. In particular, for a call option with exercise price K and maturing after n time steps, we have

$$C_n(j) = (S_n(j) - K)_+ = (S_0 u^j d^{n-j} - K)_+,$$

whereas for a similar put option, we have

$$P_n(j) = (K - S_n(j))_+ = (K - S_0 u^j d^{n-j})_+.$$

The evolution of the risky asset price and the corresponding derivative price is depicted in the following tree over three periods:

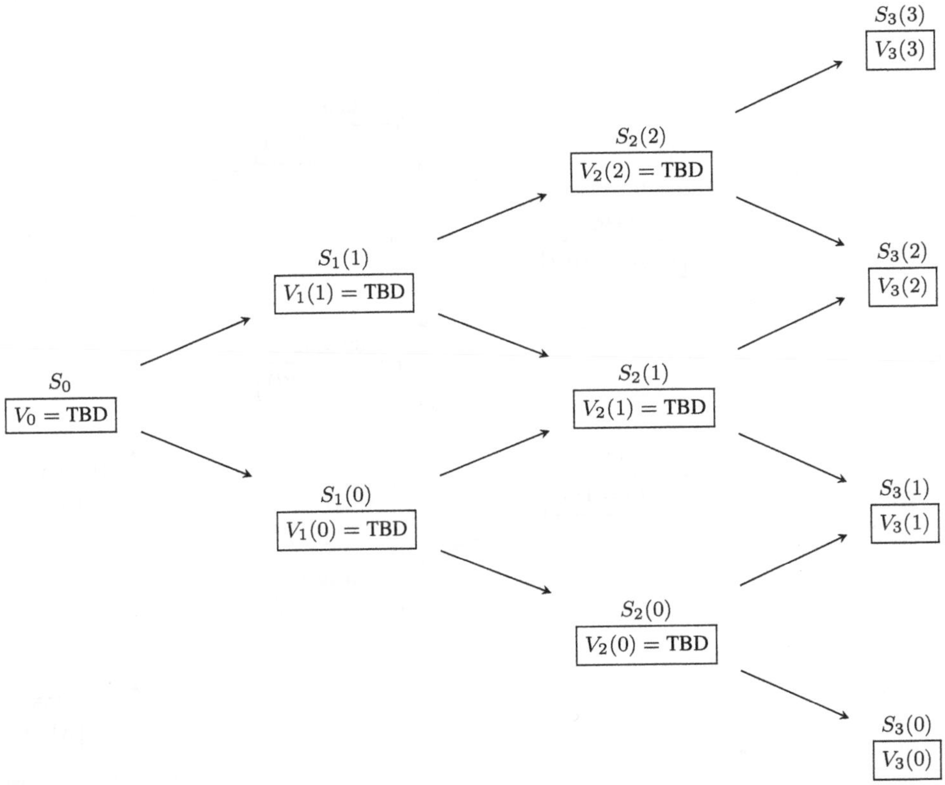

Example 11.1.3 *Call option payoff in a general tree*

The initial stock price is 32 and it may go up by 6% or down by 4% every period. Let us represent the evolution of the stock price by a three-step binomial tree along with the payoff of an at-the-money call option.

We have $S_0 = 32$ and $S_k = 32 \times 1.06^j \times 0.96^{k-j}$, for $j = 0, 1, \dots, k$ and $k = 1, 2, 3$. Moreover, the payoff of the call option is

$$C_3(0) = C_3(1) = 0, \quad C_3(2) = 2.516992 \quad \text{and} \quad C_3(3) = 6.112512.$$

Those values are illustrated in the following tree:

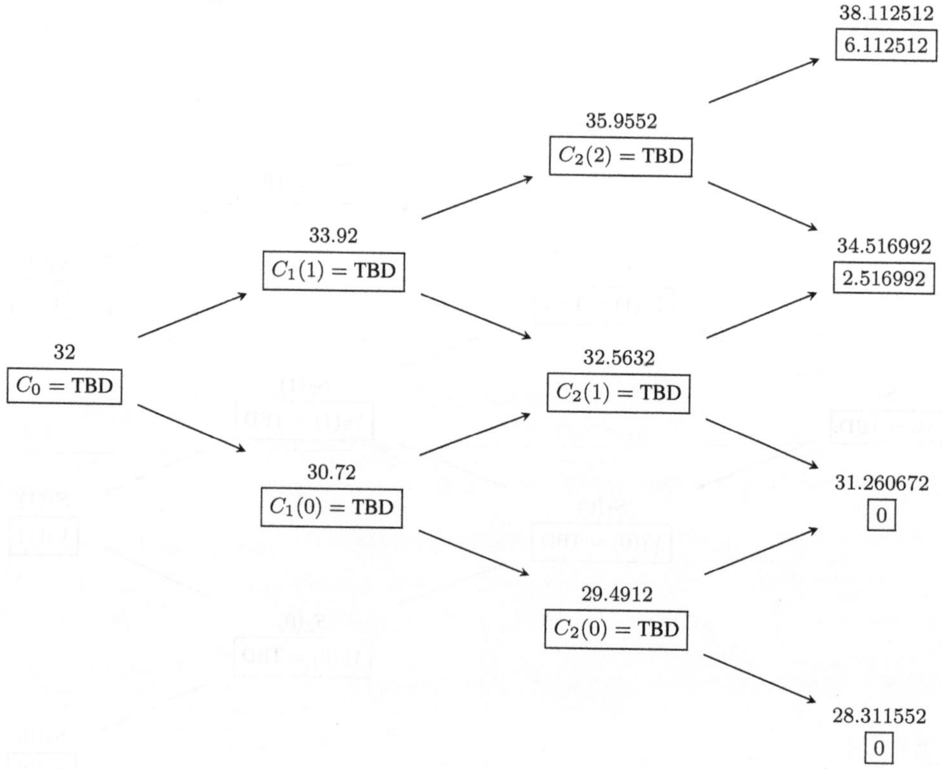

■

11.1.5 Path-dependent payoffs

Exotic or path-dependent options such as Asian or lookback options cannot be treated as easily in the n-step binomial tree. One has to be very careful with the current notation (labeling of the nodes) because it focuses on the risky asset price at any given node, not taking into account the different paths reaching that node. Unless we rely on *advanced techniques*,[1] the only choice we have when computing the final payoff of an exotic option is to consider each possible path separately. This is illustrated in the following example.

Example 11.1.4 *Payoff of a fixed-strike Asian call option*
Using the context of example 11.1.3, we now introduce a fixed-strike Asian call option with strike price $K = 31.50$. Determine the behavior of the random variable V_3.

Asian options are path-dependent derivatives so we have to be careful and consider all eight possible payoffs (eight possible trajectories) in the three-step binomial tree of example 11.1.3. For this option, we cannot use the algorithmic notation $V_k(j)$. This Asian option has a payoff given by $V_3 = (\overline{S}_3 - 31.50)_+$ whose realizations/values are computed in the next table. Note that the random variable $\overline{S}_3$ is based on an arithmetic average.

---

1 Tree or lattice methods specifically tailored for a given exotic option. This is beyond the scope of this book.

Path	S_0	S_1	S_2	S_3	Average	Payoff
uuu	32	33.92	35.9552	38.112512	35.995904	4.495904
uud	32	33.92	35.9552	34.516992	34.7973973	3.29739733
udu	32	33.92	32.5632	34.516992	33.6667307	2.16673067
udd	32	33.92	32.5632	31.260672	32.5812907	1.08129067
duu	32	30.72	32.5632	34.516992	32.600064	1.100064
dud	32	30.72	32.5632	31.260672	31.514624	0.014624
ddu	32	30.72	29.4912	31.260672	30.490624	0
ddd	32	30.72	29.4912	28.311552	29.507584	0

For example, the path corresponding to udu is highlighted in the following tree:

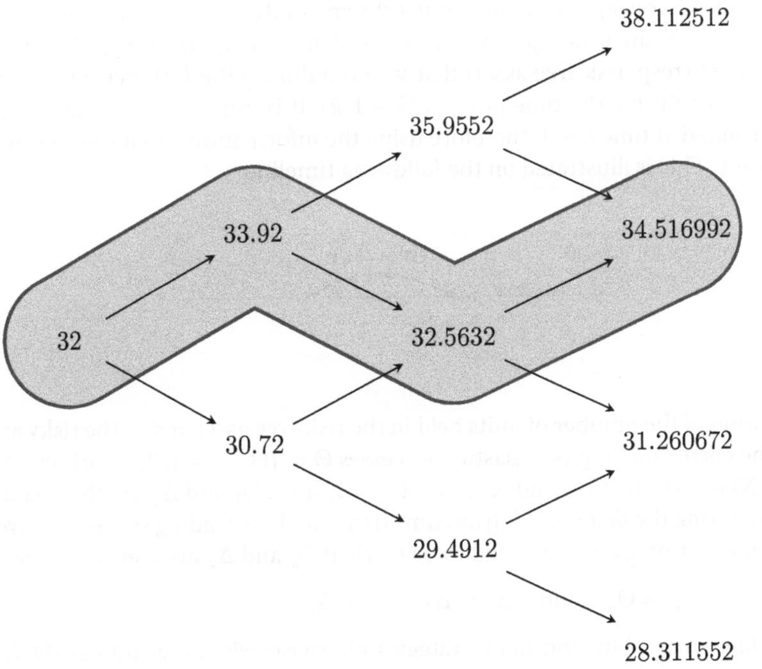

For this path, the computation of the average goes as follows:

$$\bar{S}_3(\text{udu}) = \frac{S_1(\text{udu}) + S_2(\text{udu}) + S_3(\text{udu})}{3}$$
$$= \frac{33.92 + 32.5632 + 34.516992}{3} = 33.6667307.$$

So, the payoff in this scenario is given by

$$(\bar{S}_3(\text{udu}) - 31.50)_+ = (33.6667307 - 31.50)_+ = 2.16673067.$$

Computations are done the same way for other trajectories. ■

As we saw in the previous example, pricing exotic options in an n-period binomial tree can be tedious, especially when n is large. Therefore, unless stated otherwise, the rest of this chapter will focus on simple vanilla options, even if most of the upcoming material could apply to path-dependent options with some adaptation to the notation.

11.2 Pricing by replication

Replicating a derivative in a multi-step binomial tree is similar to what we did in a two-step binomial tree. We will decompose the tree into $1 + 2 + 3 + \cdots + n = n(n + 1)/2$ one-period sub-trees and, for each, we will solve a system of two equations with two unknowns.

However, before looking at the replication procedure, we need to define and analyze trading/investment strategies, the portfolio value process, etc.

11.2.1 Trading strategies/portfolios

We now look at the concept of trading or investment strategies, also called portfolios, of which replicating strategies are a sub-group. Let us first define by Δ_k (resp. Θ_k), the number of units of the risky asset (resp. risk-free asset) that we hold during the k-th period, that is from time $k-1$ to time k or during the time interval $[k-1, k)$. It is important to note that both Δ_k and Θ_k are determined at time $k-1$, therefore using the information about the tree available up to that time point. This is illustrated on the following timeline:

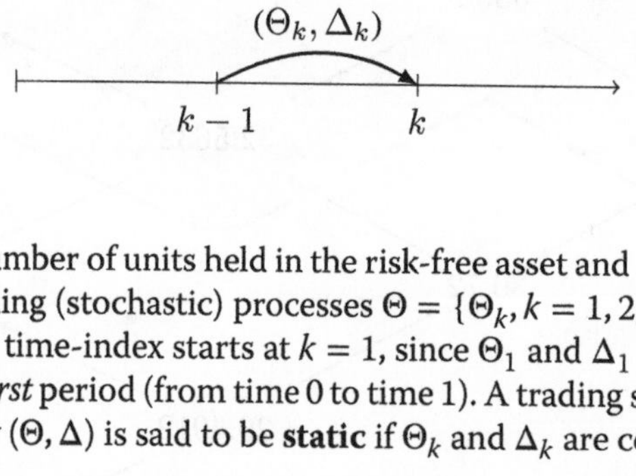

The evolution of the number of units held in the risk-free asset and in the risky asset is represented by the corresponding (stochastic) processes $\Theta = \{\Theta_k, k = 1, 2, \ldots, n\}$ and $\Delta = \{\Delta_k, k = 1, 2, \ldots, n\}$. Note that the time-index starts at $k = 1$, since Θ_1 and Δ_1 are the quantities held in the portfolio during the *first* period (from time 0 to time 1). A trading strategy is given by (Θ, Δ).

An investment strategy (Θ, Δ) is said to be **static** if Θ_k and Δ_k are constant over time, that is

$$\Theta_1 = \Theta_2 = \cdots = \Theta_n \quad \text{and} \quad \Delta_1 = \Delta_2 = \cdots = \Delta_n.$$

This is also known as a buy-and-hold strategy. Otherwise, when the strategy (Θ, Δ) is updated periodically as asset prices evolve, the strategy is said to be **dynamic**.

Example 11.2.1 *Static and dynamic investment strategies*
We know from Chapter 3 that to replicate a forward contract we need to hold one unit of stock and borrow the present value of K. Therefore, using the previously defined notation, this trading strategy is given by

$$\Delta_1 = \Delta_2 = \cdots = \Delta_n = 1$$

and

$$\Theta_1 = \Theta_2 = \cdots = \Theta_n = K/B_n.$$

This is a static strategy.

An example of a dynamic investment strategy is when we update our portfolio to maintain a fixed proportion of, say, 60%–40% (of the portfolio value) in stocks and bonds. Suppose a stock and a bond trade at \$100 each and we want/need to invest \$10,000. Therefore, at time 0, we choose $\Delta_1 = 60$ and $\Theta_1 = 40$, for a total investment of

$$60 \times 100 + 40 \times 100 = 10000.$$

After one period, if for example the stock price is worth 150 and the bond 110, then the portfolio will be worth $60 \times 150 + 40 \times 110 = 13400$. However, the position in the stock is worth 9000, which corresponds to $9000/13400 = 67.16\%$ of the portfolio value at time 1. To maintain, our 60–40 strategy, we need to sell some shares of stock and buy more bonds, so that we ultimately set

$$\Delta_2 = \frac{13400 \times 0.6}{150} = 53.6$$

and

$$\Theta_2 = \frac{13400 \times 0.4}{110} = 48.73.$$

In other words, we sell $60 - 53.6$ shares of stock and buy $48.73 - 40$ units of the bond. Those are the quantities set up at time 1 and prevailing during the second period. Because the portfolio has been rebalanced following the information observed at time 1, we say the investment strategy is dynamic.

Similar trades would need to be performed in the future, depending on the values of the basic assets at those times. ■

11.2.2 Portfolio value process

Let us now define by Π_k the value of a trading strategy at time k (before rebalancing). Thus, for each $k = 1, 2, \ldots, n$, set

$$\Pi_k = \Theta_k B_k + \Delta_k S_k. \tag{11.2.1}$$

This portfolio value is computed *before* rebalancing takes place. In particular, $\Pi_n = \Theta_n B_n + \Delta_n S_n$ is the value of the portfolio at maturity, a time where no rebalancing is needed.

At time k, based upon asset prices B_k and S_k, we will choose $(\Theta_{k+1}, \Delta_{k+1})$ for the upcoming period, i.e. the $(k+1)$-th period (from time k to time $k+1$). This is the rebalancing procedure.

To complete the definition of the *portfolio value process*, started in equation (11.2.1), we define the portfolio initial value as follows:

$$\Pi_0 = \Theta_1 B_0 + \Delta_1 S_0.$$

It is the value of the portfolio at time 0 (inception).

The notation for trading strategies and its corresponding portfolio value is complicated by the fact that the portfolio quantities Θ_k and Δ_k are set up at time $k-1$ and remain unchanged during the whole k-th period. They affect the portfolio value at both time $k-1$ (after rebalancing) and time k (before rebalancing).

Moreover, we can see that the portfolio value at time k, that is Π_k, depends on the risky asset price, so it is also a random variable. Therefore,

$$\Pi = \{\Pi_k, k = 0, 1, \ldots n\},$$

is a stochastic process and it is often called the **portfolio value process**.

Example 11.2.2 *Portfolio value process of the 60–40 portfolio*
We continue example 11.2.1 where we had $S_0 = B_0 = 100$. Recall that to hold 60% in stocks and 40% in bonds, we need to set $\Delta_1 = 60$ and $\Theta_1 = 40$ initially, so that

$$\Pi_0 = \Theta_1 B_0 + \Delta_1 S_0 = 40 \times 100 + 60 \times 100 = 10000.$$

One period later, i.e. at time 1, in the scenario that $S_1 = 150$ and $B_1 = 110$, the portfolio would be worth

$$\Pi_1 = \Theta_1 B_1 + \Delta_1 S_1 = 40 \times 150 + 60 \times 110 = 13400$$

before any rebalancing takes place.

To maintain the 60–40 proportion (of the new portfolio value of 13,400), we need to update our portfolio quantities. Based upon the information available at time 1 (in the given scenario), we set $\Delta_2 = 53.6$ and $\Theta_2 = 48.73$ for the second period (from time 1 to time 2).

Finally, at time 2, the portfolio value will be given by

$$\Pi_2 = \Theta_2 B_2 + \Delta_2 S_2 = 48.73 \times B_2 + 53.6 \times S_2,$$

which depends on the realization of the random variable S_2. ■

11.2.3 Self-financing strategies

A trading strategy (Θ, Δ) is said to be **self-financing** if, when it is rebalanced, no money is injected or withdrawn: the portfolio value, before and after rebalancing, is equal. Mathematically, this means that, at each time step $k = 1, 2, \dots, n-1$,

$$\Theta_k B_k + \Delta_k S_k = \Theta_{k+1} B_k + \Delta_{k+1} S_k \tag{11.2.2}$$

where the left-hand side is in fact Π_k as given in equation (11.2.1). Thus, a self-financing strategy does not generate any cash flows (in or out) between inception and maturity. Only at maturity or inception does it require/allow money to be injected or withdrawn from it, making it comparable to the cash flows of most derivatives (premium at inception, payoff at maturity, no other cash flow). Note that equation (11.2.2) is an equality of random variables.

The profit or loss during the k-th period, for any investment portfolio, is given by

$$\Pi_k - \Pi_{k-1} = (\Theta_k B_k + \Delta_k S_k) - (\Theta_{k-1} B_{k-1} + \Delta_{k-1} S_{k-1}).$$

For a self-financing strategy, using the self-financing condition of equation (11.2.2), we get

$$\Pi_k - \Pi_{k-1} = \Theta_k (B_k - B_{k-1}) + \Delta_k (S_k - S_{k-1}). \tag{11.2.3}$$

This provides another interpretation of the self-financing condition: the variation in the portfolio value comes from changes in values from both assets within the portfolio. This interpretation is the one to have in mind when we consider continuous-time models.

Example 11.2.3 *Self-financing condition in the 60–40 portfolio*
We continue example 11.2.2 and want to verify that the investment strategy satisfies the self-financing condition at time 1, at least in the given scenario.

Indeed, at time 1, in the given scenario, we have

$$\Pi_1 - \Pi_0 = 13400 - 10000 = 3400$$

and

$$\begin{aligned}&\Theta_1 (B_1 - B_0) + \Delta_1 (S_1 - S_0)\\ &= 40 \times (110 - 100) + 60 \times (150 - 100) = 40 \times 10 + 60 \times 50 = 3400.\end{aligned}$$

So, the self-financing condition is satisfied because the gain/loss of the portfolio corresponds to the sum of the gains/losses made on each asset. Alternatively, still in the given scenario, we have

$$\Theta_1 B_1 + \Delta_1 S_1 = \Theta_2 B_1 + \Delta_2 S_1$$
$$40 \times 110 + 60 \times 150 = 53.6 \times 110 + 48.73 \times 150 = 13400.$$

In other words, before and after rebalancing, the portfolio is worth the same. ■

11.2.4 Replicating strategy

The main objective now is to determine the initial value V_0 of a derivative having final payoff V_n. The key idea is that we need to set up a (self-financing) trading strategy such that

$$\Pi_n = V_n$$

in *each possible scenario*. Moreover, if the strategy is self-financing, which means it has no cash flows except at time 0 and time n (just like the derivative we want to replicate), the strategy is called a **replicating strategy** for this derivative.

By the no-arbitrage assumption, we can conclude that the portfolio process Π mimics the price of the derivative at each time point and in each possible scenario. Consequently, we have

$$V_k = \Pi_k$$

for all $k = 0, 1, \ldots, n$. In particular, the initial values are equal: $V_0 = \Pi_0$. The price of the derivative is equal to the cost of the replicating portfolio.

11.2.5 Backward recursive procedure

Recall that in what follows, we consider only simple options. Using the labeling of nodes presented in section 11.1.4, let

$$\Theta_k(j) \quad \text{and} \quad \Delta_k(j)$$

be the realized number of units of each asset an investor holds during the k-th period and in the scenario corresponding to the one-period sub-tree with root $(k-1, j)$. In other words, if at time $k-1$ we have observed so far j upward movements, then during the next period (the k-th), we will hold $\Theta_k(j)$ units of the risk-free asset and $\Delta_k(j)$ units of the risky asset. Then, the time-$(k-1)$ value of the corresponding portfolio in this scenario is

$$\Pi_{k-1}(j) = \Theta_k(j) B_{k-1} + \Delta_k(j) S_{k-1}(j). \tag{11.2.4}$$

Graphically, we have

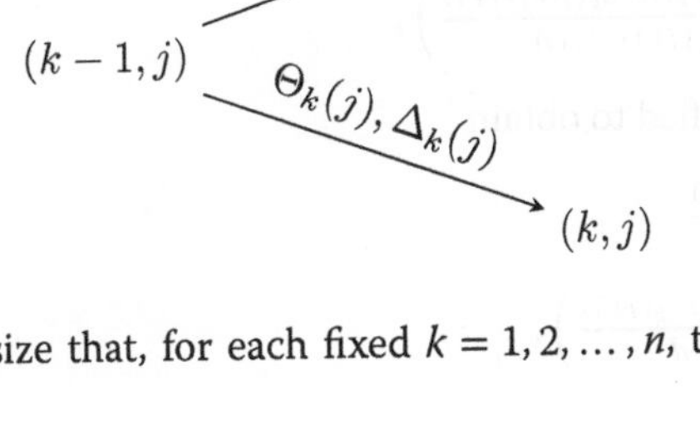

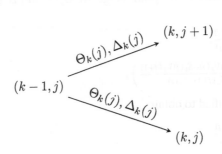

In this case, we emphasize that, for each fixed $k = 1, 2, \ldots, n$, the possible values for j are $0, 1, \ldots, k-1$.

For $k = 0$, the only possible value for j is of course 0, which is in agreement with the fact that Θ_1 and Δ_1 are deterministic values, i.e. are fixed at time 0.

Example 11.2.4 *Relating the notation from Chapters 9 and 10*
In the one-step binomial tree, the investment strategy set up at inception to replicate the realizations v^u and v^d of a payoff V_1 was the pair (x, y). Using this chapter's notation, we thus have

$$\Theta_1 = y \quad \text{and} \quad \Delta_1 = x.$$

In the two-step binomial tree, the quantities (x^d, y^d) were chosen at time 1, in the scenario $S_1 = s^d$, to replicate the values v^{du} and v^{dd}, while the quantities (x^u, y^u) were chosen, in the scenario $S_1 = s^u$, to replicate v^{uu} and v^{ud}. With the new notation, in the scenario $S_1 = s^d$ (i.e. no up-move so far), we have

$$\Theta_2(0) = y^d \quad \text{and} \quad \Delta_2(0) = x^d$$

and, in the scenario $S_1 = s^u$ (i.e. one up-move so far), we have

$$\Theta_2(1) = y^u \quad \text{and} \quad \Delta_2(1) = x^u.$$ ■

The one-period sub-tree with root at node $(k-1, j)$ is given by:

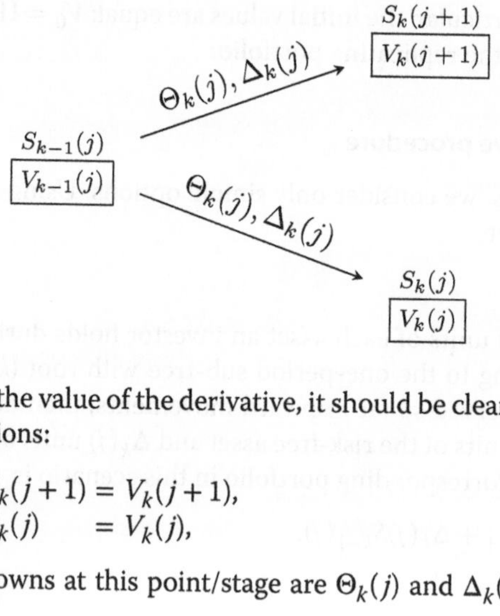

In this tree, to replicate the value of the derivative, it should be clear that we need to solve the following system of equations:

$$\begin{cases} \Theta_k(j)B_k + \Delta_k(j)S_k(j+1) = V_k(j+1), \\ \Theta_k(j)B_k + \Delta_k(j)S_k(j) \quad\;\; = V_k(j), \end{cases}$$

where the only two unknowns at this point/stage are $\Theta_k(j)$ and $\Delta_k(j)$. As in Chapter 10, the solution is

$$\begin{cases} \Delta_k(j) = \frac{V_k(j+1)-V_k(j)}{S_k(j+1)-S_k(j)}, \\ \Theta_k(j) = \frac{1}{B_k}\left(\frac{S_k(j+1)V_k(j)-S_k(j)V_k(j+1)}{S_k(j+1)-S_k(j)}\right), \end{cases}$$

which can be further simplified to obtain

$$\begin{cases} \Delta_k(j) = \frac{V_k(j+1)-V_k(j)}{(u-d)S_{k-1}(j)}, \\ \Theta_k(j) = \frac{1}{B_k}\left(\frac{uV_k(j)-dV_k(j+1)}{u-d}\right). \end{cases} \tag{11.2.5}$$

Again, to avoid arbitrage opportunities, we must have

$$V_{k-1}(j) = \Pi_{k-1}(j), \tag{11.2.6}$$

where $\Pi_{k-1}(j)$ is given in equation (11.2.4).

The full n-period binomial tree and replicating strategy for payoff V_n is obtained by pasting together all the one-period sub-trees, i.e. for each $k = 1, \dots, n$ and $j = 0, 1, \dots, k-1$.

11.2.6 Algorithm

Replication and valuation in the general n-step binomial tree are summarized by the following algorithm:

1) At maturity (time n), in each possible scenario, we must:
   a) compute the payoff for each $j = 0, 1, \dots, n$

   $$V_n(j) = g(S_n(j)).$$

   b) compute the corresponding (replicating) portfolio quantities

   $$\Theta_n(j) \quad \text{and} \quad \Delta_n(j),$$

   as given in (11.2.5) for $j = 0, 1, \dots, n-1$.
   c) compute the value of the replicating portfolio (at time $n-1$):

   $$\Pi_{n-1}(j) := \Theta_n(j)B_{n-1} + \Delta_n(j)S_{n-1}(j).$$

   d) compute the value of the derivative:

   $$V_{n-1}(j) := \Pi_{n-1}(j).$$

2) Then, we work backward and recursively: for each time $k = n-1, n-2, \dots, 1$, and in each possible scenario, i.e. for each $j = 0, 1, ..., k$, we must:
   a) from the previous step, retrieve the derivative's value $V_k(j)$;
   b) compute the corresponding (replicating) portfolio quantities: for each $j = 0, 1, \dots, k-1$, compute

   $$\Theta_k(j) \quad \text{and} \quad \Delta_k(j),$$

   as given in (11.2.5).
   c) compute the value of the replicating portfolio (at time $k-1$):

   $$\Pi_{k-1}(j) = \Theta_k(j)B_{k-1} + \Delta_k(j)S_{k-1}(j);$$

   d) compute the value of the derivative:

   $$V_{k-1}(j) = \Pi_{k-1}(j).$$

Note that in the very last step, i.e. for $k = 1$, we compute $(\Theta_1(0), \Delta_1(0))$ and then set $V_0(0) = \Pi_0(0)$ or, written as before, $V_0 = \Pi_0$.

Example 11.2.5 *Pricing a put option*
Consider a European put option on a stock with spot price 50, maturity 2 years and exercise price 52. Assume the stock price evolves according to a two-period binomial tree, where each time step corresponds to 1 year and during which the price increases or decreases by 20%. In this model, the continuously compounded annual interest rate is 5%. Let us find the replicating portfolio of this option.

We have $T = 2$, $K = 52$, $S_0 = 50$, $u = 1.2$, $d = 0.8$, $n = 2$, $h = 1$ and $r = 0.05$. We have graphically:

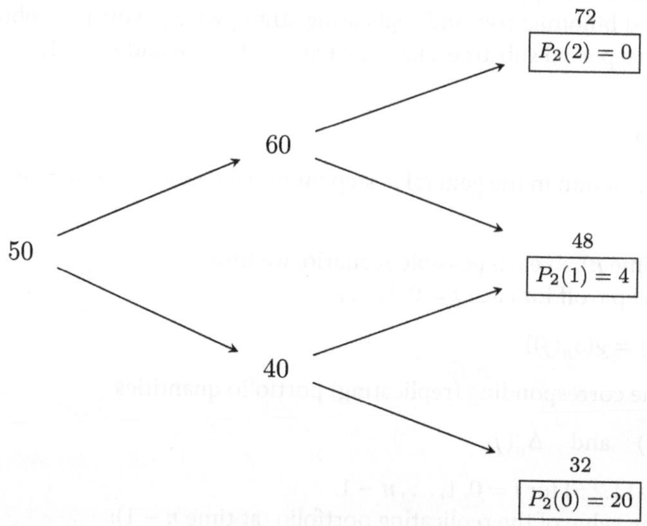

Using the expressions in (11.2.5), we find the replicating strategies for the second period: in the sub-tree with root $(1, 0)$, i.e. for $S_1 = 40$,

$$\Theta_2(0) = 52\mathrm{e}^{-0.1} \quad \text{and} \quad \Delta_2(0) = -1,$$

whereas, in the sub-tree with root $(1, 1)$, i.e. for $S_1 = 60$,

$$\Theta_2(1) = 12\mathrm{e}^{-0.1} \quad \text{and} \quad \Delta_2(1) = -\frac{1}{6}.$$

So, the option price after 1 period (at time $k = 1$) is given by

$$P_1(0) = 52\mathrm{e}^{-0.1} \times \mathrm{e}^{0.05} - 1 \times 40 = 9.46393,$$
$$P_1(1) = 12\mathrm{e}^{-0.1} \times \mathrm{e}^{0.05} - \frac{1}{6} \times 60 = 1.41475.$$

During the first period, the portfolio quantities are given by

$$\Delta_1(0) = \frac{P_1(1) - P_1(0)}{S_0(0)(u - d)} = -0.4024588,$$

$$\Theta_1(0) = \mathrm{e}^{-r}\left(\frac{uP_1(0) - dP_2(1)}{u - d}\right) = 24.315597.$$

In conclusion, the initial price of the option is

$$P_0 = 24.315597 \times 1 - 0.4024588 \times 50 = 4.192656731,$$

and more generally we have obtained

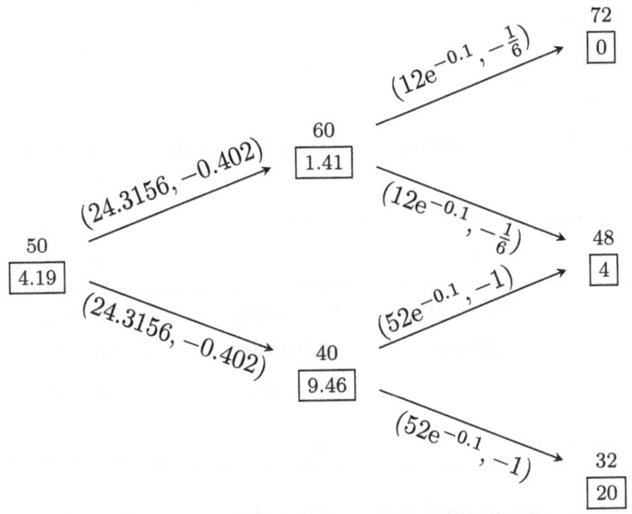

■

Example 11.2.6 *Replicating a GMMB in a binomial tree*
A GMMB was issued by your insurance company several years ago. It has an annual premium of 2% with a guaranteed minimum payout of 102. As of today, the sub-account balance is \$98.74 and the guarantee will apply 3 years from now. Given that the reference stock actually trades for 50 and increases by 7% or decreases by 5% every year, let us describe the replicating strategy covering the loss on the guarantee. Assume the risk-free rate is 3% (annually compounded) and use a binomial tree with annual time steps.

As we did in Chapter 10, replicating the loss on a variable annuity is similar to replicating options. Recall that the payoff is computed on the sub-account (after premiums and withdrawals), whereas replication is performed with the reference asset. Therefore, we will first build the binomial tree for the reference stock index and for the sub-account value. The tree for the reference stock is constructed using

$$S_k(j) = u^j d^{k-j} S_0,$$

for each $j = 0, 1, 2, 3$, with $S_0 = 50$, $u = 1.07$ and $d = 0.95$. The evolution of the reference asset is shown in the next table:

j/k	0	1	2	3
0	50	47.5	45.125	42.86875
1		53.5	50.825	48.28375
2			57.245	54.38275
3				61.25215

Then, the sub-account balance of the policyholder gets credited with the returns of the reference asset whereas premiums are withdrawn. Recursively, as we saw in Chapter 8, we have

$$A_k = A_{k-1} \times \frac{S_k}{S_{k-1}}(1 - \alpha)$$

which is equivalent to

$$A_k = A_0 \times \frac{S_k}{S_0}(1-\alpha)^k,$$

knowing that $A_0 = 98.74$. Therefore, the values $A_k(j)$ of the sub-account balance are:

j/k	0	1	2	3	Loss
0	98.74	91.92694	85.5839811	79.6786864	22.3213136
1		103.538764	96.3945893	89.7433626	12.2566374
2			108.570748	101.079366	0.92063368
3				113.847286	0

The minimum guaranteed amount is $G = 102$ and hence the loss for the insurer is computed in the last column as $\max(102 - A_3, 0)$. Therefore, the insurance company must implement an investment strategy to replicate the payoff values $\{22.3213136, 12.2566374, 0.92063368, 0\}$ in the final nodes of the tree. Indeed, when the sub-account balance is \$79.68 upon maturity, the company needs to provide for an extra \$22.32 in order for the policyholder to receive at least 102.

We will illustrate how to replicate in a single sub-tree. The whole replicating strategy and its cost are provided in the tables below.

For example, the one-period sub-tree with root $(2, 0)$, i.e. at time 2 with no up-move so far, is given by

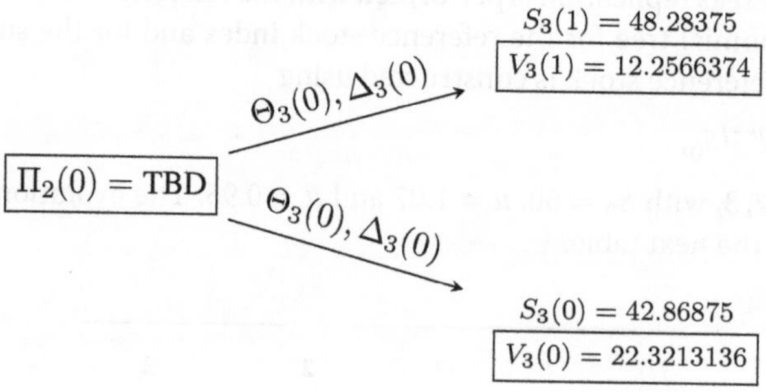

Using the formulas in (11.2.5), we get

$$\Delta_3(0) = \frac{V_3(1) - V_3(0)}{S_3(1) - S_3(0)} = \frac{12.2566374 - 22.3213136}{48.28375 - 42.86875} = -1.858665962$$

and

$$\begin{aligned}\Theta_3(0) &= \frac{1}{B_3}\left(\frac{S_3(1)V_3(0) - S_3(0)V_3(1)}{S_3(1) - S_3(0)}\right)\\ &= \frac{1}{1.03^3}\left(\frac{48.28375 \times 22.3213136 - 42.86875 \times 12.2566374}{48.28375 - 42.86875}\right)\\ &= 93.34444925.\end{aligned}$$

Hence,

$$\begin{aligned}\Pi_2(0) &= \Theta_3(0)B_2 + \Delta_3(0)S_2(0)\\ &= 93.34444925 \times 1.03^2 - 1.858665962 \times 45.125\\ &= 15.1568247.\end{aligned}$$

In order to cover a random loss of either \$22.32 or \$12.26, the insurance company must have \$15.16 to short-sell 1.86 shares of stock and invest 93.34 at the risk-free rate.

Repeating the same procedure for each sub-tree, we can determine the replicating strategy. This is given in the next table.

	$\Delta_k(j)$			$\Theta_k(j)$		
j/k	1	2	3	1	2	3
0	−1.0314657	−1.85866596	−1.85866596	55.1968642	93.3444493	93.3444493
1		−0.66425063	−1.85866596		36.1230717	93.3444493
2			−0.13401952			7.51238289

Finally, the realizations $\Pi_k(j)$ of the replicating portfolio value process are given in the last table. Note that this also corresponds to the no-arbitrage values of the insurer's loss.

j/k	0	1	2	3
0	3.62357939	7.85814956	15.1568247	22.3213136
1		1.66935538	4.56242872	12.2566374
2			0.2979397	0.92063368
3				0

■

11.3 Pricing with risk-neutral probabilities

As in the one- and two-period models, we can reorganize equation (11.2.6) (combined with the positions of equation (11.2.5)) in order to write the option price in node $(k-1, j)$ as a linear combination of its values one step ahead: for all $k = 0, 1, \dots, n-1$ and $j = 0, 1, \dots, k$,

$$V_k(j) = \frac{B_k}{B_{k+1}}(qV_{k+1}(j+1) + (1-q)V_{k+1}(j)), \tag{11.3.1}$$

where

$$\frac{B_k}{B_{k+1}} = \begin{cases} e^{-rh} & \text{if the risk-free rate is continuously compounded,}\\ \frac{1}{1+rh} & \text{if the risk-free rate is periodically compounded}\end{cases}$$

is the one-period discount factor (applying from time k to time $k-1$) and

$$q = \begin{cases} \frac{e^{rh}-d}{u-d} & \text{if the rate is continuously compounded,} \\ \frac{1+rh-d}{u-d} & \text{if the rate is periodically compounded.} \end{cases}$$

Since the model is arbitrage-free (cf. (11.1.1)), then $0 < q < 1$ and we can interpret q as a (conditional) risk-neutral probability. Again, the expression in (11.3.1) can be interpreted as a risk-neutral expectation, i.e. using risk-neutral weights. Specifically, for each $k = 0, 1, \dots, n-1$, we can write

$$\frac{V_k}{B_k} = \mathbb{E}^{\mathbb{Q}}\left[\left.\frac{V_{k+1}}{B_{k+1}}\right| S_1, \dots, S_k\right]$$

and, using iterated expectations, we further get

$$V_0 = \mathbb{E}^{\mathbb{Q}}\left[\frac{V_n}{B_n}\right]. \tag{11.3.2}$$

As in Chapters 9 and 10, it is important to recall that the (conditional) risk-neutral probability q has nothing to do with the (conditional) real-world probability p: it is only a convenient way to rewrite the value of the replicating portfolio as an expectation.

In conclusion, valuation in the general n-step binomial tree using risk-neutral formulas is summarized by the following algorithm:

1) At time n, calculate the payoff of the derivative $V_n(j) = g(S_n(j))$ for each $j = 0, 1, \dots, n$.
2) Then, proceed recursively, from time $k = n-1$ back to time $k = 0$: compute $V_k(j)$ using equation (11.3.1) for each $j = 0, 1, 2, \dots, k$.

Example 11.3.1 *Pricing a put option with risk-neutral formulas*
Let us have a second look at example 11.2.5, now using risk-neutral formulas. We start by computing the risk-neutral probability of an up-move:

$$q = \frac{e^r - d}{u-d} = \frac{e^{0.05} - 0.8}{1.2 - 0.8} = 0.628177741.$$

The values of the payoff are

$$P_2(0) = 20, \quad P_2(1) = 4 \quad \text{and} \quad P_2(2) = 0.$$

Then, for $(k,j) = (1,0)$ and $(k,j) = (1,1)$, and then for $(k,j) = (0,0)$, using the relationship

$$P_k(j) = e^{-r}(qP_{k+1}(j+1) + (1-q)P_{k+1}(j)),$$

we obtain

$$P_1(0) = 9.463930074 \quad \text{and} \quad P_1(1) = 1.414753094,$$

and finally

$$P_0 = 4.192654281.$$

■

We already know that the random variable I_k follows a binomial distribution with parameters (k,p) for each $k = 1, 2, \dots, n$. This is denoted as $I_k \overset{\mathbb{P}}{\sim} \text{Bin}(k,p)$ where we use $\overset{\mathbb{P}}{\sim}$ to emphasize that each up-move has a (conditional) probability p. The probability *measure* $\mathbb{P}$ is thus known as the real-world or actuarial probability *measure*.

But if instead each up-move had a (conditional) probability q, then we would have $I_k \overset{\mathbb{Q}}{\sim} \text{Bin}(k, q)$ where $\overset{\mathbb{Q}}{\sim}$ emphasizes the use of q instead of p. The probability *measure* $\mathbb{Q}$ is known as the risk-neutral probability *measure*. It results that, for $j = 0, 1, 2, \ldots, k$,

$$\mathbb{Q}(S_k = u^j d^{k-j} S_0) = \mathbb{Q}(I_k = j) = \binom{k}{j} q^j (1-q)^{k-j}. \tag{11.3.3}$$

Combining equation (11.3.2) with equation (11.3.3), we deduce that the initial price of a simple option with payoff $V_n = g(S_n)$ can be written as a risk-neutral binomial sum:

$$\begin{aligned} V_0 &= \frac{1}{B_n} \mathbb{E}^{\mathbb{Q}}[g(S_n)] \\ &= \frac{1}{B_n} \sum_{j=0}^{n} g(u^j d^{n-j} S_0) \mathbb{Q}(S_n = u^j d^{n-j} S_0) \\ &= \frac{1}{B_n} \sum_{j=0}^{n} g(u^j d^{n-j} S_0) \binom{n}{j} q^j (1-q)^{n-j}. \end{aligned} \tag{11.3.4}$$

Example 11.3.2 *Pricing a put option with risk-neutral formulas (continued)*
Let us use equation (11.3.4) to simplify the computations of example 11.3.1. We know that $q = 0.628177741$ and hence

$$\begin{aligned} P_0 &= \mathrm{e}^{-2r} \sum_{j=0}^{2} (52 - u^j d^{2-j} S_0)_+ \binom{2}{j} q^j (1-q)^{2-j} \\ &= \mathrm{e}^{-2r} (0 \times q^2 + 4 \times 2q(1-q) + 20(1-q)^2) \\ &= 4.192654281. \end{aligned}$$

■

Example 11.3.3 *Replicating a GMMB in a binomial tree (continued)*
We illustrate how to use risk-neutral formulas in example 11.2.6. Given the equivalence between replicating portfolios and risk-neutral formulas, we can easily combine the two approaches to determine the replicating strategy the actuary should implement during the first time period.

First, we compute the risk-neutral probability of an up-move for the reference stock. We find

$$q = \frac{1.03 - 0.95}{1.07 - 0.95} = \frac{2}{3}.$$

Second, the replicating strategy that applies in the first period requires to find the no-arbitrage price of the GMMB at time 1, i.e. corresponding to nodes (1, 0) and (1, 1). Using risk-neutral formulas, we get

$$\begin{aligned} V_1(0) &= \frac{1}{1.03^2} (V_3(0) \times (1-q)^2 + V_3(1) \times 2q(1-q) + V_3(2) q^2) \\ &= 7.858149556 \end{aligned}$$

and

$$\begin{aligned} V_1(1) &= \frac{1}{1.03^2} (V_3(1) \times (1-q)^2 + V_3(2) \times 2q(1-q) + V_3(3) q^2) \\ &= 1.669355378. \end{aligned}$$

Hence, initially the actuary must hold

$$\Delta_1 = \frac{1.669355378 - 7.858149556}{53.50 - 47.50} = -1.031465696$$

and

$$\Theta_1 = \frac{1}{1.03} \times \frac{7.858149556 \times 53.50 - 1.669355378 \times 47.50}{53.50 - 47.50} = 55.1968642,$$

as already obtained in example 11.2.6. ■

Applying equation (11.3.4) to a call option with payoff $(S_n - K)_+$, we find that the initial price can be written as

$$C_0 = \frac{1}{B_n} \sum_{j=0}^{n} (u^j d^{n-j} S_0 - K)_+ \binom{n}{j} q^j (1-q)^{n-j}.$$

However, in many scenarios, the payoff is equal to zero. Indeed, when the realization of S_n is less than K, the corresponding term in the summation is equal to zero. Therefore, if we define

$$k_n^* = \min\{i \geq 1 : u^i d^{n-i} S_0 > K\}$$

as the smallest i such that the payoff is strictly positive, then we can get rid of the *positive part* and write[2]

$$C_0 = \frac{1}{B_n} \sum_{j=k_n^*}^{n} (u^j d^{n-j} S_0 - K) \binom{n}{j} q^j (1-q)^{n-j}. \tag{11.3.5}$$

We can do similar computations for a put option. The next example shows a numerical application to a 12-step binomial tree.

Example 11.3.4 *Pricing a 1-year call option*
Assume the 1-year rate is 5% (continually compounded) and the evolution of a stock price is represented by a 12-period binomial tree: every month, the stock price can increase by 1% or decrease by 0.7%. If the current stock price is 52, let us compute the no-arbitrage price of a 1-year call option with strike 54.

Although the problem seems daunting, the call option is a vanilla option whose price will be easy to compute with equation (11.3.5). We start by computing the risk-neutral (conditional) probability of an up-move:

$$q = \frac{e^{rh} - d}{u - d} = \frac{e^{0.05/12} - 0.993}{1.01 - 0.993} = 0.657374076.$$

We compute the payoff in the *extremal upper nodes* when the price of the underlying is *big enough*, in which case the option will mature in the money. The possible outcomes

2 The result is also useful in the derivation of the Black-Scholes formula in Chapter 16.

for the payoff and the corresponding risk-neutral probabilities are given in the following table:

j	$C_{12}(j)$	$\mathbb{Q}(I_{12} = j)$
12	4.59490157	0.00651255
11	3.60865075	0.04073238
10	2.63900019	0.11676431
9	1.68567049	0.20286004
8	0.74838692	0.23789574
≤7	0	irrelevant

Consequently,

$$\begin{aligned} C_0 &= \frac{1}{B_n} \sum_{j=0}^{n} (u^j d^{n-j} S_0 - 54)_+ \mathbb{Q}(S_n = u^j d^{n-j} S_0) \\ &= e^{-0.05}(4.59490157 \times 0.00651255 \\ &\qquad + 3.60865075 \times 0.04073238 + \cdots + 0.74838692 \times 0.23789574) \\ &= 0.956030955. \end{aligned}$$

■

A risk-neutral probability measure

The probability measure $\mathbb{Q}$ is an example of what is called a risk-neutral probability measure. It is such that:

1) $\mathbb{Q}$ is *equivalent* to $\mathbb{P}$, which means here that $0 < q < 1$;
2) the stochastic process $\{S_k/B_k, k = 0, 1, 2, \ldots, n\}$ is such that:

$$\frac{S_k}{B_k} = \mathbb{E}^{\mathbb{Q}} \left[\left. \frac{S_{k+1}}{B_{k+1}} \right| S_1, S_2, \ldots, S_k \right],$$

for all $k = 0, 1, \ldots, n - 1$.

The second condition says that the process given by S_k/B_k is a $\mathbb{Q}$-martingale.

11.4 Summary

Multi-period binomial tree model

- Risk-free interest rate: r.
- Risk-free asset price: $B = \{B_k, k = 0, 1, 2, \ldots, n\}$, where $B_0 = 1$ and, for $k = 1, 2, \ldots, n$,

$$B_k = \begin{cases} B_{k-1} e^r & \text{if the risk-free rate is continuously compounded,} \\ B_{k-1}(1 + r) & \text{if the risk-free rate is periodically compounded.} \end{cases}$$

- Risky asset price: $S = \{S_k, k = 0, 1, 2, \ldots, n-1, n\}$, where S_0 is a constant and, for $k = 1, 2, \ldots, n$,

$$S_k = S_{k-1}U_k = S_0 U_1 U_2 \ldots U_k,$$

where the U_js are iid random variables taking values in $\{u, d\}$ with (real-world) probabilities $\{p, 1-p\}$. Consequently, for each $k = 1, 2, \ldots, n$,

$$\mathbb{P}(S_k = u^i d^{k-i} S_0) = \binom{k}{i} p^i (1-p)^{k-i}, \quad i = 0, 1, 2, \ldots, k.$$

- No-arbitrage condition:

$$\begin{cases} d < e^{rh} < u & \text{when interest is compounded continuously,} \\ d < 1 + rh < u & \text{when interest is compounded at each period.} \end{cases}$$

- Derivative (written on S): $V = \{V_k, k = 0, 1, 2, \ldots, n-1, n\}$, where V_0 is a constant (to be determined) and where $V_n = g(S_n)$ (simple payoff) or $V_n = g(S_1, S_2, \ldots S_n)$ (exotic path-dependent payoff).
- Important characteristics of the multi-period model:
  - at time k: there are $k+1$ possible stock prices but there are 2^k different paths leading to those prices;
  - it contains $\frac{n(n+1)}{2}$ one-period binomial sub-trees.
- Labelling the nodes:
  - the pair (k, j) corresponds to the node at time k reached by a total of j upward movements (after k steps), where $k = 0, 1, 2, \ldots, n$ and $j = 0, 1, \ldots, k$;
  - $S_k(j) = u^j d^{k-j} S_0$: realization of S_k in node (k, j);
  - for a simple payoff $V_n = g(S_n)$ with realization $V_n(j) = g(S_n(j)) = g(u^j d^{n-j} S_0)$ in node (n, j), the realizations of the time-k derivative prices V_k are given by $V_k(j)$ in node (k, j);
  - for exotic path-dependent payoffs: we cannot use this notation.

Trading strategy/portfolio

- Portfolio: a pair (Θ, Δ), where $\Theta = \{\Theta_k, k = 1, 2, \ldots, n\}$ and $\Delta = \{\Delta_k, k = 1, 2, \ldots, n\}$ are such that:
  - number of units of B held from time $k-1$ to time k (k-th period): Θ_k;
  - number of units of S held from time $k-1$ to time k (k-th period): Δ_k.
- Number of units of the risk-free asset set up in node $(k-1, j)$: $\Theta_k(j)$.
- Number of units of the risky asset set up in node $(k-1, j)$: $\Delta_k(j)$.
- Portfolio value process: $\Pi = \{\Pi_k, k = 0, 1, \ldots n\}$, where

$$\Pi_0 = \Theta_1 B_0 + \Delta_1 S_0$$

and where, for $k = 1, 2, \ldots, n$,

$$\Pi_k = \Theta_k B_k + \Delta_k S_k.$$

- Self-financing condition: (Θ, Δ) is self-financing if, at each time step $k = 1, 2, \ldots, n-1$,

$$\Theta_k B_k + \Delta_k S_k = \Theta_{k+1} B_k + \Delta_{k+1} S_k$$

or equivalently

$$\Pi_k - \Pi_{k-1} = \Theta_k (B_k - B_{k-1}) + \Delta_k (S_k - S_{k-1}).$$

Pricing by replication

- A self-financing trading strategy (Θ, Δ) is a replicating strategy for V if

$$\Pi_n = V_n.$$

Consequently, $V_k = \Pi_k$, for all $k = 0, 1, \ldots, n$.

- Replicating portfolio for V_n: for each sub-tree with root (k, j), we solve

$$\begin{cases} \Theta_k(j)B_k + \Delta_k(j)S_k(j+1) &= V_k(j+1), \\ \Theta_k(j)B_k + \Delta_k(j)S_k(j) &= V_k(j), \end{cases}$$

and find

$$\begin{cases} \Delta_k(j) = \frac{V_k(j+1)-V_k(j)}{(u-d)S_{k-1}(j)}, \\ \Theta_k(j) = \frac{1}{B_k}\left(\frac{uV_k(j)-dV_k(j+1)}{u-d}\right). \end{cases}$$

Pricing with risk-neutral probabilities

- Risk-neutral probability: $\mathbb{Q} = (q, 1-q)$, where

$$q = \begin{cases} \frac{e^{rh}-d}{u-d} & \text{if the risk-free rate is continuously compounded,} \\ \frac{1+rh-d}{u-d} & \text{if the risk-free rate is periodically compounded.} \end{cases}$$

- Risk-neutral pricing formula for a derivative with payoff V_n: for each $k = 0, 1, \ldots, n-1$,

$$\frac{V_k}{B_k} = \mathbb{E}^{\mathbb{Q}}\left[\frac{V_{k+1}}{B_{k+1}} \middle| S_1, \ldots, S_k\right]$$

or, written differently,

$$V_k(j) = \frac{B_k}{B_{k+1}}(qV_{k+1}(j+1) + (1-q)V_{k+1}(j)),$$

for all $j = 0, 1, \ldots, k$.

- Initial price: $V_0 = \mathbb{E}^{\mathbb{Q}}\left[\frac{V_n}{B_n}\right]$.
- Risk-neutral probabilities are only a mathematical convenience used for no-arbitrage pricing.

11.5 Exercises

11.1 In a four-period binomial tree, the stock price can increase by 3% or decrease by 1%. If the initial stock price is \$100 and analysts have determined that the probability that the stock price increases is 72%, compute:

(a) probability that the stock price is 108.18 after four periods (three ups, one down);

(b) probability of observing the path 100, 103, 101.97, 105.0291 and 108.18.

11.2 You model the evolution of stock prices using a three-period binomial tree. You are given:

- $u = 1.07$ and $d = 0.98$;
- $S_0 = 37$

Detail all possible scenarios/paths taken by the stock price over the next three periods. Then, compute the payoff in each possible scenario for the following derivatives:
(a) at-the-money call option;
(b) floating-strike lookback put option;
(c) fixed-strike Asian call option with strike \$35;
(d) up-and-out put option with $H = 41$ and strike \$38;
(e) asset-or-nothing call option with strike \$43.

11.3 In a 3-month CRR binomial tree, you are given:
- volatility of the stock price is 25%;
- $S_0 = 40$;
- risk-free rate is 3% (compounded continuously).

(a) Build the tree for the stock price where one time step corresponds to one month.
(b) What is the no-arbitrage price of an at-the-money 3-month put option?
(c) How many shares of stock do you need to buy/sell at time 0 to replicate the option price at time 1 (for the option in (b))?
(d) What is the no-arbitrage price of the option in (b) at time 2 given that the stock price went down twice?

11.4 A share of stock trades for \$45 and a 3-year zero-coupon bond can be bought for \$87 (face value of 100). The stock price evolves according to a three-period binomial tree with $u = 1.07$ and $d = 0.96$ ($h = 1$) and the interest rate is constant. You have \$1000 to invest.
(a) You buy 60% of your initial investment in stocks and 40% in bonds. You make no more adjustments to your portfolio over the course of time. Describe your strategy (Δ_k, Θ_k) mathematically and compute the portfolio value after 3 years.
(b) Suppose now you want to make sure that as the stock price evolves, you always hold 60% of the investment in stocks. Describe your strategy (Δ_k, Θ_k) mathematically and compute the portfolio value after 3 years.

11.5 In a 20-period binomial tree, you are given that:
- the risk-free asset is such that $B_0 = B_1 = B_2 = \cdots = B_{20} = 1$;
- the risky asset price can increase by 0.4% or decrease by 0.2% on each period;
- the initial stock price is 1000;
- one period corresponds to 1 day;
- according to analysts, the probability that the asset price goes up on any given day is 62%.

A 20-day at-the-money put option is issued. What is the no-arbitrage price of this option?

11.6 In a three-period binomial tree, you are given:
- $S_0 = 100$;
- $u = 1.04$ and $d = 0.98$;
- the risk-free asset is such that $B_k = 1.01^k, k = 0, 1, 2, 3$;
- one period is 4 months.

A 1-year floating-strike lookback call option is issued.
(a) There are eight possible paths for the stock price. Describe them all.
(b) Compute the payoff of this option on each of the possible paths. Note that the maximum/minimum is computed over S_1, S_2, S_3.

(c) You believe that over any given period, the probability that the stock price goes up is 65%. What is the expected payoff of the option?
(d) What is the risk-neutral probability of observing each path?
(e) What should be the no-arbitrage price of this option at time 0?
(f) What is the replicating strategy applying during the first period to ultimately replicate the option's payoff at maturity?

11.7 In the binomial tree of exercise 11.6, a down-and-out call option with strike \$102 and barrier \$95 is issued.
(a) Compute the payoff of the option on each of the possible paths.
(b) Using risk-neutral probabilities, compute the option's initial price.
(c) At each node, determine the replicating portfolio. What do you observe whenever the stock price is close to the barrier (over or below)?

11.8 The price of a stock can increase by 5% or decrease by 2%. A dividend of \$2.50 will be paid two periods from now and the initial stock price is \$30. You build a four-period binomial tree.
(a) Build the binomial tree for the ex-dividend stock price.
(b) Compute the payoff of a four-period 28-strike put option.
(c) Given that the stock price went down twice after two periods, compute the put option price at time 2 if the periodic interest rate is 1%.

11.9 In a three-period binomial tree, you are given:
- each period is 3 months;
- the initial stock price is \$57;
- the stock price can increase by a factor of $u = 1.03$ or decrease by a factor of $d = 0.99$;
- the 3-month interest rate observed at time 0 is 2% (annualized, compounded quarterly) ($r_0^{0.25} = 0.02$);
- future 3-month interest rate, i.e. $r_k^{0.25}, k = 1, 2$, is such that $r_k^{0.25} = r_{k-1}^{0.25} \pm 0.5\%$;
- when the interest rate goes up, the stock price goes down.

(a) Build the binomial tree for the stock, for the interest rate and for a simple bank account with $B_0 = 1$.
(b) Verify that the no-arbitrage condition holds in each sub-tree.
(c) Using replicating portfolios, compute the initial no-arbitrage price of an at-the-money call option.
(d) Using risk-neutral probabilities, verify the option price with the replicating portfolio value in each node of your work in (c).

11.10 Equity-indexed annuities (EIAs)
You are using the binomial tree of exercise 11.6.
(a) Compute the payoff a three-period compound annual ratchet EIA whose participation rate is 80%.
(b) Find the participation rate β such that the no-arbitrage value of the benefit is equal to the initial investment.

11.11 Guaranteed minimum maturity benefit (GMMB)
A 5-year GMMB is issued to a policyholder on a reference portfolio whose value can increase by 7% annually or decrease by 3% over the same period. The annual premium

is 1.75% whereas the initial investment is 100,000. Calculate the loss in each possible scenario if the guaranteed amount is 85% of the capital. Assume one period corresponds to a year.

11.12 Guaranteed minimum withdrawal benefit (GMWB)
There are 3 years remaining on a GMWB whose sub-account balance is currently \$26,000. The policyholder is allowed to withdraw 10,000 at the end of each year. If the reference index has a volatility of 30% and you are using a CRR binomial tree, calculate the loss in each possible scenario when the annual premium is 2.25%. Assume one period corresponds to a year.

12
Further topics in the binomial tree model

In Chapter 11, we presented the general binomial tree model. In that market model, we determined replicating portfolios and computed no-arbitrage prices for vanilla European options, from which we derived risk-neutral pricing formulas. As mentioned in Section 6.5, in practice, most traded options are American while many underlying assets pay a regular stream of income such as dividends.

The main objective of this chapter is to build upon the knowledge gained in Chapter 11 to extend the binomial tree to more realistic situations. The specific objectives are to:

- determine replicating portfolios;
- compute no-arbitrage prices;
- derive risk-neutral formulas,

for

- American-style options (especially American put options);
- options on stocks paying continuous dividends;
- currency options;
- futures options.

12.1 American options

Until now, in the binomial model, we have only considered European-style options. We will now analyze American-style options, which can be exercised at any time prior to maturity, and determine how an issuer can manage this additional risk when the dynamics of the underlying asset comes from a multi-period binomial tree.

12.1.1 American put options

Recall from Chapter 6 that it is never optimal to early-exercise an American call whereas it could be optimal to early-exercise an American put option, when both derivatives are written on a non-dividend-paying stock. These two conclusions being model-free, they hold in the binomial model.

One question remains: when should we exercise an American put option? Now that we have a model for the underlying asset, we can quantify the impact of early exercise on the price of a put option.

In what follows, let us denote the American put option price by $P = \{P_k, k = 0, 1, \dots, n\}$. Suppose that at time $n-1$ the put is still *alive*, i.e. it has not been exercised yet at the previous

Actuarial Finance: Derivatives, Quantitative Models and Risk Management, First Edition.
Mathieu Boudreault and Jean-François Renaud.

times $n-2, n-3, \ldots, 1, 0$. We want to determine whether a *rational investor*[1] should exercise the option or hold on to it for the last time period. To do so, we need to compare the values of those two possibilities: at time $n-1$,

- if it is exercised, the value is called the **early-exercise value** (intrinsic value) and it corresponds to what the option holder will receive from exercising early, which here means she will receive $K - S_{n-1}$;
- if it is not exercised immediately, the value is called the **holding value** or **continuation value** and it corresponds to the value of holding on to the American put option for one last time period, so there are no cash flows at that time.

At time $n-1$, the holding/continuation value is denoted by H_{n-1} and it is equal to the no-arbitrage price of a one-period European put option.

Of course, at time $n-1$, the investor will early-exercise the option only if she observes that $K - S_{n-1} > 0$. Furthermore, if acting rationally, she will early-exercise the option only if the exercise value is larger than the holding/continuation value, i.e. if $K - S_{n-1} > H_{n-1}$. Therefore, the value of an American put option at time $n-1$, which we denote here by P_{n-1}, is given by

$$P_{n-1} = \max(K - S_{n-1}, H_{n-1}). \tag{12.1.1}$$

At time $n-1$, any other value for this option would introduce an arbitrage opportunity.

A similar reasoning applies at each time point: the American put holder decides whether it is better to early-exercise or to hold on to the option for (at least) one more period. Mathematically, we have to compare, at each time $k = 0, 1, \ldots, n-1$:

- the early-exercise value (intrinsic value) $K - S_k$, that is what the option holder will receive from exercising early;
- the holding/continuation value given by H_k, that is the value of holding on to the American put option for (at least) an additional time period.

Again, a rational investor will only early-exercise the option at time k if the exercise value is larger than the continuation value, i.e. if $K - S_k > H_k$. Consequently, for each time $k = 0, 1, \ldots, n-1$, we have

$$P_k = \max(K - S_k, H_k). \tag{12.1.2}$$

Again, any other time-k value for this option would introduce an arbitrage opportunity.

As a result, the value of an American put option is always (in all scenarios) greater than or equal to its early-exercise value at any time during the life of the option. This is in agreement with the model-free relationships obtained in Chapter 6.

It should be noted that:

- the formula in equation (12.1.2) is valid in all possible scenarios, as it is an equality of random variables;
- the time-k continuation value H_k contains the value of potentially early exercising later (at times $k+1, k+2, \ldots, n-1$), so it is a positive quantity;
- if $K - S_k \leq 0$, then the option is out of the money at time k, so exercising is irrational and we have $P_k = H_k$.

All is left to do now is to compute $H = \{H_k, k = 0, 1, \ldots, n-1\}$, the **holding/continuation value process**. We must determine the value $H_k(j)$ for each node (k, j), which is equivalent to

1 A rational investor will want to maximize her gains/minimize her losses.

finding the no-arbitrage price of a derivative that will be worth either $P_{k+1}(j+1)$ or $P_{k+1}(j)$, depending if we observe an up-move or a down-move, over the next time period:

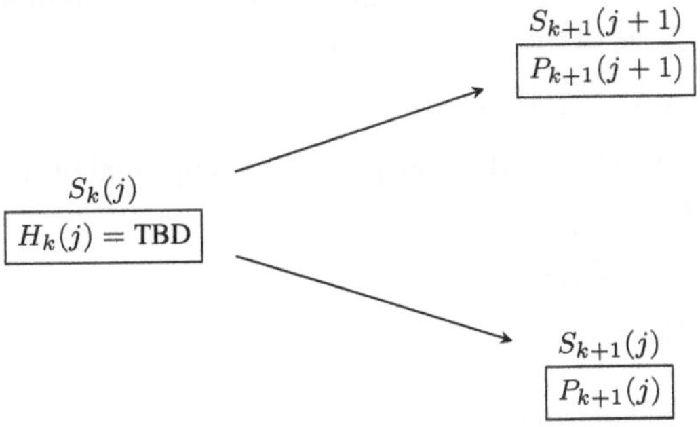

This is like pricing in a one-period binomial tree with $H_k(j)$ at the root, and $P_{k+1}(j+1)$ and $P_{k+1}(j)$ at the leaves. Using the risk-neutral formula of equation (11.3.1) in Chapter 11, we can write

$$H_k(j) = \frac{B_k}{B_{k+1}}(qP_{k+1}(j+1) + (1-q)P_{k+1}(j)), \tag{12.1.3}$$

where

$$\frac{B_k}{B_{k+1}} = \begin{cases} e^{-rh} & \text{if the risk-free rate is continuously compounded,} \\ \frac{1}{1+rh} & \text{if the risk-free rate is periodically compounded} \end{cases}$$

and

$$q = \begin{cases} \frac{e^{rh}-d}{u-d} & \text{if the rate is continuously compounded,} \\ \frac{1+rh-d}{u-d} & \text{if the rate is periodically compounded.} \end{cases}$$

Finally, now that we have a value for the American put at every time point, we can build a replicating portfolio just as in Chapter 11. More precisely, for each time $k = 1, 2, \dots, n$ and then for each $j = 0, 1, \dots, k-1$, we have

$$\begin{cases} \Theta_k(j) = \frac{P_k(j+1)-P_k(j)}{(u-d)S_{k-1}(j)}, \\ \Delta_k(j) = \frac{1}{B_k}\left(\frac{uP_k(j)-dP_k(j+1)}{u-d}\right). \end{cases} \tag{12.1.4}$$

Of course, these expressions are similar to those obtained in (11.2.5) of Chapter 11.

Again, to avoid arbitrage opportunities, we must have $P_{k-1}(j) = \Pi_{k-1}(j)$, where as before

$$\Pi_{k-1}(j) = \Theta_k(j)B_{k-1} + \Delta_k(j)S_{k-1}(j).$$

Backward recursive algorithm

The algorithm for pricing and replicating an American put option in an n-period binomial tree is as follows:

1) At maturity time n, in each possible scenario, i.e. for each $j = 0, 1, ..., n$, we set:

$$P_n(j) = (K - S_n(j))_+.$$

2) Then, we work backward and recursively: for each time $k = n-1, n-2, \dots, 1, 0$, and then in each possible scenario, i.e. for each $j = 0, 1, \dots, k$:
   (a) We compute the holding/continuation value $H_k(j)$ as given in Equation (12.1.3) above (or we use the replicating portfolio quantities for the corresponding one-period sub-tree).
   (b) We compute the value of the American put:
   $$P_k(j) = \max(K - S_k(j), H_k(j)),$$
   allowing us to determine whether early exercise is optimal or not in node (k, j).
   (c) If needed, we compute
   $$\Theta_{k+1}(j) \quad \text{and} \quad \Delta_{k+1}(j),$$
   as given by (12.1.4).

Example 12.1.1 *Pricing an American put option*
A stock whose current price is \$50 may increase by 20% or decrease by 20% each year. Assume the annual interest rate is 5% continuously compounded. Using a binomial tree and risk-neutral formulas, find the no-arbitrage price of a 52-strike American put option maturing in 2 years.

We have $S_0 = 50$, $u = 1.2$, $d = 0.8$ and $B_k = e^{0.05k}$, for $k = 0, 1, 2$. We have graphically:

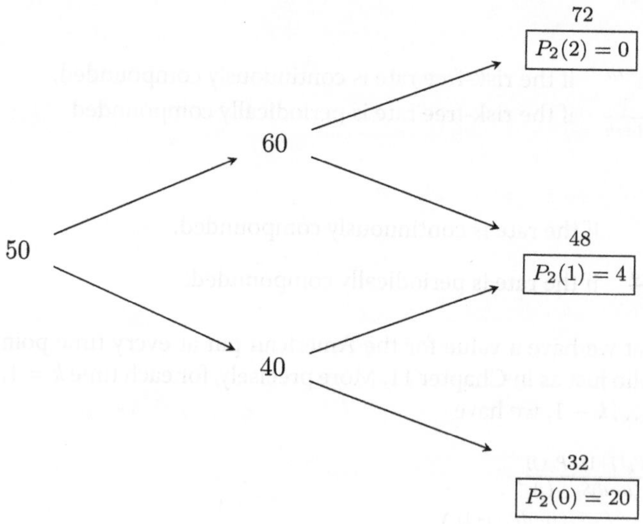

Let us follow closely the above algorithm. At maturity, i.e. at time 2, the possible values of the payoff are

$$P_2(0) = 20, \quad P_2(1) = 4 \quad \text{and} \quad P_2(2) = 0.$$

At time 1, we compute the continuation values:

$$H_1(1) = e^{-0.05}(q \times 0 + (1-q) \times 4) = 1.41475$$
$$H_1(0) = e^{-0.05}(q \times 4 + (1-q) \times 20) = 9.46393,$$

where

$$q = \frac{B_1 - d}{u - d} = \frac{e^{0.05} - 0.8}{1.2 - 0.8} = 0.628177741.$$

The next step is to compare the continuation values and the exercise values. At node (1, 1), we have

$$K - S_1(1) = 52 - 60 < 1.41475 = H_1(1),$$

so $P_1(1) = \max(K - S_1(1), H_1(1)) = 1.41475$, and, at node $(1, 0)$, we have

$$K - S_1(0) = 52 - 40 = 12 > 9.46393 = H_1(0),$$

so $P_1(0) = \max(K - S_1(0), H_1(0)) = 12$. Therefore, the put option is exercised early in node $(1, 0)$, i.e. at time 1 in the scenario corresponding to *no up-move so far*.

At time 0, we repeat the same procedure. The continuation value at time 0 is the no-arbitrage price of a derivative paying either 1.41475 or 12 (over the next period). Thus, we have

$$H_0(0) = e^{-0.05}(q \times 1.41475 + (1 - q) \times 12) = 5.089632474.$$

The exercise value at time 0 is $K - S_0 = 52 - 50 = 2$, so it is less than the continuation value: the option is not immediately exercised at time 0 (right after issuance). So the initial price of this American put option is \$5.0896.

The tree corresponding to the valuation of this American put option is the following:

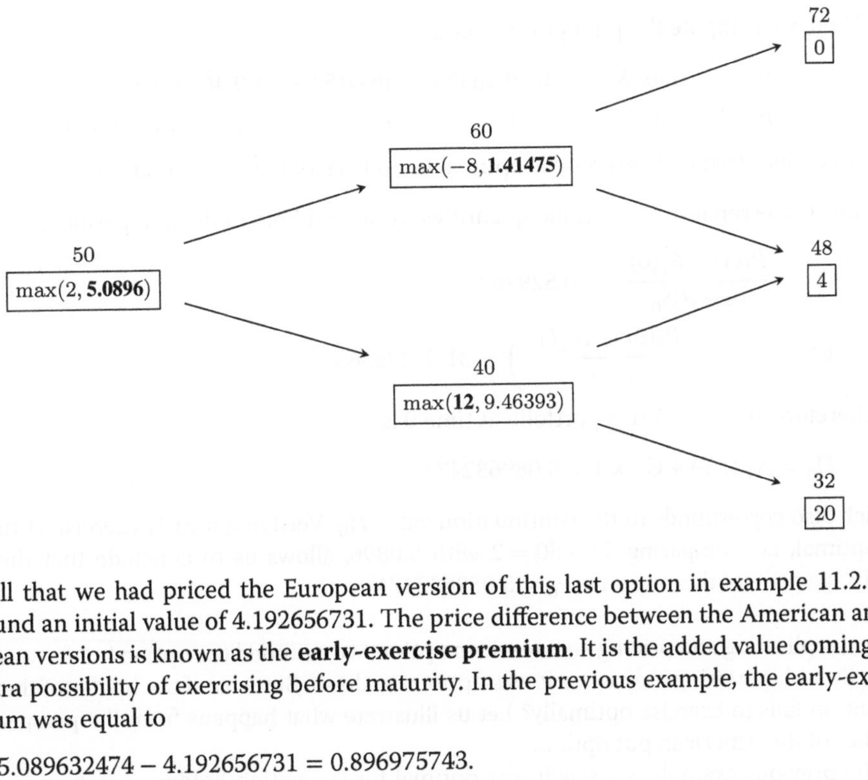

■

Recall that we had priced the European version of this last option in example 11.2.5. We had found an initial value of 4.192656731. The price difference between the American and the European versions is known as the **early-exercise premium**. It is the added value coming from the extra possibility of exercising before maturity. In the previous example, the early-exercise premium was equal to

$$5.089632474 - 4.192656731 = 0.896975743.$$

Example 12.1.2 *Replication of an American put option*
Let us find the replicating portfolio for the American put option of the previous example.

We know already that

$$P_2(0) = 20, \quad P_2(1) = 4 \quad \text{and} \quad P_2(2) = 0.$$

At time 1, we have:

1) The replicating portfolio quantities we should hold during the second period are given by: in node $(1,0)$,

$$\Delta_2(0) = \frac{P_2(1) - P_2(0)}{(u-d)S_1(0)} = -1$$

$$\Theta_2(0) = e^{-0.1}\left(\frac{uP_2(0) - dP_2(1)}{u-d}\right) = 52e^{-0.1}$$

and, in node $(1,1)$,

$$\Delta_2(1) = \frac{P_2(2) - P_2(1)}{(u-d)S_1(1)} = -1/6$$

$$\Theta_2(1) = e^{-0.1}\left(\frac{uP_2(1) - dP_2(2)}{u-d}\right) = 12e^{-0.1}.$$

2) The corresponding continuation values are equal to

$$H_1(0) = \Delta_2(0)S_1(0) + \Theta_2(0)B_1 = 9.46393,$$
$$H_1(1) = \Delta_2(1)S_1(1) + \Theta_2(1)B_1 = 1.41475.$$

3) Then, we compute the put option prices:

$$P_1(0) = \max(K - S_1(0), 9.46393) = \max(52-40, 9.46393) = 12,$$
$$P_1(1) = \max(K - S_1(1), 1.41475) = \max(52-60, 1.41475) = 1.41475.$$

We deduce that early exercise is optimal in node $(1,0)$ but not in node $(1,1)$.

At time 0, the replicating portfolio quantities we should hold in the first period are:

$$\Delta_1 = \frac{P_1(1) - P_1(0)}{(u-d)S_0} = -0.5292625,$$

$$\Theta_1 = e^{-0.1}\left(\frac{uP_1(0) - dP_1(1)}{u-d}\right) = 31.55275563.$$

Therefore, the cost of this portfolio at time 0 is

$$\Pi_0 = \Delta_1 \times 50 + \Theta_1 \times 1 = 5.089632474,$$

which also corresponds to the continuation value H_0. Verifying if early exercise at time 0 is optimal, i.e. comparing $52 - 50 = 2$ with 5.0896, allows us to conclude that the no-arbitrage price of this option is $P_0 = 5.089632474$. ■

When replicating and pricing an American option, we assumed the option holder was acting rationally and exercising only when it was optimal to do so. But what if the holder of the American option fails to exercise optimally? Let us illustrate what happens from the perspective of the seller of the American put option.

In the previous example, we saw it was optimal for the option holder to early-exercise her American put at node $(1,0)$, that is at time 1 after having observed no up-move. From a risk management point of view, i.e. from the seller's perspective, the replicating portfolio is defined in such a way that it has enough value, at any time and in any scenario, to cover an optimal exercise of the put. If, in node $(1,0)$, the holder does exercise her option and therefore sells the stock (to the option seller) for 52, even if it is worth only 40, then the corresponding value of the replicating portfolio will also be equal to 12. Consequently, the option issuer will offset its *temporary loss* of 12 with the replicating portfolio value.

A *less informed* investor could instead decide, in node $(1, 0)$, to hold on to the option for (at least) one more period. In this case, the investor would be acting sub-optimally. We have computed that this holding value, in node $(1, 0)$, is equal to 9.46, while the (replicating) portfolio is worth 12. Consequently, now that the holder has missed the opportunity to optimally exercise her option, the option seller needs only \$9.46393 out of the \$12 he already owns to continue replicating the American put for one more period (they will offset each other at maturity). In other words, there is a positive cash flow for the option seller: he can invest that difference of $12 - 9.46 > 0$ at the risk-free rate and cash $(12 - 9.46393)e^{0.05} = 2.666097$ at maturity.

The seller of an American option always benefits from sub-optimal exercise if he sets up the replicating portfolio at inception and given that rational behavior is accounted for in this replicating portfolio. In this case, the seller is prepared for the worst-case scenario, i.e. optimal early exercise of the put.

12.1.2 American call options

The previous algorithm is valid for any American derivative in the binomial model as long as the exercice value is computed accordingly. In particular, for an American call written on a non-dividend-paying asset, we could illustrate that it is not optimal to exercise early: at every time point, the continuation value is greater than the exercise value.

Example 12.1.3 *Pricing an American call option*
Let us consider the call version of the American put option of example 12.1.1. Recall that we have $S_0 = 50$, $u = 1.2$, $d = 0.8$ and $B_k = e^{0.05k}$, for $k = 0, 1, 2$. We have already computed $q = 0.628177741$.

We will use risk-neutral valuation to obtain the no-arbitrage price of this call option. At time 1, the continuation values are given by

$$H_1(0) = \frac{B_1}{B_2}(qC_2(1) + (1-q)C_2(0)) = 0,$$

$$H_1(1) = \frac{B_1}{B_2}(qC_2(2) + (1-q)C_2(1)) = 11.95082302,$$

and the corresponding call option prices by

$$C_1(0) = \max(S_1(0) - K, 0) = \max(40 - 52, 0) = 0,$$

$$C_1(1) = \max(S_1(1) - K, 11.95082302) = \max(60 - 52, 11.95082302)$$
$$= 11.95082302.$$

Finally,

$$H_0(0) = \frac{1}{B_1}(qC_1(1) + (1-q)C_1(0)) = 7.141108544$$

and therefore

$$C_0(0) = \max(S_0 - K, 7.141108544) = \max(50 - 52, 7.141108544) = 7.141108544.$$

Hence, the American call option is never exercised early and its value is the same as an otherwise equivalent European call option. The entire tree for the valuation of the American call option is illustrated here:

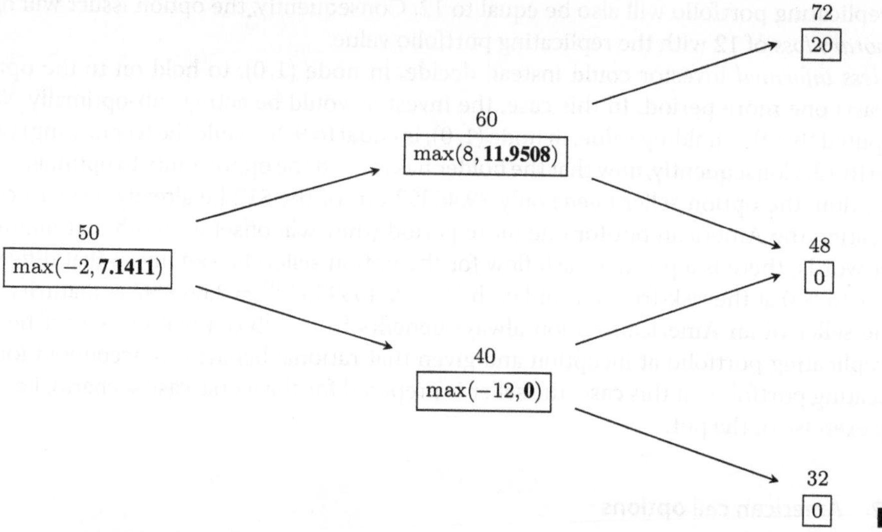

■

12.2 Options on dividend-paying stocks

Recall that, in Chapter 10, we looked at the impact of discrete (dollar) dividends, paid by a stock following a two-period binomial tree, on the pricing of an option.

Now assume that the annual continuously compounded dividend yield is given by γ and that dividends are continuously reinvested in the stock, as described in Chapter 2. In this case, an investment worth S_0 at time 0 will grow to a value of $S_t e^{\gamma t}$ after t years. In a binomial setting, this means that, at time $k-1$, an investment worth S_{k-1} in a stock will accumulate to $S_k e^{\gamma h}$ at time k, where $h = T/n$ is the duration (in years) of one period.

Recall that the payoff of an option is computed on the ex-dividend price, which is the stock price observed on the market. Assume the ex-dividend price S follows a general binomial tree with factors u and d. In this case, the one-period sub-tree with root at node $(k-1, j)$ is given by:

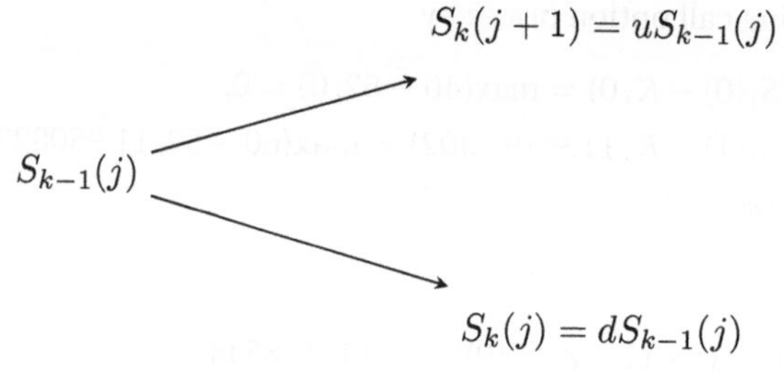

The no-arbitrage assumption between the basic assets, i.e. the stock and the risk-free bond, becomes

$$de^{\gamma h} < e^{rh} < ue^{\gamma h}$$

meaning that investing $\$S_0$ in the stock, and reinvesting dividends, cannot always (in all scenarios) yield more or yield less than investing $\$S_0$ at the risk-free rate.

In this binomial setting, to identify the replicating strategy of an option with payoff V_n written on this dividend-paying stock, we need to use the fact that investing $S_{k-1}(j)$ in a stock will accumulate to either $S_k(j)e^{\gamma h} = dS_{k-1}(j)e^{\gamma h}$ or $S_k(j+1)e^{\gamma h} = uS_{k-1}(j)e^{\gamma h}$. Therefore, the system of equations we need to solve, for the one-period sub-tree with root positioned at node $(k-1, j)$, is

$$\begin{cases} \Theta_k(j)B_k + \Delta_k(j)S_k(j+1)e^{\gamma h} = V_k(j+1), \\ \Theta_k(j)B_k + \Delta_k(j)S_k(j)e^{\gamma h} = V_k(j). \end{cases}$$

It is the same system as in Chapter 11, except that the realizations $S_k(j+1)$ and $S_k(j)$ are multiplied now by $e^{\gamma h}$. Consequently, we easily find

$$\begin{cases} \Delta_k(j) = e^{-\gamma h} \times \frac{V_k(j+1)-V_k(j)}{S_k(j+1)-S_k(j)}, \\ \Theta_k(j) = \frac{1}{B_k}\left(\frac{S_k(j+1)V_k(j)-S_k(j)V_k(j+1)}{S_k(j+1)-S_k(j)}\right). \end{cases} \tag{12.2.1}$$

Note that, compared with the replicating portfolio given in (11.2.5) for the non-dividend-paying case, only the expression for the number of shares of the risky asset has changed.

Then, the time-$(k-1)$ price of the derivative must be equal to the corresponding value of the replicating portfolio

$$V_{k-1}(j) = \Pi_{k-1}(j),$$

with $\Pi_{k-1}(j) = \Theta_k(j)B_{k-1} + \Delta_k(j)S_{k-1}(j)$.

Example 12.2.1 *Replicating an option on a dividend-paying stock*
A stock currently trades at 35 and the (annual) dividend yield on that stock is 2%. Assume the stock price can only increase by 7% per year or decrease by 4% per year, according to a binomial tree in which each step corresponds to 1 year. Assume also that the (annual) risk-free interest rate is 3%, i.e. $B_k = 1.03^k$, for $k = 0, 1, 2$. Let us find the strategy that replicates the payoff of a 2-year at-the-money European call option on this dividend-paying stock.

The stock price evolves from $S_0 = 35$ with factors $u = 1.07$ and $d = 0.96$ and investing in the stock generates an additional yield of $\gamma = 0.02$. Note that there are no arbitrage opportunities in this market because

$$0.96e^{0.02} = 0.979393286 < 1.03 < 1.091615434 = 1.07e^{0.02}.$$

The evolution of the (ex-dividend) stock price, together with the option payoff, is given in the following tree:

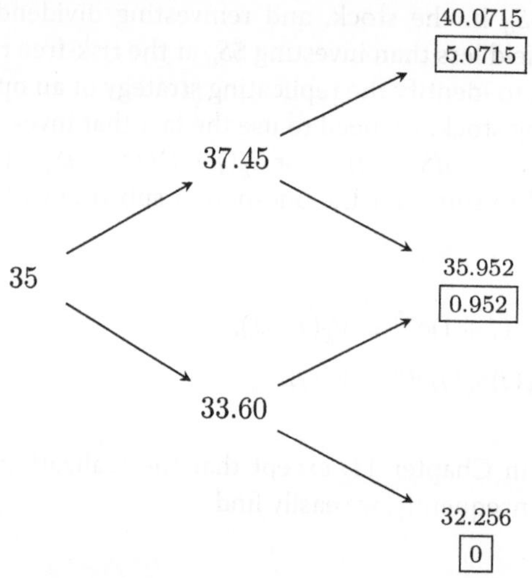

We are now ready to replicate this call option. In the upper one-period sub-tree, i.e. with root at node $(1, 1)$, using the expressions given in (12.2.1), we find

$$\Delta_2(1) = e^{-0.02} \times \frac{5.0715 - 0.952}{40.0715 - 35.952} = 0.980198673$$

$$\Theta_2(1) = \frac{1}{1.03^2}\left(\frac{40.0715 \times 0.952 - 35.952 \times 5.0715}{40.0715 - 35.952}\right) = -32.99085682$$

whereas in the lower one-period sub-tree, i.e. with root at node $(1, 0)$, we find

$$\Delta_2(0) = e^{-0.02} \times \frac{0.952 - 0}{35.952 - 32.256} = 0.252475416$$

$$\Theta_2(0) = \frac{1}{1.03^2}\left(\frac{35.952 \times 0 - 32.256 \times 0.952}{35.952 - 32.256}\right) = -7.831429575.$$

The value of the option at time 1 is thus equal to Π_1 and then

$$\begin{aligned} C_1(1) &= \Theta_2(1)B_1 + \Delta_2(1)S_1(1) \\ &= -32.99085682 \times 1.03 + 0.980198673 \times 37.45 = 2.727857791 \\ C_1(0) &= \Theta_2(0)B_1 + \Delta_2(0)S_1(0) \\ &= -7.831429575 \times 1.03 + 0.252475416 \times 33.60 = 0.41680151. \end{aligned}$$

Finally, replicating the option value at time 1, we get

$$\Delta_1(0) = e^{-0.02} \times \frac{2.727857791 - 0.41680151}{37.45 - 33.60} = 0.58838813$$

$$\Theta_1(0) = \frac{1}{1.03^2}\left(\frac{37.45 \times 0.41680151 - 33.60 \times 2.727857791}{37.45 - 33.60}\right) = -19.17710383$$

with

$$C_0 = \Theta_1(0)B_0 + \Delta_1(0)S_0 = -19.17710383 \times 1 + 0.58838813 \times 35 = 1.416480723.$$

To illustrate the effect of dividends for an investor, we will *run* this example forward in time for one period, from time 0 to time 1. After one period, one share of stock will be worth either 37.45 or 33.60 in this market model. An investor buying one share of stock is also entitled to dividends. Therefore, buying one share of stock worth 35 at time 0 will grow to either $37.45e^{0.02}$ or $33.60e^{0.02}$ (capital gain/loss and reinvested dividends). Hence, after one period, the replicating portfolio will be worth either

$$-19.17710383 \times 1.03 + 0.58838813 \times (37.45e^{0.02}) = 2.727857791$$

or

$$-19.17710383 \times 1.03 + 0.58838813 \times (33.60e^{0.02}) = 0.41680151,$$

which corresponds to either $C_1(0)$ or $C_1(1)$.

We can provide a similar argument for the next time period to illustrate that we need only a smaller number of shares to replicate the same payoffs. ∎

As in Chapter 11, the value of the replicating portfolio $\Pi_{k-1}(j)$ can be rewritten as a risk-neutral expectation:

$$V_{k-1}(j) = \Pi_{k-1}(j) = \frac{B_{k-1}}{B_k}(qV_k(j+1) + (1-q)V_k(j)), \tag{12.2.2}$$

where the risk-neutral probability (of an up-move) is again given by

$$q = \frac{e^{(r-\gamma)h} - d}{u - d},$$

when the risk-free rate r is constant and continuously compounded.

Example 12.2.2 *Pricing an option on a dividend-paying stock using risk-neutral formulas*
In the context of Example 12.2.1, let us use risk-neutral formulas to obtain the no-arbitrage price of the option.

The risk-neutral probability of an up-move is

$$q = \frac{e^{-0.02} \times 1.03 - 0.96}{1.07 - 0.96} = 0.450951214.$$

Using (12.2.2), we easily get

$$C_1(1) = \frac{1}{1.03}(q \times 5.0715 + (1-q) \times 0.952) = 2.727857791$$

$$C_1(0) = \frac{1}{1.03}(q \times 0.952 + (1-q) \times 0) = 0.41680151$$

and finally,

$$C_0 = \frac{1}{1.03}(q \times 2.727857791 + (1-q) \times 0.41680151) = 1.416480723,$$

as obtained by replication. ∎

Finally, as mentioned in Chapter 3 and Chapter 6, options on stock indices can be treated like options on a dividend-paying stock, where γ is then interpreted as the average dividend yield on the stocks that compose the index.

12.3 Currency options

Suppose you have Euros in your portfolio and will need to convert them into U.S. dollars some time in the future. In other words, you are exposed to the *risky* exchange rate. You could fix the selling price of Euros right now by shorting *Euro futures* or, as we will see below, by buying put options on the Euro. This section focuses on pricing currency options in the binomial model.

Let S_k be the time-k price in the domestic currency, chosen to be the U.S. dollar, of *one unit* of a foreign currency, say the Euro. This is known as the foreign exchange rate, a quantity evolving randomly over time. As seen in Chapter 3, holding a foreign currency earns interest at the foreign risk-free rate, denoted by r_f, whereas the domestic currency earns interest at a risk-free rate of r in a domestic bank account.

Similar to currency forwards, currency futures and currency swaps, a **currency option** is an option to buy (or sell) *one unit* of the foreign currency at the maturity date. It is important to note that all prices (including the strike price) are expressed in the domestic currency (U.S. dollar).

We will now price and replicate currency options in the binomial model. We will use the similarities with dividend-paying stocks: indeed, the interest earned in the foreign bank account (r_f) is similar to a dividend yield paid by a stock. Therefore, the binomial model gives the dynamic of the foreign currency with respect to the domestic currency.

Assume the price (in U.S. dollars) S for one Euro follows a general binomial tree with factors u and d. It is the *exchange rate*. In this case, the one-period sub-tree with root at node $(k-1, j)$ is given by:

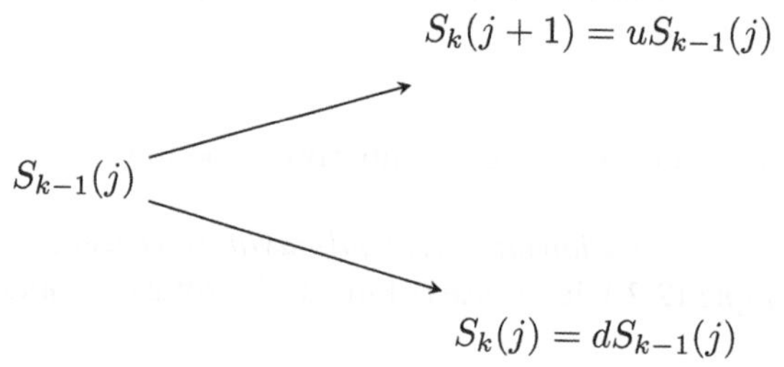

The no-arbitrage assumption in this setup is given by

$$de^{r_f h} < e^{rh} < ue^{r_f h}.$$

To replicate the payoff of a currency option, we will need to periodically adjust the number of units held in both bank accounts according to the *exchange rate*. Given that one Euro earns the risk-free rate r_f in between two portfolio adjustments, then the set of replicating equations at node $(k-1, j)$ is given by

$$\begin{cases} \Theta_k(j)B_k + \Delta_k(j)S_k(j+1)e^{r_f h} = V_k(j+1), \\ \Theta_k(j)B_k + \Delta_k(j)S_k(j)e^{r_f h} = V_k(j) \end{cases}$$

where both sides of both equations are expressed in U.S. dollars. Note the similarity of these equations to those for the replication of options written on dividend-paying stocks.

As before, the solution is given by

$$\Delta_k(j) = e^{-r_f h} \times \frac{V_k(j+1) - V_k(j)}{(u-d)S_{k-1}(j)},$$

$$\Theta_k(j) = \frac{1}{B_k}\left(\frac{uV_k(j) - dV_k(j+1)}{u-d}\right).$$

Consequently, using the value of the replicating portfolio, we can obtain the following risk-neutral pricing formula:

$$V_{k-1}(j) = \Pi_{k-1}(j) = \frac{B_{k-1}}{B_k}(qV_k(j+1) + (1-q)V_k(j)),$$

where the risk-neutral probability (of an up-move) is given by

$$q = \frac{e^{(r-r_f)h} - d}{u - d}.$$

Example 12.3.1 *Pricing a currency option*
A U.S.-based company will receive 1 million Canadian dollars in two months. The current exchange rate is quoted as 0.75 USD per CAD and is subject to monthly movements of ±5%. The (continuously compounded) risk-free rate in Canada is 1% and it is 0.75% in the U.S. The company benefits if the exchange rate becomes larger than \$0.75 USD/CAD, and loses otherwise. The American company decides to buy 1 million at-the-money put options to benefit from changes in the exchange rate while limiting losses. Using a two-period binomial tree, let us find the no-arbitrage price of one such currency option.

We have $h = 1/12$, $S_0 = K = 0.75$, $r_f = 0.01$, $r = 0.0075$, $u = 1.05$ and $d = 0.95$. Therefore, we have

$$q = \frac{e^{(r-r_f)h} - d}{u - d} = \frac{e^{(0.0075-0.01)\times 1/12} - 0.95}{1.05 - 0.95} = 0.497916884$$

and the realizations of the payoff are

$$P_2(2) = (0.75 - 0.75(1.05)^2)_+ = 0$$
$$P_2(1) = (0.75 - 0.75(1.05)(0.95))_+ = 0.001875$$
$$P_2(0) = (0.75 - 0.75(0.95)^2)_+ = 0.073125.$$

The put option price is therefore

$$P_0 = e^{-3r/12}(q^2 \times P_2(2) + 2q(1-q) \times P_2(1) + (1-q)^2 \times P_2(0))$$

and we find $P_0 = 0.01933509$. ■

Example 12.3.2 *Replication of a currency option*
Another U.S.-based company owes 1 million Euros to an important provider, an amount that needs to be delivered in 2 months. The current exchange rate is 1.15 USD per EUR and is subject to monthly movements of ±10%. The risk-free rate in Europe is 1% whereas in the US it is at 0%. Using a two-period binomial tree, let us describe how to protect the American company against adverse changes in the exchange rate, while retaining the benefits.

The US company's obligation is currently worth 1.15 million U.S. dollars but it could be more if the exchange rate goes up over the next 2 months. The company could replicate a call option on the Euro: the gains from the call option would offset the increase in the obligation's value, but would not affect it otherwise. Let us describe the replication of an at-the-money call option on the Euro.

Since $r = 0$, we have $B_0 = B_1 = B_2 = 1$. At time 1, the exchange rate will be either 1.265 USD per EUR or 1.035 USD per EUR.

- If we end up in node $(1, 1)$, i.e. if $S_1 = 1.265$, then we need to replicate, over the next period, a payoff taking values 0.2415 or 0. The system of equations we need to solve is

$$\begin{cases} \Theta_2(1) + 1.3915e^{0.01/12}\Delta_2(1) = 0.2415, \\ \Theta_2(1) + 1.1385e^{0.01/12}\Delta_2(1) = 0. \end{cases}$$

Therefore, we find $\Delta_2(1) = 0.953750331$ and $\Theta_2(1) = -1.08675$ and the value of the call option at time 1 (in the upper node) is

$$V_1(1) = \Theta_2(1) \times 1 + \Delta_2(1) \times 1.265 = 0.119744169.$$

- If we end up in node (1, 0), i.e. if $S_1 = 1.035$, then we need to replicate, over the next period, a payoff of 0 in both scenarios. We deduce that $\Theta_2(0) = \Delta_2(0) = 0$ and thus $V_1(0) = 0$.

Finally, back at time 0, the system of equations is

$$\begin{cases} \Theta_1 + 1.265e^{0.01/12}\Delta_1 = 0.119744169, \\ \Theta_1 + 1.035e^{0.01/12}\Delta_1 = 0 \end{cases}$$

and we find $\Delta_1 = 0.520193147$ and $\Theta_1 = -0.538848761$.

As usual, the risk management strategy is as follows:

- At time 0, the company needs $\Theta_1 \times B_0 + \Delta_1 \times S_0 = -0.538848761 + 0.520193147 \times 1.15 = 0.059373358$.
- At time 1, and depending on the exchange rate, the company updates its portfolio to either $(-1.08675, 0.953750331)$ or $(0, 0)$.
- At time 2, the exchange rate will be either 1.3915, 1.1385 or 0.9315 USD per EUR. In the first case, the obligation to the provider goes up to close to 1.4 million USD (1.3915 precisely) which is offset by a replicating portfolio worth 241,500. Overall, its obligation is frozen at 1.15 million USD.

  In the other cases, the replicating portfolio is worth 0 and the obligation to the provider is worth either 1.1385 or 0.9315 million USD. The company has thus retained the benefits of a decreasing exchange rate while protecting against an increase in the exchange rate. ■

12.4 Options on futures

In this section, we are interested in *futures options*. An **option on futures** or a **futures option** is an option written on a futures contract. It is an agreement that gives the possibility to enter the underlying futures contract at maturity for a predetermined price. Recall from Section 3.5 in Chapter 3 that a futures contract is marked-to-market, which means that the daily gains/losses are deposited/withdrawn from a margin account accumulating interest. Consequently, there is an important difference between holding a futures contract or an option on a futures contract: the holder of a futures contract is entitled to the interest on the margin account whereas the holder of the futures option will not receive any of that interest during the life of the option. This is similar to options on dividend-paying stocks.

More precisely, a **futures call option** provides the right, but not the obligation, to enter the long position of a futures contract upon the option maturity. If exercised, the option holder enters the long position of a futures contract at no cost and receives the difference between the futures price and the strike price. Similarly, a **futures put option** provides the right, but not the obligation, to enter the short position of a futures contract upon the option maturity. If exercised, the option holder enters the short position of a futures contract at no cost and receives the difference between the strike price and the futures price. Even if in practice futures options are generally American, in what follows we will only consider European futures options. The decisions to enter or not the futures contract is based on the following event: for a futures call (resp. put) option, the futures contract is entered into if the underlying asset futures price at maturity F_T is larger (resp. less) than the strike price K specified in the option's contract.

At the maturity of the option, the investor can decide to immediately offset its position into the futures contracts. Consequently, we can assume that the payoff of a futures call option and that of a futures put option, with maturity T and strike price K, are given by

$$(F_T - K)_+ \qquad \text{and} \qquad (K - F_T)_+,$$

respectively, where F_T is the futures price at time T of the underlying asset (spot contract). The delivery date corresponding to the futures prices F_T is not mentioned here, but it must be equal to or greater than the option maturity time T. When these two maturities coincide, the futures price and the spot price also coincide, meaning that $F_T = S_T$, as seen in Chapter 3.

12.4.1 Futures price in a binomial environment

For a dividend-paying stock, the option payoff is directly computed on the ex-dividend price. This is the price observed by investors on the market and it represents the price at which we can buy the stock at any time. As a result, the binomial tree of Section 12.2 was needed to represent the evolution of that price.

In a similar fashion, we will use a binomial tree to model the evolution of the futures price (of the underlying asset). This is because the payoff of a futures option is directly tied to the futures price of the underlying asset and not to the spot price of this asset. Using notation as above, let us define by $F_k(j)$ the futures price in node (k,j), i.e. at time k, in the scenario that we have observed j upward movements so far.

Let us look at the futures price in node $(k-1,j)$. Over the next period, the futures price can increase by a factor u or decrease by a factor d:

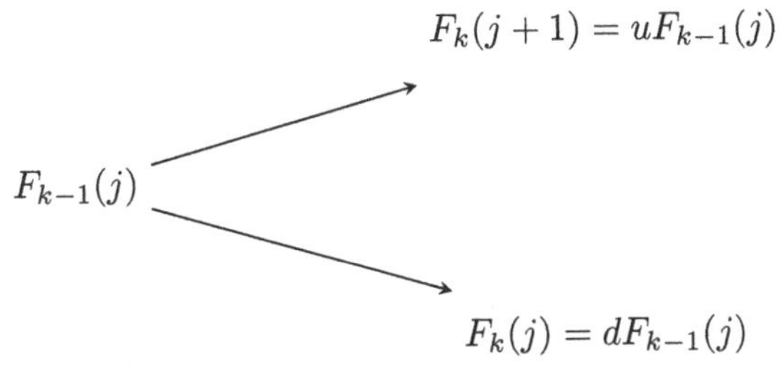

Therefore, the gain or loss between time $k-1$ and time k, is either

$$F_k(j) - F_{k-1}(j) = F_{k-1}(j)d - F_{k-1}(j) = F_{k-1}(j)(d-1),$$

in a downward market, or

$$F_k(j+1) - F_{k-1}(j) = F_{k-1}(j)u - F_{k-1}(j) = F_{k-1}(j)(u-1),$$

in an upward market. Due to the marking-to-market process, the gain (or loss) is then deposited (or withdrawn) at time k in (from) the margin account and will earn interest at the risk-free rate only *one period later* at time $k+1$.

12.4.2 Replication and risk-neutral pricing

We will replicate a futures option using futures contracts, as a basic asset, rather than the underlying asset. As a result, we will be using the periodic gain and loss of the futures contract to replicate the payoff of the option. Therefore, the marking-to-market process of futures contracts has to be appropriately accounted for. This is explained below.

We seek to replicate the futures option in node $(k-1, j)$, just before the k-th period. We know that, one period later, the option can take only two values: $V_k(j)$ or $V_k(j+1)$. Correspondingly, the gain or loss on the futures contract will be either $F_{k-1}(j)(d-1)$ or $F_{k-1}(j)(u-1)$. As before, let $\Theta_k(j)$ be the number of units of the risk-free asset that we need in the replicating portfolio, during the k-th period, to replicate this option. But now, let $\Delta_k(j)$ be the number of units of futures contracts that we need in the replicating portfolio, during the k-th period, to replicate the option.

We are now ready to set up the system of equations for replication:

$$\begin{cases} \Theta_k(j)B_k + \Delta_k(j)F_{k-1}(j)(u-1) = V_k(j+1), \\ \Theta_k(j)B_k + \Delta_k(j)F_{k-1}(j)(d-1) = V_k(j). \end{cases}$$

Solving for $\Delta_k(j)$, we find that

$$\Delta_k(j) = \frac{V_k(j+1) - V_k(j)}{F_{k-1}(j)(u-d)}$$

and we obtain

$$\Theta_k(j) = \frac{1}{B_k(u-d)}((uV_k(j) - dV_k(j+1)) - (V_k(j+1) - V_k(j))).$$

As usual, because of the no-arbitrage assumption, the cost of the replicating portfolio in node $(k-1, j)$ coincides with the option price in that same node, so we have the following equality:

$$V_{k-1}(j) = \Theta_k(j) \times B_{k-1} + \Delta_k(j) \times 0 = \Theta_k(j) \times B_{k-1}.$$

Note that the 0 comes from the fact that we can buy or sell as many futures contracts as we want at no cost. Even if futures contracts are costless to enter into, we must still compute $\Delta_k(j)$ because it has an impact on gains and losses in the margin account and ultimately our ability to replicate the option. The following example will help clarify the process.

Example 12.4.1 *Replication of a futures call option*
The current futures price of a given asset is \$35 and we think the futures price can only increase by 7% or decrease by 5% over each upcoming period. One dollar deposited in a bank account earns 1% per period. Let us determine how to replicate an at-the-money call option on a futures contract maturing in two periods.

We have $u = 1.07$, $d = 0.95$ and $B_k = 1.01^k$, for $k = 0, 1, 2$. The binomial tree for the futures price is given by:

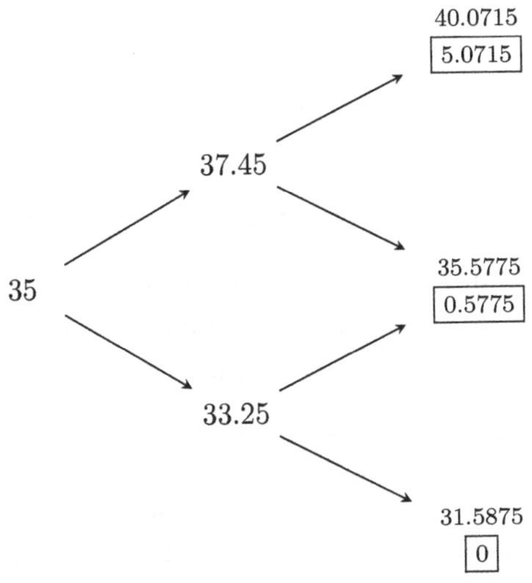

The payoff of this option at maturity is such that

$$C_2(2) = 5.0715, \quad C_2(1) = 0.5775 \quad \text{and} \quad C_2(0) = 0.$$

If we consider the one-period sub-tree with root $F_1 = 37.45$, then we must replicate the values 5.0715 and 0.5775. The corresponding system of equations for replication is

$$\begin{cases} \Theta_2(1) \times (1.01)^2 + \Delta_2(1) \times 37.45 \times (1.07 - 1) = 5.0715, \\ \Theta_2(1) \times (1.01)^2 + \Delta_2(1) \times 37.45 \times (0.95 - 1) = 0.5775. \end{cases}$$

We deduce that $\Delta_2(1) = 1$ and $\Theta_2(1) = 2.401725321$, and therefore

$$C_1(1) = \Theta_2(1) \times B_1 = 2.401725321 \times 1.01 = 2.425742574.$$

Similarly, if we consider the one-period sub-tree with root $F_1 = 33.25$, then we must replicate the values 0.5775 and 0. The system of equations for replication is

$$\begin{cases} \Theta_2(0) \times (1.01)^2 + \Delta_2(0) \times 33.25 \times (0.07) = 0.5775, \\ \Theta_2(0) \times (1.01)^2 + \Delta_2(0) \times 33.25 \times (-0.05) = 0. \end{cases}$$

We deduce that $\Delta_2(0) = 0.144736842$ and $\Theta_2(0) = 0.235883737$, and therefore

$$C_1(0) = \Theta_2(0) \times B_1 = 0.235883737 \times 1.01 = 0.238242574.$$

Finally, at time 0, the system of equations is given by

$$\begin{cases} \Theta_1 \times 1.01 + \Delta_1 \times 35 \times (0.07) = 2.425742574, \\ \Theta_1 \times 1.01 + \Delta_1 \times 35 \times (-0.05) = 0.238242574. \end{cases}$$

We deduce that $\Delta_1 = 0.5208333333$ and $\Theta_1 = 1.13831773$, and therefore the futures option's price is

$$C_0 = \Theta_1 \times B_0 = 1.13831773 \times 1.00 = 1.13831773.$$

To illustrate how the above dynamic replicating strategy works to replicate the payoff of the futures option, the next table gives the cash flows of the strategy for two periods for each possible path.

	Margin balance				
Scenario/Time	**0**	**1**	**2**	**r-f invest.**	**Total**
uu	0	1.27604167	3.91030208	1.16119792	5.0715
ud	0	1.27604167	−0.58369792	1.16119792	0.5775
du	0	−0.91145833	−0.58369792	1.16119792	0.5775
dd	0	−0.91145833	−1.16119792	1.16119792	0

At time 0, according to the replicating strategy, an amount of \$1.13831773 needs to be invested at the risk-free rate whereas 0.52083 futures contracts must be entered (at no cost). The amount of \$1.13831773 accumulates to \$1.16119792 after two periods, as can be seen in the fifth column.

In the meantime, a margin is set up with initial balance at \$0 (second column). At time 1, the futures price will be either 37.45 or 33.25, so the gain/loss on each single futures contract will be 2.45 or −1.75. We have a long position of 0.52083 futures contracts so the total gain/loss is either 1.276034 or −0.9114525, which represents the margin balance at time 1 (third column).

The replicating strategy needs to be rebalanced at time 1 by adjusting the number of futures contracts to hold. Hence, the margin balance at time 2 represents the accumulated value of the margin balance at time 1 plus the gain or loss on the futures contracts position over the second period. For example,

$$3.91030208 = 1.27604167 \times 1.01 + 1 \times (40.0715 - 37.45)$$
$$-1.16119792 = -0.91145833 \times 1.01 + 0.144736842 \times (31.5875 - 33.25).$$

Finally, adding up the margin balance and the risk-free investment at time 2 perfectly matches the payoff of the futures option in every possible scenario.

The marking-to-market process illustrated above is also equivalent to:

1) cancelling (clearing) out the long position in the 0.52083 unit of the futures contract at time 1;
2) investing/borrowing the proceeds at the risk-free rate until the option's maturity;
3) entering the appropriate number of futures contracts for the last period. ■

Finally, as we have done previously, if we set

$$q = \frac{1-d}{u-d},$$

then we can rewrite the price of the futures option as follows:

$$\begin{aligned} V_{k-1}(j) &= \Theta_k(j) \times B_{k-1} \\ &= \mathrm{e}^{-rh}\left(V_k(j)\frac{u-1}{u-d} + V_k(j+1)\frac{1-d}{u-d}\right). \end{aligned}$$

Therefore, we can write

$$V_{k-1}(j) = e^{-rh}(qV_k(j+1) + (1-q)V_k(j)),$$

which can be interpreted again as a discounted (conditional) risk-neutral expectation, where q can be viewed as the risk-neutral probability of an up-move.

Example 12.4.2 *Risk-neutral pricing of a futures option*
In the context of the previous example, we can use the above risk-neutral pricing formula to compute the no-arbitrage price of the futures option.

First, we have

$$q = \frac{1-d}{u-d} = \frac{1-0.95}{1.07-0.95} = \frac{5}{12}.$$

The option price is thus

$$\begin{aligned} V_0 &= \frac{1}{(1.01)^2}(q \times C_2(2) + 2q(1-q) \times C_2(1) + (1-q)^2 \times C_2(2)) \\ &= 1.13831773, \end{aligned}$$

as expected. ■

12.5 Summary

American options

- Options that can be exercised any time before or at maturity.
- Early-exercise value: amount received upon exercise of an American option.
- Holding value: value of the option if it is not exercised.
- An American option is exercised early only if

  Exercise value > Holding value.

- American put:
  - put option price at time k: P_k;
  - early-exercise value at time k: $K - S_k$;
  - holding value at time k in node (k, j):

    $$H_k(j) = \frac{B_k}{B_{k+1}}(qP_{k+1}(j+1) + (1-q)P_{k+1}(j));$$

  - put option value at time k: $P_k = \max(K - S_k, H_k)$;
  - otherwise, replication and risk-neutral formulas are obtained as before (backward, recursively).
- American call:
  - never optimal to early-exercise an American call option written on a non-dividend-paying asset;
  - an American call and a European call written on a non-dividend-paying asset have the same price.

Options on dividend-paying stocks

- Dividend rate: γ.
- Dividends are continuously reinvested in the stock.

- The ex-dividend price follows a binomial tree.
- No-arbitrage condition: $de^{\gamma h} < e^{rh} < ue^{\gamma h}$.
- Replicating strategy (Θ, Δ) for the payoff V_n:
  - for the sub-tree with root $(k-1, j)$, we must solve:

$$\begin{cases} \Theta_k(j)B_k + \Delta_k(j)S_k(j+1)e^{\gamma h} = V_k(j+1), \\ \Theta_k(j)B_k + \Delta_k(j)S_k(j)e^{\gamma h} = V_k(j); \end{cases}$$

  - the solution is given by:

$$\begin{cases} \Delta_k(j) = e^{-\gamma h} \times \frac{V_k(j+1)-V_k(j)}{S_k(j+1)-S_k(j)}, \\ \Theta_k(j) = \frac{1}{B_k}\left(\frac{S_k(j+1)V_k(j)-S_k(j)V_k(j+1)}{S_k(j+1)-S_k(j)}\right). \end{cases}$$

- The time-$(k-1)$ price of V_n is such that

$$V_{k-1}(j) = \Pi_{k-1}(j) = \Theta_k(j)B_{k-1} + \Delta_k(j)S_{k-1}(j).$$

- The time-$(k-1)$ price of V_n (and its replicating portfolio) is also equal to

$$V_{k-1}(j) = \frac{B_{k-1}}{B_k}(qV_k(j+1) + (1-q)V_k(j)),$$

where $q = \dfrac{e^{(r-\gamma)h} - d}{u-d}$.

Currency options

- Time-k price (number of units of domestic currency) to buy/sell one unit of the foreign currency: S_k.
- Foreign-denominated bank account earns interest at rate r_f.
- Domestic-denominated bank account earns interest at rate r.
- All other formulas (replication and pricing) are equal to those for options on dividend-paying stocks with γ replaced by r_f.

Futures options

- Options to buy or sell futures contracts.
- Payoff is computed with the futures price:

$$\text{call: } C_n = (F_T - K)_+ \text{ and put: } P_n = (K - F_T)_+.$$

- The futures price follows a binomial tree.
- Basic assets for the replication of futures options: futures contracts (periodic gain/loss) and risk-free asset.

- Remember that futures contracts are:
  - costless to enter (long or short) at any time;
  - marked-to-market: gain/loss deposited/withdrawn in the margin, earns interest.
- Replicating strategy (Θ, Δ) for the payoff V_n:
  - for the sub-tree with root $(k-1, j)$, we must solve:

$$\begin{cases} \Theta_k(j)B_k + \Delta_k(j)F_{k-1}(j)(u-1) = V_k(j+1), \\ \Theta_k(j)B_k + \Delta_k(j)F_{k-1}(j)(d-1) = V_k(j); \end{cases}$$

  - the solution is given by:

$$\begin{cases} \Delta_k(j) = \frac{V_k(j+1)-V_k(j)}{F_{k-1}(j)(u-d)} \\ \Theta_k(j) = \frac{1}{B_k(u-d)}((uV_k(j) - dV_k(j+1)) - (V_k(j+1) - V_k(j))). \end{cases}$$

- The time-$(k-1)$ price of V_n is such that

$$V_{k-1}(j) = \Theta_k(j) \times B_{k-1}.$$

- The time-$(k-1)$ price of V_n (and its replicating portfolio) is also equal to

$$V_{k-1}(j) = \frac{B_{k-1}}{B_k}(qV_k(j+1) + (1-q)V_k(j)),$$

where $q = \dfrac{1-d}{u-d}$.

12.6 Exercises

12.1 A share of stock currently trades for \$42 and its volatility is 25%. The risk-free rate is 3% (continuously compounded) and the evolution of the stock price is modeled with a three-period Cox-Ross-Rubinstein (CRR) tree. A 3-month 40-strike American put option has just been issued.

(a) Using risk-neutral probabilities, calculate the no-arbitrage price of this option.

(b) Describe in details the replicating strategy at each node.

12.2 A dividend-paying stock sells for \$50 and a single dividend of \$7 will be paid in one month. In a two-period binomial tree where one period corresponds to a month, the stock price can increase or decrease by 4% periodically whereas the risk-free rate is 2.4% (annual, compounded monthly). Using risk-neutral probabilities, find the no-arbitrage price of an:

(a) American call option with strike \$51 maturing in 2 months;

(b) American put option with strike \$49 maturing in 2 months.

12.3 A dividend-paying stock trades for \$78 and continuously pays a dividend yield of 1% annually. The (ex-dividend) price can move up by 6% or down by 3% once every 6 months. Moreover, the risk-free rate is 2% annually and continuously compounded. Using risk-neutral probabilities, find the no-arbitrage price of:

(a) 80-strike European call option;

(b) 80-strike American call option;

(c) 76-strike European put option;

(d) 76-strike American put option.

All options mature in a year.

12.4 A European investment bank issues a 6-month USD put option on the USD-EUR exchange rate. The risk-free rate in Europe is 3% and 1% in the U.S. (both rates are continuously compounded). The current exchange rate is 1.25 USD per EUR and the strike price is 1.35 USD per EUR. The evolution of the USD-EUR exchange rate is modeled by a two-period binomial tree with $u = 1.15$ and $d = 1/u$.

(a) Express the current exchange rate and strike in EUR per USD.

(b) Compute the no-arbitrage price of this put option (in EUR).

(c) Describe in detail how the investment bank can exactly offset the cash flows from the put option it has issued.

12.5 The current price of 6-month S&P futures is 2400. It is well known that every month, the price of this specific futures goes up by 3% or down 1%. The risk-free rate is 1.2% compounded monthly. A 3-month at-the-money put option on S&P futures is issued.

(a) Using risk-neutral probabilities, calculate the no-arbitrage price of this option.

(b) Describe in detail the replicating strategy at each node.

13

Market incompleteness and one-period trinomial tree models

Market incompleteness is everywhere in actuarial science, as typical insurable risks embedded in life insurance or homeowner's insurance cannot be (exactly) replicated using tradable assets. Public policy and tight regulations in the insurance industry prevent insurers from trading their insurance policies to exactly offset a liability. It may also be impossible to (exactly) replicate certain financial risks. For example, prior to the creation of credit default swaps (see Chapter 4), it was nearly impossible to hedge against losses resulting from credit risk.

The risk management implications of an *incomplete market* are rather important. It basically means that some risks cannot be *synthetically replicated*. We illustrate this in the next example.

Example 13.0.1 *Impact of default risk on replicating portfolios*[1]
Suppose that on the financial market, a share of ABC inc. currently trades for \$40 and it is assumed that, in a year, its value will be either \$50 or \$30. You also know that \$100 invested at the risk-free rate will accumulate to \$104 in a year. Your company sells a 1-year at-the-money investment guarantee on ABC inc.; this is marketed as an investment that provides the upside on the stock return with a protection against the downside. We know from Chapter 6 that the investment guarantee can be replicated by combining the stock and a put option on this stock. As the risk manager, you decide to replicate this put option and then add one share of the stock to the resulting replicating portfolio.

As we have done before, we could model the price of this stock using a binomial tree. However, there is a probability of $1/100$ that ABC inc. will go bankrupt during the next year, which means its stock price will plunge to zero. As this probability is very small, you decide to ignore it and go through with *replication* assuming the stock price follows a binomial tree model.

In a binomial model, we can easily *replicate* the put option with the following system of equations:

$$\begin{aligned} 0 &= 50x + 1.04y \\ 10 &= 30x + 1.04y. \end{aligned}$$

We obtain $x = -0.5$ and $y = 25/1.04 = 24.038461538$. Therefore, the no-arbitrage price of the put option is $40x + 1y = 4.038461538$.

In conclusion, to *replicate* the investment guarantee, the company needs to buy $1 + x = 1 + (-0.5) = 0.5$ share of the stock and $y = 24.0385$ units of the risk-free asset, which can

1 This example is inspired by example 2 in [18].

Actuarial Finance: Derivatives, Quantitative Models and Risk Management, First Edition.
Mathieu Boudreault and Jean-François Renaud.

be done by selling the investment guarantee for $44.038461538 = 40 + 4.038461538$ to the customer. In 99 out of 100 cases, this will be a riskless venture for the company.

However, in the unlikely event that the stock price goes to 0, i.e. if ABC inc. bankrupts, then your company will have to pay the guaranteed amount of \$40 to the customer, while its assets in the portfolio will be worth only

$$0.5 \times 0 + 24.038461538 \times 1.04 = 25.$$

In 1 out of 100 cases, your company will suffer an important loss of $40 - 25 = 15$. Consequently, there is real danger to assume the stock price follows a binomial model in this situation. ■

This example illustrates that a risk such as bankruptcy or the possibility of a financial crisis, if not properly hedged, can have a very important impact on a risk management strategy. This risk might have been ignored by the risk manager because default risk was considered insignificant (as in the above example) or because no additional assets were available in the market to eliminate this risk. We will come back to this example later in Section 13.4.

In this chapter, we will introduce the concept of *market incompleteness* and we will present a few one-period trinomial tree models with the main objective of understanding the mathematical, financial and risk management implications of the impossibility to replicate specific risks in the financial and insurance markets. The model is very simple and tractable as it only adds a third possible outcome to a one-period tree model. This seemingly simple extension has numerous implications that we will analyze in various contexts.

One-period trinomial tree models also allow us to easily explain and illustrate some of the most complex concepts in financial mathematics, namely complete and incomplete markets, super and sub-replicating portfolios, the existence of multiple risk-neutral probabilities and the Fundamental Theorem of Asset Pricing (FTAP).

The specific objectives are to:

- differentiate complete and incomplete markets;
- build sub-replicating and super-replicating portfolios;
- understand that in incomplete markets, there is a range of prices that prevents arbitrage opportunities;
- derive bounds on the admissible risk-neutral probabilities and relate the resulting prices to sub- and super-replicating portfolios;
- replicate a derivative in a one-period trinomial tree model with three traded assets;
- examine the risk management implications of ignoring possible outcomes when replicating a derivative;
- analyze how actuaries cope with the incompleteness of insurance markets.

The beginning of this chapter has the same structure as Chapters 9–11, i.e. we first introduce the trinomial model for a risky asset, then we attempt to replicate payoffs and retrieve the risk-neutral probabilities. The last sections of this chapter will analyze more specific consequences of market incompleteness in actuarial contexts.

13.1 Model

Just like the one-period binomial model, the (one-asset) **one-period *trinomial* model** presented in this chapter is based upon two time points: the beginning of the period is time 0 and the end of the period is time 1. Again, time 0 represents inception or issuance of a derivative while time 1 stands for its maturity. It is a *frictionless* financial market composed of the following two assets:

- a risk-free asset (a bank account or a bond) which evolves according to the risk-free interest rate r;
- a risky asset (a stock or an index) with a known initial value and a random future value.

Note that in Section 13.4, the *two-asset trinomial model* will be presented to include an additional risky asset like a second stock.

13.1.1 Risk-free asset

The risk-free asset $B = \{B_0, B_1\}$ is an asset for which capital and interest are repaid with certainty at maturity. In a one-step trinomial tree model, we assume there exists a constant interest rate $r \geq 0$ for which we can either write

$$B_1 = \begin{cases} e^r & \text{if the risk-free rate is continuously compounded,} \\ (1+r) & \text{if the risk-free rate is periodically compounded,} \end{cases}$$

with $B_0 = 1$.

13.1.2 Risky asset

The evolution of the price of the risky asset is denoted by $S = \{S_0, S_1\}$ where S_0 is the price today which is a known (observed) quantity. Moreover, S_1 represents the risky asset price at the end of the period which is random. This random variable S_1 is taking values in $\{s^u, s^m, s^d\}$ with probability p_u, p_m and p_d respectively, where $0 < p_u, p_m, p_d < 1$ are fixed values such that $p_u + p_m + p_d = 1$ and where $s^d < s^m < s^u$. In other words, S_1 has the following probability distribution (with respect to the real-world (or actuarial) probability $\mathbb{P}$):

$$\mathbb{P}(S_1 = s^u) = p_u, \quad \mathbb{P}(S_1 = s^m) = p_m, \quad \mathbb{P}(S_1 = s^d) = p_d.$$

Note that if we specify the values of p_u and p_m, then we must have $p_d = 1 - p_u - p_m$, and vice versa. Therefore, the one-period trinomial tree model is a seven-parameter model, namely B_1(or r), $S_0, s^d, s^m, s^u, p_u, p_m$, two more than the corresponding one-period binomial model.

The one-period trinomial tree is often represented graphically as follows:

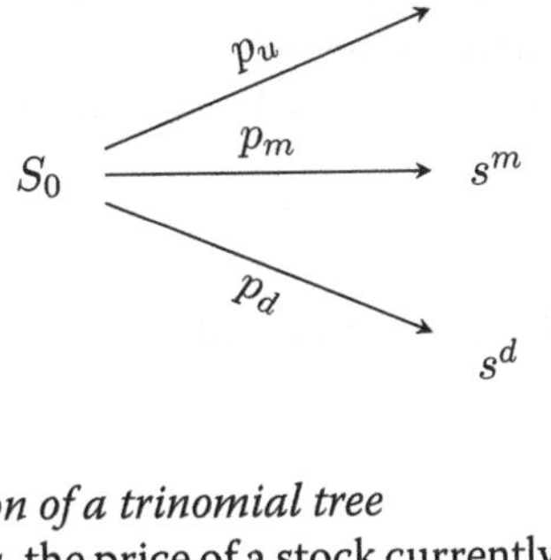

Example 13.1.1 *Construction of a trinomial tree*
According to financial analysts, the price of a stock currently selling for \$52 will be equal to either \$55 or \$50 in 3 months if the markets behave *normally*. However, it will dramatically drop to \$26 if a financial crisis occurs. These analysts have determined there is only a 4% chance for a crisis to occur, while the other two scenarios are equally probable. A 3-month Treasury zero-coupon bond, with a face value of \$100, currently trades for \$99.

Let us build the corresponding one-step trinomial tree for the stock price in 3 months.

First, we have $B_0 = 1$ and $B_1 = \frac{1}{0.99} = 1.0101$. From the above information, $p_d = 0.04$ while $p_u = p_m = 0.48$, which means that the trinomial tree for the stock is

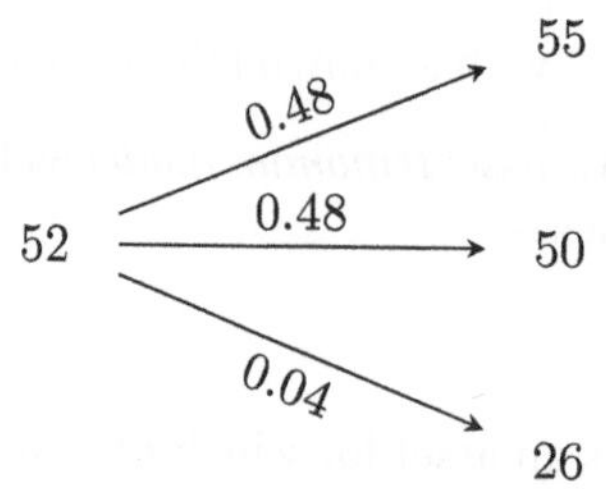

■

To make sure a trinomial model is arbitrage-free, none of the two assets should be *doing better* than the other one in all three scenarios. In other words, investing S_0\$ in either the risk-free asset B or the risky asset S should be such that $S_1 > S_0B_1$ and $S_1 < S_0B_1$ are both inadmissible. Mathematically, the arbitrage-free condition reads as follows:

$$s^d < S_0B_1 < s^u \tag{13.1.1}$$

which looks exactly like the condition in equation (9.1.1) for the binomial model. One must realize here that this no-arbitrage condition also says that it does not matter whether S_0B_1 is smaller or larger than s^m, as long as it is not bigger than s^u or smaller than s^d.

Example 13.1.2 *Arbitrage between the stock and the bond*
The model depicted in example 13.1.1 is arbitrage-free because investing S_0 to buy a stock may yield more or less than lending at the risk-free rate. The same can be said when lending S_0 at the risk-free rate: it may earn more or less than buying a stock. Mathematically,

$$s^d = 26 < S_0\frac{B_1}{B_0} = 52\frac{1}{0.99} = 52.\overline{52} < 55 = s^u.$$

Suppose now that the risk-free asset is such that $B_1 = 1.07$. Then $S_0B_1 = 52 \times 1.07 = 55.64 > 55 = s^u$, which means the no-arbitrage condition in (13.1.1) is violated.

To exploit this arbitrage opportunity, the key is always to buy the cheapest asset/portfolio and sell the most expensive i.e., we invest (borrow) in the asset that yields the most (least). Therefore, we sell the stock for \$52 and invest the proceeds at the risk-free rate. At time 0, the net cash flow is 0. At maturity, the risk-free investment earns 55.64 and we need to close our short position on the stock. We buy the stock at either \$55, \$50 or \$26. The arbitrage profit is

$$55.64 - S_1 > 0,$$

in all three scenarios. The arbitrage opportunity is thus exploited in exactly the same manner as in the binomial model. ■

13.1.2.1 Simplified tree

Again, we can represent the stock price at maturity using an *up factor u*, a *middle factor m*, and a *down factor d* as follows:

$$u = \frac{s^u}{S_0}, \qquad m = \frac{s^m}{S_0} \qquad \text{and} \qquad d = \frac{s^d}{S_0},$$

which implies that we can write

$$s^u = uS_0, \qquad s^m = mS_0 \qquad \text{and} \qquad s^d = dS_0.$$

Clearly, we have that $d < m < u$, and the tree:

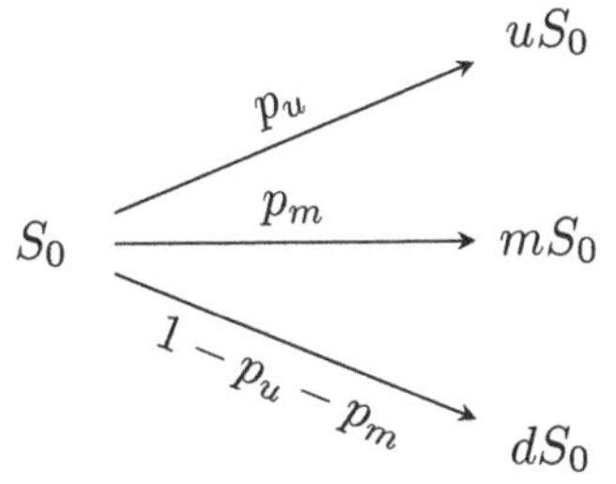

With this new notation, the no-arbitrage condition (13.1.1) is equivalent to

$$d < B_1 < u,$$

as in equation (11.1.1). Recall that B_1 is either equal to e^r or $1 + r$.

The underlying probability model

For the one-period trinomial tree, one can choose its sample space Ω_1 to be

$$\Omega_1 = \{\omega : \omega = \mathrm{u}, \ \mathrm{m} \text{ or } \mathrm{d}\}.$$

It is then clear that each state of nature $\omega \in \Omega_1$, that is each scenario, corresponds to a branch in the tree. Clearly, the possible prices at maturity are

$$S_1(\mathrm{u}) = s^u \equiv uS_0, \quad S_1(\mathrm{m}) = s^m \equiv mS_0 \quad \text{and} \quad S_1(\mathrm{d}) = s^d \equiv dS_0.$$

13.1.3 Derivatives

We now introduce a third asset, more precisely a derivative written on the risky asset S, which is completely characterized by its payoff. The evolution of the price of this third asset is denoted by $V = \{V_0, V_1\}$, where V_0 is today's price while V_1 is its (random) value at the end of the period. As in the binomial model, V_1 is a random variable representing the payoff.

As there are three possible outcomes in the trinomial model, the payoff V_1 can take three values, denoted by v^u, v^m and v^d. More precisely, the random variable V_1 takes the value v^u when S_1 takes the value s^u, it takes the value v^m when S_1 takes the value s^m, and finally it takes the value v^d when S_1 takes the value s^d. Consequently, V_1 has the following probability distribution (with respect to the real-world probability $\mathbb{P}$):

$$\mathbb{P}(V_1 = v^u) = p_u, \qquad \mathbb{P}(V_1 = v^m) = p_m \\ \text{and} \qquad \mathbb{P}(V_1 = v^d) = p_d = 1 - p_u - p_m.$$

As in the binomial model, we can now superimpose the evolution of the derivative price V (in squares) over the evolution of the stock price S:

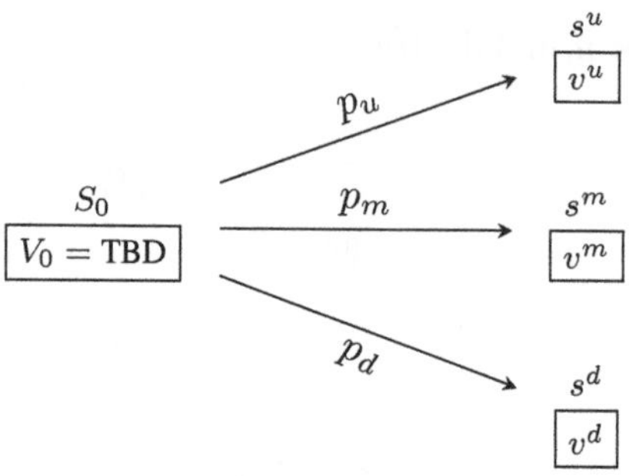

Finally, we will use the notation $C = \{C_0, C_1\}$ and $P = \{P_0, P_1\}$ for a call option price and a put option price respectively, where the payoffs are

$$C_1 = (S_1 - K)_+$$
$$P_1 = (K - S_1)_+,$$

if they both mature at time 1 and both have strike price K.

13.2 Pricing by replication

The key principle underlying replication in the one-period trinomial model is the same as in the binomial model: we seek to find a strategy that replicates the payoff in all scenarios. But as we will see below, allowing a third possible outcome for the risky asset will complicate matters.

13.2.1 (In)complete markets

For a given payoff V_1, we want to find the time-0 price V_0 of this derivative in a one-period trinomial model. In the one-period binomial model, we were able to find a replicating portfolio mimicking the payoff in each and every scenario of that model, using only the basic assets B and S.

Again, let us define a trading strategy (or portfolio) as a pair (x, y), where x (resp. y) is the number of units of S (resp. B) held in the portfolio from time 0 to time 1. Then, the value of this strategy over time is given by $\Pi = \{\Pi_0, \Pi_1\}$, where

$$\Pi_0 = xS_0 + yB_0$$

is the time-0 value, while

$$\Pi_1 = xS_1 + yB_1$$

is the (random) time-1 value. If our goal is to replicate the payoff V_1, then we should choose the pair (x, y) to make sure that

$$\Pi_1 = V_1$$

in each of the three possible scenarios, i.e. we must have

$$\begin{cases} xs^u + yB_1 = v^u, \\ xs^m + yB_1 = v^m, \\ xs^d + yB_1 = v^d. \end{cases} \tag{13.2.1}$$

So, we need to solve (13.2.1), a linear system of three equations with two unknowns, where B_1, s^d, s^m, s^u are given model parameters and v^d, v^m, v^u are given by the derivative payoff at maturity. The values of x and y are the two unknowns.

Example 13.2.1 Let us consider the following trinomial model.[2] Assume the risk-free asset is such that $B_1 = 1.1$ while the risky asset can take three possible values in one period: $s^u = 7.7$, $s^m = 5.5$ and $s^d = 3.3$. Note that the stock currently trades for $S_0 = 6$. We want to find the price of a 1-period 7.70-strike put option written on S. Graphically, we have

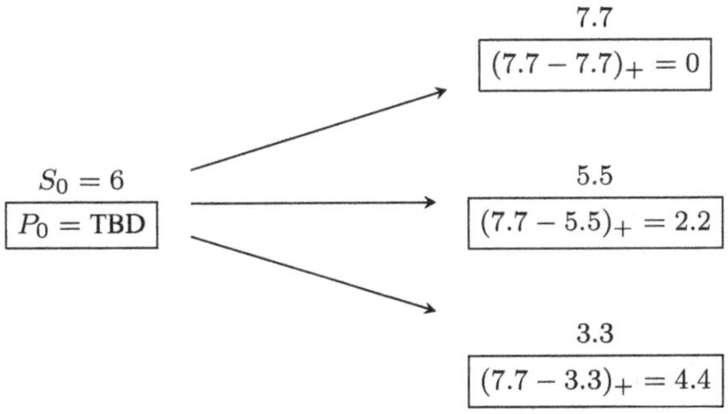

We can easily verify that this trinomial tree is free of arbitrage. Note that we do not know what the values of the actuarial probabilities p^u, p^m and p^d are.

Let us find the replicating portfolio, i.e. let us find (x, y) such that

$$\begin{cases} x7.7 + y1.1 = 0, \\ x5.5 + y1.1 = 2.2, \\ x3.3 + y1.1 = 4.4. \end{cases}$$

From the first two equations, we get

$$x = -1 \quad \text{and} \quad y = 7.$$

The couple (x, y) is also a solution to the third equation:

$$3.3 \times (-1) + 1.1 \times 7 = 4.4.$$

In other words, the trading strategy $(x, y) = (-1, 7)$ is a replicating strategy for this put option. By the no-arbitrage principle, we then have $P_0 = \Pi_0$, which yields

$$P_0 = xS_0 + yB_0 = (-1) \times 6 + 7 \times 1 = 1.$$ ■

2 The framework of this example and the ones that follow is inspired by an illustration presented in Anda Gadidov's lecture notes.

In general, it is not possible to solve a linear system of three equations with two unknowns. This means that the last example was a (deliberate) coincidence.

Example 13.2.2 In the previous example, if we had instead considered a put option with strike price $K = 6$, the system of equations resulting from replication (in each scenario) would have been:

$$\begin{cases} x7.7 + y1.1 = 0, \\ x5.5 + y1.1 = 0.5, \\ x3.3 + y1.1 = 2.7. \end{cases}$$

Unfortunately, this linear system does not admit a solution. Indeed, we can find the only x and y solving the first two equations:

$$x = -0.22727273 \quad \text{and} \quad y = 1.5909091.$$

However, these values do not satisfy the third equation:

$$3.3 \times (-0.22727273) + 1.1 \times (1.5909091) \neq 2.7.$$

In other words, there is no replicating portfolio for this put option. ■

The last two examples illustrate that in the one-period trinomial model, it is not always possible to replicate the payoff of a derivative. Indeed, the 7.70-strike put option could be replicated whereas the 6-strike put option could not.

A derivative is said to be **attainable** if we can find a self-financing strategy to replicate its payoff. In the one-period trinomial model, it is easy to determine whether a security is attainable. If one can find a unique solution to the system of equations in (13.2.1), then the derivative is attainable. But in most cases, we know that we cannot get a unique solution to a system of three equations with two unknowns. Therefore, most derivatives in the one-period trinomial tree model are not attainable. In the previous example, the 7.70-put option was attainable whereas the 6-strike put option was not.

A financial market is said to be **complete** if there are no arbitrage opportunities in this market and all derivatives are attainable. From this definition, we can deduce that the binomial model is a complete model because we can easily replicate all derivatives in this market by solving systems of two equations with two unknowns. The trinomial model, composed of a risk-free asset and a risky asset, is an **incomplete** model because we cannot replicate all derivatives. In fact, it suffices not to be able to replicate only one derivative for a model to be incomplete.

13.2.2 Intervals of no-arbitrage prices

The question now is: how do we find the no-arbitrage price of a derivative in a trinomial tree model if we cannot exactly replicate its payoff (in all scenarios)? But first, does a unique no-arbitrage price even exist? The answer to the latter question is no, not in the trinomial model. Instead of finding a unique no-arbitrage price for a given derivative, we will find an *interval of no-arbitrage prices* by excluding prices for which arbitrage opportunities would exist.

Here is an example on how to exclude *non-valid* prices in a trinomial model.

Example 13.2.3 *Sub-replication*
Consider a one-period trinomial tree where a stock evolves as follows:

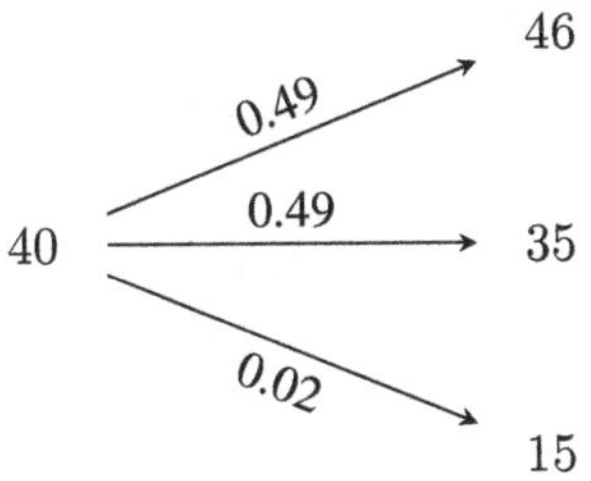

The one-period risk-free rate is at 4%. You have sold a put option with strike price \$37 that expires in one period. You want to manage the risk of your financial position. The payoff of the put option is superimposed to the evolution of the stock price in the tree below.

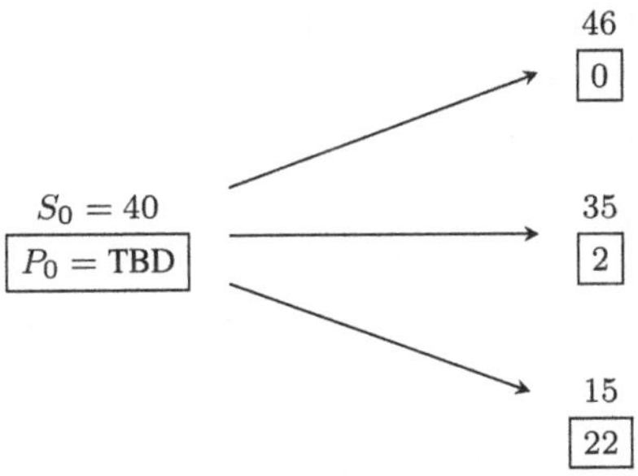

We do not expect to be able to replicate this put option. However, from our own analyses and experience, we think that the probability of a financial crisis and hence that the stock price crashes to 15\$ is only 2%. Therefore, let us ignore this scenario and replicate the put only in the other two *normal* scenarios, as if it were a binomial model. In other words, let us find the pair $(\underline{x}, \underline{y})$ solving

$$\begin{cases} \underline{x}46 + \underline{y}(1.04) = (37 - 46)_+ = 0, \\ \underline{x}35 + \underline{y}(1.04) = (37 - 35)_+ = 2. \end{cases}$$

We know that the solution to this system is given by

$$\underline{x} = \frac{0 - 2}{46 - 35} = -0.1818182,$$

$$\underline{y} = \frac{1}{1.04}\left(\frac{2 \times 46 - 0 \times 35}{46 - 35}\right) = 8.041958.$$

The initial value of this strategy is then given by

$$\underline{\Pi}_0 = \underline{x}S_0 + \underline{y}B_0 = -0.1818182 \times 40 + 8.041958 = 0.76923.$$

By construction, the value of the strategy at time 1 in the *normal* scenarios is such that

$$\begin{cases} \underline{\pi}_1^u = \underline{x}s^u + \underline{y}(1.04) = 0, \\ \underline{\pi}_1^m = \underline{x}s^m + \underline{y}(1.04) = 2. \end{cases}$$

Based upon our judgment, there should be a 98% chance that the portfolio will exactly replicate the payoff of the option at maturity.

But what will happen to this portfolio if a financial crisis occurs? In that case, the time-1 value of the strategy will be equal to

$$\underline{\pi}_1^d = \underline{x}s^d + \underline{y}(1.04) = -0.1818182 \times 15 + 8.041958 \times 1.04 = 5.6363$$

while the put option payoff will be $(37 - 15)_+ = 22$. In other words, the portfolio will be short by $22 - 5.636363 = 16.36364$, with probability 0.02.

The portfolio given by $(\underline{x}, \underline{y}) = (-0.1818182, 8.041958)$ is what we call a *sub-replicating portfolio* because

$$\underline{\Pi}_1 = \underline{x}S_1 + \underline{y}B_1 \leq P_1,$$

in all three scenarios (see also next table).

	Stock	Put	Sub-rep ptf
Value if bull market	46	0	0
Value if bear market	35	2	2
Value if financial crisis	15	22	5.6363
Initial price	40	TBD	0.76923

Consequently, to avoid arbitrage opportunities, this put option must not be sold for a value less than $\underline{\Pi}_0 = 0.76923$. ■

Let us illustrate how to exploit an arbitrage opportunity in this setting.

Example 13.2.4 *Exploiting an arbitrage opportunity*
In the previous example, we built a sub-replicating portfolio, i.e. a portfolio such that $\underline{\Pi}_1 \leq P_1$, in all three scenarios. Suppose now that the put option can be bought for only \$0.50. Let us explain how to exploit this arbitrage opportunity.

As the option is underpriced, we will buy it and we will short the sub-replicating portfolio. We buy the put option (−0.50 at inception) and sell the sub-replicating portfolio (+0.76923 at inception), for a net initial cash flow of 0.26923. At maturity, if the stock price is either 46 or 35, the net proceeds are zero. However, if the stock market crashes, the strategy yields a profit of $22 - 5.6363 = 16.3637$. Therefore, you receive 0.269 at inception, and you will receive a non-negative amount at maturity, thus exploiting this arbitrage opportunity. ■

It is also possible to create *super-replicating portfolios* whose terminal values dominate that of the corresponding option payoff in all scenarios. We illustrate this in the next example.

Example 13.2.5 *Super-replication*
Your colleague thinks that you should instead replicate the put in the bull market scenario and in the financial crisis scenario only, and therefore ignore the bear market scenario. In other words, she suggests to find the pair $(\overline{x}, \overline{y})$ solving

$$\begin{cases} \overline{x}46 + \overline{y}(1.04) = (37 - 46)_+ = 0, \\ \overline{x}15 + \overline{y}(1.04) = (37 - 15)_+ = 22. \end{cases}$$

We know that the solution to this system of equations is given by

$$\overline{x} = \frac{0-22}{46-15} = -0.7096774,$$

$$\overline{y} = \frac{1}{1.04}\left(\frac{2\times 46 - 0\times 15}{46-15}\right) = 31.38958.$$

The initial value of this strategy is then given by

$$\overline{\Pi}_0 = \overline{x}S_0 + \overline{y}B_0 = -0.7096774\times 40 + 31.38958 = 3.002484.$$

By construction, its value at time 1 is such that

$$\begin{cases} \overline{\pi}_1^u = \overline{x}s^u + \overline{y}(1.04) = 0, \\ \overline{\pi}_1^d = \overline{x}s^d + \overline{y}(1.04) = 22. \end{cases}$$

But what happens if we set up this portfolio and then the bear market scenario is realized? In that case, the time-1 value of the strategy would be equal to

$$\overline{\pi}_1^m = \overline{x}s^m + \overline{y}(1.04) = -0.7096774\times 35 + 31.38958\times 1.04 = 7.806454$$

while the put option payoff would be $(37-35)_+ = 2$. In other words, selling the put and holding this portfolio will generate an extra amount of $7.806454 - 2 = 5.806454$, with probability 0.49, and break even the rest of the time.

The portfolio given by $(\overline{x}, \overline{y}) = (-0.7096774, 31.38958)$ is a *super-replicating portfolio* because

$$\overline{\Pi}_1 = \overline{x}S_1 + \overline{y}B_1 \geq P_1,$$

in all three scenarios. ■

Let us now illustrate another arbitrage opportunity.

Example 13.2.6 *Exploiting an arbitrage opportunity (continued)*
In the previous example, we built a super-replicating portfolio, that is a portfolio such that $\overline{\Pi}_1 \geq P_1$ in all three scenarios. Suppose now that the put option can be sold for \$4. Explain how to exploit an arbitrage opportunity, if any.

The following table shows the value of the put option and of the super-replicating portfolio in all scenarios.

	Stock	Put	Super-rep ptf
Value if bull market	46	0	0
Value if bear market	35	2	7.806454
Value if financial crisis	15	22	22
Initial price	40	TBD	3.002484

Selling the put option and buying the super-replicating portfolio will yield an important profit in the bear market scenario. Now if the put option sells for \$4 on the market, we can simply sell it and buy the super-replicating portfolio, yielding a profit of nearly \$1 at inception ($4 - 3.002 = 0.998$) and an additional gain when the second scenario realizes. This is an arbitrage opportunity since there is no possibility of loss.

If the put option was sold below \$3 (say even \$2.99), then the preceding strategy would be a bet that the bear market will occur at maturity. This is no longer an arbitrage opportunity. Therefore, any price below 3.002484 will prevent arbitrage opportunities. ■

In the previous examples, there were three pairs of scenarios we could choose from to set up portfolios inspired by our work in the binomial model. There is one such pair we have not looked at. More precisely, what happens if we now decide to ignore the bull market scenario?

We are essentially replicating the scenarios where the put matures in the money. We will be short-selling the stock and investing 35.57692308 at the risk-free rate for a negative initial cost (−4.423076923).[3] This is shown in the next table.

	Stock	Put	This ptf
Value if bull market	46	0	−9
Value if bear market	35	2	2
Value if financial crisis	15	22	22
Initial price	40	TBD	−4.423076923

This strategy is sub-replicating the payoff of the put option because $\underline{\Pi}_1 \leq P_1$, in all three scenarios. What this says is that the no-arbitrage price of the put option should be at least −4.423076923, which is not informative since we already found that it should be larger than 0.76923 to avoid arbitrage.

To conclude the above examples, the initial price of the put option should be:

- larger than 0.76923;
- lower than 3.002484;
- larger than −4.4231.

Taking the intersection of the latter three conditions, we find that any price in the interval (0.76923, 3.002484) will not introduce an arbitrage opportunity in this market model.

13.2.3 Super- and sub-replicating strategies

We are now ready to formalize the ideas seen in the previous section. In a one-period trinomial model, we can build three strategies associated to the three pairs of scenarios: (u, m), (u, d) and (m, d). The idea is to *ignore* one scenario, (temporarily) treat the model as a binomial model and build the corresponding *replicating* portfolio. These strategies are not replicating strategies for the trinomial model as we are not mimicking the payoffs in all three scenarios.

Out of these three strategies, there always exists at least one super-replicating strategy and at least one sub-replicating strategy. Figure 13.1 illustrates various payoffs that are convex or concave functions of S_1. A line that is located at or above (below) the three points represents a super- (sub-) replicating strategy. The case of call (upper left panel) and put (upper right panel) options is depicted in the upper panels of Figure 13.1. We see that depending on the shape of the payoff function, we might have two sub-replicating strategies and one super-replicating strategies or vice versa.

3 If you wonder why we obtained a negative cost, this is because the binomial tree built with stock prices of 35 or 15 creates arbitrage opportunities between the risky and the risk-free assets.

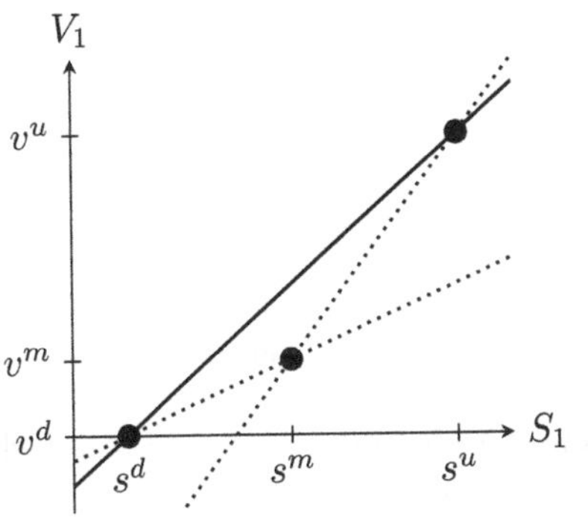

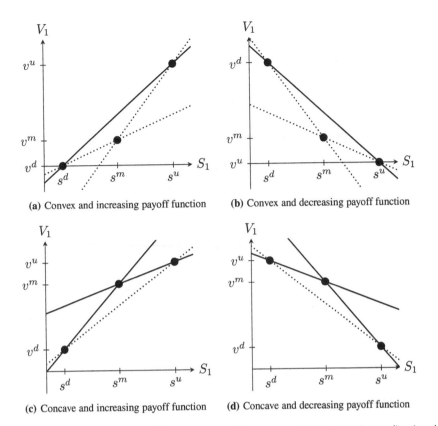

(a) Convex and increasing payoff function

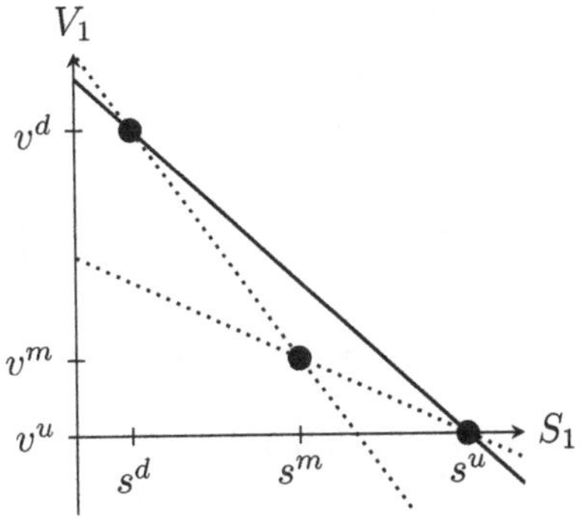

(b) Convex and decreasing payoff function

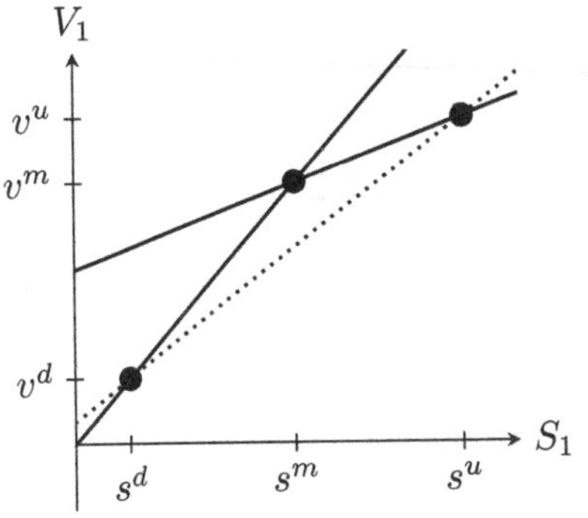

(c) Concave and increasing payoff function

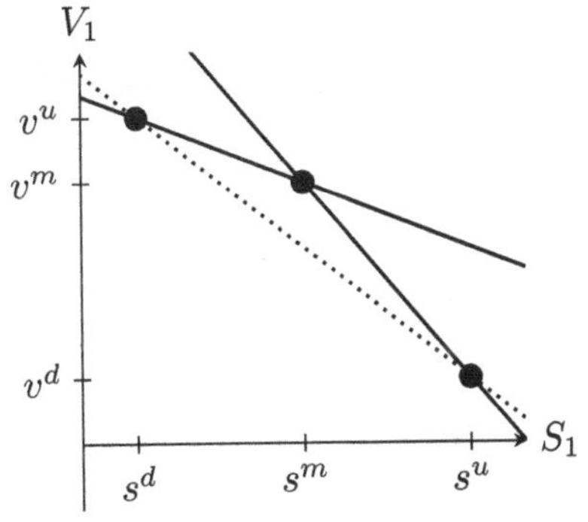

(d) Concave and decreasing payoff function

Figure 13.1 Different payoff functions and their corresponding super-replicating (continuous lines) and sub-replicating (dotted lines) strategies

A **super-replicating strategy** (or portfolio) for a derivative with payoff $V_1 = g(S_1)$ is a pair $(\tilde{x}, \tilde{y})$ such that

$$\tilde{x}S_1 + \tilde{y}B_1 \geq V_1 = g(S_1)$$

in all three scenarios, i.e.

$$\begin{cases} \tilde{x}s^u + \tilde{y}B_1 \geq v^u = g(s^u), \\ \tilde{x}s^m + \tilde{y}B_1 \geq v^m = g(s^m), \\ \tilde{x}s^d + \tilde{y}B_1 \geq v^d = g(s^d). \end{cases}$$

The *best* super-replicating strategy $(\overline{x}, \overline{y})$ is the *cheapest* in the set of all super-replicating strategies. In other words, there might exist many super-replicating strategies $(\tilde{x}, \tilde{y})$ for V_1 and the best is simply the one with the lowest cost. We denote the initial value of $(\overline{x}, \overline{y})$ by $\overline{\Pi}_0$.

In the case of call and put options, a quick plot of the payoff function, as in Figure 13.1, suffices to see that there is only one super-replicating portfolio: the one based upon u and d (see the upper panels of Figure 13.1). Therefore, the *best* super-replicating strategy for a call option is obtained by solving

$$\begin{cases} \overline{x}s^u + \overline{y}B_1 = (s^u - K)_+ \\ \overline{x}s^d + \overline{y}B_1 = (s^d - K)_+, \end{cases} \tag{13.2.2}$$

whose unique solution is given by

$$\bar{x} = \frac{(s^u - K)_+ - (s^d - K)_+}{s^u - s^d}$$

$$\bar{y} = \frac{1}{B_1}\left(\frac{(s^d - K)_+ s^u - (s^u - K)_+ s^d}{s^u - s^d}\right).$$

We can proceed similarly for a put option to find its super-replicating portfolio.

Example 13.2.7 Let us consider again the example where $B_1 = 1.1, S_0 = 6$ and where the possible values for S_1 are $s^u = 7.7$, $s^m = 5.5$ and $s^d = 3.3$.

In this model, the super-replicating portfolio of a put option with strike $K = 6$ is given by

$$\bar{x} = \frac{(6 - s^u)_+ - (6 - s^d)_+}{s^u - s^d} = \frac{0 - 2.7}{7.7 - 3.3} = -0.6136364,$$

$$\bar{y} = \frac{1}{B_1}\left(\frac{(6 - s^d)_+ s^u - (6 - s^u)_+ s^d}{s^u - s^d}\right) = \frac{1}{1.1}\left(\frac{2.7 \times 7.7 - 0 \times 3.3}{7.7 - 3.3}\right)$$

$$= 4.295455.$$

And then its initial value is

$$\overline{\Pi}_0^{\text{put}} = \bar{x}S_0 + \bar{y}B_0 = (-0.6136364) \times 6 + 4.295455 \times 1 = 0.613636.$$ ■

A **sub-replicating strategy** (or portfolio) for a derivative with payoff $V_1 = g(S_1)$ is a pair $(\tilde{x}, \tilde{y})$ such that

$$\tilde{x}S_1 + \tilde{y}B_1 \leq V_1 = g(S_1)$$

in all three scenarios, i.e.

$$\begin{cases} \tilde{x}s^u + \tilde{y}B_1 \leq v^u = g(s^u), \\ \tilde{x}s^m + \tilde{y}B_1 \leq v^m = g(s^m), \\ \tilde{x}s^d + \tilde{y}B_1 \leq v^d = g(s^d). \end{cases}$$

The *best* sub-replicating strategy $(\underline{x}, \underline{y})$ is the *most expensive* in the set of all sub-replicating strategies. In other words, there might exist many sub-replicating strategies $(\tilde{x}, \tilde{y})$ for V_1 and the best is simply the most expensive. We denote the initial value of $(\underline{x}, \underline{y})$ by $\underline{\Pi}_0$.

Due to the convexity of the payoff of call and put options, there are always two sub-replicating strategies: one based upon scenarios u and m, and another based upon scenarios m and d. However, determining the pair of scenarios yielding the best sub-replicating strategy depends on whether $S_0B_1 > s_m$ or $S_0B_1 < s_m$. If $S_0B_1 < s_m$, we should consider m and d.

The methodology is still very similar: select the pair of scenarios u and m, replicate the payoff in those two scenarios and compute the cost of the corresponding sub-replicating portfolio. Then, repeat for the pair of scenarios m and d. This is illustrated in the next example.

Example 13.2.8 Let us consider again Example 13.2.2 where $B_1 = 1.1, S_0 = 6$ and where the possible values for S_1 are $s^u = 7.7$, $s^m = 5.5$ and $s^d = 3.3$. We know there are two sub-replicating portfolios: one obtained by replicating the payoff with scenarios d and m, and another with scenarios m and u.

The first sub-replicating portfolio is given by

$$\tilde{x} = \frac{(6-s^m)_+ - (6-s^d)_+}{s^m - s^d} = \frac{0.5-2.7}{5.5-3.3} = -1,$$

$$\tilde{y} = \frac{1}{B_1}\left(\frac{(6-s^d)_+ s^m - (6-s^m)_+ s^d}{s^m - s^d}\right) = \frac{1}{1.1}\left(\frac{2.7\times 5.5 - 0.5\times 3.3}{5.5-3.3}\right)$$
$$= 5.454545.$$

The initial cost of this portfolio is

$$\Pi_0 = \tilde{x}S_0 + \tilde{y}B_0 = (-1)\times 6 + 5.454545\times 1 = -0.54545.$$

The second sub-replicating portfolio is given by

$$\underline{x} = \frac{(6-s^u)_+ - (6-s^m)_+}{s^u - s^m} = \frac{0-0.5}{7.7-5.5} = -0.2272727,$$

$$\underline{y} = \frac{1}{B_1}\left(\frac{(6-s^m)_+ s^u - (6-s^u)_+ s^m}{s^u - s^m}\right) = \frac{1}{1.1}\left(\frac{0.5\times 7.7 - 0\times 5.5}{7.7-5.5}\right)$$
$$= 1.590909$$

with initial cost

$$\underline{\Pi}_0 = \underline{x}S_0 + \underline{y}B_0 = (-0.2272727)\times 6 + 1.590909\times 1 = 0.2272728.$$

As we used suggestive notation, we conclude that the best sub-replicating portfolio is indeed the second one since $0.2272728 > -0.54545$. ∎

What we should remember from this section is that replicating a payoff V_1 in the one-period trinomial tree model almost never leads to a solution (a unique replicating portfolio). Consequently, there is no such thing as a unique price V_0. Most of the time, we will find a range of prices that prevents arbitrage opportunities: any choice of V_0 in the (open) interval $(\underline{\Pi}_0, \overline{\Pi}_0)$ will not give rise to an arbitrage opportunity in the trinomial tree. This is also true of incomplete market models in general.

In the rare cases where we can find a unique replicating portfolio, the interval will collapse to a single value

$$\Pi_0 = \underline{\Pi}_0 \equiv \overline{\Pi}_0$$

and then we have a unique no-arbitrage price $V_0 = \Pi_0$ corresponding to this strategy's initial value.

Moreover, from a risk management perspective, it is important to understand that the seller of a derivative can no longer be indifferent between:

- buying the equivalent derivative elsewhere on the market (offsetting positions); or
- replicating the liability with basic traded assets.

This is because, in an incomplete market, these two alternatives are not exactly equivalent. Risk management with the second alternative entails some risk.

Which price to use (in an incomplete market)?

In a complete market, there exists a unique replicating strategy for all derivatives. Therefore, risk management is easy: issuing a derivative and holding the corresponding replicating portfolio bears no risk. Moreover, the no-arbitrage price of a derivative is unique.

In an incomplete market, for a given derivative, we find a range of no-arbitrage prices $(\underline{\Pi}_0, \overline{\Pi}_0)$: each price in that interval prevents arbitrage opportunities. This is not very useful: we cannot buy or sell a security using an interval of prices. We need to find *a* price.

However, the largest price at which a buyer is willing to buy the derivative is $\underline{\Pi}_0$ whereas the smallest price at which the seller is willing to sell is $\overline{\Pi}_0$. This is because the best sub- (super-) replicating portfolio assures the buyer (seller) cannot lose. It is therefore impossible to obtain a unique price at which both the seller and buyer will agree. At any price in between ($\underline{\Pi}_0$ and $\overline{\Pi}_0$), the buyer and/or seller will assume some level of risk. Thus, in order to find *a* price, we will need additional assumptions on the market and the behavior of the buyer(s) and seller(s).

13.3 Pricing with risk-neutral probabilities

Remember that in the one-period binomial model, we rewrote expressions for the cost of replicating portfolios and interpreted them as discounted expectations using risk-neutral probabilities. It is possible to carry out a similar exercise in the one-period trinomial tree model by working with sub- and super-replicating portfolios. In doing so, we will also introduce a very important theorem in the theory of derivatives pricing, which is known as the *Fundamental Theorem of Asset Pricing*.

13.3.1 Maximal and minimal probabilities

Let us recall that in the one-period binomial model of Chapter 9, the price of the replicating portfolio for an arbitrary derivative with payoff V_1 is given by

$$\Pi_0 = xS_0 + yB_0$$

with

$$x = \frac{v^u - v^d}{s^u - s^d},$$

$$y = \frac{1}{B_1}\left(\frac{v^d s^u - v^u s^d}{s^u - s^d}\right).$$

By the no-arbitrage principle, we must have that $V_0 = \Pi_0$.

As we saw, it turns out that there exists *unique* risk-neutral probabilities q and $1 - q$, given by

$$q = \frac{S_0 B_1 - s^d}{s^u - s^d},$$

such that

$$V_0 = \frac{1}{B_1}\mathbb{E}^{\mathbb{Q}}[V_1] = \frac{1}{B_1}(qv^u + (1-q)v^d),$$

for any derivative with payoff V_1. The risk-neutral probabilities belong to the one-period binomial model, not to a particular derivative.

In the trinomial model, we found that the cost of the best sub-replicating portfolio $\underline{\Pi}_0$ provides the smallest no-arbitrage price for the corresponding derivative. Even if this portfolio is built to mimick the payoff in only two scenarios, it is still possible to rewrite its cost in terms of a discounted risk-neutral expectation. This will attribute fictitious weights to scenarios u, m and d, similar to what happened in the binomial model. The same can be done for the best super-replicating portfolio.

Let us illustrate the process with call and put options. The cheapest super-replicating portfolio is obtained by considering scenarios u and d, for both options. Hence, we can write

$$\overline{\Pi}_0^{\text{call}} = \frac{1}{B_1}(\overline{q}_u(s^u - K)_+ + (1-\overline{q}_u)(s^d - K)_+),$$

$$\overline{\Pi}_0^{\text{put}} = \frac{1}{B_1}(\overline{q}_u(K - s^u)_+ + (1-\overline{q}_u)(K - s^d)_+),$$

where

$$\overline{q}_u = \frac{S_0B_1 - s^d}{s^u - s^d}.$$

Recalling the no-arbitrage condition of (13.1.1), it is clear that $0 < \overline{q}_u < 1$. As a consequence, we can also write

$$\overline{\Pi}_0^{\text{call}} = \frac{1}{B_1}\mathbb{E}^{\overline{\mathbb{Q}}}[(S_1 - K)_+],$$

$$\overline{\Pi}_0^{\text{put}} = \frac{1}{B_1}\mathbb{E}^{\overline{\mathbb{Q}}}[(K - S_1)_+],$$

where the notation $\overline{\mathbb{Q}}$ means we are attributing probabilities $\{\overline{q}_u, 0, 1-\overline{q}_u\}$ to scenarios $\{u, m, d\}$, respectively.

The best sub-replicating portfolio is obtained by considering scenarios m and d, or scenarios m and u. To illustrate the methodology, suppose that $S_0B_1 < s_m$, in which case we should consider scenarios m and d. Therefore, we obtain

$$\underline{\Pi}_0^{\text{call}} = \frac{1}{B_1}\left(\underline{q}_m(s^m - K)_+ + (1-\underline{q}_m)(s^d - K)_+\right)$$

$$\underline{\Pi}_0^{\text{put}} = \frac{1}{B_1}\left(\underline{q}_m(K - s^m)_+ + (1-\underline{q}_m)(K - s^d)_+\right),$$

where

$$\underline{q}_m = \frac{S_0B_1 - s^d}{s^m - s^d}.$$

Since we assumed that $S_0B_1 < s_m$, it is clear that $0 < \underline{q}_m < 1$. As a consequence, we can also write

$$\underline{\Pi}_0^{\text{call}} = \frac{1}{B_1}\mathbb{E}^{\underline{\mathbb{Q}}}[(S_1 - K)_+],$$

$$\underline{\Pi}_0^{\text{put}} = \frac{1}{B_1}\mathbb{E}^{\underline{\mathbb{Q}}}[(K - S_1)_+],$$

where the notation $\underline{\mathbb{Q}}$ means we are attributing probabilities $\{0, \underline{q}_m, 1-\underline{q}_m\}$ to scenarios $\{u, m, d\}$, respectively. The methodology is similar when $S_0B_1 > s_m$, in which case we should consider scenarios m and u.

For convenience, let us define $\overline{C}_0 = \overline{\Pi}_0^{\text{call}}$ and $\overline{P}_0 = \overline{\Pi}_0^{\text{put}}$, as well as $\underline{C}_0 = \underline{\Pi}_0^{\text{call}}$ and $\underline{P}_0 = \underline{\Pi}_0^{\text{put}}$. Then, the range of call option prices such that there are no arbitrage opportunities is given by

$$(\underline{C}_0, \overline{C}_0),$$

while, for the put, it is given by

$$(\underline{P}_0, \overline{P}_0).$$

Risk-neutral probabilities

You might have noticed that we did not use the terminology *risk-neutral probabilities* for $\underline{q}_m$ or $\overline{q}_u$.

This is because formally, for a probability measure $\mathbb{Q}$ to be a risk-neutral probability measure, it has to be *equivalent* to $\mathbb{P}$. Two probability measures $\mathbb{P}$ and $\mathbb{Q}$ are equivalent when outcomes that are certain or impossible are the same under both measures.

We know that $0 < p_d, p_m, p_u < 1$ and hence all three scenarios are possible. However, both $\underline{\mathbb{Q}}$ and $\overline{\mathbb{Q}}$ assign a zero probability to one of the three possible outcomes and therefore both $\underline{\mathbb{Q}}$ and $\overline{\mathbb{Q}}$ are not equivalent to $\mathbb{P}$. For example, we have $\overline{\mathbb{Q}}(S_1 = s^m) = 0$ while $\mathbb{P}(S_1 = s^m) = p^m > 0$.

13.3.2 Fundamental Theorem of Asset Pricing

The approach we just presented can be framed into a more general setting based upon the *Fundamental Theorem of Asset Pricing*, which is valid for many financial market models such as the binomial model and the trinomial model, as well as the Black-Scholes-Merton model of Chapter 16.

It is beyond the scope of this book to present a formal statement of the FTAP. First, we will present an accessible formulation for binomial and trinomial trees. In Chapter 17, we will present another version for the Black-Scholes-Merton model.

Loosely speaking, the **Fundamental Theorem of Asset Pricing** states that:

1) a market model is free of arbitrage opportunities if and only if there exists at least one set of risk-neutral probabilities;
2) an arbitrage-free market model is complete if and only if there exists a unique set of risk-neutral probabilities.

In a one-period market model, such as the binomial and trinomial models, we obtain risk-neutral probabilities through the following equation:

$$S_0 B_1 = \mathbb{E}^{\mathbb{Q}}[S_1]. \tag{13.3.1}$$

Equation (13.3.1) is known as the **risk-neutral condition.**[4] In other words, risk-neutral probabilities are *artificial probabilities* such that the risky asset S earns the risk-free rate r on average. Of course, this is not true using real-world probabilities: there is a *real* rate of return μ for S that is given implicitly by the model parameters. Let us see how the FTAP works in models we already know.

4 Also known as the martingale condition, see e.g. Chapter 17.

13.3.2.1 One-period binomial model

Let us first illustrate how the FTAP can be applied in the now well-known one-period binomial tree model. First of all, let us assume that the market is free of arbitrage opportunities, i.e. that $s^d < S_0 B_1 < s_u$. In a binomial environment, the risk-neutral condition of equation (13.3.1) is equivalent to

$$S_0 B_1 = \mathbb{E}^{\mathbb{Q}}[S_1] = q s^u + (1-q) s^d.$$

Finding risk-neutral probabilities thus amounts to finding $q \in (0, 1)$.

Clearly, we have a unique solution given by

$$q = \frac{S_0 B_1 - s^d}{s^u - s^d}.$$

Since there is no arbitrage in this market, we have $0 < q < 1$ and also $0 < 1 - q < 1$. As expected from the FTAP, under the no-arbitrage assumption, we were able to find a risk-neutral probability. Finally, the risk-neutral probability q is unique and therefore the one-period binomial tree model is a complete market model.

13.3.2.2 One-period trinomial model

In the one-period trinomial model, the risk-neutral condition of equation (13.3.1) is equivalent to

$$S_0 B_1 = \mathbb{E}^{\mathbb{Q}}[S_1] = q_u s^u + q_m s^m + q_d s^d.$$

Finding risk-neutral probabilities amounts to finding $0 < q_d, q_m, q_u < 1$ such that

$$q_d + q_m + q_u = 1. \tag{13.3.2}$$

We have three unknowns, namely q_d, q_m and q_u, but only two conditions, namely the risk-neutral condition of (13.3.1) and the probability condition of (13.3.2). This means there are infinitely many solutions or, said differently, infinitely many different sets of probabilities $\{q_d, q_m, q_u\}$.

As expected from the FTAP, since we are assuming the trinomial model is free of arbitrage, we are able to find at least one set of risk-neutral probabilities $\{q_d, q_m, q_u\}$.

The converse of the FTAP can be illustrated as follows. Assume there exists a set of risk-neutral probabilities $\{q_d, q_m, q_u\}$. This means that $0 < q_u < 1$, $0 < q_m < 1$ and $0 < q_d = 1 - q_u - q_m < 1$, and that

$$S_0 B_1 = q_u s^u + q_m s^m + q_d s^d.$$

Since $\{q_d, q_m, q_u\}$ are probabilities, we have that $S_0 B_1$ is a weighted average of the values $\{s^d, s^m, s^u\}$, so we have

$$s^d < S_0 B_1 < s^u.$$

This is the no-arbitrage condition for the one-period trinomial model.

13.3.3 Finding risk-neutral probabilities

We always assume that the trinomial tree is free of arbitrage, meaning that equation (13.1.1) is verified. Then, the FTAP guarantees the existence of at least one set of risk-neutral probabilities $\{q_d, q_m, q_u\}$.

As discussed above, we will find infinitely many risk-neutral triplets $\{q_u, q_m, q_d\}$ satisfying conditions (13.3.1) and (13.3.2). We have that $0 < q_d, q_m, q_u < 1$ are such that

$$S_0 B_1 = q_u s^u + q_m s^m + q_d s^d$$
$$1 = q_u + q_m + q_d.$$

Clearly, we can reduce this to a system of one equation with two unknowns:

$$S_0 B_1 = q_u s^u + q_m s^m + (1 - q_u - q_m) s^d$$

or, written differently,

$$S_0 B_1 - s^d = q_u (s^u - s^d) + q_m (s^m - s^d). \tag{13.3.3}$$

In other words, we can parameterize q_u in terms of q_m, or vice versa.[5] However, we have to make sure that

$$0 < q_u < 1,$$
$$0 < q_m < 1,$$
$$0 < 1 - q_u - q_m < 1.$$

These last conditions will restrict the range of values for q_u and q_m, i.e. they will not be allowed to vary in the whole interval $(0, 1)$.

The algebra needed to determine all possible risk-neutral triplets $\{q_u, q_m, q_d\}$ can be tedious but it is elementary, as we will see in the next example.

Example 13.3.1 Let us consider again the example where $B_1 = 1.1$, $S_0 = 6$ and where the possible values for S_1 are $s^u = 7.7$, $s^m = 5.5$ and $s^d = 3.3$. As verified earlier, this trinomial model is free of arbitrage, so according to the FTAP, there are risk-neutral probabilities in this model.

The risk-neutral condition in (13.3.1) can be written here as

$$6 = S_0 = \mathbb{E}^{\mathbb{Q}}\left[\frac{S_1}{B_1}\right] = \frac{1}{1.1}(q_u(7.7) + q_m(5.5) + (1 - q_u - q_m)(3.3)).$$

Collecting the terms for q_u and q_m, we get

$$q_u + (0.5)q_m = 0.75.$$

If we decide to express q_u in terms of q_m, then we have

$$q_u = -(0.5)q_m + 0.75$$

and

$$q_d = 1 - q_u - q_m = 1 - (-(0.5)q_m + 0.75) - q_m = 0.25 - (0.5)q_m.$$

Our free variable q_m might not be allowed to take any value in $(0, 1)$. We must keep in mind that q_u and q_d, as functions of q_m, must take values in $(0, 1)$:

$$0 < q_u = 0.75 - (0.5)q_m < 1,$$
$$0 < 1 - q_u - q_m = 0.25 - (0.5)q_m < 1.$$

5 It is also possible to pick any other pair: for example, we could have decided to parameterize q_d in terms of q_u.

These two inequalities are equivalent to q_m satisfying both

$$1.5 > q_m > -0.5,$$
$$0.5 > q_m > -1.5.$$

This means that q_m must be chosen such that

$$0 < q_m < 0.5,$$

or q_u and q_d might not belong to the interval $(0, 1)$.

In conclusion, we have fully characterized the set of risk-neutral triplets (in terms of q_m). Indeed, for any fixed and arbitrary value of q_m taken in $(0, 0.5)$, the corresponding values for q_u and q_d are given by

$$q_u = -(0.5)q_m + 0.75 \qquad \text{and} \qquad q_d = 0.25 - (0.5)q_m.$$

For example, if we take $q_m = 0.4$, then

$$q_u = -(0.5) \times 0.4 + 0.75 = 0.55 \qquad \text{and} \qquad q_d = 0.25 - (0.5) \times 0.4 = 0.05.$$

and we obtain the following risk-neutral triplet $\mathbb{Q} = \{q_u, q_m, q_d\} = \{0.55, 0.4, 0.05\}$. ■

13.3.4 Risk-neutral pricing

Now that we know how to obtain risk-neutral probabilities $\{q_u, q_m, q_d\}$ in a one-period trinomial tree, the next step is to find the interval of no-arbitrages prices $(\underline{V}_0, \overline{V}_0)$ as discounted risk-neutral expectations.

Let us denote by $\mathbb{Q}$ a set of risk-neutral probability weights $\{q_u, q_m, q_d\}$ (meeting the risk-neutral condition). Then, for a random payoff V_1, the discounted risk-neutral expectation

$$V_0(\mathbb{Q}) = \frac{1}{B_1}\left(q_u v^u + q_m v^m + q_d v^d\right),$$

lies within the interval of no-arbitrage prices meaning that

$$V_0(\mathbb{Q}) \in (\underline{V}_0, \overline{V}_0).$$

We will not provide a proof of this statement, but the discussion leading to both $\underline{V}_0$ and $\overline{V}_0$ should provide the necessary intuition.

This means that for a given risk-neutral probability triplet $\mathbb{Q} = \{q_u, q_m, q_d\}$, the value $V_0(\mathbb{Q})$ is a valid no-arbitrage price for the payoff V_1. The notation $V_0(\mathbb{Q})$ thus emphasizes that this price depends on the specific choice of risk-neutral triplet $\mathbb{Q} = \{q_u, q_m, q_d\}$.

If the values of q_u and q_d are expressed as functions of q_m, then we will write $V_0(q_m)$ instead of $V_0(\mathbb{Q})$ to emphasize that the price is a function of q_m. In other words, we will obtain a parameterized (in terms of q_m) interval of arbitrage-free prices, just as in the super-replication and sub-replication procedure. The next example illustrates the process.

Example 13.3.2 Let us go back to the previous example where $B_1 = 1.1$, $S_0 = 6$, $s^u = 7.7$, $s^m = 5.5$ and $s^d = 3.3$. We have fully characterized the set of risk-neutral triplets in terms of q_m: for each $0 < q_m < 0.5$, if we set

$$q_u = -(0.5)q_m + 0.75 \qquad \text{and} \qquad q_d = 0.25 - (0.5)q_m,$$

then the resulting triplet $\mathbb{Q} = \{q_u, q_m, q_d\}$ is a set of risk-neutral probabilities.

We would like to find the interval of no-arbitrage prices for a put option with strike price $K = 6$ and repeat the procedure for a strike price of $K = 7.7$.

First, if $K = 6$, then $v^u = (6 - s^u)_+ = 0$, $v^m = (6 - s^m)_+ = 0.5$ and $v^d = \left(6 - s^d\right)_+ = 2.7$, and, for a given $\mathbb{Q} = \{q_u, q_m, q_d\}$, we have

$$P_0(\mathbb{Q}) = \frac{1}{1.1}[q_u \times 0 + q_m \times 0.5 + q_d \times 2.7].$$

As we have chosen to express q_u and q_d as functions of q_m, we further obtain

$$\begin{aligned}
P_0(q_m) &= \frac{1}{1.1}[0 + q_m \times 0.5 + \left(0.25 - (0.5)q_m\right) \times 2.7] \\
&= \frac{1}{1.1}\left[0.675 - (0.85)q_m\right] \\
&= 0.613636 - (0.7727273)q_m.
\end{aligned}$$

Clearly, when q_m covers the interval $(0, 0.5)$, the corresponding values of $P_0(q_m)$, which decreases in terms of q_m, will be such that

$$0.2272723 < P_0(q_m) < 0.613636.$$

As we now know, any put option price located within this interval will prevent arbitrage opportunities. We notice that those also correspond to the cost of the best super- and sub-replicating portfolios (see examples 13.2.7 and 13.2.8).

Second, if $K = 7.7$, then $v^u = (7.7 - s^u)_+ = 0$, $v^m = (7.7 - s^m)_+ = 2.2$ and $v^d = (7.7 - s^d)_+ = 4.4$, and, for a given $\mathbb{Q} = \{q_u, q_m, q_d\}$, we have

$$\begin{aligned}
P_0(\mathbb{Q}) &= \frac{1}{1.1}[q_u \times 0 + q_m \times 2.2 + q_d \times 4.4] \\
&= \frac{1}{1.1}[0 + q_m \times 2.2 + (0.25 - (0.5)q_m) \times 4.4] \\
&= \frac{1}{1.1}[1.1 + q_m \times 0] \\
&= 1.
\end{aligned}$$

In other words, no matter what is the triplet $\mathbb{Q} = \{q_u, q_m, q_d\}$, the interval of no-arbitrage prices for this put collapses to a single value, more precisely $P_0 = 1$. This should not be a surprise since we were able to exactly replicate this put option. ■

Maximal and minimal no-arbitrage prices

It is possible to prove that

$$\underline{V}_0 = \inf_{\mathbb{Q}} V_0(\mathbb{Q}) \leq \sup_{\mathbb{Q}} V_0(\mathbb{Q}) = \overline{V}_0,$$

where $\inf_{\mathbb{Q}}$ and $\sup_{\mathbb{Q}}$ mean taking the *infimum* and the *supremum* over all possible risk-neutral triplets $\mathbb{Q} = \{q_u, q_m, q_d\}$, respectively.

In particular, if q_u and q_d are expressed as functions of q_m, then clearly we have

$$\inf_{\mathbb{Q}} V_0(\mathbb{Q}) = \inf_{q_m} V_0(q_m),$$

where $\inf_{q_m}$ means taking the *infimum* within the possible values for q_m. It is now a minimization problem for a one-variable function. Similarly, we have

$$\sup_{\mathbb{Q}} V_0(\mathbb{Q}) = \sup_{q_m} V_0(q_m).$$

13.4 Completion of a trinomial tree

The reader might have noticed that the incompleteness of the trinomial tree model comes from an insufficient number of basic assets to perfectly hedge (to replicate) all risks in the model. Mathematically, this means trying to solve a system of three equations, associated to the three possible outcomes, with only two unknowns, associated to the two basic assets B and S.

If three basic assets were traded in this trinomial market, then replication would mean solving a system of three equations with three unknowns, making replication achievable. Adding a third basic asset, i.e. another risky asset such as another stock or index, will *complete the trinomial market model*, as we will see below.

13.4.1 Trinomial tree with three basic assets

If we add a third basic asset to the previous trinomial tree model, this asset must be such that:

- it cannot be replicated with the first two basic assets, i.e. it is not a *redundant* asset;
- it does not introduce any arbitrage opportunities into the market.

From a financial point of view, the first condition amounts to adding a *new* asset and not just a combination of the previous two basic assets. From a mathematical point of view, we want this asset to help us solve the system of equations associated to the replication problem. Let us illustrate this with an example.

Example 13.4.1 *Trinomial model with two stocks*
Consider a trinomial market model where two stocks are traded, that of ABC inc. and of XYZ inc., each evolving according to a 3-month trinomial tree. This is illustrated in the trees below.

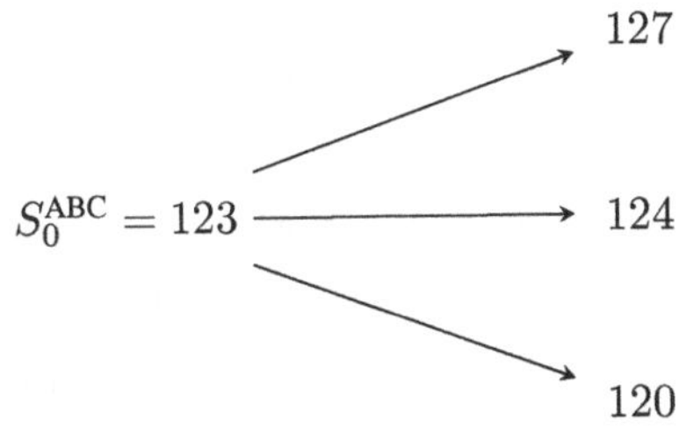

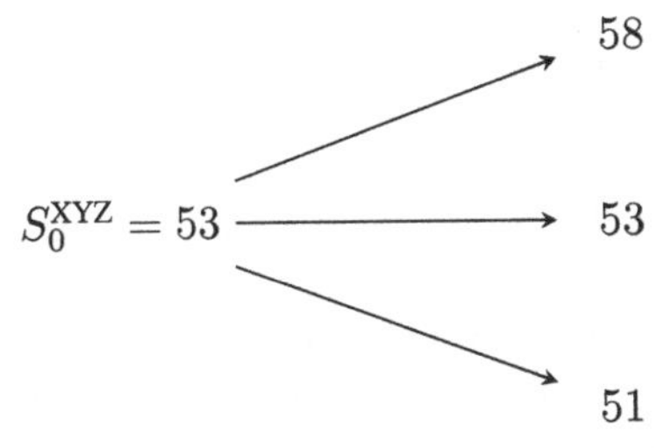

We should note that the up scenario occurs at the same time for both stocks. We then say the events $\{S_1^{ABC} = 127\}$ and $\{S_1^{XYZ} = 58\}$ are the same. This also holds true for the other two scenarios, meaning that the events $\{S_1^{ABC} = 124\}$ and $\{S_1^{XYZ} = 53\}$ are also the same.

Assume also that the annual interest rate, compounded each 3 months, is at 4%. Thus, we have $B_1 = 1.01$. We want to find the no-arbitrage price of a 3-month exotic derivative with payoff

$$V_1 = \left(\frac{S_1^{ABC} + S_1^{XYZ}}{2} - 86 \right)_+ .$$

In each scenario, the value of this payoff is given by

	V_1
u	6.50
m	2.50
d	0

In this *modified* trinomial model, a trading strategy is a triplet (x^{ABC}, x^{XYZ}, y), where x^{ABC} (resp. x^{XYZ} and y) is the number of units of S^{ABC} (resp. S^{XYZ} and B) held in the portfolio from time 0 to time 1. If we want this strategy to replicate the payoff, we must solve:

$$\begin{cases} (127)x^{ABC} + (58)x^{XYZ} + (1.01)y = 6.50 \\ (124)x^{ABC} + (53)x^{XYZ} + (1.01)y = 2.50 \\ (120)x^{ABC} + (51)x^{XYZ} + (1.01)y = 0. \end{cases}$$

Subtracting the first two equations and the last two in order to eliminate $(1.01)y$, we get

$$\begin{cases} 3x^{ABC} + 5x^{XYZ} = 4 \\ 4x^{ABC} + 2x^{XYZ} = 2.5. \end{cases}$$

Consequently,

$$x^{ABC} = 0.32142857 \qquad \text{and} \qquad x^{XYZ} = 0.60714286.$$

Putting these values back into one of the three original equations, we obtain

$$(120)0.32142857 + (51)0.60714286 + (1.01)y = 0,$$

which yields $y = -68.847242$. Therefore, the initial value of this replicating portfolio (for this exotic derivative) is given by

$$\begin{aligned} &(123)x^{ABC} + (53)x^{XYZ} + (1)y \\ &= 123 \times 0.32142857 + 53 \times 0.60714286 - 1 \times 68.847242 \\ &= 2.8670437. \end{aligned}$$

By the no-arbitrage principle, this is also the derivative's initial price. ■

In the previous example, we saw a modified version of the trinomial tree model, now with three basic assets, namely B, S^{ABC} and S^{XYZ}. This time, the fact that we were able to replicate the payoff V_1 in a trinomial environment is not coincidental. Since none of the basic assets is redundant, it allowed us to replicate all the risks in the model. Mathematically, this meant solving a system of three equations, associated to the three possible outcomes, with now three unknowns, associated to the three basic assets B, S^{ABC} and S^{XYZ}.

Here is another example where the third asset will be a credit default swap, as seen in Section 4.4 of Chapter 4.

Example 13.4.2 *Completing a trinomial market by adding a CDS*[6]
Let us have another look at Example 13.0.1. We will introduce a third basic asset, namely a credit default swap (CDS) paying 1$ upon ABC inc.'s default, and 0 otherwise. As we are in a one-period model, assume the CDS requires a single premium at time 0. The following table illustrates the possible realizations of the three basic assets in the three scenarios, along with the payoff of the investment guarantee (IG).

Scenario	Bond B_1	Stock S_1	CDS CDS_1	IG
u	1.04	50	0	50
m	1.04	30	0	40
d (default)	1.04	0	1	40
Initial price	1	40	0.07	TBD

From the table, we have that $S_0 = 40$ and that $\text{CDS}_0 = 0.07$. Let us see how to use the CDS to manage credit risk and replicate the IG.

Let x^S, x^{CDS}, y be the number of shares of the stock, the number of units of CDS and the number of bonds we need to buy to exactly replicate the payoff of the investment guarantee. We need to solve a system of three equations along with three unknowns:

$$\begin{cases} 50 = 1.04y + 50x^S + 0x^{CDS}, \\ 40 = 1.04y + 30x^S + 0x^{CDS}, \\ 40 = 1.04y + 0x^S + 1x^{CDS}. \end{cases}$$

From the first two equations, we directly find $x^S = 0.5$ and $y = 25/1.04 = 24.038461538$. This is the same as in example 13.0.1. Now, from the third equation, we find

$$40 - 1.04y = 15 = x^{CDS}.$$

Therefore, we need to add 15 units of this CDS to the portfolio.

The initial cost of this replicating portfolio (for the IG) is thus

$$y + 40x^S + 0.07x^{CDS} = 44.038461538 + 0.07 \times 15 = 45.08846154.$$

Compared with the strategy deployed in example 13.0.1, for an extra dollar (1.05 to be more precise) the company can completely hedge against credit risk and assure to meet its obligations in all three scenarios.

This is interesting because replicating the first two scenarios might yield a loss of \$15 in case of bankruptcy of ABC inc. Therefore, we need to buy the number of CDS required to protect against a possible loss of \$15. ■

The previous example is pretty realistic in the sense that credit default swaps became commonly traded in the financial markets in the early 2000s. They were one of the many securities that appeared in the markets following the deregulation of financial markets between 1970 and

6 This example is inspired by Example 2 in [18].

2000. *Securitization* is a process commonly used to create tradable securities and it became increasingly important during the same period to help complete the market.

In the insurance business, catastrophe and longevity derivatives, as seen in Chapter 7, can also be viewed as securities helping toward the *completion* of the insurance market. Natural catastrophe risk and longevity risk were traditionally managed by setting aside an appropriate reserve or with reinsurance. The emergence of these insurance derivatives helped risk managers transfer a significant part of these risks to other investors willing to bear them.

13.5 Incompleteness of insurance markets

Market incompleteness arises because there are risks that cannot be perfectly hedged using traded assets. Insurance markets are incomplete by nature: an insurer cannot replicate the payoff of a life insurance policy issued on the life of an individual by trading on the markets, e.g. by buying an equivalent policy from another company. Recall that this issue was discussed in Chapter 1.

The same holds true for most risks covered by life insurance, homeowner's and car insurance, health insurance, etc. Therefore, the risk arising from issuing a call option on a stock needs to be managed differently than the risk coming from selling car insurance.

Example 13.5.1 *Selling car insurance*
Imagine a very simple insurance market: there is one car owner seeking loss protection in case of an accident and there is an insurance company willing to sell such an insurance policy. Over the next year, there is a 2% probability of a car accident with damage \$5000 and a 98% probability of no loss at all. The expected loss is thus \$100.

The customer faces a risk and she seeks to transfer this risk in exchange for a premium. Unfortunately, with only one customer in the market, the insurance company cannot do much to manage this risk. The company cannot resell the policy to another entity, nor can it hedge this risk on the financial market. This over-simplified insurance market is incomplete. ■

Example 13.5.2 *Incompleteness and mortality risk*
Consider an insured investing \$100 in the following insurance product: the initial investment grows with a financial index while the capital is guaranteed upon death or maturity, whichever comes first. Assume the guaranteed amount at time 1 (maturity) is:

- \$112 if the investor dies prior to maturity;
- \$108 if the investor is still alive at maturity.

The current value of the financial index is $S_0 = 2400$ and, at time 1, the index has possible terminal values of $s^u = 3000$ and $s^d = 2100$, which means we assume the price of the financial index follows this tree:

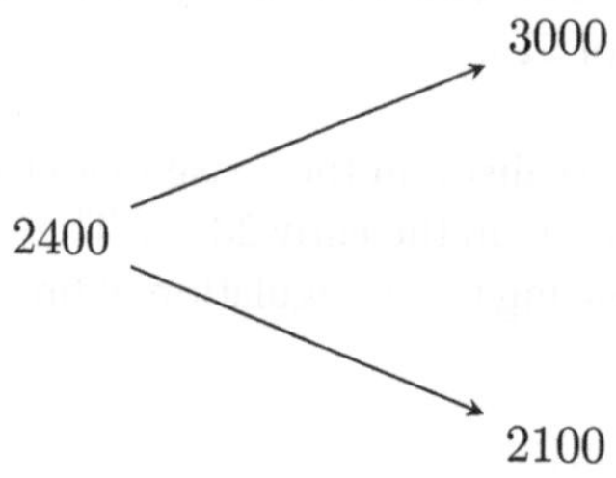

Even if the financial index has two possible values at maturity, there is a total of four outcomes: the policyholder is either dead or alive at maturity, while the index has two possible values. For example, at maturity, the policyholder could be alive and the index could be worth $2100, or the policyholder could have died during the period and the index could be worth $2100, and so on. There are four possibilities, as shown in the next table.

Scenario	Index	Insured	Benefit
1	3000	Alive	125
2	3000	Dead	125
3	2100	Alive	108
4	2100	Dead	112

However, as we can see in this table, benefits only take three different values in these four states because the death and maturity benefits are the same when the index is at 3000.

Consequently, if we are only interested in pricing and risk managing this insurance product, we can model this situation with a one-period trinomial tree model:

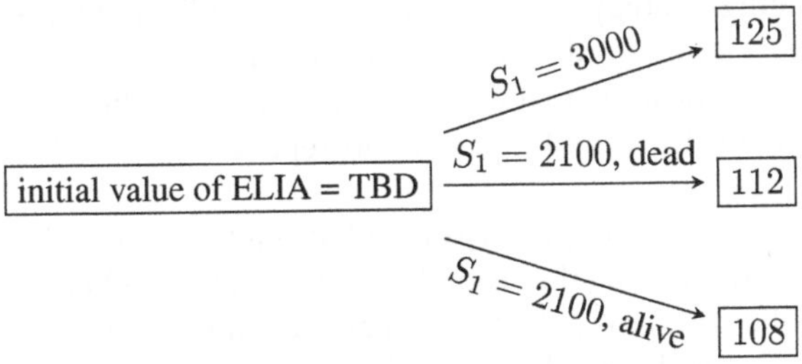

There are two basic assets traded in this market and three possible outcomes driven by the survival of the individual. Market incompleteness arises from the (legal) impossibility to trade individual mortality risk. ∎

We have illustrated how market incompleteness appears in the insurance market. We now illustrate how insurers deal with market incompleteness when they issue typical insurance policies to individuals.

Example 13.5.3 *Selling car insurance (continued)*
Car insurance works well when there are many similar individuals whose risk is independent from one to another. Now assume the company has issued such a policy to 10,000 of these individuals and that if one ends up in a car accident then the likelihood of another insured ending up in another accident is not impacted (statistical independence). Each of these insureds faces an important potential loss, but here insurance will work: the company can write a car insurance policy to each of these individuals and charge a premium of about $100 due to diversification.

It is important to note that even if the insurance market is composed of thousands of i.i.d. individuals, it is still incomplete because it is impossible for an insurance company to trade assets to replicate the risks arising from its car insurance policies. The insurer can *manage* this risk by issuing a sufficient number of policies, thanks to diversification. As seen in Chapter 1, for such a large portfolio of independent and identically distributed risks, it can predict with *almost certainty* the total number of claims. ■

Example 13.5.4 *Incompleteness and mortality risk (continued)*
In the context of the previous example, assume that $1 invested at the risk-free rate accumulates to 1.10 over that period. Assume also that mortality risk (from the insured) is independent of the evolution of the financial index.

Let us describe the risk management implications of selling this equity-linked insurance product with an initial investment of $100 to:

1) a single 60-year-old policyholder having a 1-year death probability of 1%;
2) 10,000 (independent) 60-year-old policyholders each having a 1-year death probability of 1%.

There are two risks embedded in this contract: the financial risk from the random variations of the financial index used to determine the benefit, while the actuarial risk stems from the uncertainty of the survival of the policyholder(s).

1) With a single policyholder in the insurance market, the insurer assumes entirely the uncertainty on the life of the policyholder as it cannot *trade the mortality* of this investor.

One possible hedging strategy could be built as if the policyholder were to survive until maturity, i.e. by ignoring the third scenario (the one where the index goes down to $2100 and the policyholder is dead at maturity, in which case the benefit would be equal to $112). As we now know, this would lead to a sub-replicating portfolio for this investment guarantee.

One more conservative hedging strategy could be built as if the policyholder were to die prior to maturity, i.e. by ignoring the second scenario (the one where the index goes down to $2100 and the policyholder is still alive at maturity, in which case the benefit would be equal to $108). As we now know, this would lead to a super-replicating portfolio for this investment guarantee.

2) Across the 10,000 insureds, the financial risk is systematic, i.e. the financial index will go up or will go down for all 10,000 contracts, while the mortality risk is diversifiable. Now, let us see how the insurance company can manage the risks underlying these contracts, as we did in Section 8.5 of Chapter 8. In particular, we refer to example 8.5.2.

Due to the law of large numbers, we can expect that 100 policyholders will die by the end of the year, whereas 9900 of these will survive. In other words, we expect that 100 contracts will require the payment at maturity of a benefit worth either $125 or $112 depending on the index value at that time, while 9900 contracts will require the payment of a benefit worth either $125 or $108 depending on the index value at that time.

Said differently, selling this investment guarantee to those 10,000 insureds is, on average, equivalent to selling instead:

- 100 investment guarantees with a guaranteed amount of 112;
- 9900 investment guarantees with a guaranteed amount of 108.

As mentioned before, as the number of deaths will not be exactly 100, the insurance company will not be perfectly hedged. In practice, companies set aside a provision for adverse deviations of mortality. ■

These examples have shown that to manage typical insurable risks that cannot be replicated in the markets, insurance companies need to diversify these risks across a large number of policyholders. We have also illustrated that systematic and diversifiable risks need to be treated differently. Ideally, systematic risks should be hedged as much as possible, or even replicated when possible, whereas diversifiable risks should be mitigated by underwriting additional *independent* policyholders.

13.6 Summary

One-period trinomial model

- Risk-free interest rate: r.
- Risk-free asset price: $B = \{B_0, B_1\}$, where $B_0 = 1$ and

$$B_1 = \begin{cases} B_0 e^r & \text{if the risk-free rate is continuously compounded,} \\ B_0(1+r) & \text{if the risk-free rate is periodically compounded.} \end{cases}$$

- Risky asset price: $S = \{S_0, S_1\}$, where S_1 is a random variable such that

$$S_1 = \begin{cases} s^u & \text{with probability } p_u, \\ s^m & \text{with probability } p_m, \\ s^d & \text{with probability } 1 - p_u - p_m. \end{cases}$$

- No-arbitrage condition: $s^d < S_0 B_1 < s^u$.
- Derivative (written on S): $V = \{V_0, V_1\}$ where V_1 is a random variable such that

$$V_1 = \begin{cases} v^u & \text{if } S_1 = s^u, \\ v^m & \text{if } S_1 = s^m, \\ v^d & \text{if } S_1 = s^d. \end{cases}$$

- Simplified tree: up and down factors $u > m > d$ such that

$$s^u = uS_0, \quad s^m = mS_0 \quad \text{and} \quad s^d = dS_0.$$

- Alternative no-arbitrage condition: $d < B_1 < u$.

Trading strategy/portfolio

- Portfolio: a pair (x, y).
- Number of units of S, the risky asset: x.
- Number of units of B, the risk-free asset: y.
- Portfolio's value: $\Pi = \{\Pi_0, \Pi_1\}$ where

$$\Pi_0 = xS_0 + yB_0 \quad \text{and} \quad \Pi_1 = xS_1 + yB_1.$$

Pricing by replication

- A strategy (x, y) replicates V_1 if its time-1 value is such that $\Pi_1 = V_1$, i.e. if

$$\begin{cases} xs^u + yB_1 = v^u, \\ xs^m + yB_1 = v^m, \\ xs^d + yB_1 = v^d. \end{cases}$$

- For most payoffs V_1, a (unique) solution/portfolio cannot be found.
- Attainable payoff: V_1 is attainable if it admits a replicating portfolio.
- Complete market model: all payoffs are attainable.
- One-period trinomial model (with one risky asset) is an incomplete market.
- A strategy (x, y) super-replicates V_1 if its time-1 value is such that $\Pi_1 \geq V_1$, i.e. if

$$\begin{cases} xs^u + yB_1 \geq v^u, \\ xs^m + yB_1 \geq v^m, \\ xs^d + yB_1 \geq v^d. \end{cases}$$

- The *best* super-replicating strategy $(\overline{x}, \overline{y})$ is the *cheapest*. Its initial value is given by $\overline{\Pi}_0$.
- A strategy (x, y) sub-replicates V_1 if its time-1 value is such that $\Pi_1 \leq V_1$, i.e. if

$$\begin{cases} xs^u + yB_1 \leq v^u, \\ xs^m + yB_1 \leq v^m, \\ xs^d + yB_1 \leq v^d. \end{cases}$$

- The *best* sub-replicating strategy $(\underline{x}, \underline{y})$ is the *most expensive*. Its initial value is given by $\underline{\Pi}_0$.
- No-arbitrage price V_0 of a derivative is not unique: there exists an interval of no-arbitrage prices $(\underline{\Pi}_0, \overline{\Pi}_0)$.
- When a security is attainable, the interval collapses to a single value $V_0 = \underline{\Pi}_0 = \overline{\Pi}_0$.

Pricing with risk-neutral probabilities

- Risk-neutral probabilities: triplet $\mathbb{Q} = \{q_d, q_m, q_u\}$ such that

$$\begin{aligned} S_0 B_1 &= q_u s^u + q_m s^m + q_d s^d && \text{(risk-neutral condition)}, \\ 1 &= q_u + q_m + q_d && \text{(probability condition)}. \end{aligned}$$

- For each triplet $\mathbb{Q} = \{q_d, q_m, q_u\}$, there is a corresponding no-arbitrage price given by the discounted risk-neutral expectation:

$$V_0(\mathbb{Q}) = \frac{1}{B_1}\left(q_u v^u + q_m v^m + q_d v^d\right).$$

- In trinomial model:
  - exists infinitely many risk-neutral triplets;
  - no-arbitrage price of a derivative is not unique: there exists an interval of no-arbitrage prices $(\underline{V}_0, \overline{V}_0)$.
- Fundamental Asset Pricing Theorem:
  - a market model is free of arbitrage opportunities if and only if there exists at least one triplet of risk-neutral probabilities;
  - an arbitrage-free market model is complete if and only if there exists a unique triplet of risk-neutral probabilities.

Completion of a trinomial model

- Add another basic risky asset such that:
  - not redundant, i.e. cannot be replicated with the first two basic assets;
  - does not introduce any arbitrage opportunities.
- Consequence of completion: replication is possible (exists unique replicating portfolio) for any payoff. Market model is complete.

Incompleteness of insurance markets

- Incompleteness in general stems from a lack of tradable assets.
- Insurance markets are incomplete by nature: it is not possible to replicate most typical insurable risks using tradable assets. Laws and regulations prevent this.
- Diversification is a mean to attenuate the adverse effects of the incompleteness of insurance markets.

13.7 Exercises

13.1 Suppose there are three states of the economy, namely $\Omega = \{\omega_1, \omega_2, \omega_3\}$, and there exist three securities $S_T^{(i)}$, for $i = 1, 2, 3$, paying \$1 at time T if scenario/state i is observed. Such a market is known as an *Arrow-Debreu market* (or economy). The payoff at time T and the current prices for these securities are shown in the table.

ω_i	$S_0^{(i)}$	$S_T^{(i)}$
ω_1	0.45	1
ω_2	0.40	1
ω_3	0.10	1

Assume that $T = 1$ for simplicity.

(a) To avoid arbitrage opportunities, what is the current price of a security paying \$1 with certainty?

(b) What is the risk-free rate?

(c) What is the risk-neutral probability of observing state ω_2?

(d) To avoid arbitrage opportunities, what is the current price of a stock that is worth 100 in state ω_1, 75 in state ω_2 and 0 otherwise?

13.2 In a financial market, it is possible to buy a share of stock whose value in one period will be in $\{5, 10, 15\}$. Assume all derivatives mature in one period and the risk-free rate is 0%. Determine whether the following derivatives are attainable:

(a) a 10-strike call option;

(b) a 5-strike call option;

(c) a 10-strike put option;

(d) a 15-strike put option.

In (a)–(d), would you change your answer if $B_0 = 1$ and $B_1 > 1$? What is (are) the condition(s) on the payoff to ensure a derivative is attainable? Is this market model complete? Justify your answer.

13.3 Suppose that for \$27, you can buy an asset whose value in one period will be in $\{23, 28, 32\}$. Moreover, the risk-free asset is such that $B_0 = 1$ and $B_1 = 1.1$.

(a) Verify that this model is arbitrage-free.

(b) Is this market model complete? Justify your answer.

(c) Using sub-replication and super-replication, find the interval of no-arbitrage prices for a call option with strike price $K = 30$.

(d) Find all risk-neutral probabilities.
(e) Using risk-neutral pricing, find the interval of no-arbitrage prices for a call option with strike price $K = 30$.

13.4 As in example 13.2.1, a risky asset will take one of the three following values one period later: 7.70, 5.50 or 3.30. Moreover, the risk-free rate is 10% and the initial stock price is 6.60.
(a) Compute the risk-neutral weights of observing u and d if we ignore the existence of scenario m. Is this a probability? Compute the price of a 6-strike put option and relate to the cost of the sub- or super-replicating portfolio.
(b) Compute the risk-neutral weights of observing u and m if we ignore the existence of scenario d. Is this a probability? Compute the price of a 6-strike put option and relate to the cost of the sub- or super-replicating portfolio.
(c) Compute the risk-neutral weights of observing m and d if we ignore the existence of scenario u. Is this a probability? Compute the price of a 6-strike put option and relate to the cost of the sub- or super-replicating portfolio.
(d) In light of (a), (b) and (c), what do you observe?

13.5 Consider the following one-period market model with three scenarios and three assets:

Scenario	$S_1^{(1)}$	$S_1^{(2)}$	$S_1^{(3)}$
1	85	0	0
2	100	7	0
3	125	31	18

with $S_0^{(1)} = 100$, $S_0^{(2)} = 9$ and $S_0^{(3)} = 6$. Build a replicating portfolio using the above three assets only in order to find the value of a risk-free asset paying \$1 in all three scenarios. Then, deduce the interest rate in this market model.

13.6 In the context of example 13.0.1, replicate the investment guarantee assuming the underlying stock follows a binomial model, by ignoring the *bankruptcy scenario*, and set aside an amount corresponding to the average loss in the *bankruptcy scenario*. Compare this hedging strategy to the one using a CDS as presented in example 13.4.2. Does it change your answer if you set aside twice or five times the average loss?

13.7 Verify that the model considered in example 13.4.1 is free of arbitrage opportunities.

13.8 In the context of example 13.5.4 where mortality is managed with a large portfolio, describe the cash flows of the hedging strategy in the scenario that:
- 93 insureds die during the year (instead of 100);
- 109 insureds die during the year (instead of 100).

13.9 In example 13.3.1, we expressed q_u and q_d in terms of q_m. Redo the example and express q_m and q_d in terms of q_u. Verify that the intervals you obtain are consistent with those already obtained and that setting $q_u = 0.55$ also yields the triplet $\mathbb{Q} = \{q_u, q_m, q_d\} = \{0.55, 0.4, 0.05\}$.

13.10 Consider the following two-period market model:

scenario	S_0	S_1	S_2	C_2
1	2	2	2	0
2	2	2	3	0
3	2	4	3	0
4	2	4	4	1
5	2	4	6	3

with $B_0 = (1.1)^{-2}$, $B_1 = (1.1)^{-1}$ and $B_2 = 1$.

An investor believes that the risky asset will be worth less than \$3 at time 2, so she is selling right now the call option with payoff C_2, i.e. written on S and with a strike price of $K = 3$.

(a) Use a tree to describe the dynamics of S.

(b) Verify that this model is free of arbitrage opportunities.

(c) Find a two-period replicating portfolio for this call option and deduce the initial value of the option.

(d) Find the risk-neutral probabilities in this model by looking at each one-period subtree separately.

(e) For each risk-neutral probability, compute the corresponding (risk-neutral) value of the option. Is there a pattern?

(f) With the above computations in hand, would you say that this model is complete? Justify your answer.

13.11 Consider the following two-period market model where B is the risk-free asset and S the risky asset:

scenario	S_0	S_1	S_2	B_0	B_1	B_2
1	1	1	1	1	1.1	1.2
2	1	1	2	1	1.1	1.2
3	1	2	1	1	1.1	1.3
4	1	2	2	1	1.1	1.3
5	1	2	3	1	1.1	1.3

At time 0, Mr. Brown has the following financial position: he holds three units of the risk-free asset and four units of the risky asset. Another investor, Mrs. McFly, is short two units of the risk-free asset and long nine units of the risky asset. They will both hold on to these positions until time two.

Mrs. McFly has an offer for Mr. Brown: at time 2, he can choose to exchange their financial positions, or not.

How much should Mr. Brown give to/receive from Mrs. McFly, at time 0, for this offer to be *fair*? To provide an answer to this question, follow these steps:

(a) Use a tree to describe the dynamics of S and compute the payoff of the above *exchange derivative*.

(b) Verify that this model is free of arbitrage opportunities.

(c) Find the risk-neutral probabilities in this model by looking at each one-period subtree separately.

(d) Is the exchange derivative an attainable derivative? Is this model complete? Justify your answer.

(e) For each risk-neutral probability, compute the corresponding (risk-neutral) value of the exchange derivative.

(f) So, should Mr. Brown give something to Mrs. McFly to enter this agreement? Or is it the other way around?

Exercises 13.10 and 13.11 have been inspired by two exercises from Geneviève Gauthier, with her permission.

Part III

Black-Scholes-Merton model

14

Brownian motion

Brownian motion is a very important stochastic process widely studied in probability and has been applied with success in many areas such as physics, economics and finance. Its name originates from botanist Robert Brown, who studied moving pollen particles in the early 1800s.

The first application of Brownian motion in finance can be traced back to Louis Bachelier in 1900 in his doctoral thesis titled *Théorie de la spéculation*. Then, physicist Albert Einstein and mathematician Norbert Wiener studied Brownian motion from a mathematical point of view. Later, in the 1960s and 1970s, (financial) economists Paul Samuelson, Fischer Black, Myron Scholes and Robert Merton all used this stochastic process for asset price modeling.

This part of the book (Chapters 14–20) is dedicated to the Black-Scholes-Merton model, a framework that laid the foundations of modern finance and largely contributed to the development of options markets in the early 1970s. The cornerstone of the model is a stochastic process known as Brownian motion and whose randomness drives stock prices.

Therefore, this chapter aims at providing the necessary background on Brownian motion to understand the Black-Scholes-Merton model and how to price and manage (hedge) options in that model. We also focus on simulation and estimation of this process, which are very important in practice.

Even if Brownian motion is a sophisticated mathematical object with many interesting properties, the role of this chapter is not to get overly technical but rather to understand the main properties and gain as much intuition as possible. More specifically, the objectives are to:

- provide an introduction to the lognormal distribution;
- compute truncated expectations and the stop-loss transform of a lognormally distributed random variable;
- obtain a standard Brownian motion as the limit of random walks;
- define standard Brownian motion and understand its basic properties;
- relate the construction and properties of linear and geometric Brownian motions to standard Brownian motion;
- simulate standard, linear and geometric Brownian motions to generate scenarios;
- estimate a geometric Brownian motion from a given data set.

14.1 Normal and lognormal distributions

We begin this chapter with a primer on the normal distribution and the lognormal distribution.

Actuarial Finance: Derivatives, Quantitative Models and Risk Management, First Edition.
Mathieu Boudreault and Jean-François Renaud.

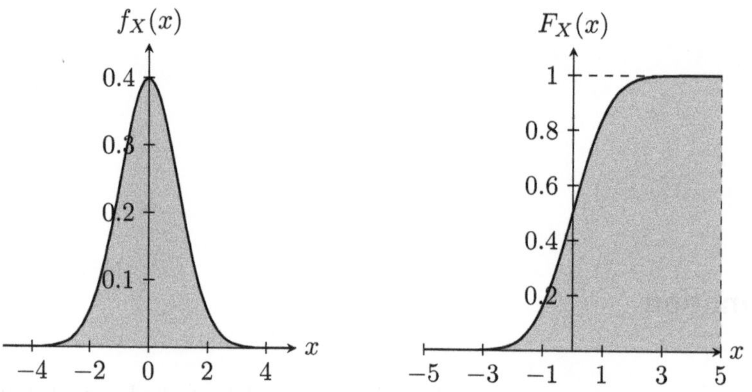

(a) Probability density function **(b)** Cumulative distribution function

Figure 14.1 Standard normal distribution ($\mu = 0$ and $\sigma^2 = 1$)

14.1.1 Normal distribution

A random variable X is said to follow a **normal distribution** with mean $\mu \in \mathbb{R}$ and variance[1] σ^2 if its probability density function (p.d.f.) is given by

$$f_X(x) = \frac{1}{\sqrt{2\pi\sigma^2}} \mathrm{e}^{-\frac{(x-\mu)^2}{2\sigma^2}},$$

for all $x \in \mathbb{R}$, or equivalently if its cumulative distribution function (c.d.f.) is given by

$$\mathbb{P}(X \leq x) = \int_{-\infty}^{x} \frac{1}{\sqrt{2\pi\sigma^2}} \mathrm{e}^{-\frac{(y-\mu)^2}{2\sigma^2}} \mathrm{d}y,$$

for all $x \in \mathbb{R}$. In both cases, this is denoted by

$$X \overset{\mathbb{P}}{\sim} \mathcal{N}(\mu, \sigma^2).$$

In particular, a random variable Z is said to follow a *standard* normal distribution if $\mu = 0$ and $\sigma^2 = 1$ and this is denoted by $Z \overset{\mathbb{P}}{\sim} \mathcal{N}(0, 1)$. The p.d.f. and the c.d.f. of the standard normal distribution are given by

$$\phi(x) = \frac{1}{\sqrt{2\pi}} \mathrm{e}^{-\frac{x^2}{2}} \tag{14.1.1}$$

and

$$N(x) = \int_{-\infty}^{x} \phi(y) \mathrm{d}y,$$

respectively. Because of the symmetry (around zero) of ϕ, we have that $N(x) = 1 - N(-x)$, for all $x \in \mathbb{R}$. The functions $\phi(\cdot)$ and $N(\cdot)$ will be used very frequently in the following chapters. Figure 14.1 shows the p.d.f. (left panel) and c.d.f. (right panel) of the standard normal distribution.

1 It is assumed that $\sigma > 0$.

The notation $\overset{\mathbb{P}}{\sim}$ emphasizes that probabilities are computed with the *(actuarial) probability measure* $\mathbb{P}$. This level of precision about the *probability measure* used is not crucial at this point but will become important for option pricing, as it was in the binomial model.

The normal distribution is also uniquely determined by its moment generating function (m.g.f.): if $X \overset{\mathbb{P}}{\sim} \mathcal{N}(\mu, \sigma^2)$, then its m.g.f. (w.r.t. $\mathbb{P}$) is given by

$$M_X(\lambda) = \mathbb{E}[e^{\lambda X}] = e^{\lambda\mu + \frac{\lambda^2\sigma^2}{2}}, \tag{14.1.2}$$

for all $\lambda \in \mathbb{R}$, and vice versa. Recall that the m.g.f. is a function of λ, while μ and σ^2 are (fixed) parameters.

The normal distribution also has the following properties:

1) If $Z \overset{\mathbb{P}}{\sim} \mathcal{N}(0,1)$, then $X = \mu + \sigma Z \overset{\mathbb{P}}{\sim} \mathcal{N}(\mu, \sigma^2)$. Conversely, if $X \overset{\mathbb{P}}{\sim} \mathcal{N}(\mu, \sigma^2)$, then $Z = (X - \mu)/\sigma \overset{\mathbb{P}}{\sim} \mathcal{N}(0,1)$. This last transformation, from X to Z, is known as *standardization*. Hence, we deduce that

$$\mathbb{P}(X \leq x) = \mathbb{P}\left(Z \leq \frac{x-\mu}{\sigma}\right) = N\left(\frac{x-\mu}{\sigma}\right), \tag{14.1.3}$$

for all $x \in \mathbb{R}$.

2) The normal distribution is *additive*: if $X_1 \overset{\mathbb{P}}{\sim} \mathcal{N}(\mu_1, \sigma_1^2)$ and $X_2 \overset{\mathbb{P}}{\sim} \mathcal{N}(\mu_2, \sigma_2^2)$ are independent, then $X_1 + X_2 \overset{\mathbb{P}}{\sim} \mathcal{N}(\mu_1 + \mu_2, \sigma_1^2 + \sigma_2^2)$.

Bivariate and multivariate normal distributions

We say that the couple (X_1, X_2) follows a *bivariate normal distribution*, or *bivariate Gaussian distribution*, if the bivariate p.d.f. of (X_1, X_2) is given by

$$f_{X_1,X_2}(x_1, x_2) = \frac{1}{2\pi\sigma_1\sigma_2\sqrt{1-\rho^2}} \exp\left(-\frac{z}{2(1-\rho^2)}\right)$$

where

$$z = \frac{(x_1-\mu_1)^2}{\sigma_1^2} + \frac{(x_2-\mu_2)^2}{\sigma_2^2} - \frac{2\rho(x_1-\mu_1)(x_2-\mu_2)}{\sigma_1\sigma_2},$$

for all $x_1, x_2 \in \mathbb{R}$.

The bivariate normal distribution is specified by five parameters: the means (μ_1, μ_2), the variances (σ_1^2, σ_2^2) and the *correlation* $-1 \leq \rho \leq 1$. Indeed, we can deduce that $X_1 \overset{\mathbb{P}}{\sim} \mathcal{N}(\mu_1, \sigma_1^2)$ and $X_2 \overset{\mathbb{P}}{\sim} \mathcal{N}(\mu_2, \sigma_2^2)$, and also that $\mathrm{Cov}(X_1, X_2) = \rho\sigma_1\sigma_2$.

A multivariate normal distribution is defined similarly for a random vector $(X_1, X_2, \ldots, X_n)$. It is characterized by a vector of means and a covariance matrix. The multivariate normal distribution is also very important in modern portfolio theory to determine asset allocation.

14.1.2 Lognormal distribution

A random variable X is said to be lognormally distributed if $Y = \ln(X)$ follows a normal distribution. In other words, if Y follows a normal distribution then $X = e^Y$ is said to follow a

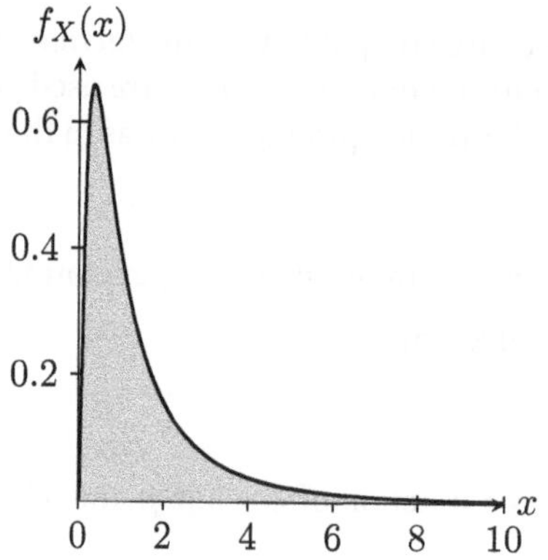

(a) Probability density function

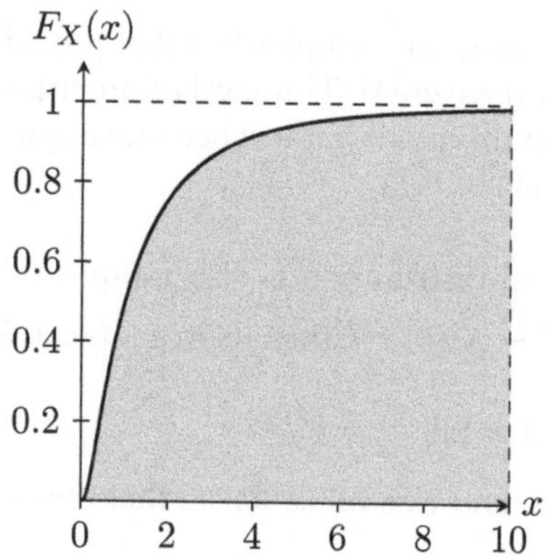

(b) Cumulative distribution function

Figure 14.2 Lognormal distribution with parameters $\mu = 0$ and $\sigma^2 = 1$

lognormal distribution. Note that this means that a lognormal random variable is a positive random variable.

More precisely, we say that X follows a **lognormal distribution** with parameters μ and σ^2, which we will denote by

$$X \overset{\mathbb{P}}{\sim} \mathcal{LN}(\mu, \sigma^2),$$

if $Y = \ln(X) \overset{\mathbb{P}}{\sim} \mathcal{N}(\mu, \sigma^2)$. Then, its p.d.f. is given by

$$f_X(x) = \frac{1}{x\sqrt{2\pi\sigma^2}} e^{-\frac{(\ln(x)-\mu)^2}{2\sigma^2}} \mathbb{1}_{(0,\infty)}(x) = \begin{cases} \frac{1}{x\sqrt{2\pi\sigma^2}} e^{-\frac{(\ln(x)-\mu)^2}{2\sigma^2}} & \text{if } x > 0, \\ 0 & \text{if } x \leq 0. \end{cases} \tag{14.1.4}$$

Figure 14.2 shows the p.d.f. (left panel) and c.d.f. (right panel) of a lognormal distribution. It should be clear that the lognormal distribution is not a symmetric distribution as is the normal distribution. Also, it puts no probability mass on the negative part of the real line.

Example 14.1.1 Assume $X \overset{\mathbb{P}}{\sim} \mathcal{LN}(-1, 2)$. Compute the following probability: $\mathbb{P}(X > 9)$.

First, taking the log on both sides, we can write

$$\{X > 9\} = \{\ln(X) > \ln(9)\}.$$

We know that $Y = \ln(X) \overset{\mathbb{P}}{\sim} \mathcal{N}(-1, 2)$. Then, standardizing Y, i.e. subtracting $\mu = -1$ and dividing by $\sigma = \sqrt{2}$ to obtain a standard normal random variable $Z = \frac{Y-(-1)}{\sqrt{2}}$, we can write

$$\begin{aligned} \mathbb{P}(X > 9) &= \mathbb{P}\left(\frac{Y-(-1)}{\sqrt{2}} > \frac{\ln(9)-(-1)}{\sqrt{2}}\right) \\ &= \mathbb{P}(Z > 2.26) \\ &= 1 - N(2.26) = 0.011910625, \end{aligned}$$

where we used the symmetry (around zero) of the standard normal distribution. ∎

We can generalize the calculations of the last example and identify the c.d.f. of a lognormally distributed random variable $X \overset{\mathbb{P}}{\sim} \mathcal{LN}(\mu, \sigma^2)$. Using equation (14.1.3), we see that it is simply given by

$$\mathbb{P}(X \leq x) = \begin{cases} 0 & \text{if } x \leq 0, \\ N\left(\frac{\ln(x)-\mu}{\sigma}\right) & \text{if } x > 0, \end{cases}$$

since, for $x > 0$, we have $\mathbb{P}(X \leq x) = \mathbb{P}(\ln(X) \leq \ln(x))$, where $\ln(X) \overset{\mathbb{P}}{\sim} \mathcal{N}(\mu, \sigma^2)$.

The lognormal distribution is not *additive*: if X_1 and X_2 are independent and lognormally distributed (possibly with different parameters), then we can *not* say that $X_1 + X_2$ is also lognormally distributed. However, it inherits the following two properties from the normal distribution:

1) If $X \overset{\mathbb{P}}{\sim} \mathcal{LN}(0, 1)$, then $e^{\mu} X^{\sigma} = \exp(\mu + \sigma \ln(X)) \overset{\mathbb{P}}{\sim} \mathcal{LN}(\mu, \sigma^2)$.
2) The lognormal distribution is *multiplicative*: if X_1 and X_2 are independent and lognormally distributed (possibly with different parameters), then $X_1 \times X_2$ is also lognormally distributed.

Let us verify the second property. Assume that $X_1 \overset{\mathbb{P}}{\sim} \mathcal{LN}(\mu_1, \sigma_1^2)$ and $X_2 \overset{\mathbb{P}}{\sim} \mathcal{LN}(\mu_2, \sigma_2^2)$ are independent. This implies that $Y_1 = \ln(X_1) \overset{\mathbb{P}}{\sim} \mathcal{N}(\mu_1, \sigma_1^2)$ and $Y_2 = \ln(X_2) \overset{\mathbb{P}}{\sim} \mathcal{N}(\mu_2, \sigma_2^2)$ are also independent. Since the normal distribution is additive, we have that $Y_1 + Y_2 \overset{\mathbb{P}}{\sim} \mathcal{N}(\mu_1 + \mu_2, \sigma_1^2 + \sigma_2^2)$. Hence,

$$X_1 \times X_2 = e^{Y_1} \times e^{Y_2} = e^{Y_1+Y_2} \overset{\mathbb{P}}{\sim} \mathcal{LN}\left(\mu_1 + \mu_2, \sigma_1^2 + \sigma_2^2\right).$$

This last property will be very useful when dealing with geometric Brownian motion in Section 14.5.

Mean and variance

Computing the mean and variance of the lognormal distribution follows readily from the m.g.f. of the normal distribution as given in equation (14.1.2). Indeed, if $X \overset{\mathbb{P}}{\sim} \mathcal{LN}(\mu, \sigma^2)$, then $Y = \ln(X) \overset{\mathbb{P}}{\sim} \mathcal{N}(\mu, \sigma^2)$ and we can write

$$\mathbb{E}[X] = \mathbb{E}[e^Y] = M_Y(1) = e^{\mu+\frac{\sigma^2}{2}} \tag{14.1.5}$$

and

$$\mathbb{E}[X^2] = \mathbb{E}[e^{2Y}] = M_Y(2) = e^{2\mu+2\sigma^2}.$$

Finally, we deduce that

$$\mathbb{V}\text{ar}(X) = \mathbb{E}[X^2] - \mathbb{E}[X]^2 = e^{2\mu+2\sigma^2} - e^{2\mu+\sigma^2} = e^{2\mu+\sigma^2}\left(e^{\sigma^2} - 1\right).$$

Note that we have not used the p.d.f. (nor the c.d.f.) of the lognormal distribution to compute these first two moments.

High order moments and moment generating function

It is easy to deduce that a lognormally distributed random variable X, with parameters μ and σ^2, has moments of all orders: for any $n \geq 1$, we have

$$\mathbb{E}[X^n] = e^{n\mu + n^2\sigma^2/2}.$$

However, the m.g.f. of X does not exist, which means that we cannot write $\mathbb{E}[e^{\lambda X}]$ for a positive value of λ (this expectation does not exist). In fact, no matter the value of $\lambda > 0$, the integral

$$\mathbb{E}[e^{\lambda X}] = \int_0^{\infty} e^{\lambda x} \frac{1}{x\sqrt{2\pi\sigma^2}} e^{-\frac{(\ln(x)-\mu)^2}{2\sigma^2}} dx$$

diverges. This means that the lognormal distribution is not uniquely determined by its moments.

Truncated expectations

In the following chapters, we will often compute truncated expectations of lognormally distributed random variables. This will be useful for option pricing in the Black-Scholes-Merton model.

Let us consider the following *truncated expectation*

$$\mathbb{E}\left[X \mathbb{1}_{\{X \leq a\}}\right],$$

where $X \overset{\mathbb{P}}{\sim} \mathcal{LN}(\mu, \sigma^2)$ and where $a > 0$.

It is very tempting to immediately write down an integral using the p.d.f. of the lognormal distribution. Instead, let us use the fact that we can write $X = e^{\mu + \sigma Z}$, where

$$Z = \frac{\ln(X) - \mu}{\sigma} \overset{\mathbb{P}}{\sim} \mathcal{N}(0, 1).$$

Then, the truncated expectation can be rewritten as

$$\mathbb{E}\left[X \mathbb{1}_{\{X \leq a\}}\right] = e^{\mu} \mathbb{E}\left[e^{\sigma Z} \mathbb{1}_{\left\{Z \leq \frac{\ln(a)-\mu}{\sigma}\right\}}\right]. \tag{14.1.6}$$

To ease notation, let us set $c = (\ln(a) - \mu)/\sigma$ and compute

$$\mathbb{E}\left[e^{\sigma Z} \mathbb{1}_{\{Z \leq c\}}\right] = \int_{-\infty}^{c} e^{\sigma z} \frac{e^{-\frac{z^2}{2}}}{\sqrt{2\pi}} dz.$$

The trick here is to *complete the square*: since we have that

$$\sigma z - \frac{z^2}{2} = \frac{\sigma^2}{2} - \frac{(z-\sigma)^2}{2},$$

then

$$\mathbb{E}\left[e^{\sigma Z} \mathbb{1}_{\{Z \leq c\}}\right] = e^{\sigma^2/2} \int_{-\infty}^{c} \frac{e^{-\frac{(z-\sigma)^2}{2}}}{\sqrt{2\pi}} dz = e^{\sigma^2/2} N(c - \sigma).$$

Using this last equality together with (14.1.6), we can conclude that

$$\mathbb{E}\left[X \mathbb{1}_{\{X \leq a\}}\right] = e^{\mu + \frac{\sigma^2}{2}} N\left(\frac{\ln(a) - \mu}{\sigma} - \sigma\right).$$

Moreover, since $\mathbb{1}_{\{X>a\}} = 1 - \mathbb{1}_{\{X\leq a\}}$, we can also write

$$\begin{aligned}
\mathbb{E}\left[X\mathbb{1}_{\{X>a\}}\right] &= \mathbb{E}\left[X\left(1-\mathbb{1}_{\{X\leq a\}}\right)\right] \\
&= \mathbb{E}[X] - \mathbb{E}\left[X\mathbb{1}_{\{X\leq a\}}\right] \\
&= e^{\mu+\frac{\sigma^2}{2}} - e^{\mu+\frac{\sigma^2}{2}} N\left(\frac{\ln(a)-\mu}{\sigma} - \sigma\right) \\
&= e^{\mu+\frac{\sigma^2}{2}} \left(1 - N\left(\frac{\ln(a)-\mu}{\sigma} - \sigma\right)\right) \\
&= e^{\mu+\frac{\sigma^2}{2}} N\left(-\frac{\ln(a)-\mu}{\sigma} + \sigma\right),
\end{aligned}$$

where in the last step we used the symmetry of the standard normal distribution.

Stop-loss transforms

Recall that we defined in Chapter 6 the stop-loss function $(x-a)_+$. Using its properties, we can write

$$(X-a)_+ = (X-a)\mathbb{1}_{\{X>a\}} = X\mathbb{1}_{\{X>a\}} - a\mathbb{1}_{\{X>a\}}.$$

The *stop-loss transform*[2] of X, defined by $\mathbb{E}[(X-a)_+]$, is thus

$$\begin{aligned}
\mathbb{E}[(X-a)_+] &= \mathbb{E}\left[X\mathbb{1}_{\{X>a\}}\right] - a\mathbb{E}\left[\mathbb{1}_{\{X>a\}}\right] \\
&= \mathbb{E}\left[X\mathbb{1}_{\{X>a\}}\right] - a\mathbb{P}(X>a).
\end{aligned}$$

Finally, using the above calculations, we get the following identity:

$$\mathbb{E}[(X-a)_+] = e^{\mu+\frac{\sigma^2}{2}} N\left(-\frac{\ln(a)-\mu}{\sigma} + \sigma\right) - aN\left(-\frac{\ln(a)-\mu}{\sigma}\right). \tag{14.1.7}$$

This is the Black-Scholes formula in disguise. We will come back to this in Chapter 16.

14.2 Symmetric random walks

In this section, we provide the necessary background on (symmetric) random walks. It is the only discrete-time stochastic process we will consider in this chapter.

Let us consider a sequence of independent and identically distributed (iid) random variables $\{\epsilon_i, i = 1, 2, \dots\}$ whose common distribution is given by

$$\mathbb{P}(\epsilon_i = 1) = \mathbb{P}(\epsilon_i = -1) = \frac{1}{2}, \quad i = 1, 2, \dots.$$

We define the corresponding **symmetric random walk (SRW)** $X = \{X_n, n \geq 0\}$ by $X_0 = 0$ and, for $n \geq 1$, by

$$X_n = \sum_{i=1}^{n} \epsilon_i = \epsilon_1 + \epsilon_2 + \cdots + \epsilon_n. \tag{14.2.1}$$

It is a discrete-time stochastic process. Figure 14.3 shows a path of a symmetric random walk up to time $n = 10$.

---

2 The stop-loss transform is also known as the *stop-loss premium* in actuarial mathematics when a is a deductible and X is a loss random variable.

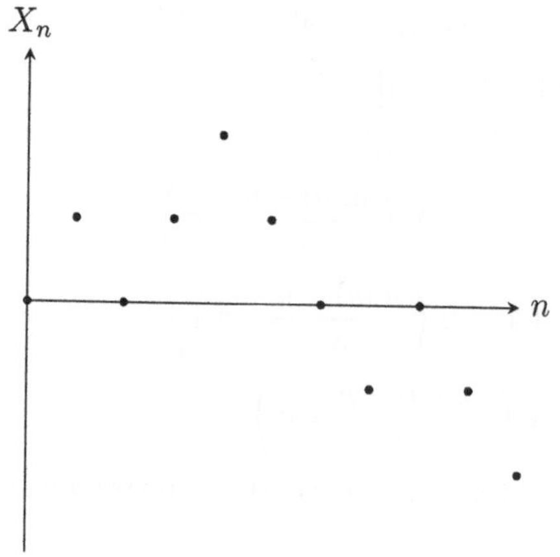

Figure 14.3 A symmetric random walk. At each time step, it moves upward or downward by 1 with equal probability 0.5

The term *symmetric* refers to the symmetry in the distribution of the ϵs: at each time step, the random walk moves up or down by a jump of magnitude 1, with equal probability 1/2. Indeed, for any $n \geq 1$, we have $X_n - X_{n-1} = \epsilon_n$.

If the common distribution of the ϵs is not symmetric, then the random walk $X = \{X_n, n \geq 0\}$ is said to be a **non-symmetric random walk.**

More generally, for times m and n such that $0 \leq m < n$, the corresponding **increment** $X_n - X_m$ is given by

$$X_n - X_m = \sum_{i=1}^{n} \epsilon_i - \sum_{i=1}^{m} \epsilon_i = \sum_{i=m+1}^{n} \epsilon_i = \epsilon_{m+1} + \epsilon_{m+2} + \cdots + \epsilon_n.$$

We deduce easily that the increments of a SRW $X = \{X_n, n \geq 0\}$ are independent (over non-overlapping time intervals) and stationary. More precisely,

1) for any $m \geq 1$ and $0 \leq n_1 < n_2 < \cdots < n_m$, the random variables

$$X_{n_2} - X_{n_1}, \ldots, X_{n_m} - X_{n_{m-1}}$$

are independent;
2) for any $k, n \geq 1$, the random variables $X_{n+k} - X_k$ and X_n have the same distribution.

These two properties rely heavily on the fact that the ϵs are iid. Indeed, in the first case each increment is based on different ϵs whereas in the second case, both $X_{n+k} - X_k$ and X_n are the sum of the same number of ϵs.

A SRW is a *good* model for the evolution of your wealth if you play a coin-toss game. Indeed, assume a *fair* coin is tossed repeatedly. Each time the coin ends up on heads (with probability 1/2), you win \$1, otherwise you lose \$1. Your gain/loss for the first toss (± 1) is modelled by ϵ_1, the second one by ϵ_2, etc. Therefore, X_n is the cumulative amount of gains/losses after n throws, i.e. your wealth after n tosses.

Random walks also play a key role in finance in the weak form of the efficient market hypothesis (EMH). If asset prices abide by this theory, an investor should not be consistently making profits by using past prices. As a result, asset prices should follow a (non-symmetric) random walk.

14.2.1 Markovian property

A SRW possesses the **Markovian property**, i.e. at each time step it depends only on the last known value, not on the preceding ones. Mathematically, for $0 \le m < n$,

$$\mathbb{P}(X_n \le x \mid X_0, X_1, \dots, X_m) = \mathbb{P}(X_n \le x \mid X_m),$$

for any $x \in \mathbb{R}$, or equivalently

$$\mathbb{P}(X_n = z \mid X_0, X_1, \dots, X_m) = \mathbb{P}(X_n = z \mid X_m),$$

for any $z \in \mathbb{Z} = \{\dots, -2, -1, 0, 1, 2, \dots\}$. Clearly, this comes from the fact that

$$X_n = X_m + \sum_{i=m+1}^{n} \epsilon_i = X_m + \epsilon_{m+1} + \dots + \epsilon_n.$$

No extra knowledge about $\epsilon_1, \epsilon_2, \dots, \epsilon_m$ is needed, except for the cumulative value X_m.

Said differently, if we want to predict the value of X_n (in the future) knowing the whole random walk up to time m (the present), only the knowledge of X_m is useful.

Consequently, for a function $g(\cdot)$, we have

$$\mathbb{E}[g(X_n) \mid X_0, X_1, \dots, X_m] = \mathbb{E}[g(X_n) \mid X_m]. \tag{14.2.2}$$

In particular, when $m = n - 1$, we have

$$\mathbb{E}[g(X_n) \mid X_0, X_1, \dots, X_{n-1}] = \mathbb{E}[g(X_n) \mid X_{n-1}].$$

Note that we do not need the random walk to be symmetric for it to be a Markov process.

14.2.2 Martingale property

A SRW possesses the **martingale property**, i.e. the prediction of a process in the future is given by its last known value. Mathematically, for $0 \le m < n$,

$$\mathbb{E}[X_n \mid X_0, X_1, \dots, X_m] = X_m. \tag{14.2.3}$$

Since the SRW is also a Markov process, using (14.2.2), we can further write

$$\mathbb{E}[X_n \mid X_0, X_1, \dots, X_m] = \mathbb{E}[X_n \mid X_m].$$

Again, using the fact that $X_n = X_m + \sum_{i=m+1}^{n} \epsilon_i$ and the linearity property of the conditional expectation, we further have

$$\begin{aligned}
\mathbb{E}[X_n \mid X_0, X_1, \dots, X_m] &= \mathbb{E}\left[X_m + \sum_{i=m+1}^{n} \epsilon_i \,\middle|\, X_m\right] \\
&= \mathbb{E}[X_m \mid X_m] + \mathbb{E}\left[\sum_{i=m+1}^{n} \epsilon_i \,\middle|\, X_m\right] \\
&= X_m + \sum_{i=m+1}^{n} \mathbb{E}[\epsilon_i] \\
&= X_m,
\end{aligned}$$

since $\sum_{i=m+1}^{n} \epsilon_i$ is independent of $X_m = \sum_{j=1}^{m} \epsilon_j$ and $\mathbb{E}[\epsilon_i] = 0$ for each i. The fact that the random walk is symmetric is crucial for the martingale property.

Being a martingale, the coin toss game is said to be a *fair* game because the expected gain/loss at each coin toss (each one-step increment) is equal to zero.

It is important to understand that a stochastic process with the Markov property does not necessarily possess the martingale property. In particular, we know from equation (14.2.2) with $g(x) = x$ that a Markov process X is such that

$$\mathbb{E}[X_n \mid X_0, X_1, \dots, X_m] = \mathbb{E}[X_n \mid X_m],$$

but this conditional expectation is not necessarily equal to X_m as in (14.2.3). We will encounter such a process in Section 14.5.3, namely *geometric Brownian motion.* The converse is also not true, meaning that a stochastic process that has the martingale property does not necessarily have the Markov property.

14.3 Standard Brownian motion

Brownian motion arises naturally as the limit of symmetric random walks. This section presents the construction of (standard) Brownian motion on that basis in addition to studying its properties. We will conclude the section by illustrating how to simulate a Brownian motion.

14.3.1 Construction as the limit of symmetric random walks

The symmetric random walk $X = \{X_n, n \geq 0\}$ defined in equation (14.2.1) is a discrete-time stochastic process. To make it a continuous-time process, we can interpolate the trajectories or, even simpler, keep them constant in between time points. We will choose the latter option. More precisely, for each $t \geq 0$, set

$$X_t = X_k \quad \text{if } k \leq t < k+1,$$

where k is a non-negative integer. Note that for each $t \geq 0$, there is a unique such k: it is called the *integer part* of t and is often written as $\lfloor t \rfloor$. In other words, we now have a continuous-time version of the SRW:

$$X_t = \sum_{i=1}^{\lfloor t \rfloor} \epsilon_i,$$

for each $t \geq 1$, while $X_t = 0$ for each $0 \leq t < 1$. A sample path of this process is shown in Figure 14.4.

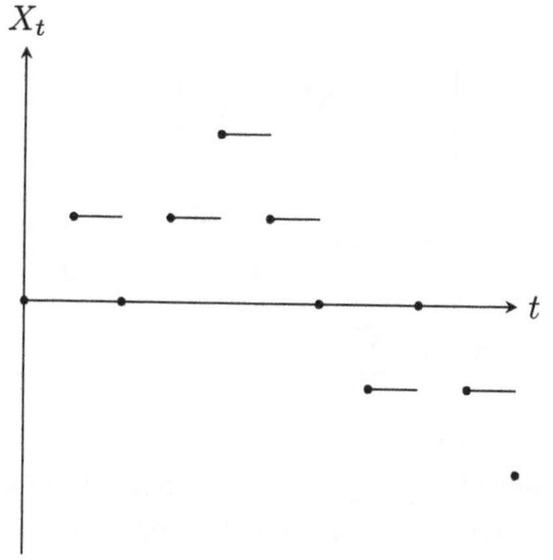

Figure 14.4 A symmetric random walk where the process is kept constant in between time steps

Our objective now is to see what happens (and what we would obtain) if we *speed up* this process, i.e. if we allow for more than one movement per unit time interval. If we want to have n movements per unit time interval, we can simply consider the process $X^{(n)}$ defined by

$$X_t^{(n)} = X_{nt} = \sum_{i=1}^{\lfloor nt \rfloor} \epsilon_i.$$

Indeed, between time 0 and time 1, the process makes n jumps:

$$X_1^{(n)} = \sum_{i=1}^{n} \epsilon_i.$$

In fact, this is the case for any unit time interval: for $t \geq 0$,

$$X_{t+1}^{(n)} - X_t^{(n)} = \sum_{i=\lfloor nt \rfloor+1}^{\lfloor n(t+1) \rfloor} \epsilon_i,$$

where $\lfloor n(t+1) \rfloor - \lfloor nt \rfloor = n$. A sample path of the *accelerated* random walk $X^{(2)}$ is shown in Figure 14.5.

However, those n jumps are all of magnitude 1. We could keep on increasing n and see what happens at the limit, but we know from the Central Limit Theorem (CLT) that this is going nowhere. We must normalize the ϵs if we want to obtain some sort of convergence. Let us make this normalization depend on the number of jumps per unit time interval, that is let the process have jumps of magnitude $1/\sqrt{n}$.

We are now ready to identify the sequence of symmetric random walks that will converge to a standard Brownian motion. For each $n \geq 1$, we define the process $W^{(n)} = \{W_t^{(n)}, t \geq 0\}$ by

$$W_t^{(n)} = \frac{X_t^{(n)}}{\sqrt{n}} = \sum_{i=1}^{\lfloor nt \rfloor} \frac{\epsilon_i}{\sqrt{n}}.$$

See the panels of Figure 14.6 for sample paths of $W^{(1)}$, $W^{(2)}$, $W^{(4)}$ and $W^{(8)}$.

For each fixed n, the continuous-time process $W^{(n)} = \{W_t^{(n)}, t \geq 0\}$ is a sort of (continuous-time) *accelerated* and *rescaled* symmetric random walk. When it moves, the process $W^{(n)}$

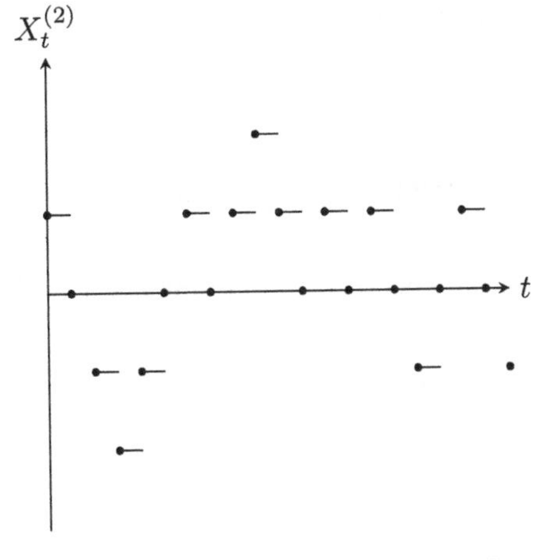

Figure 14.5 An accelerated symmetric random walk with $n = 2$

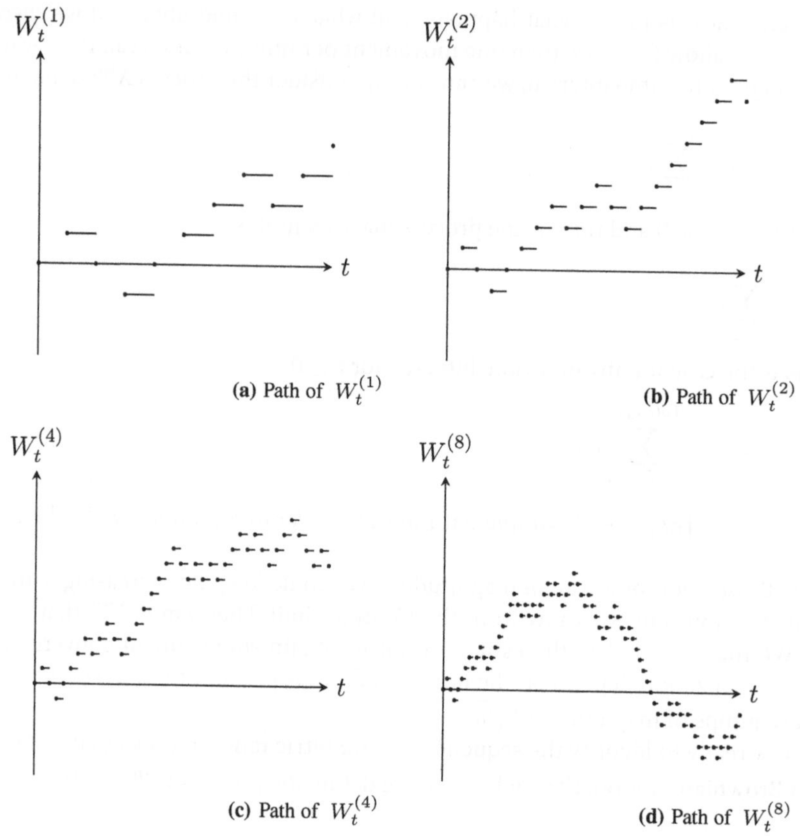

(a) Path of $W_t^{(1)}$ **(b)** Path of $W_t^{(2)}$ **(c)** Path of $W_t^{(4)}$ **(d)** Path of $W_t^{(8)}$

Figure 14.6 Accelerated and rescaled symmetric random walks

moves upward or downward with magnitude $1/\sqrt{n}$ and equal probability $1/2$. Consequently, for each fixed n, the increments of $W^{(n)} = \{W_t^{(n)}, t \geq 0\}$ are independent (over non-overlapping time intervals) and stationary. These properties are inherited from the underlying SRW X.

More precisely, we have:

1) for any $m \geq 1$ and $0 \leq t_1 < t_2 < \cdots < t_m$, the random variables

$$W_{t_2}^{(n)} - W_{t_1}^{(n)}, W_{t_3}^{(n)} - W_{t_2}^{(n)}, \ldots, W_{t_m}^{(n)} - W_{t_{m-1}}^{(n)}$$

are independent;

2) for any $s, t \geq 0$, the random variables $W_{t+s}^{(n)} - W_t^{(n)}$ and $W_s^{(n)}$ have the same distribution.

Note the similarities with the increments of X.

It is now time to take the limit when n goes to infinity. Since $\mathbb{E}[\epsilon_i] = 0$ and $\mathbb{V}\mathrm{ar}(\epsilon_i) = 1$, for each $i \geq 1$, we have that

$$\mathbb{E}\left[W_t^{(n)}\right] = 0 \quad \text{and} \quad \mathbb{V}\mathrm{ar}\left(W_t^{(n)}\right) = \frac{\lfloor nt \rfloor}{n},$$

for any $n \geq 1$ and $t \geq 0$.

From the CLT, we know that

$$W_1^{(n)} = \sum_{i=1}^{n} \frac{\epsilon_i}{\sqrt{n}}$$

converges to a $\mathcal{N}(0,1)$-distributed random variable, as n goes to infinity. Let us denote this random variable obtained at the limit by W_1. In other words, we have obtained

$$W_1^{(n)} \xrightarrow[n\to\infty]{} W_1.$$

For any integer time $t = k \geq 1$, as

$$\mathbb{E}\left[W_k^{(n)}\right] = 0 \qquad \text{and} \qquad \mathbb{V}\text{ar}\left(W_k^{(n)}\right) = k,$$

then, from the CLT, we have that

$$W_k^{(n)} = \sum_{i=1}^{nk} \frac{\epsilon_i}{\sqrt{n}} = \sqrt{k} \sum_{i=1}^{nk} \frac{\epsilon_i}{\sqrt{nk}}$$

converges to a $\mathcal{N}(0,k)$-distributed random variable, as n goes to infinity. Let us denote this random variable obtained at the limit by W_k, for each integer $k \geq 1$.

In general, at any *real* time $t > 0$, we will have that

$$W_t^{(n)} = \sum_{i=1}^{\lfloor nt \rfloor} \frac{\epsilon_i}{\sqrt{n}} = \frac{\sqrt{\lfloor nt \rfloor}}{\sqrt{n}} \sum_{i=1}^{\lfloor nt \rfloor} \frac{\epsilon_i}{\sqrt{\lfloor nt \rfloor}}$$

converges to a $\mathcal{N}(0,t)$-distributed random variable, as n goes to infinity, since

$$\frac{\lfloor nt \rfloor}{n} \longrightarrow t,$$

as n goes to infinity.[3] Let us denote this random variable obtained at the limit by W_t, for each $t > 0$. Therefore, we have obtained: for all $t > 0$,

$$W_t^{(n)} \xrightarrow[n\to\infty]{} W_t.$$

In conclusion, if we further set $W_0 = 0$ and if we regroup the normal random variables W_t just obtained, then we have a continuous-time stochastic process $W = \{W_t, t \geq 0\}$. This process is called a *standard Brownian motion*. Luckily enough, this new process will also have independent and stationary increments.

14.3.2 Definition

Formally, the process $W = \{W_t, t \geq 0\}$ we have just obtained is a **standard Brownian motion (SBM)**, also known as a **Wiener process**. Mathematically, a standard Brownian motion is a continuous-time stochastic process issued from zero ($W_0 = 0$), with independent and normally distributed (stationary) increments:

- for all $n \geq 1$ and for any choice of time points $0 \leq t_1 < t_2 < \cdots < t_n$, the following random variables are independent:

$$W_{t_2} - W_{t_1}, W_{t_3} - W_{t_2}, \ldots, W_{t_n} - W_{t_{n-1}};$$

- for all $s < t$, the random variable $W_t - W_s$ has the same probability distribution as $W_{t-s} \overset{\mathbb{P}}{\sim} \mathcal{N}(0, t-s)$.

3 Indeed, by definition, we have $\lfloor nt \rfloor \leq nt < \lfloor nt \rfloor + 1$. Dividing by n, we get $\lfloor nt \rfloor / n \leq t < (\lfloor nt \rfloor + 1)/n$. The second inequality yields $t - 1/n < \lfloor nt \rfloor / n$ and, together with the first inequality, we get $t - 1/n < \lfloor nt \rfloor / n \leq t$. This concludes the proof.

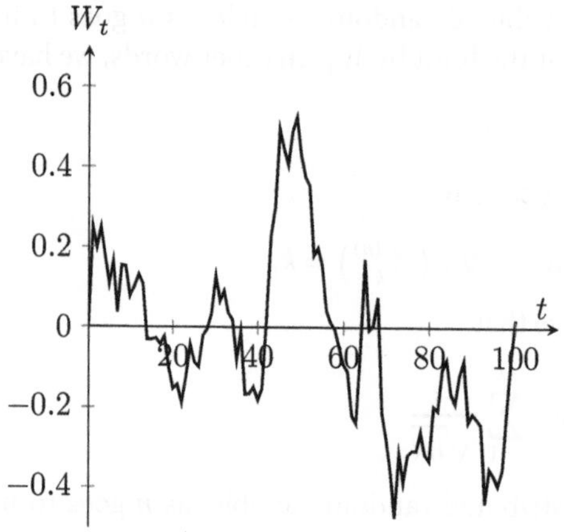

Figure 14.7 Sample path of a standard Brownian motion

Figure 14.7 shows a sample path of a standard Brownian motion. We will explain in Section 14.3.6 how to generate such trajectories.

This definition of standard Brownian motion follows directly from its construction as the limit of random walks, namely its increments over non-overlapping time intervals are independent and normally distributed.

Poisson process

The definition of a Brownian motion should be reminiscent of the definition of a Poisson process, well known in actuarial science. Indeed, a continuous-time stochastic process $N = \{N_t, t \geq 0\}$ is said to be a Poisson process with rate λ if $N_0 = 0$ and if it has the following additional properties:

1) for all $n \geq 1$ and for any choice of time points $0 \leq t_1 < t_2 < \cdots < t_n$, the following random variables are independent:

$$N_{t_2} - N_{t_1}, N_{t_3} - N_{t_2}, \ldots, N_{t_n} - N_{t_{n-1}};$$

2) for all $s < t$, the random variable $N_t - N_s$ has the same probability distribution as N_{t-s}, i.e. a Poisson distribution with mean $\lambda(t - s)$.

14.3.3 Distributional properties

From the definition of Brownian motion, we deduce that for each $t > 0$, the random variable W_t is normally distributed with mean 0 and variance t, i.e. for all $a \in \mathbb{R}$,

$$\mathbb{P}(W_t \leq a) = \int_{-\infty}^{a} \frac{1}{\sqrt{2\pi t}} \mathrm{e}^{-\frac{y^2}{2t}} \, \mathrm{d}y,$$

where $\mathbb{E}[W_t] = 0$ and $\mathbb{V}\mathrm{ar}(W_t) = \mathbb{E}[(W_t)^2] = t$. Note that the variance (and the second moment) increases linearly with time.

Example 14.3.1 What is the probability that a Brownian motion is below −1 at time 3.5?

We want to compute the following probability: $\mathbb{P}(W_{3.5} < -1)$. Since $W_{3.5} \overset{\mathbb{P}}{\sim} \mathcal{N}(0, 3.5)$, then

$$\mathbb{P}(W_{3.5} < -1) = \mathbb{P}\left(\frac{W_{3.5}}{\sqrt{3.5}} < \frac{-1}{\sqrt{3.5}}\right) = N(-1/\sqrt{3.5}) = 0.296490049. \qquad \blacksquare$$

It is not enough to know the distribution of W_t at each time t to fully characterize a Brownian motion. We also need to specify the *dependence structure*. In fact, a standard Brownian motion is a *Gaussian process*, meaning that the random vector $(W_{t_1}, W_{t_2}, \dots, W_{t_n})$, extracted from this process, follows a multivariate normal distribution, for any choice of fixed times $t_1, t_2, \dots, t_n$. In particular, for fixed times s and t, the joint normal distribution of (W_s, W_t) is characterized by its means, variances and its covariance function $\mathbb{C}\text{ov}(W_s, W_t)$.

We already know that $\mathbb{E}[W_t] = 0$, for all $t \geq 0$. So, all that is left is to compute $\mathbb{C}\text{ov}(W_s, W_t)$, for all $s, t \geq 0$, to fully specify the distribution of W. To compute the covariance function, we will rely on the properties of the increments.

First, note that

$$\mathbb{C}\text{ov}(W_s, W_t) = \mathbb{E}[W_s W_t] - \mathbb{E}[W_t]\mathbb{E}[W_s],$$

can be simplified to

$$\mathbb{C}\text{ov}(W_s, W_t) = \mathbb{E}[W_s W_t]$$

since $\mathbb{E}[W_t] = \mathbb{E}[W_s] = 0$. Now, if we assume that $s < t$, then $W_t - W_s$ is independent of $W_s - W_0 = W_s$ and we can write

$$\begin{aligned}
\mathbb{C}\text{ov}(W_s, W_t) &= \mathbb{E}[W_s W_t] \\
&= \mathbb{E}[W_s(W_t - W_s + W_s)] \\
&= \mathbb{E}[W_s(W_t - W_s)] + \mathbb{E}[(W_s)^2] \\
&= \mathbb{E}[W_s]\mathbb{E}[W_t - W_s] + s \\
&= s.
\end{aligned}$$

By symmetry, if we assume that $s > t$, then we get $\mathbb{C}\text{ov}(W_t, W_s) = t$. Consequently, we have obtained that

$$\mathbb{C}\text{ov}(W_s, W_t) = \min\{s, t\}, \tag{14.3.1}$$

for all $s, t \geq 0$.

As a conclusion, the values taken by a Brownian motion at two different time points are not independent (the covariance is not equal to zero). That should have been expected since W_s is somewhat *included in* W_t. Indeed, by definition of the approximating sequence of symmetric random walks, we had

$$W_t^{(n)} = W_s^{(n)} + \left(W_t^{(n)} - W_s^{(n)}\right) = \sum_{i=1}^{\lfloor ns \rfloor} \frac{\epsilon_i}{\sqrt{n}} + \sum_{i=\lfloor ns \rfloor + 1}^{\lfloor nt \rfloor} \frac{\epsilon_i}{\sqrt{n}}.$$

Clearly, $W_s^{(n)}$ and $W_t^{(n)}$ are not independent. Since W_s and W_t can be obtained as the limits of $W_s^{(n)}$ and $W_t^{(n)}$, they are also expected to be dependent, which is the case.

Example 14.3.2 What is the probability that a Brownian motion is below −1 at time 3.5, knowing that it was equal to 1 at time 1.25?

We want to compute the following conditional probability: $\mathbb{P}(W_{3.5} < -1 \mid W_{1.25} = 1)$. Using the independence between $W_{3.5} - W_{1.25}$ and $W_{1.25}$, we can write

$$\begin{aligned}\mathbb{P}(W_{3.5} < -1 \mid W_{1.25} = 1) &= \mathbb{P}((W_{3.5} - W_{1.25}) + W_{1.25} < -1 \mid W_{1.25} = 1)\\ &= \mathbb{P}((W_{3.5} - W_{1.25}) + 1 < -1)\\ &= \mathbb{P}(W_{3.5} - W_{1.25} < -2)\\ &= N\left(\frac{-2-0}{\sqrt{2.25}}\right)\\ &= N(-1.33) = 0.091759136,\end{aligned}$$

where we used the fact that the increment $W_{3.5} - W_{1.25}$ follows a normal distribution with mean 0 and variance 2.25. ■

14.3.4 Markovian property

A standard Brownian motion is a Markov process, i.e. for any $0 \leq t < T$, we have

$$\mathbb{P}(W_T \leq x \mid W_s, 0 \leq s \leq t) = \mathbb{P}(W_T \leq x \mid W_t),$$

for all $x \in \mathbb{R}$. In words, this means that the conditional distribution of W_T given the history of the process up to time t is the same as its conditional distribution given W_t. Saying that a standard Brownian motion is a Markovian process means that for any fixed time t, knowing the value of the random variable W_t provides the same information as knowing the whole (truncated) trajectory $W_s, 0 \leq s \leq t$ for the prediction of future values, in particular that of W_T.

The Markovian property of a standard Brownian motion $W = \{W_t, t \geq 0\}$ is also inherited from the same property for each $W^{(n)} = \{W_t^{(n)}, t \geq 0\}$ in the approximating sequence of accelerated and rescaled symmetric random walks.

Consequently, for a *sufficiently well-behaved* function $g(\cdot)$, we can write

$$\mathbb{E}[g(W_T) \mid W_s, 0 \leq s \leq t] = \mathbb{E}[g(W_T) \mid W_t].$$

Note that the conditional expectation on the right-hand side is a function of W_t, so it is a random variable.

Building on the independence between $W_T - W_t$ and W_t, we can further write

$$\mathbb{E}[g(W_T) \mid W_t] = \mathbb{E}[g(W_T - W_t + W_t) \mid W_t] = \mathbb{E}[g(W_T - W_t + x)]|_{x=W_t}. \tag{14.3.2}$$

Here is how to compute/understand the last expectation:

1) Consider x as a *dummy* variable and compute $\mathbb{E}[g(W_T - W_t + x)]$ using the fact that $W_T - W_t \overset{\mathbb{P}}{\sim} \mathcal{N}(0, T-t)$, which will generate an expression in terms of x.
2) Set $f(x)$ to be this expression computed in step 1).
3) Set $\mathbb{E}[g(W_T) \mid W_t] = f(W_t)$.

Of course, the *temporary* function f and then the final expectation both depend on g. The following two examples will illustrate these steps.

Example 14.3.3 Let us compute $\mathbb{E}[W_T^2 \mid W_t]$ using equation (14.3.2) with $g(x) = x^2$.

The first step is to compute

$$\mathbb{E}[(W_T - W_t + x)^2],$$

where x is a *dummy* variable. Since $W_T - W_t \overset{\mathbb{P}}{\sim} \mathcal{N}(0, T-t)$, expanding the square, we get

$$\begin{aligned}\mathbb{E}[(W_T - W_t + x)^2] &= \mathbb{E}[(W_T - W_t)^2] + 2x\mathbb{E}[W_T - W_t] + x^2 \\ &= (T-t) + 2x \times 0 + x^2 \\ &= T - t + x^2,\end{aligned}$$

which is, as announced, a function of x.

Now, we set $f(x) = (T-t) + x^2$ and, finally, we have

$$\mathbb{E}\left[W_T^2 \mid W_t\right] = f(W_t) = (T-t) + W_t^2.$$

As expected, this random variable is a function of W_t. ■

It turns out that equation (14.3.2) is not *vital* in the previous example, but it will be quite *handy* in the following example.

Example 14.3.4 Let us compute $\mathbb{E}[(W_T)_+ \mid W_t]$ using equation (14.3.2) with $g(x) = (x)_+ = \max(x, 0)$.

The first step is to compute

$$\mathbb{E}[(W_T - W_t + x)_+],$$

where x is a *dummy* variable. Since $W_T - W_t \overset{\mathbb{P}}{\sim} \mathcal{N}(0, T-t)$, then

$$\begin{aligned}\mathbb{E}[(W_T - W_t + x)_+] &= \int_{-\infty}^{\infty} (y+x)_+ \frac{e^{-\frac{y^2}{2(T-t)}}}{\sqrt{2\pi(T-t)}} dy \\ &= \int_{-x}^{\infty} (y+x) \frac{e^{-\frac{y^2}{2(T-t)}}}{\sqrt{2\pi(T-t)}} dy \\ &= \int_{-x}^{\infty} y \frac{e^{-\frac{y^2}{2(T-t)}}}{\sqrt{2\pi(T-t)}} dy + x(1 - N(-x)),\end{aligned}$$

since $y + x = (y+x)_+ > 0$ if and only if $y > -x$. Finally, since

$$\frac{d}{dy}\left[e^{-\frac{y^2}{2(T-t)}}\right] = -\frac{y}{T-t} e^{-\frac{y^2}{2(T-t)}},$$

we deduce that

$$\int_{-x}^{\infty} y \frac{e^{-\frac{y^2}{2(T-t)}}}{\sqrt{2\pi(T-t)}} dy = -\sqrt{\frac{T-t}{2\pi}} e^{-\frac{y^2}{2(T-t)}} \Bigg|_{y=-x}^{\infty} = \sqrt{\frac{T-t}{2\pi}} e^{-\frac{x^2}{2(T-t)}}.$$

We also have that $1 - N(-x) = N(x)$.

Now, we set

$$f(x) = \sqrt{\frac{T-t}{2\pi}} e^{-\frac{x^2}{2(T-t)}} + xN(x)$$

and, finally, we have

$$\mathbb{E}[(W_T)_+ \mid W_t] = \sqrt{\frac{T-t}{2\pi}} e^{-\frac{W_t^2}{2(T-t)}} + W_t N(W_t).$$ ■

14.3.5 Martingale property

Standard Brownian motion also inherits the *martingale property* from the approximating sequence of accelerated and rescaled symmetric random walks. More precisely, a standard Brownian motion $W = \{W_t, t \geq 0\}$ is a martingale, i.e. for any $0 \leq t < T$, we have

$$\mathbb{E}[W_T \mid W_s, 0 \leq s \leq t] = W_t.$$

Note the resemblance with the condition in (14.2.3).

As $W = \{W_t, t \geq 0\}$ is also a Markov process, we only need to show that

$$\mathbb{E}[W_T \mid W_t] = W_t$$

to verify that it is a martingale.

Once again, we will use the fact that $W_T = W_t + (W_T - W_t)$ and the independence between $W_T - W_t$ and W_t. Using the linearity of conditional expectations, we can write

$$\begin{aligned}\mathbb{E}[W_T \mid W_t] &= \mathbb{E}[(W_T - W_t) + W_t | W_t] \\ &= \mathbb{E}[W_T - W_t | W_t] + \mathbb{E}[W_t | W_t] \\ &= \mathbb{E}[W_T - W_t] + W_t \\ &= W_t.\end{aligned}$$

So, a standard Brownian motion is indeed a martingale.

Example 14.3.5 Let us verify that $M_t = W_t^2 - t$, defined for each $t \geq 0$, is a martingale *with respect to Brownian motion.* In other words, we want to verify the following martingale property: for $0 \leq t < T$, we want to have

$$\mathbb{E}\left[W_T^2 - T \mid W_s, 0 \leq s \leq t\right] = W_t^2 - t.$$

Using the Markov property of Brownian motion, we already know that

$$\mathbb{E}\left[W_T^2 - T \mid W_s, 0 \leq s \leq t\right] = \mathbb{E}\left[W_T^2 - T \mid W_t\right]$$

and, by linearity of conditional expectations, we further have

$$\mathbb{E}\left[W_T^2 - T \mid W_t\right] = \mathbb{E}\left[W_T^2 \mid W_t\right] - T.$$

Finally, from example 14.3.3, we already know that

$$\mathbb{E}\left[W_T^2 \mid W_t\right] = (T - t) + W_t^2.$$

Putting the pieces together yields the result. ■

Sample path properties

Figure 14.7 seems to suggest that the trajectories of a Brownian motion are continuous (as functions of time) and it is indeed the case. However, it is beyond the scope of this book to provide a formal proof of the continuity of Brownian motion's paths.

Even though trajectories of a Brownian motion are continuous functions of time, they are nowhere differentiable. In other words, for each state of nature ω, the function $t \mapsto W_t(\omega)$ is continuous but so irregular (it has spikes everywhere once we zoom in) that we cannot make sense of something like $\frac{\mathrm{d}}{\mathrm{d}t} W_t(\omega)$. This fact has huge mathematical consequences. From a modeling point of view, it means that we should be very careful when handling this stochastic process. In Chapter 15, we will come back to sample path properties of Brownian motion, as they motivate the definition of Ito's stochastic integral.

14.3.6 Simulation

For many applications in finance and actuarial science, simulations of Brownian motion trajectories over a given time interval $[0, T]$ help generate stochastic scenarios of useful economic variables such as stock prices and interest rates (see also Chapter 19).

Since Brownian motion is a continuous-time stochastic process, it is impossible to simulate every W_t for every $t \in [0, T]$ as this time interval is (uncountably) infinite. Instead, we choose n time points $0 < t_1 < t_2 < \cdots < t_n = T$ and simulate the random vector $(W_{t_1}, W_{t_2}, \ldots, W_{t_n})$. This random vector is a discretized version of a Brownian motion trajectory over the time interval $[0, T]$. Of course, one should take n as large as possible. Simulation methods for continuous-time stochastic processes are often called *discretization schemes*.

To simulate a (discretized) path $(W(0), W(h), W(2h), \ldots, W((n-1)h), W(T))$ of a standard Brownian motion over the time interval $[0, T]$, we make use of its definition. The algorithm is as follows:

1) We first choose $n \geq 1$ to divide the time interval $[0, T]$ into smaller sub-intervals of the same size $h = T/n$: $0 < h < 2h < 3h < \ldots < (n-1)h < T$.
2) We define $W(0) = 0$ and then, for each $i = 1, 2, \ldots, n$, we:
   a) generate $Z \sim \mathcal{N}(0, 1)$ (see also section 19.1.3);
   b) compute

$$W(ih) = W((i-1)h) + \sqrt{h}Z.$$

3) The output is $(W(0), W(h), W(2h), \ldots, W((n-1)h), W(T))$, a sampled discretized trajectory of a standard Brownian motion, which is a synthetically generated realization of the random vector

$$(W_0, W_h, W_{2h}, \ldots, W_{(n-1)h}, W_T).$$

Figure 14.8 gives two examples of such a discretized Brownian motion trajectory: the grey line is a trajectory with 100 time steps while the dotted line is a trajectory with 20 time steps.

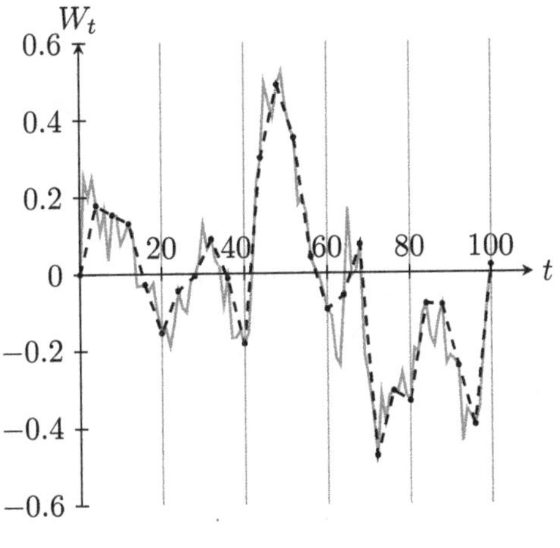

Figure 14.8 Sample path of a discretized standard Brownian motion (SBM)

Even if we know that $W_{ih} \sim \mathcal{N}(0, ih)$, for each $i = 1, 2, \ldots, n$, we cannot simulate successively $W(h) \sim \mathcal{N}(0, h)$, then $W(2h) \sim \mathcal{N}(0, 2h)$, and so on, with finally $W(T) \sim \mathcal{N}(0, T)$. Doing so would generate a sample of independent random variables and we know, from the above discussion, that the random vector $(W_{t_1}, W_{t_2}, \ldots, W_{t_n})$ has a multivariate normal distribution with dependence structure given by the covariance function in equation (14.3.1).

Example 14.3.6 *Simulation of a standard Brownian motion*
Using a computer, you have generated the following four realizations from the standard normal distribution:

1.3265 −0.8355 1.3672 0.8012.

Using the random numbers in this order, let us generate a discretized sample path for a standard Brownian motion over the time interval $[0, \frac{1}{3}]$.

We have $n = 4$ and $T = \frac{1}{3}$ and thus $h = \frac{1}{12}$. Therefore, we define:

$$\begin{aligned}
W(0) &= 0,\\
W(1/12) &= W(0) + \sqrt{1/12} \times 1.3265 = 0.382927566,\\
W(2/12) &= W(1/12) + \sqrt{1/12} \times -0.8355 = 0.141739491,\\
W(3/12) &= W(2/12) + \sqrt{1/12} \times 1.3672 = 0.536416135,\\
W(4/12) &= W(3/12) + \sqrt{1/12} \times 0.8012 = 0.767702653.
\end{aligned}$$

Then,

$$\begin{aligned}
&(0, W(1/12), W(2/12), W(3/12), W(4/12))\\
&= (0, 0.382927566, 0.141739491, 0.536416135, 0.767702653)
\end{aligned}$$

is the corresponding discretized trajectory of a standard Brownian motion. It is shown in Figure 14.9. ■

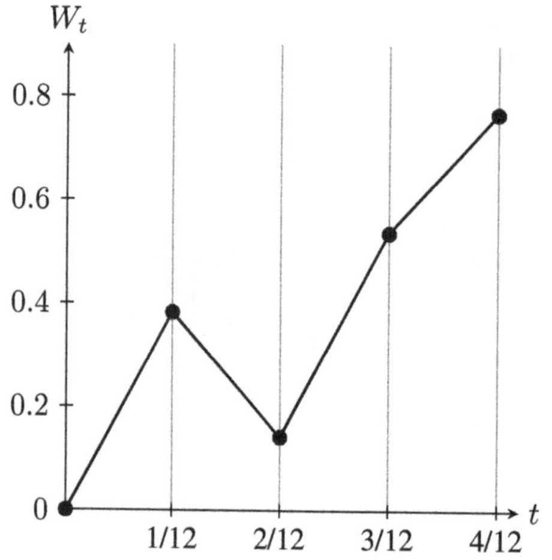

Figure 14.9 Sample path of the standard Brownian motion depicted in example 14.3.6

14.4 Linear Brownian motion

Linear Brownian motion, also known as *arithmetic Brownian motion* or *Brownian motion with drift*, is obtained by transforming a standard Brownian motion using an affine function. In other words, a linear Brownian motion is a translated, *tilted* and *stretched/dilated* Brownian motion.

More precisely, for two constants $\mu \in \mathbb{R}$ and $\sigma > 0$, we define the corresponding **linear Brownian motion**, issued from $X_0 \in \mathbb{R}$, by $X = \{X_t, t \geq 0\}$, where

$$X_t = X_0 + \mu t + \sigma W_t.$$

We call μ the **drift coefficient** and σ the **volatility coefficient** or **diffusion coefficient** of X.

The drift coefficient μ adds a *trend*, upward if $\mu > 0$ or downward if $\mu < 0$, while the volatility coefficient *dilates* (if $\sigma > 1$) or *compresses* (if $\sigma < 1$) the movements of the underlying standard Brownian motion. The effect of μ and σ on the Brownian motion is illustrated in Figure 14.10.

14.4.1 Distributional properties

Because X_t is obtained as a linear transformation of W_t which is normally distributed, X_t is also normally distributed. For each $t > 0$, we have

$$X_t \overset{\mathbb{P}}{\sim} \mathcal{N}(X_0 + \mu t, \sigma^2 t).$$

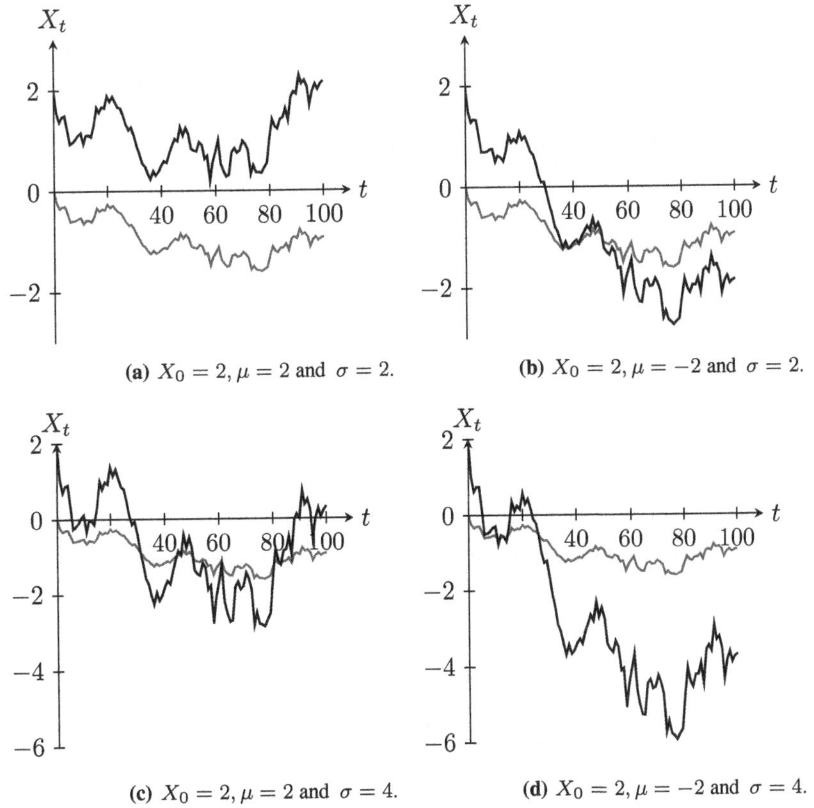

(a) $X_0 = 2, \mu = 2$ and $\sigma = 2$.

(b) $X_0 = 2, \mu = -2$ and $\sigma = 2$.

(c) $X_0 = 2, \mu = 2$ and $\sigma = 4$.

(d) $X_0 = 2, \mu = -2$ and $\sigma = 4$.

Figure 14.10 Sample paths of linear Brownian motions (black) along with the corresponding standard Brownian motion trajectories (grey)

We see that the mean increases (resp. decreases), as time goes by, at rate $\mu > 0$ (resp. $\mu < 0$) while the variance increases at rate σ^2.

Example 14.4.1 What is the probability that a linear Brownian motion, issued from $X_0 = 0$, with drift $\mu = -0.25$ and with volatility $\sigma = 0.1$ is below -1 at time 3.5?

We want to compute the following probability: $\mathbb{P}(X_{3.5} < -1)$, where

$$X_{3.5} = -0.25 \times 3.5 + 0.1 \times W_{3.5}.$$

Since $X_{3.5} \overset{\mathbb{P}}{\sim} \mathcal{N}(-0.25 \times 3.5, (0.1)^2 \times 3.5)$, we can write

$$\begin{aligned}\mathbb{P}(X_{3.5} < -1) &= \mathbb{P}\left(\frac{X_{3.5} + 0.25 \times 3.5}{\sqrt{(0.1)^2 \times 3.5}} < \frac{-1 + 0.25 \times 3.5}{\sqrt{(0.1)^2 \times 3.5}}\right)\\ &= N\left(\frac{-1 + 0.25 \times 3.5}{\sqrt{(0.1)^2 \times 3.5}}\right)\\ &= N(-0.668153105) = 0.252017933.\end{aligned}$$

■

For fixed times $0 \leq s < t$, we have

$$\begin{aligned}X_t - X_s &= (x + \mu t + \sigma W_t) - (x + \mu s + \sigma W_s)\\ &= \mu(t - s) + \sigma(W_t - W_s).\end{aligned}$$

As the increment $W_t - W_s$ is normally distributed with mean 0 and variance $t - s$, the increment $X_t - X_s$ is also normally distributed but with mean $\mu(t - s)$ and variance $\sigma^2(t - s)$. Also, since the increments of a Brownian motion (over disjoint time intervals) are independent, and because a linear Brownian motion is just one affine transformation away from a standard Brownian motion, its increments are also independent.[4]

More precisely, a linear Brownian motion is (also) a stochastic process with independent and normally distributed (stationary) increments:

- for all $n \geq 1$ and for all choice of time points $0 \leq t_1 < t_2 < \cdots < t_n$, the following random variables are independent:

$$X_{t_2} - X_{t_1}, X_{t_3} - X_{t_2}, \ldots, X_{t_n} - X_{t_{n-1}};$$

- for all $s < t$, the random variable $X_t - X_s$ has the same probability distribution as $X_{t-s} - X_0 \overset{\mathbb{P}}{\sim} \mathcal{N}(\mu(t - s), \sigma^2(t - s))$.

Example 14.4.2 For a linear Brownian motion starting at $X_0 = -2$, with drift $\mu = 1$ and diffusion $\sigma = 3$, calculate the probability that the process is below 10 at time 5 if it is already at 6.5 at time 2.

We need to compute $\mathbb{P}(X_5 \leq 10|X_2 = 6.5)$. Subtracting X_2 on both sides of the inequality, we obtain

$$\begin{aligned}\mathbb{P}(X_5 \leq 10|X_2 = 6.5) &= \mathbb{P}(X_5 - X_2 \leq 10 - 6.5|X_2 = 6.5)\\ &= \mathbb{P}(X_5 - X_2 \leq 3.5)\end{aligned}$$

as $X_5 - X_2$ is independent from X_2.

4 Recall that if the random variables X and Y are independent, then, for given functions $f(\cdot)$ and $g(\cdot)$, the random variables $f(X)$ and $g(Y)$ are also independent.

Moreover, the increment $X_5 - X_2$ follows a normal distribution with mean $1 \times (5-2) = 3$ and variance $3^2 \times (5-2) = 27$. Therefore, the desired probability is

$$\mathbb{P}(X_5 - X_2 \leq 3.5) = N\left(\frac{3.5-3}{\sqrt{27}}\right) = N(0.096225045) = 0.53832908. \qquad \blacksquare$$

Finally, we can compute the covariance of a linear Brownian motion at two different time points. Indeed, we have

$$\begin{aligned}\mathrm{Cov}(X_s, X_t) &= \mathrm{Cov}(\mu s + \sigma W_s, \mu t + \sigma W_t)\\ &= \sigma^2 \mathrm{Cov}(W_s, W_t)\\ &= \sigma^2 \min\{s, t\}.\end{aligned}$$

14.4.2 Markovian property

Standard Brownian motion also transfers its Markovian property to linear Brownian motion $X = \{X_t, t \geq 0\}$. For any fixed times t and T such that $0 \leq t < T$, we have

$$\mathbb{P}(X_T \leq x \mid X_s, 0 \leq s \leq t) = \mathbb{P}(X_T \leq x \mid X_t),$$

for all $x \in \mathbb{R}$.

From the definition of linear Brownian motion, we know that $W_s = \sigma^{-1}(X_s - X_0 - \mu s)$, for all $0 \leq s \leq t$. As a result, we quickly realize that conditioning on $X_s, 0 \leq s \leq t$ is equivalent to conditioning on $W_s, 0 \leq s \leq t$.

In other words, saying that the linear Brownian motion $X = \{X_t, t \geq 0\}$ is a Markov process also means that, for any $0 \leq t < T$, we have

$$\mathbb{P}(X_T \leq x \mid W_s, 0 \leq s \leq t) = \mathbb{P}(X_T \leq x \mid W_t),$$

for all $x \in \mathbb{R}$.

Consequently, to compute an expectation of the form

$$\mathbb{E}[g(X_T) \mid X_s, 0 \leq s \leq t] = \mathbb{E}[g(X_T) \mid W_s, 0 \leq s \leq t],$$

where $t \leq T$ and where $g(\cdot)$ is a function, we can rely on the algorithm behind equation (14.3.2). Indeed, since we can write

$$\mathbb{E}[g(X_T) \mid W_t] = \mathbb{E}[g(X_0 + \mu T + \sigma W_T) \mid W_t],$$

we are dealing with the same type of expectation.

Example 14.4.3 Let us compute $\mathbb{E}[X_T^2 \mid X_t]$, where $0 \leq t < T$.

First, let us write

$$\mathbb{E}\left[X_T^2 \mid X_t\right] = \mathbb{E}[(X_0 + \mu T + \sigma W_T)^2 \mid W_t].$$

Using the linearity property of conditional expectations and *expanding the square*, we further have

$$\begin{aligned}\mathbb{E}\left[X_T^2 \mid X_t\right] &= (X_0 + \mu T)^2 + 2(X_0 + \mu T)\sigma \mathbb{E}[W_T \mid W_t] + \sigma^2 \mathbb{E}[(W_T)^2 \mid W_t]\\ &= (X_0 + \mu T)^2 + 2(X_0 + \mu T)\sigma W_t + \sigma^2\left((T-t) + W_t^2\right),\end{aligned}$$

where, in the last step, we used previously computed conditional expectations for standard Brownian motion (see the previous section). $\blacksquare$

14.4.3 Martingale property

In general, linear Brownian motions are *not* martingales because they have a *trend* coming from the drift coefficient μ. However, when $\mu = 0$, the corresponding linear Brownian motion takes the form $X_t = X_0 + \sigma W_t$, for all $t \geq 0$. In this case, it is a martingale.

Formally, for $t < T$, we have

$$\begin{aligned}\mathbb{E}[X_T \mid X_s, 0 \leq s \leq t] &= \mathbb{E}[X_T \mid W_t] \\ &= X_0 + \mu T + \sigma\mathbb{E}[W_T | W_t] \\ &= X_0 + \mu T + \sigma W_t \\ &= \mu(T - t) + X_t,\end{aligned}$$

where, in the second last step, we used the fact that a standard Brownian motion is a martingale.

In conclusion, a linear Brownian motion X possesses the martingale property, i.e.

$$\mathbb{E}[X_T \mid X_s, 0 \leq s \leq t] = X_t,$$

for all $0 \leq t < T$, if and only if $\mu = 0$.

14.4.4 Simulation

To simulate a (discretized) path $(X(0), X(h), X(2h), \dots, X((n-1)h), X(T))$ of a linear Brownian motion with coefficients μ and σ over the time interval $[0, T]$, there are two equivalent algorithms.

The first algorithm mimics the one for a standard Brownian motion:

1) We first choose $n \geq 1$ to divide the time interval $[0, T]$ into smaller sub-intervals of the same size $h = T/n$: $0 < h < 2h < 3h < \dots < (n-1)h < T$.
2) We define $X(0) = X_0$ and then, for each $i = 1, 2, \dots, n$, we:
   a) generate $Z \sim \mathcal{N}(0, 1)$;
   b) compute

$$X(ih) = X((i-1)h) + \mu h + \sigma\sqrt{h}Z.$$

3) The output is $(X(0), X(h), X(2h), \dots, X((n-1)h), X(T))$, a sampled discretized trajectory of a linear Brownian motion, which is a synthetically generated realization of the random vector

$$(X_0, X_h, X_{2h}, \dots, X_{(n-1)h}, X_T).$$

The second algorithm relies on the fact that we might have already simulated a standard Brownian motion, i.e. that we have generated

$$(W(0), W(T/n), W(2T/n), \dots, W((n-1)T/n), W(T)).$$

Then, we apply the corresponding transformation: for each $i = 0, 1, \dots, n$, we set

$$X(ih) = X_0 + \mu \times ih + \sigma \times W(ih).$$

Again, the output is $(X(0), X(T/n), X(2T/n), \dots, X((n-1)T/n), X(T))$ which is illustrated in Figure 14.11.

Example 14.4.4 *Simulation of a linear Brownian motion*
Using the same random numbers as in example 14.3.6 (in the same order), we can generate a discretized sample path for a linear Brownian motion with drift $\mu = 2$ and diffusion $\sigma = 10$, over the time interval $[0, \frac{1}{3}]$. Assume the process starts at 0, i.e. that $X_0 = 0$.

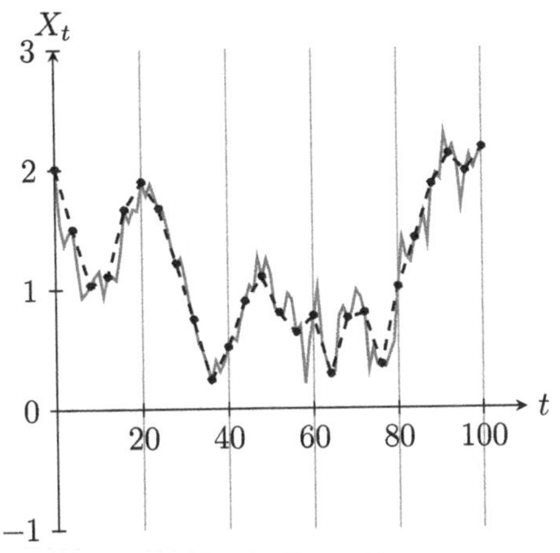

Figure 14.11 Sample path of a discretized linear Brownian motion

Again, we have $n = 4$, $T = \frac{1}{3}$ and $h = \frac{1}{12}$. Therefore,

$$\begin{aligned}
X(0) &= 0, \\
X(1/12) &= X(0) + 2 \times 1/12 + 10\sqrt{1/12} \times 1.3265 = 3.995942327, \\
X(2/12) &= X(1/12) + 2 \times 1/12 + 10\sqrt{1/12} \times -0.8355 = 1.750728244, \\
X(3/12) &= X(2/12) + 2 \times 1/12 + 10\sqrt{1/12} \times 1.3672 = 5.864161351, \\
X(4/12) &= X(3/12) + 2 \times 1/12 + 10\sqrt{1/12} \times 0.8012 = 8.343693196.
\end{aligned}$$

Then,

$$\begin{aligned}
&(0, X(1/12), X(2/12), X(3/12), X(4/12)) \\
&= (0, 3.995942327, 1.750728244, 5.864161351, 8.343693196)
\end{aligned}$$

is the corresponding discretized trajectory of this linear Brownian motion. See Figure 14.12. ■

14.5 Geometric Brownian motion

A *geometric Brownian motion* is obtained by modifying a linear Brownian motion with an exponential function. More precisely, for two constants $\mu \in \mathbb{R}$ and $\sigma > 0$, we define the corresponding **geometric Brownian motion (GBM)**, issued from $S_0 > 0$, by $S = \{S_t, t \geq 0\}$, where

$$S_t = S_0 \exp(\mu t + \sigma W_t) = e^{X_t},$$

where X is a linear Brownian motion issued from $X_0 = \ln(S_0)$, with drift coefficient μ and volatility coefficient σ. Therefore, a GBM is a continuous-time stochastic process taking only positive values, which was not the case for standard and linear Brownian motions. Figure 14.13 illustrates three sample paths of geometric Brownian motions (with different parameters).

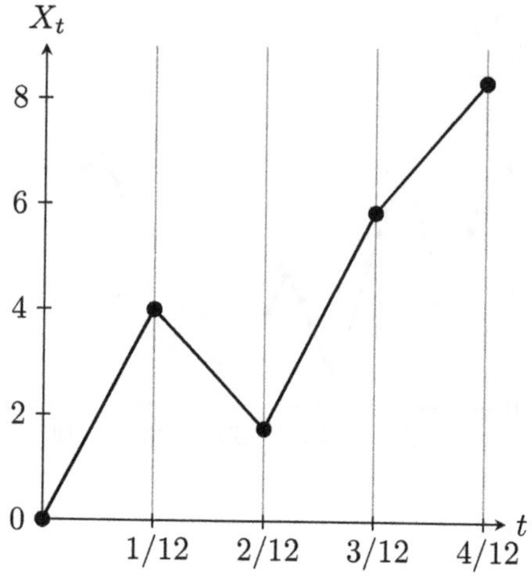

Figure 14.12 Sample path generated in example 14.4.4

14.5.1 Distributional properties

From the definition, we have that S_t is lognormally distributed, for each $t > 0$. Indeed, since we have

$$S_t = \exp(\ln(S_0) + \mu t + \sigma W_t),$$

where

$$\ln(S_0) + \mu t + \sigma W_t \overset{\mathbb{P}}{\sim} \mathcal{N}(\ln(S_0) + \mu t, \sigma^2 t),$$

then

$$S_t \overset{\mathbb{P}}{\sim} \mathcal{LN}(\ln(S_0) + \mu t, \sigma^2 t).$$

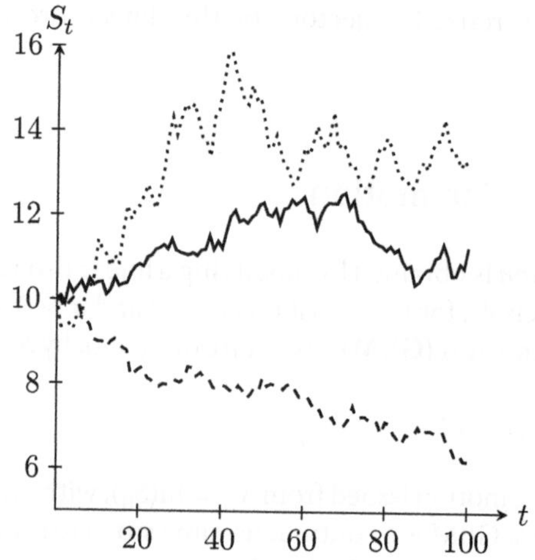

Figure 14.13 Sample paths of geometric Brownian motions (GBMs) with different parameters

We deduce, from equation (14.1.5), that the mean (function) of a GBM is given by

$$\mathbb{E}[S_t] = S_0 e^{\left(\mu+\frac{\sigma^2}{2}\right)t}.$$

We see that this mean increases (geometrically) with time if and only if $\mu + \frac{\sigma^2}{2} > 0$.

Similarly, using the results of Section 14.1.2, we can compute the variance of a GBM (or any higher order moments):

$$\mathbb{V}\mathrm{ar}(S_t) = S_0^2 e^{2\mu t+\sigma^2 t}\left(e^{\sigma^2 t} - 1\right).$$

Geometric Brownian motions have been used for decades as a mathematical model for the price of a risky asset. For example, GBM serves as a model for the stock price in the framework proposed by Black, Scholes and Merton (BSM). They are also used to generate economic scenarios in the banking and insurance industry.

Example 14.5.1 Assume a stock price evolves according to a GBM with parameters $\mu = 0.07$ and $\sigma = 0.3$. The current stock price is $S_0 = 100$. Assume the time unit is a calendar year. What is the probability that the stock price will be greater than \$120 in 3 years from now?

We want to compute $\mathbb{P}(S_3 > 120)$. We have

$$\begin{aligned}\mathbb{P}(S_3 > 120) &= \mathbb{P}(\ln(S_3) > \ln(120)) \\ &= \mathbb{P}(\ln(S_3/S_0) > \ln(120/S_0)) \\ &= \mathbb{P}(\mathcal{N}(3\mu, 3\sigma^2) > \ln(1.2)) \\ &= \mathbb{P}(\mathcal{N}(0.21, 0.27) > \ln(1.2)) \\ &= 1 - N\left(\frac{\ln(1.2) - 0.21}{\sqrt{0.27}}\right) = 0.52124.\end{aligned}$$

■

From the definition, for $0 \le t < T$, we can write

$$\begin{aligned}\frac{S_T}{S_t} &= \frac{S_0 \exp(\mu T + \sigma W_T)}{S_0 \exp(\mu t + \sigma W_t)} \\ &= \exp(\mu(T-t) + \sigma(W_T - W_t)) \\ &= \exp(X_T - X_t).\end{aligned}$$

As $X_T - X_t$ is normally distributed, then S_T/S_t is lognormally distributed.

As opposed to standard and linear Brownian motions, the increments of a geometric Brownian motion are not independent nor stationary. However, since the increments of a linear Brownian motion (over disjoint time intervals) are independent and because a GBM is just an exponential transformation of a linear Brownian motion, then the *relative* increments of the GBM are independent.

More generally,

- for all $n \ge 1$ and for all choice of time points $0 \le t_1 < t_2 < \cdots < t_n$, the following random variables are independent:

$$\frac{S_{t_1}}{S_0}, \frac{S_{t_2}}{S_{t_1}}, \frac{S_{t_3}}{S_{t_2}}, \ldots, \frac{S_{t_n}}{S_{t_{n-1}}};$$

- for all $t < T$, the random variable S_T/S_t has the same probability distribution as $S_{T-t}/S_0 \overset{\mathbb{P}}{\sim} \mathcal{LN}(\mu(T-t), \sigma^2(T-t))$.

From the properties of the relative increments of GBM, we can easily compute a conditional expectation of the form $\mathbb{E}[S_T \mid S_t]$, where $t < T$. Indeed, we can write

$$\begin{aligned}\mathbb{E}[S_T \mid S_t] &= \mathbb{E}\left[S_t \frac{S_T}{S_t} \middle| S_t\right] \\ &= S_t \mathbb{E}\left[\frac{S_T}{S_t} \middle| S_t\right] \\ &= S_t \mathbb{E}\left[\frac{S_T}{S_t}\right]\end{aligned}$$

because $\frac{S_T}{S_t}$ is independent of S_t (or equivalently of $\frac{S_t}{S_0}$). Consequently, since

$$\mathbb{E}\left[\frac{S_T}{S_t}\right] = e^{(\mu+\sigma^2/2)(T-t)}$$

because $S_T/S_t \overset{\mathbb{P}}{\sim} \mathcal{LN}(\mu(T-t), \sigma^2(T-t))$, we have that

$$\mathbb{E}[S_T \mid S_t] = S_t e^{(\mu+\sigma^2/2)(T-t)}. \tag{14.5.1}$$

As geometric Brownian motions are widely used to model asset prices, it is important to interpret this model from a financial standpoint. For $t_2 > t_1$, the random variable S_{t_2}/S_{t_1} is the *accumulation factor* of \$1 invested in S over the time interval $[t_1, t_2]$. Taking the logarithm of this accumulation factor, we get

$$\ln\left(\frac{S_{t_2}}{S_{t_1}}\right),$$

the log-return of this asset, between time t_1 and time t_2.

Since the log-return has a normal distribution with mean $\mu(t_2 - t_1)$ and variance $\sigma^2(t_2 - t_1)$, μ can be interpreted as the *mean annual log-return* whereas σ is the *annual volatility* of the asset's log-returns.

14.5.2 Markovian property

Since geometric Brownian motion is one deterministic transformation away from standard Brownian motion (or linear Brownian motion), it is also a Markov process. This means that for any $0 \le t < T$, we have

$$\mathbb{P}(S_T \le x \mid S_u, 0 \le u \le t) = \mathbb{P}(S_T \le x \mid S_t),$$

for all $x \in \mathbb{R}$.

As before, for a fixed $t > 0$, conditioning on $S_u, 0 \le u \le t$ is equivalent to conditioning on $X_u, 0 \le u \le t$ or even $W_u, 0 \le u \le t$. Also, note that if we know the value of S_t, then we know the value of W_t, and vice versa. Consequently, for a GBM, the Markovian property can be restated as follows:

$$\mathbb{P}(S_T \le x \mid S_u, 0 \le u \le t) = \mathbb{P}(S_T \le x \mid W_t),$$

for all $x \in \mathbb{R}$.

To compute an expectation of the form

$$\mathbb{E}[g(S_T) \mid S_u, 0 \le u \le t] = \mathbb{E}[g(S_T) \mid W_t],$$

where $g(\cdot)$ is a function, we can rely on the algorithm based on equation (14.3.2). Again, since we can write

$$\mathbb{E}[g(S_T) \mid W_t] = \mathbb{E}[g(\exp(\ln(S_0) + \mu T + \sigma W_T)) \mid W_t],$$

we are dealing with the same type of expectation.

We can also use more explicitly the fact that S_T/S_t is independent of S_t and write

$$\mathbb{E}[g(S_T) \mid W_t] = \mathbb{E}[g((S_T/S_t) \times S_t) \mid W_t].$$

In this case, since knowing W_t is the same as knowing the value of S_t, we can compute this last expectation using the algorithm based on equation (14.3.2). This is illustrated in the following example.

Example 14.5.2 Let us compute $\mathbb{E}[(S_T - 1)_+ \mid S_t]$, where $t < T$.

Following the previous methodology, we first compute

$$\mathbb{E}\left[\left(x\frac{S_T}{S_t} - 1\right)_+\right],$$

where x is a dummy variable. Since $x > 0$, using a property of the lognormal distribution, we have that $x(S_T/S_t) \overset{\mathbb{P}}{\sim} \mathcal{LN}(\ln(x) + \mu(T-t), \sigma^2(T-t))$. Therefore, we can apply the formula for the stop-loss transform of a lognormal distribution, as obtained in (14.1.7), and deduce that

$$\mathbb{E}\left[\left(x\frac{S_T}{S_t} - 1\right)_+\right] = \mathrm{e}^{m+\frac{b^2}{2}} N\left(-\frac{\ln(1) - m}{b} + b\right) - N\left(-\frac{\ln(1) - m}{b}\right),$$

with $m = \ln(x) + \mu(T-t)$ and $b^2 = \sigma^2(T-t)$. To conclude, we replace x by S_t in the last expression and then we obtain

$$\begin{aligned}
&\mathbb{E}[(S_T - 1)_+ \mid S_t] \\
&= S_t \mathrm{e}^{(\mu+\sigma^2/2)(T-t)} N\left(\frac{\ln(S_t) + \mu(T-t)}{\sigma\sqrt{T-t}} + \sigma\sqrt{T-t}\right) \\
&\quad - N\left(\frac{\ln(S_t) + \mu(T-t)}{\sigma\sqrt{T-t}}\right).
\end{aligned}$$

■

14.5.3 Martingale property

In general, geometric Brownian motions are not martingales because they have an *exponential trend*. Indeed, from equation (14.5.1), for $0 \leq t < T$, we have

$$\mathbb{E}[S_T \mid S_u, 0 \leq u \leq t] = \mathbb{E}[S_T \mid S_t] = S_t \mathrm{e}^{(\mu+\sigma^2/2)(T-t)}.$$

In other words, S satisfies the martingale property, i.e.

$$\mathbb{E}[S_T \mid S_u, 0 \leq u \leq t] = S_t,$$

for all $0 \leq t < T$, if and only if the parameters μ and σ are such that $\mu + \sigma^2/2 = 0$.

14.5.4 Simulation

As for linear Brownian motion, to simulate a (discretized) path $(S(0), S(h), S(2h), \dots, S((n-1)h), S(T))$ of a geometric Brownian motion with coefficients μ and σ over the time interval $[0, T]$, there are two equivalent algorithms.

The first algorithm is based on the properties of the relative increments:

1) We first choose $n \geq 1$ to divide the time interval $[0, T]$ into smaller sub-intervals of the same size $h = T/n$: $0 < h < 2h < 3h < \dots < (n-1)h < T$.
2) We define $S(0) = S_0$ and then, for each $i = 1, 2, \dots, n$, we:
   a) generate $Z \sim \mathcal{N}(0, 1)$;
   b) compute

$$S(ih) = S((i-1)h) \times \exp\left(\mu h + \sigma\sqrt{h}Z\right).$$

3) The output is $(S(0), S(h), S(2h), \dots, S((n-1)h), S(T))$, a sampled discretized trajectory of a geometric Brownian motion, which is a synthetically generated realization of the random vector

$$(S_0, S_h, S_{2h}, \dots, S_{(n-1)h}, S_T).$$

The second algorithm relies on the fact that we might have already simulated a standard Brownian motion (or a linear Brownian motion), i.e. that we have generated

$$(W(0), W(h), W(2h), \dots, W((n-1)h), W(T)).$$

Then, we apply the corresponding exponential transformation: for each $i = 0, 1, \dots, n$, we set

$$S(ih) = S_0 \exp(\mu \times ih + \sigma \times W(ih)).$$

Again, the output is a realization of the random vector $(S_0, S_h, S_{2h}, \dots, S_{(n-1)h}, S_T)$.

Example 14.5.3 *Simulation of a geometric Brownian motion*
A stock currently trades for \$100 and its mean annual log-return is 7% whereas its volatility is 25%. Using the random numbers from example 14.3.6 (in the same order), we can generate a discretized sample path for this geometric Brownian motion over the time interval $[0, \frac{1}{3}]$.

Again, we have $n = 4$, $T = \frac{1}{3}$ and $h = \frac{1}{12}$. Furthermore, we have $S_0 = 100$, $\mu = 0.1$ and $\sigma = 0.25$. Therefore,

$$\begin{aligned}
S(0) &= 100, \\
S(1/12) &= S(0) \times \exp\left(0.1 \times 1/12 + 0.25\sqrt{1/12} \times 1.3265\right) = 110.9672831, \\
S(2/12) &= S(1/12) \times \exp\left(0.1 \times 1/12 + 0.25\sqrt{1/12} \times -0.8355\right) = 105.3482707, \\
S(3/12) &= S(2/12) \times \exp\left(0.1 \times 1/12 + 0.25\sqrt{1/12} \times 1.3672\right) = 117.2459916, \\
S(4/12) &= S(3/12) \times \exp\left(0.1 \times 1/12 + 0.25\sqrt{1/12} \times 0.8012\right) = 125.2647105.
\end{aligned}$$

Then,

$$\begin{aligned}
&(100, S(1/12), S(2/12), S(3/12), S(4/12)) \\
&= (100, 110.9672831, 105.3482707, 117.2459916, 125.2647105)
\end{aligned}$$

is the corresponding discretized trajectory of this geometric Brownian motion. It is shown in Figure 14.14. ■

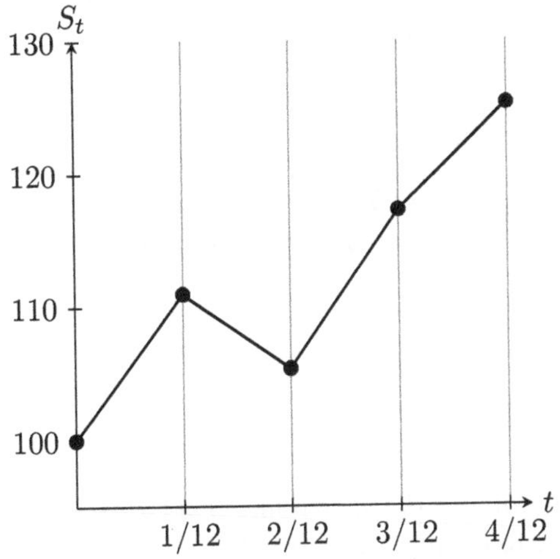

Figure 14.14 Sample path of the geometric Brownian motion depicted in example 14.5.3

14.5.5 Estimation

Given the importance of GBM in financial and actuarial applications, a natural question arises: how should/can we determine the values of μ and σ given asset price data?

Assume that the time unit is a calendar year and that data is collected periodically (weekly, daily,[5] hourly, etc.) at each time step of length h for a total of T years. Overall, we will have $n = T/h$ observations. For example, if $h = 1/12$ and we collect $T = 10$ years of data, then we have a total of $n = 120$ observations.

The idea is to capitalize on the fact that the relative increments of a GBM are independent and identically distributed according to a lognormal distribution. Said differently, the random variables

$$\ln\left(\frac{S_h}{S_0}\right), \ln\left(\frac{S_{2h}}{S_h}\right), \dots, \ln\left(\frac{S_{nh}}{S_{(n-1)h}}\right)$$

are independent and distributed according to the same normal distribution: for all $j = 1, 2, \dots, n$, we have

$$\ln\left(\frac{S_{jh}}{S_{(j-1)h}}\right) \overset{\mathbb{P}}{\sim} \mathcal{N}(\mu h, \sigma^2 h).$$

For example, if μ and σ are annual parameters and if we collect monthly data for n consecutive months, i.e. if we have realized values for $\ln(S_{j/12}/S_{(j-1)/12})$, where $j = 1, 2, \dots, n$, then these log-returns are drawn from a normal distribution with mean $\mu \times 1/12$ and variance $\sigma^2 \times 1/12$.

Fortunately, estimating parameters of a normal distribution is straightforward. We will estimate the parameters μ and σ^2 using *maximum likelihood estimation*. Recall that the **maximum likelihood estimators** (MLEs) of the mean and the variance of a normal distribution are given by the sample mean and the sample variance. More precisely, if we have the following observations

$$S_0 = s_0, S_h = s_1, S_{2h} = s_2, \dots, S_{(n-1)h} = s_{n-1}, S_{nh} = s_n$$

5 Instead of using time steps of 1/365 for the number of *calendar* days in a year, financial applications generally use the number of *trading* days per year, which is about 250–252.

and if we set $x_j = \ln(s_j/s_{j-1})$, for each $j = 1, 2, \dots, n$, then the MLEs $\hat{\mu}_n$ and $\hat{\sigma}_n$, of μ and σ respectively, are given by

$$\hat{\mu}_n = \frac{\overline{x}}{h} \quad \text{and} \quad \hat{\sigma}_n = \frac{s_x}{\sqrt{h}},$$

where $\overline{x} = (1/n)\sum_{i=1}^n x_i$ and $s_x^2 = (1/n)\sum_{i=1}^n (x_i - \overline{x})^2$. Recall that s_x^2 is a biased estimator for the variance of the x_is but this bias decreases quickly for large samples. If one is concerned with such bias, one could replace s_x^2 by

$$\frac{1}{n-1}\sum_{i=1}^{n}(x_i - \overline{x})^2.$$

Example 14.5.4 *Maximum likelihood estimation of GBM*
Assume that you have the following monthly data for the stock price of ABC inc.:

Date	Price
December 31st, 2015	51
January 31st, 2015	54
February 28th, 2015	61
March 31st, 2015	53
April 30th, 2015	49

In practice, the analyst would use the closing price of the asset at the end of the last trading day of any given month. Let us compute the MLE estimates of μ and σ corresponding to these observations.

We have $n = 4$ and $h = 1/12$. From our sample $(s_0, s_1, s_2, s_3, s_4) = (51, 54, 61, 53, 49)$ of prices, we must compute the corresponding sample of continuously compounded monthly log-returns: $x_j = \ln(s_j/s_{j-1})$, for each $j = 1, 2, 3, 4$. And then we must compute the sample mean and sample variance.

The log-returns observed over each month are given in the following table:

Month	Log-return
January	$\ln\left(\frac{54}{51}\right) = 0.05715841$
February	$\ln\left(\frac{61}{54}\right) = 0.12188982$
March	$\ln\left(\frac{53}{61}\right) = -0.14058195$
April	$\ln\left(\frac{49}{53}\right) = -0.07847162$

The sample mean and variance are

$$\overline{x} = -0.01 \quad \text{and} \quad s_x^2 = 0.0109113.$$

Then, we easily obtain the following MLE estimates for the parameters:

$$\widehat{\mu}_n = \frac{-0.01}{1/12} = -0.12,$$

$$\widehat{\sigma}_n = \frac{\sqrt{0.0109}}{\sqrt{1/12}} = 0.361850245.$$

■

From the properties of the log function, we know that

$$\sum_{j=1}^{n} \ln(s_j/s_{j-1}) = \sum_{j=1}^{n} [\ln(s_j) - \ln(s_{j-1})] = \ln(s_n/s_0)$$

and thus the MLE of μ can be further simplified as

$$\widehat{\mu}_n = \frac{\ln(s_n/s_0)}{nh}.$$

Precision of the MLE for the drift

Although the computation of the MLE for μ is straightforward, it remains that $\widehat{\mu}_n$ is an imprecise estimator. Given that $T = nh$, then

$$\begin{aligned}\mathbb{V}\text{ar}\left(\frac{\ln(S_T/S_0)}{T}\right) &= \frac{1}{T^2}\mathbb{V}\text{ar}(\ln(S_T/S_0)) \\ &= \frac{1}{T^2} \times \sigma^2 T \\ &= \frac{\sigma^2}{T}.\end{aligned}$$

No matter how frequently we collect data (how small h is), the variability of $\widehat{\mu}_n$ only decreases linearly with T, the number of years of data. For example, if we know that $\sigma = 0.25$, then 10 years of data implies that the 95% confidence interval around μ is $\widehat{\mu}_n \pm 0.155$ which is very large when the mean annual log-return is typically $\mu \in [-0.2, 0.2]$. The lesson here is that T needs to be large for the estimator of the drift to be precise.

14.6 Summary

Normal distribution

- Notation: $X \overset{\mathbb{P}}{\sim} \mathcal{N}(\mu, \sigma^2)$, with $\mathbb{E}[X] = \mu$ and $\mathbb{V}\text{ar}(X) = \sigma^2$.
- Probability density function:

$$f_X(x) = \frac{1}{\sqrt{2\pi\sigma^2}} e^{-\frac{(x-\mu)^2}{2\sigma^2}}.$$

- Cumulative distribution function of the standard normal distribution: $N(z)$.
- Standardization: if $X \overset{\mathbb{P}}{\sim} \mathcal{N}(\mu, \sigma^2)$, then $Z = \frac{X-\mu}{\sigma} \overset{\mathbb{P}}{\sim} \mathcal{N}(0, 1)$.
- Cumulative distribution function of $X \overset{\mathbb{P}}{\sim} \mathcal{N}(\mu, \sigma^2)$:

$$\mathbb{P}(X \leq x) = N\left(\frac{x - \mu}{\sigma}\right).$$

- Moment generating function: $M_X(\lambda) = e^{\lambda\mu + \frac{\lambda^2\sigma^2}{2}}$.

Lognormal distribution

- Notation: $X \overset{\mathbb{P}}{\sim} \mathcal{LN}(\mu, \sigma^2)$.
- Representation: $X = e^Y$ where $Y \overset{\mathbb{P}}{\sim} \mathcal{N}(\mu, \sigma^2)$.
- Cumulative distribution function:

$$\mathbb{P}(X \leq x) = N\left(\frac{\ln x - \mu}{\sigma}\right), \quad x > 0.$$

- Expectation and variance:

$$\mathbb{E}[X] = \exp\left(\mu + \frac{1}{2}\sigma^2\right) \quad \text{and} \quad \mathbb{V}\text{ar}(X) = e^{2\mu+\sigma^2}\left(e^{\sigma^2} - 1\right).$$

- Truncated expectations: for $a > 0$,

$$\mathbb{E}\left[X\mathbb{1}_{\{X\leq a\}}\right] = e^{\mu+\frac{\sigma^2}{2}} N\left(\frac{\ln(a) - \mu}{\sigma} - \sigma\right).$$

- Stop-loss transforms: for $a > 0$,

$$\mathbb{E}[(X-a)_+] = e^{\mu+\frac{\sigma^2}{2}} N\left(-\frac{\ln(a) - \mu}{\sigma} + \sigma\right) - aN\left(-\frac{\ln(a) - \mu}{\sigma}\right).$$

Symmetric random walk

- Jumps: independent ϵ_is such that $\mathbb{P}(\epsilon_i = 1) = \mathbb{P}(\epsilon_i = -1) = \frac{1}{2}$.
- Symmetric random walk: $\{X_n, n \geq 0\}$ such that $X_0 = 0$ and $X_n = \sum_{i=1}^n \epsilon_i$.
- Properties of the increments:
  1) for $m \geq 1$ and $0 \leq n_1 < n_2 < \cdots < n_m$, the random variables

  $$X_{n_2} - X_{n_1}, \ldots, X_{n_m} - X_{n_{m-1}}$$

  are independent;
  2) for $k, n \geq 1$, the random variables $X_{n+k} - X_k$ and X_n have the same distribution.
- Markov property: for $1 \leq m < n$,

$$\mathbb{P}(X_n = x \mid X_0, X_1, \ldots, X_m) = \mathbb{P}(X_n = x \mid X_m).$$

- Martingale property: for $0 \leq m < n$,

$$\mathbb{E}[X_n \mid X_0, X_1, \ldots, X_m] = X_m.$$

Standard Brownian motion

- Construction: limit of accelerated and rescaled symmetric random walks.
- Definition: a process $W = \{W_t, t \geq 0\}$ is a standard Brownian motion if
  - $W_0 = 0$;
  - for all $n \geq 1$ and $0 \leq t_1 < t_2 < \cdots < t_n$,

  $$W_{t_2} - W_{t_1}, W_{t_3} - W_{t_2}, \ldots, W_{t_n} - W_{t_{n-1}}$$

  are independent;
  - for all $s < t$, $W_t - W_s$ has the same distribution as $W_{t-s} \overset{\mathbb{P}}{\sim} \mathcal{N}(0, t-s)$.
- Mean function: $\mathbb{E}[W_t] = 0$, for all $t \geq 0$.
- Variance function: $\mathbb{V}\text{ar}(W_t) = t$, for all $t \geq 0$.
- Dependence structure: $\mathbb{C}\text{ov}(W_s, W_t) = \min\{s, t\}$, for all $s, t \geq 0$.
- Markov property: for $0 \leq t < T$,

$$\mathbb{P}(W_T \leq x \mid W_s, 0 \leq s \leq t) = \mathbb{P}(W_T \leq x \mid W_t).$$

- Martingale property: for $0 \le t < T$,

$$\mathbb{E}[W_T \mid W_s, 0 \le s \le t] = W_t.$$

- Simulation: choose n, set $h = T/n$ and set $W(0) = 0$, and then, for each $i = 1, 2, \dots, n$, generate $Z \sim \mathcal{N}(0,1)$ and compute $W(ih) = W((i-1)h) + \sqrt{h}Z$.

Linear Brownian motion

- Definition: a linear Brownian motion with drift μ and diffusion σ is defined by

$$X_t = X_0 + \mu t + \sigma W_t.$$

- Distribution: $X_t \overset{\mathbb{P}}{\sim} \mathcal{N}(X_0 + \mu t, \sigma^2 t)$.
- Mean function: $\mathbb{E}[X_t] = X_0 + \mu t$, for all $t \ge 0$.
- Variance function: $\mathbb{V}\mathrm{ar}(X_t) = \sigma^2 t$, for all $t \ge 0$.
- Increments: independent and stationary, i.e.
  - for all $n \ge 1$ and $0 \le t_1 < t_2 < \cdots < t_n$,

  $$X_{t_2} - X_{t_1}, X_{t_3} - X_{t_2}, \dots, X_{t_n} - X_{t_{n-1}}$$

  are independent;
  - for all $s < t$, $X_t - X_s$ has the same distribution as $X_{t-s} \overset{\mathbb{P}}{\sim} \mathcal{N}(\mu(t-s), \sigma^2(t-s))$.
- Markov property (with respect to W): for $0 \le t < T$,

$$\mathbb{P}(X_T \le x \mid W_s, 0 \le s \le t) = \mathbb{P}(X_T \le x \mid W_t).$$

- A linear Brownian motion is a martingale if and only if $\mu = 0$.
- Simulation: choose n, set $h = T/n$ and set $X(0) = X_0$, and then, for each $i = 1, 2, \dots, n$, generate $Z \sim \mathcal{N}(0,1)$ and compute $X(ih) = X((i-1)h) + \mu h + \sigma\sqrt{h}Z$.

Geometric Brownian motion

- Definition: a geometric Brownian motion with parameters μ and σ is defined by

$$S_t = S_0 \exp(\mu t + \sigma W_t).$$

- Distribution: $S_t \overset{\mathbb{P}}{\sim} \mathcal{LN}(\ln(S_0) + \mu t, \sigma^2 t)$.
- Mean function: $\mathbb{E}[S_t] = S_0 e^{(\mu + \frac{\sigma^2}{2})t}$, for all $t \ge 0$.
- Variance function: $\mathbb{V}\mathrm{ar}(S_t) = S_0^2 e^{2\mu t + \sigma^2 t}(e^{\sigma^2 t} - 1)$, for all $t \ge 0$.
- Relative increments are independent and stationary, not the usual increments.
- Markov property (with respect to W): for $0 \le t < T$,

$$\mathbb{P}(S_T \le x \mid W_s, 0 \le s \le t) = \mathbb{P}(S_T \le x \mid W_t).$$

- A geometric Brownian motion is a martingale if and only if $\mu + \frac{1}{2}\sigma^2 = 0$.
- Simulation: choose n, set $h = T/n$ and set $S(0) = S_0$, and then, for each $i = 1, 2, \dots, n$, generate $Z \sim \mathcal{N}(0,1)$ and compute $S(ih) = S((i-1)h)\exp(\mu h + \sigma\sqrt{h}Z)$.
- Estimation: for a time step h and a sample $\{s_0, s_1, s_2, \dots, s_n\}$, then the MLEs for μ and σ are given by

$$\widehat{\mu}_n = \frac{\overline{x}}{h} = \frac{\ln(s_n/s_0)}{nh} \quad \text{and} \quad \widehat{\sigma}_n = \frac{s_x}{\sqrt{h}},$$

where $\overline{x}$ and s_x^2 are the sample mean and the sample variance of $x_j = \ln(s_j/s_{j-1})$, $j = 1, 2, \dots, n$.

14.7 Exercises

14.1 For $X \overset{\mathbb{P}}{\sim} \mathcal{LN}(\mu, \sigma)$ and a constant a, compute $\mathbb{E}[(a - X)_+]$ using the properties of stop-loss functions.

14.2 A particle is randomly moving over one dimension according to a standard Brownian motion.

(a) On average after ten periods, where will the particle be?

(b) After five periods, you find the particle is located at -3. Where do you expect the particle to be five periods later?

(c) In (b), what is the probability that the particle is above zero five periods later (given that after five periods, it is located at -3)?

14.3 Compute the following quantities:

(a) $\mathbb{E}[W_{\sqrt{2}}]$;

(b) $\mathbb{V}\mathrm{ar}(W_{8/9})$;

(c) $\mathbb{E}[W_{17/2}^2]$;

(d) $\mathbb{E}[\mathrm{e}^{0.2+W_{0.1}}]$;

(e) $\mathbb{P}(0.5\mathrm{e}^{0.2+W_{0.1}} \leq 0.5\mathrm{e}^{0.2})$.

14.4 Consider a random variable $Z \overset{\mathbb{P}}{\sim} \mathcal{N}(0,1)$. For each $t \geq 0$, set $X_t = \sqrt{t}Z$. Argue that the stochastic process $\{X_t, t \geq 0\}$ has continuous trajectories and verify that for each fixed $t \geq 0$, the random variable X_t follows a $\mathcal{N}(0,t)$ distribution. Is $\{X_t, t \geq 0\}$ a standard Brownian motion? Justify your answer.

14.5 Let $\{W_t, t \geq 0\}$ and $\{\widetilde{W}_t, t \geq 0\}$ be two independent standard Brownian motions and let ρ be a fixed number between 0 and 1. For each $t \geq 0$, set $X_t = \rho W_t + \sqrt{1-\rho^2}\widetilde{W}_t$. Is $\{X_t, t \geq 0\}$ a standard Brownian motion? Justify your answer.

14.6 Fix $\lambda > 0$. Verify that the stochastic process $\{B_t, t \geq 0\}$, defined by

$$B_t = \frac{1}{\sqrt{\lambda}} W_{\lambda t},$$

is also a standard Brownian motion.

14.7 Verify that $\{M_t, t \geq 0\}$, defined by

$$M_t = \frac{\mathrm{e}^{\sigma W_t}}{\mathbb{E}[\mathrm{e}^{\sigma W_t}]},$$

is a martingale.

14.8 You just bought a *vintage car* for \$20,000. Assume its future value can be modeled by a geometric Brownian motion with parameters $\mu = -0.14$ and $\sigma = 0.07$.

(a) What is its value expected to be in 4 years from now? What about its median value?

(b) What is the probability that its price in 4 years will be greater than \$20,000?

(c) How much time must elapse so that the expected market value of the car corresponds to \$1,000?

14.9 You have observed the following (annual) values for what you assume is a geometric Brownian motion: $S_0 = 100$, $S_1 = 98$, $S_2 = 100$, $S_3 = 101$, $S_4 = 105$ and $S_5 = 104$.

(a) Compute the corresponding estimates for μ and σ^2, given on an annual basis.

(b) If your data were observed monthly instead of annually, how would your estimates be affected?

14.10 Using a computer, you have sampled the following normal random numbers: 0.9053, 1.4407, −1.0768, −1.3102, 0.0302. Using a time step $h = 0.2$, generate a sample a path for:

(a) a standard Brownian motion;

(b) a linear Brownian motion with $X_0 = 0, \mu = 0.07, \sigma = 0.3$;

(c) a geometric Brownian motion with $S_0 = 100, \mu = 0.07, \sigma = 0.3$.

14.11 Consider a geometric Brownian motion given by $S_t = S_0 \exp(\mu t + \sigma W_t)$, for each $t \geq 0$.

(a) For a fixed number a, verify that

$$\mathbb{E}\left[S_t^a\right] = S_0^a e^{a\mu t} \exp\left(t\frac{a^2\sigma^2}{2}\right).$$

(b) For fixed numbers $0 < t < T$, verify that

$$\mathbb{E}[S_t S_T] = S_0^2 e^{\mu(t+T)} e^{\frac{1}{2}(\sigma^2(3t+T))}.$$

(c) Identify the probability distribution of the following random variable:

$$(S_1 S_2 S_3)^{1/3}.$$

Exercises 14.4 and 14.5 have been inspired by two exercises from Geneviève Gauthier, with her permission.

15

Introduction to stochastic calculus***

Usually, in most textbooks and research papers, the evolution of the stock price in the Black-Scholes-Merton (BSM) model is given by a so-called *stochastic differential equation*:

$$\mathrm{d}S_t = \mu S_t \mathrm{d}t + \sigma S_t \mathrm{d}W_t$$

(see also equation (17.1.8) in Chapter 17). This last equation has an equivalent *stochastic integral form*:

$$S_T - S_0 = \int_0^T \mathrm{d}S_t = \int_0^T \mu S_t \mathrm{d}t + \int_0^T \sigma S_t \mathrm{d}W_t.$$

In other words, in the BSM framework, the dynamic of the stock price S is the sum of two *random* integrals: the integral of the process $\{\mu S_t, t \geq 0\}$ with respect to the time variable and an integral of the process $\{\sigma S_t, t \geq 0\}$ with respect to Brownian motion.

Before making sense of these integrals, recall that we mentioned in Chapter 14 that the stock price in the BSM framework is represented by a geometric Brownian motion. Therefore, the latter two representations, in terms of a stochastic differential equation and of the sum of random integrals, should be two equivalent representations of the same stochastic process, namely a GBM. To better understand these concepts, we will provide below a heuristic introduction to *stochastic calculus*.

Stochastic calculus is a set of tools, in the field of probability theory, to work with continuous-time stochastic processes, just like we do with functions in classical differential and integral calculus. One of the main mathematical objects is the so-called *stochastic integral*. Stochastic calculus is widely used in mathematical finance, but also in mathematical biology and physics. Stochastic calculus arises naturally in continuous-time actuarial finance whenever, for example, we need to consider dynamic (replicating) portfolios requiring continuous trading or if we need to determine the dynamic of a stochastic interest rate.

The overall objective of this chapter is to provide a heuristic introduction to stochastic calculus based on Brownian motion by defining Ito's stochastic integral and stochastic differential equations. This is a rather complex topic so the presentation focuses on providing a working knowledge of the material. We aim at an understanding suitable for the pricing and hedging of options in the Black-Scholes-Merton model. This chapter is intended for readers with a stronger mathematical background and is *not* mandatory to understand the upcoming chapters.

More specifically, the learning objectives are to:

- understand the definition of a stochastic integral and its basic properties;
- compute the mean and variance of a given stochastic integral;
- apply Ito's lemma to simple situations;
- understand how a stochastic process can be the solution to a stochastic differential equation;

Actuarial Finance: Derivatives, Quantitative Models and Risk Management, First Edition.
Mathieu Boudreault and Jean-François Renaud.

- recognize the SDEs for linear and geometric Brownian motions, the Ornstein-Uhlenbeck process and the square-root process, and understand the role played by their coefficients;
- solve simple SDEs such as those corresponding to linear and geometric Brownian motions, and the Ornstein-Uhlenbeck process.

Before we begin, we emphasize there are two types of stochastic integrals that are of interest in line with the specific objectives:

- stochastic Riemann integrals, i.e. classical Riemann integrals of the path of a stochastic process with respect to the time variable (e.g. $\int_0^T \mu S_t \mathrm{d}t$);
- stochastic integrals with respect to Brownian motion, i.e. integrals of a stochastic process with respect to a Brownian motion (e.g. $\int_0^T \sigma S_t \mathrm{d}W_t$).

In both cases, the resulting integral will be a random variable, thus the name *stochastic* integral.

15.1 Stochastic Riemann integrals

Let us consider a stochastic process $H = \{H_t, 0 \leq t \leq T\}$ with continuous trajectories, such as for example a linear Brownian motion or a geometric Brownian motion. If we draw a trajectory of the process H, i.e. if we look at one scenario ω of the underlying experiment, then the corresponding path $t \mapsto H_t(\omega)$ is a function of the time variable. For the function $f(t) = H_t(\omega)$, we could compute the following Riemann integral:

$$\int_0^t f(s)\mathrm{d}s = \int_0^t H_s(\omega)\mathrm{d}s.$$

Since a Riemann integral is obtained as the limit of *Riemann sums*, i.e.

$$\int_0^t f(s)\mathrm{d}s = \lim_{n\to\infty} \sum_{i=0}^{n-1} f(t_i) \times (t_{i+1} - t_i), \tag{15.1.1}$$

where the time points given by $t_i = it/n$ partition the interval $[0, t]$, then we can write

$$\int_0^t H_s(\omega)\mathrm{d}s = \lim_{n\to\infty} \sum_{i=0}^{n-1} H_{t_i}(\omega) \times (t_{i+1} - t_i).$$

Note that even if the notation does not make it explicit, the t_is depend on the number of time points n.

For each n, the realization $\sum_{i=0}^{n-1} H_{t_i}(\omega) \times (t_{i+1} - t_i)$ is the (stochastic) total area of n rectangles, where the (random) height of the i-th rectangle is given by the value $H_{t_i}(\omega)$ and the length of its base by the value $t_{i+1} - t_i$. This is illustrated in Figure 15.1.

As we can repeat this for every scenario $\omega \in \Omega$, we have defined a random variable which we will denote simply by

$$\int_0^t H_s \mathrm{d}s.$$

In some sense, it is the *(random) area* under the curves given by the trajectories of H.

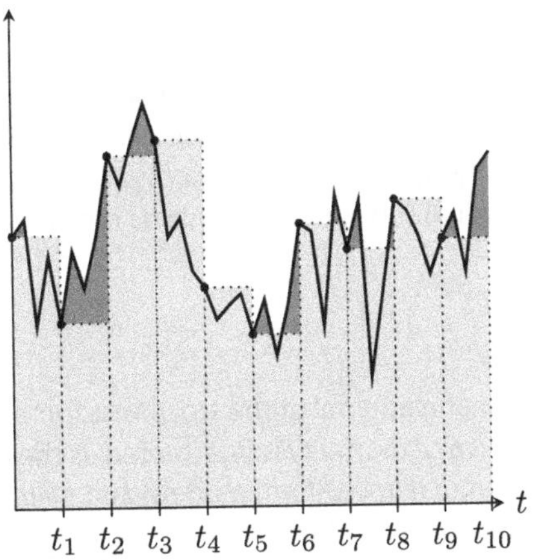

Figure 15.1 Riemann integral of a given path of a stochastic process using Riemann sums

It is possible to show that[1] the *average random area* is such that:

$$\mathbb{E}\left[\int_0^T H_t \mathrm{d}t\right] = \int_0^T \mathbb{E}[H_t]\mathrm{d}t,$$

i.e. we can interchange the order of the expectation and the Riemann integral (as long as each quantity is well defined). The integral on the right-hand-side is a *genuine* Riemann integral of the deterministic function $t \mapsto \mathbb{E}[H_t]$.

Example 15.1.1 *Riemann integral of standard Brownian motion*
Let us compute the average random area below a Brownian trajectory over a unit time interval, i.e. the expectation of the random variable $\int_0^1 W_t \mathrm{d}t$.

As Brownian motion starts from zero and is highly symmetrical, we should expect this expectation to be equal to zero. Indeed, we have

$$\mathbb{E}\left[\int_0^1 W_t \mathrm{d}t\right] = \int_0^1 \mathbb{E}[W_t]\mathrm{d}t = 0,$$

as the (deterministic) function $t \mapsto \mathbb{E}[W_t]$ is identically equal to zero, i.e. $\mathbb{E}[W_t] = 0$ for all values of t. ■

Example 15.1.2 *Riemann integral of a stochastic process*
What is the expectation of the random variable given by $\int_1^{1.5} W_t^2 \mathrm{d}t$?

We have

$$\mathbb{E}\left[\int_1^{1.5} W_t^2 \mathrm{d}t\right] = \int_1^{1.5} \mathbb{E}\left[W_t^2\right] \mathrm{d}t = \int_1^{1.5} t\mathrm{d}t = \frac{(1.5)^2}{2} - \frac{1^2}{2} = \frac{5}{8},$$

where we used the fact that $\mathbb{E}[W_t^2] = t$ for all values of t. ■

1 However, the proof is beyond the scope of this book.

Stochastic force of mortality

Stochastic Riemann integrals appear in various areas of actuarial science, actuarial finance and mathematical finance. To quantify longevity risk, there exists a string of literature that models the (stochastic) force of mortality by a stochastic process $H = \{H_t, t \geq 0\}$. In such models, the remaining lifetime τ of an individual subject to a stochastic mortality intensity $H = \{H_t, t \geq 0\}$ is such that

$$\mathbb{P}(\tau > u) = \mathbb{E}\left[\exp\left(-\int_0^u H_t \mathrm{d}t\right)\right],$$

which involves a stochastic Riemann integral of the stochastic force of mortality H.

In credit risk modeling, the force of mortality is interpreted as the *default intensity* to represent the time until default (in the class of *reduced-form* models) and *failure/hazard rate* in engineering to characterize the time until failure of an object.

15.2 Ito's stochastic integrals

Now, we want to define the stochastic integral of a stochastic process $H = \{H_t, 0 \leq t \leq T\}$ with respect to Brownian motion, i.e. give a meaning to a random variable denoted by

$$\int_0^T H_t \mathrm{d}W_t.$$

As in a stochastic Riemann integral, the integrand is a stochastic process H, but now the *integrator* is a standard Brownian motion. Symbolically, $\mathrm{d}t$ needs to be replaced by $\mathrm{d}W_t$. Intuitively, we want to compute the area under the trajectories of H using a Brownian motion W, meaning that both the height (H_t) and the width ($\mathrm{d}W_t$) of the rectangles will be random.

As for other integrals, a stochastic integral over the time interval $[0, t]$ is obtained as the limit of Riemann sums as the partition of the interval $[0, t]$ becomes finer and finer. More precisely, the random variable called **Ito's stochastic integral of H with respect to Brownian motion** is defined by

$$\int_0^t H_s \mathrm{d}W_s = \lim_{n\to\infty} \sum_{i=0}^{n-1} H_{t_i}\left(W_{t_{i+1}} - W_{t_i}\right), \tag{15.2.1}$$

where the time points are given by $t_i = it/n$, for each $i = 0, 1, \ldots, n$. Again, let us keep in mind that the t_is depend on n, so they change when n goes to infinity. The *limiting random variable* is denoted by $\int_0^t H_s \mathrm{d}W_s$ because, as we will see below, it behaves like other types of integrals in many ways; see Section 15.2.4.

Example 15.2.1 Fix a real number c. If $H_t = c$ for all t, i.e. if it is not affected by the scenario ω or by the time index t, then

$$\sum_{i=0}^{n-1} H_{t_i}\left(W_{t_{i+1}} - W_{t_i}\right) = \sum_{i=0}^{n-1} c\left(W_{t_{i+1}} - W_{t_i}\right) = c\left(W_{t_n} - W_{t_0}\right).$$

Since $t_n = nt/n = t$, we have $W_{t_n} - W_{t_0} = W_t - W_0 = W_t$ and, by the definition of the stochastic integral given above, we have

$$\int_0^t c\,\mathrm{d}W_s = \lim_{n\to\infty} cW_t = cW_t.$$

In particular, this means that a linear Brownian motion $X_t = X_0 + \mu t + \sigma W_t$ can also be written as

$$X_t = X_0 + \int_0^t \mu \, ds + \int_0^t \sigma \, dW_s.$$

■

Limit of random variables

The limit of a sequence of random variables is a topic often overlooked in a first probability course. But in fact, both the Law of Large Numbers and the Central Limit Theorem rely on limits of random variables. The results are: given a sequence of independent and identically distributed random variables $(X_n, n \geq 1)$ with common mean m and common variance σ^2, there exists a random variable X^{LLN} and a random variable X^{CLT} such that

$$\frac{1}{n} \sum_{i=1}^{n} X_i \underset{n \to \infty}{\longrightarrow} X^{LLN}$$

and

$$\frac{1}{\sqrt{n}} \sum_{i=1}^{n} \left(\frac{X_i - m}{\sigma} \right) \underset{n \to \infty}{\longrightarrow} X^{CLT}$$

where $X^{LLN} = m$ is a constant (not affected by the scenario ω) and X^{LLN} is a normally distributed random variable with mean 0 and variance 1.

Similar conclusions will be obtained for stochastic integrals: the properties of the random variable $\int_0^t H_s dW_s$, given by the limit of the random variables $\sum_{i=0}^{n-1} H_{t_i}(W_{t_{i+1}} - W_{t_i})$, will be derived from the properties of those Riemann sums.

15.2.1 Riemann sums

Intuitively, if we fix a scenario $\omega \in \Omega$, then the corresponding paths $H_t(\omega)$ and $W_t(\omega)$ are both functions of t and then we can write down the non-random Riemann sum:

$$\sum_{i=0}^{n-1} H_{t_i}(\omega) \left(W_{t_{i+1}}(\omega) - W_{t_i}(\omega) \right).$$

For each n, this is the total area of n rectangles, where the height of the i-th rectangle is given by $H_{t_i}(\omega)$ and the width of its base is given by $W_{t_{i+1}}(\omega) - W_{t_i}(\omega)$.

Example 15.2.2 *Realization of a Riemann sum*
Let us fix $n = 4$ and $t = \frac{1}{3}$, yielding $t_i = i/12$, and let us choose $H_t = |W_t|$. Consider a scenario ω in which we have:

$$W_{1/12}(\omega) = 0.382927566,$$
$$W_{2/12}(\omega) = -0.141739491,$$
$$W_{3/12}(\omega) = 0.536416135,$$
$$W_{4/12}(\omega) = 0.767702653.$$

Note that $W_0(\omega) = 0$ by definition, no matter which scenario ω occurs. Since $H_t = |W_t|$ at any time t, we have

$$\begin{aligned}
&\sum_{i=0}^{3} H_{i/12}(\omega)(W_{(i+1)/12}(\omega) - W_{i/12}(\omega)) \\
&\quad = |W_0(\omega)|(W_{1/12}(\omega) - W_0(\omega)) + |W_{1/12}(\omega)|(W_{2/12}(\omega) - W_{1/12}(\omega)) \\
&\qquad + |W_{2/12}(\omega)|(W_{3/12}(\omega) - W_{2/12}(\omega)) + |W_{3/12}(\omega)|(W_{4/12}(\omega) - W_{3/12}(\omega)) \\
&\quad = 0 + |0.382927566| \times (-0.141739491 - 0.382927566) \\
&\qquad + |-0.141739491| \times (0.536416135 - (-0.141739491)) \\
&\qquad + |0.536416135| \times (0.767702653 - 0.536416135) \\
&\quad = 0.01927777.
\end{aligned}$$

One should remark that in this scenario, the area of the first (non-null) rectangle is negative: $|0.382927566| \times (-0.141739491 - 0.382927566) = -0.2009095$. ∎

As we can repeat this for every scenario $\omega \in \Omega$, then for each n we have a *doubly random Riemann sum* that we denote by

$$\sum_{i=0}^{n-1} H_{t_i}\left(W_{t_{i+1}} - W_{t_i}\right). \tag{15.2.2}$$

It is a random variable that we will call the n**-th (random) Riemann sum** (of H).

Example 15.2.3 *Riemann sums of geometric Brownian motion*
Let us compute the expectation of the n-th Riemann sum of the process $H_t = \exp(W_t)$, i.e.

$$\sum_{i=0}^{n-1} \exp(W_{t_i})\left(W_{t_{i+1}} - W_{t_i}\right).$$

Recall that since increments of Brownian motion are independent, we have that $\exp(W_{t_i})$ is independent of $W_{t_{i+1}} - W_{t_i}$, for each i. Consequently,

$$\begin{aligned}
\mathbb{E}\left[\sum_{i=0}^{n-1} \exp(W_{t_i})\left(W_{t_{i+1}} - W_{t_i}\right)\right] &= \sum_{i=0}^{n-1} \mathbb{E}\left[\exp\left(W_{t_i}\right)\left(W_{t_{i+1}} - W_{t_i}\right)\right] \\
&= \sum_{i=0}^{n-1} \mathbb{E}\left[\exp\left(W_{t_i}\right)\right]\mathbb{E}\left[W_{t_{i+1}} - W_{t_i}\right] \\
&= 0,
\end{aligned}$$

as increments of Brownian motion are zero-mean random variables. ∎

15.2.2 Elementary stochastic processes

Depending on the integrand, the limiting procedure involved in defining the stochastic integral as given in (15.2.1) is not always necessary. We have seen such a trivial illustration in example 15.2.1. In this section, we introduce a class of stochastic processes called **elementary stochastic processes** whose stochastic integrals do not require the limiting procedure. Hence, each such stochastic integral is equal to a Riemann sum.

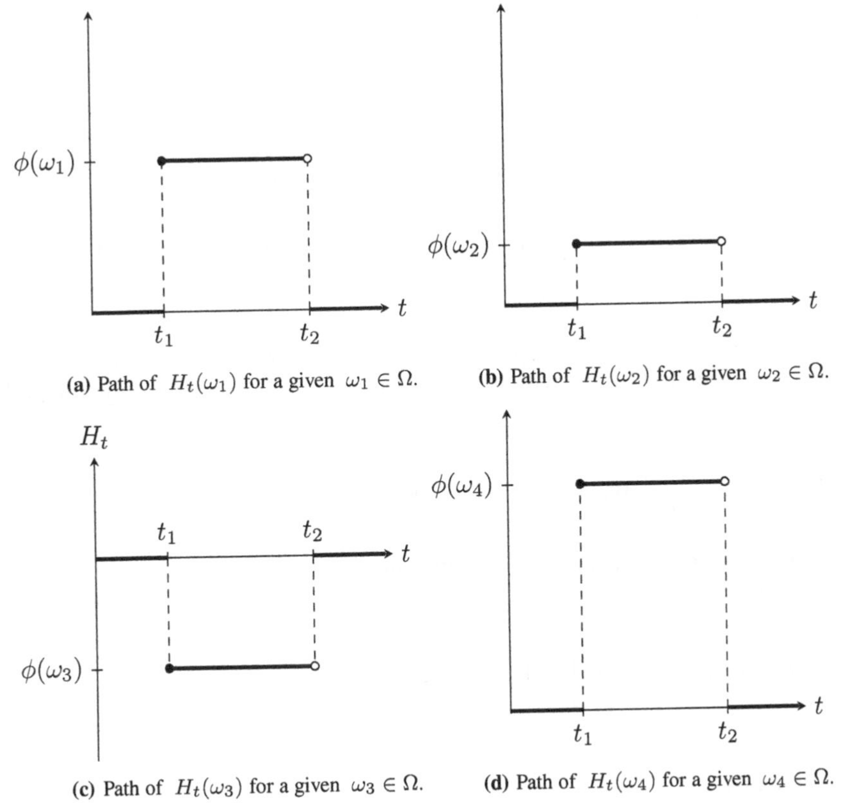

(a) Path of $H_t(\omega_1)$ for a given $\omega_1 \in \Omega$.

(b) Path of $H_t(\omega_2)$ for a given $\omega_2 \in \Omega$.

(c) Path of $H_t(\omega_3)$ for a given $\omega_3 \in \Omega$.

(d) Path of $H_t(\omega_4)$ for a given $\omega_4 \in \Omega$.

Figure 15.2 Four paths of the same elementary one-step stochastic process H_t as defined in (15.2.3) with a random variable ϕ

The simplest stochastic process one can imagine[2] is a one-step *elementary* stochastic process, i.e. a process of the form

$$H_t = \phi \times \mathbb{1}_{(t_1,t_2]}(t), \tag{15.2.3}$$

where ϕ is a random variable[3] and where $0 \leq t_1 < t_2 \leq T$. Here, ϕ, t_1 and t_2 specify the elementary process: they are the parameters of H.

Figure 15.2 shows four different trajectories of this process H associated to four different scenarios.

As we can see in Figure 15.2, the trajectories of the process H given in (15.2.3) consist of a single step of height $\phi(\omega)$ when the scenario is ω. Consequently, for such a process, its stochastic integral does not require the limiting procedure: it is directly equal to the Riemann sum $\phi \times (W_{t_2} - W_{t_1})$. In other words, the stochastic integral of the process H given in (15.2.3), over the time interval $[0, T]$, is given by the following random variable:

$$\int_0^T H_t \mathrm{d}W_t = \int_0^T \phi \times \mathbb{1}_{(t_1,t_2]}(t)\mathrm{d}W_t = \phi \times \left(W_{t_2} - W_{t_1}\right). \tag{15.2.4}$$

---

2 Except maybe for a deterministic function.

3 For modeling and mathematical purposes, this random variable must be determined with (at most) the information accumulated up to time t_1.

Example 15.2.4 If $H_t = \sqrt{3}\mathbb{1}_{(t_1,t_2]}(t)$, with $0 \leq t_1 < t_2 \leq T$, then it is a one-step elementary process as given in (15.2.3) with $\phi = \sqrt{3}$ a constant. Consequently, the stochastic integral of H is given by

$$\int_0^T H_t \mathrm{d}W_t = \int_0^T \sqrt{3}\mathbb{1}_{(t_1,t_2]}(t)\mathrm{d}W_t = \sqrt{3}\left(W_{t_2} - W_{t_1}\right).$$

In this case, the stochastic integral is normally distributed with mean 0 and variance $3 \times (t_2 - t_1)$.

■

Example 15.2.5 If $H_t = W_{t_1}\mathbb{1}_{(t_1,t_2]}(t)$, with $0 \leq t_1 < t_2 \leq T$, then it is a one-step elementary process as given in (15.2.3) with $\phi = W_{t_1}$. Consequently, the stochastic integral of H is given by

$$\int_0^T H_t \mathrm{d}W_t = \int_0^T W_{t_1}\mathbb{1}_{(t_1,t_2]}(t)\mathrm{d}W_t = W_{t_1}\left(W_{t_2} - W_{t_1}\right).$$

In this case, the stochastic integral is the product of two independent zero-mean normal random variables. In particular, we have

$$\begin{aligned}\mathbb{E}\left[\int_0^T W_{t_1}\mathbb{1}_{(t_1,t_2]}(t)\mathrm{d}W_t\right] &= \mathbb{E}\left[W_{t_1}\left(W_{t_2} - W_{t_1}\right)\right] \\ &= \mathbb{E}\left[W_{t_1}\right]\mathbb{E}\left[\left(W_{t_2} - W_{t_1}\right)\right] = 0\end{aligned}$$

and consequently

$$\begin{aligned}\mathbb{V}\mathrm{ar}\left(\int_0^T W_{t_1}\mathbb{1}_{(t_1,t_2]}(t)\mathrm{d}W_t\right) &= \mathbb{E}\left[\left(\int_0^T W_{t_1}\mathbb{1}_{(t_1,t_2]}(t)\mathrm{d}W_t\right)^2\right] \\ &= \mathbb{E}\left[\left(W_{t_1}\right)^2\left(W_{t_2} - W_{t_1}\right)^2\right] = \mathbb{E}\left[\left(W_{t_1}\right)^2\right]\mathbb{E}\left[\left(W_{t_2} - W_{t_1}\right)^2\right] \\ &= t_1(t_2 - t_1).\end{aligned}$$

■

More generally, a stochastic process $H = \{H_t, 0 \leq t \leq T\}$ is said to be an **elementary stochastic process** if it is such that

$$H_t = \sum_{i=0}^{n-1} \phi_i \mathbb{1}_{(t_i,t_{i+1}]}(t), \tag{15.2.5}$$

where $0 = t_0 < t_1 < \cdots < t_{n-1} < t_n = T$ generate non-overlapping time intervals and where, for each i, the random variable ϕ_i is completely determined by the Brownian motion up to time t_i; this is a mathematical technicality we have to deal with. Note that we have $H_0 = 0$ for any scenario ω.

For a given $\omega \in \Omega$, the trajectory $H_t(\omega)$ is made of steps with various heights determined by the draws of $\phi_1(\omega), \phi_2(\omega), \ldots, \phi_n(\omega)$. This is illustrated in Figure 15.3 showing trajectories of $H_t(\omega)$ for two scenarios.

Again, as we can see in Figure 15.3, since the trajectories of the process H given in (15.2.5) consist of steps of height $\phi_i(\omega)$ when the scenario is ω, then its stochastic integral does not require

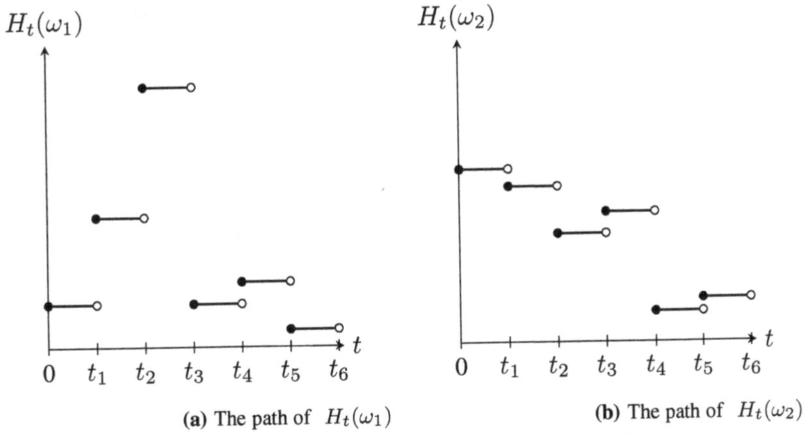

Figure 15.3 Two different paths of the same elementary stochastic process H

the limiting procedure. The stochastic integral of H given in (15.2.5) over the time interval $[0, T]$ is the extension of equation (15.2.4): it is given by the following random variable:

$$\int_0^T H_t \mathrm{d}W_t = \sum_{i=0}^{n-1} \phi_i (W_{t_{i+1}} - W_{t_i}). \tag{15.2.6}$$

Example 15.2.6 For the elementary process $H_t = \mathrm{e}^{-1}\mathbb{1}_{(2,4]}(t) + 4\mathbb{1}_{(7,10]}(t) + 12\mathbb{1}_{(15,21]}(t)$, the corresponding stochastic integral is simply

$$\int_0^T H_t \mathrm{d}W_t = \mathrm{e}^{-1}(W_4 - W_2) + 4(W_{10} - W_7) + 12(W_{21} - W_{15}).$$

Since increments of Brownian motion over non-overlapping intervals are independent, this is a sum of independent normally distributed random variables. Hence, the stochastic integral is normally distributed with mean

$$\mathbb{E}\left[\int_0^T H_t \mathrm{d}W_t\right] = \mathrm{e}^{-1}\mathbb{E}[W_4 - W_2] + 4\mathbb{E}[W_{10} - W_7] + 12\mathbb{E}[W_{21} - W_{15}] = 0$$

and variance

$$\begin{aligned}\mathbb{V}\mathrm{ar}\left(\int_0^T H_t \mathrm{d}W_t\right) &= \mathrm{e}^{-2}\mathbb{V}\mathrm{ar}(W_4 - W_2) + 4^2\mathbb{V}\mathrm{ar}(W_{10} - W_7) + 12^2\mathbb{V}\mathrm{ar}(W_{21} - W_{15}) \\ &= \mathrm{e}^{-2} \times (4-2) + 4^2 \times (10-7) + 12^2 \times (21-15) = 912.2707.\end{aligned}$$

■

As observed in the last example, if H is an elementary stochastic process as in (15.2.5) with non-random ϕ_i, then its stochastic integral is normally distributed with mean zero and variance equal to

$$\sum_{i=0}^{n-1} \phi_i^2 (t_{i+1} - t_i).$$

Note that in other/most cases, we cannot say much about the distribution of a stochastic integral.

Example 15.2.7 For the process $H_t = 15\mathbb{1}_{(2,4]}(t) + W_6\mathbb{1}_{(7,10]}(t)$, its stochastic integral is the random variable given by

$$\int_0^T H_t \mathrm{d}W_t = 15(W_4 - W_2) + W_6(W_{10} - W_7).$$

Using the properties of Brownian motion, we can compute its expectation, which is equal to zero. However, the variance is more challenging to compute because $W_4 - W_2$ and W_6 are not independent. First, since the mean of the integral is 0, we have

$$\mathbb{V}\mathrm{ar}\left(\int_0^T H_t \mathrm{d}W_t\right) = \mathbb{E}\left[(15(W_4 - W_2) + W_6(W_{10} - W_7))^2\right].$$

We then need to expand the square within the expectation, which yields three expectations. The first expectation is

$$\mathbb{E}\left[15^2(W_4 - W_2)^2\right] = 225 \times 2 = 450.$$

The third expectation is

$$\mathbb{E}\left[W_6^2(W_{10} - W_7)^2\right] = \mathbb{E}\left[W_6^2\right]\mathbb{E}\left[(W_{10} - W_7)^2\right] = 6 \times (10 - 7) = 18$$

since we are taking the expectation of independent random variables (Brownian increments over non-overlapping intervals). The second expectation is more complicated. We have

$$\mathbb{E}[(W_4 - W_2)W_6(W_{10} - W_7)]$$

which is the expectation of the product of three random variables, some of them being dependent. However, since we can write

$$W_6 = (W_6 - W_4) + (W_4 - W_2) + (W_2 - W_0),$$

then

$$\begin{aligned}(W_4 - W_2)W_6(W_{10} - W_7) &= (W_4 - W_2) \\ &\times ((W_6 - W_4) + (W_4 - W_2) + (W_2 - W_0)) \times (W_{10} - W_7)\end{aligned}$$

and its expectation is equal to zero. Combining everything, we obtain

$$\mathbb{V}\mathrm{ar}\left(\int_0^T H_t \mathrm{d}W_t\right) = 450 + 18 = 468.$$

■

15.2.3 Ito-integrable stochastic processes

We would like to analyze the stochastic integral of a more general class of stochastic processes, which we will call **Ito-integrable stochastic processes**. Such processes need to have the following properties.

First, the process $H = \{H_t, 0 \leq t \leq T\}$ should be an **adapted stochastic process**, that is a stochastic process such that for each time t, the random variable H_t is completely determined by the Brownian trajectory up to time t, i.e. by the random variables W_s, for all $0 \leq s \leq t$. In some sense, an adapted process follows the *flow of information* provided by the Brownian motion W: it does not look into the future. Second, the process H also needs to be *sufficiently integrable* for the variance of its stochastic integral to be finite; more on this below. These are mathematical technicalities we have to deal with, but we will not say more than we already have on this matter. Elementary stochastic processes defined in (15.2.5) are Ito-integrable processes by definition.

The stochastic integral of an Ito-integrable process H, as defined in (15.2.1), requires the limiting procedure, meaning that

$$\int_0^t H_s \mathrm{d}W_s = \lim_{n\to\infty} \sum_{i=0}^{n-1} H_{t_i}\left(W_{t_{i+1}} - W_{t_i}\right),$$

is obtained as the limit of the Riemann sums where $t_i = it/n$, for each $i = 0, 1, \dots, n$. We can also say that each n-th Riemann sum is the stochastic integral of an elementary stochastic process $\widetilde{H}$ given by

$$\widetilde{H}_t = \sum_{i=0}^{n-1} H_{t_i} \mathbb{1}_{(t_i, t_{i+1}]}(t)$$

implying that

$$\int_0^t \widetilde{H}_s \mathrm{d}W_s = \sum_{i=0}^{n-1} H_{t_i}\left(W_{t_{i+1}} - W_{t_i}\right).$$

Said differently, the stochastic integral of a process H is obtained by approximating H with elementary stochastic processes $\widetilde{H}$ having an increasing number of time steps n. This is illustrated in Figure 15.4.

Example 15.2.8 *Stochastic integrals of functions of Brownian motion*
First, let us have a look at the following stochastic integral: $\int_0^t W_s \mathrm{d}W_s$. In this case, we have $H = W$ and then, by definition,

$$\int_0^t W_s \mathrm{d}W_s = \lim_{n\to\infty} \sum_{i=0}^{n-1} W_{t_i}\left(W_{t_{i+1}} - W_{t_i}\right).$$

Similarly, the stochastic integral $\int_0^t \exp(W_s)\mathrm{d}W_s$ is obtained by the following limiting procedure:

$$\int_0^t \exp(W_s)\mathrm{d}W_s = \lim_{n\to\infty} \sum_{i=0}^{n-1} \exp(W_{t_i})\left(W_{t_{i+1}} - W_{t_i}\right).$$

Recall that these last Riemann sums have been studied in example 15.2.3. ■

Convergence in quadratic mean

The type of convergence involved in the definition of the stochastic integral given in (15.2.1) is the *convergence in quadratic mean*. Mathematically, it means that

$$\mathbb{E}\left[\left(\int_0^t H_s \mathrm{d}W_s - \sum_{i=0}^{n-1} H_{t_i}\left(W_{t_{i+1}} - W_{t_i}\right)\right)^2\right] \underset{n\to\infty}{\longrightarrow} 0.$$

In other words, as n increases, the *mean-square error* between the Riemann sums and the stochastic integral gets closer to zero.

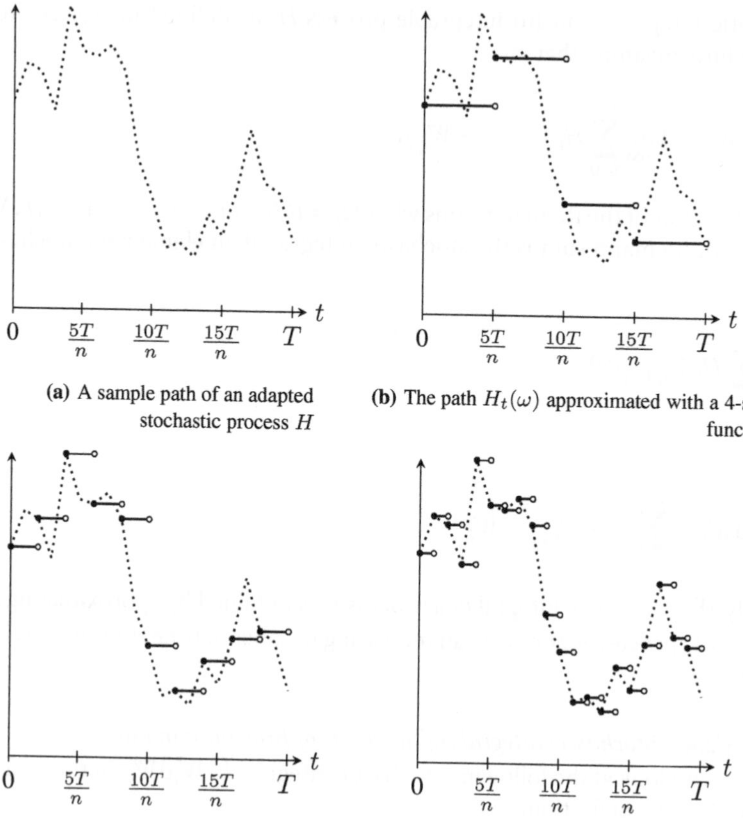

(a) A sample path of an adapted stochastic process H

(b) The path $H_t(\omega)$ approximated with a 4-step function

(c) The path $H_t(\omega)$ approximated with a 10-step function

(d) The path $H_t(\omega)$ approximated with a 20-step function

Figure 15.4 A sample path $H_t(\omega)$ approximated by different step functions. For n large, the path of the step function gets closer to the path $H_t(\omega)$

15.2.4 Properties

Let us now analyze some of the properties of Ito's stochastic integral. It must be noted that since a stochastic integral is a random variable, there will be two types of properties:

1) properties related to its *random nature*, i.e. its mean and variance, and whenever possible its probability distribution;
2) properties related to its *integral nature*, i.e. its approximation by Riemann sums and its linearity.

As mentioned before, the properties of stochastic integrals are mostly inherited from those of the approximating Riemann sums.

Let $H = \{H_t, 0 \leq t \leq T\}$ and $G = \{G_t, 0 \leq t \leq T\}$ be Ito-integrable stochastic processes and let α_1, α_2 be real numbers. Then, for any time $T > 0$, we have:

1) Stochastic integrals are zero-mean random variables:

$$\mathbb{E}\left[\int_0^T H_t \mathrm{d}W_t\right] = 0.$$

2) Stochastic integrals have a variance given by:

$$\mathbb{V}\mathrm{ar}\left(\int_0^T H_t \mathrm{d}W_t\right) = \int_0^T \mathbb{E}\left[H_t^2\right] \mathrm{d}t, \tag{15.2.7}$$

which is also known as **Ito's isometry.**

3) If H is a non-random function of time, i.e. if it is not affected by the scenario ω, then

$$\int_0^T H_t \mathrm{d}W_t \sim \mathcal{N}\left(0, \int_0^T H_t^2 \mathrm{d}t\right).$$

4) Stochastic integrals are linear:

$$\int_0^T (\alpha_1 H_t + \alpha_2 G_t)\mathrm{d}W_t = \alpha_1 \int_0^T H_t \mathrm{d}W_t + \alpha_2 \int_0^T G_t \mathrm{d}W_t.$$

Let us show all these properties and how they are inherited from the Riemann sums. For any Ito-integrable process $H = \{H_t, 0 \le t \le T\}$, its n-th Riemann sum

$$\sum_{i=0}^{n-1} H_{t_i}\left(W_{t_{i+1}} - W_{t_i}\right),$$

where $t_i = iT/n$ for each i, is a zero-mean random variable. Indeed, we have

$$\begin{aligned}\mathbb{E}\left[\sum_{i=0}^{n-1} H_{t_i}\left(W_{t_{i+1}} - W_{t_i}\right)\right] &= \sum_{i=0}^{n-1} \mathbb{E}\left[H_{t_i}\left(W_{t_{i+1}} - W_{t_i}\right)\right] \\ &= \sum_{i=0}^{n-1} \mathbb{E}\left[H_{t_i}\right]\mathbb{E}\left[W_{t_{i+1}} - W_{t_i}\right] = 0,\end{aligned}$$

where in the second last equality we used the fact that, for each i, the random variable H_{t_i} is independent of the random variable $W_{t_{i+1}} - W_{t_i}$; this is because H_{t_i} is completely determined by the Brownian trajectory up to time t_i.

Given that the mean of the Riemann sum is equal to zero, its variance is such that

$$\mathbb{V}\mathrm{ar}\left(\sum_{i=0}^{n-1} H_{t_i}\left(W_{t_{i+1}} - W_{t_i}\right)\right) = \mathbb{E}\left[\left(\sum_{i=0}^{n-1} H_{t_i}\left(W_{t_{i+1}} - W_{t_i}\right)\right)^2\right].$$

Distributing the square and using the linearity property of the expectation, we get

$$\begin{aligned}&\mathbb{V}\mathrm{ar}\left(\sum_{i=0}^{n-1} H_{t_i}\left(W_{t_{i+1}} - W_{t_i}\right)\right) \\ &= \mathbb{E}\left[\sum_{i=0}^{n-1}\sum_{j=0}^{n-1} H_{t_i}H_{t_j}\left(W_{t_{i+1}} - W_{t_i}\right)\left(W_{t_{j+1}} - W_{t_j}\right)\right] \\ &= \sum_{i=0}^{n-1}\sum_{j=0}^{n-1} \mathbb{E}\left[H_{t_i}H_{t_j}\left(W_{t_{i+1}} - W_{t_i}\right)\left(W_{t_{j+1}} - W_{t_j}\right)\right].\end{aligned}$$

When $i = j$, again by independence, we have

$$\mathbb{E}\left[H_{t_i}^2\left(W_{t_{i+1}} - W_{t_i}\right)^2\right] = \mathbb{E}\left[H_{t_i}^2\right]\mathbb{E}\left[\left(W_{t_{i+1}} - W_{t_i}\right)^2\right] = \mathbb{E}\left[H_{t_i}^2\right](t_{i+1} - t_i)$$

When $i < j$, using the formula of iterated expectations, we get

$$\begin{aligned}
&\mathbb{E}[H_{t_i}H_{t_j}(W_{t_{i+1}} - W_{t_i})(W_{t_{j+1}} - W_{t_j})] \\
&= \mathbb{E}[\mathbb{E}[H_{t_i}H_{t_j}(W_{t_{i+1}} - W_{t_i})(W_{t_{j+1}} - W_{t_j}) \mid W_u, u \leq t_j]] \\
&= \mathbb{E}[H_{t_i}H_{t_j}(W_{t_{i+1}} - W_{t_i})\mathbb{E}[(W_{t_{j+1}} - W_{t_j}) \mid W_u, u \leq t_j]] \\
&= \mathbb{E}[H_{t_i}H_{t_j}(W_{t_{i+1}} - W_{t_i})\mathbb{E}[(W_{t_{j+1}} - W_{t_j})]] = 0,
\end{aligned}$$

where, once more, we used the independence properties of Brownian increments. Of course, changing the role played by i and j, we get a symmetric result when $j < i$. In conclusion,

$$\mathbb{V}\mathrm{ar}\left(\sum_{i=0}^{n-1} H_{t_i}(W_{t_{i+1}} - W_{t_i})\right) = \sum_{i=0}^{n-1} \mathbb{E}[H_{t_i}^2](t_{i+1} - t_i). \tag{15.2.8}$$

Informally speaking, since by definition

$$\sum_{i=0}^{n-1} H_{t_i}(W_{t_{i+1}} - W_{t_i}) \underset{n\to\infty}{\longrightarrow} \int_0^T H_t \mathrm{d}W_t,$$

we can expect that[4]

$$\begin{aligned}
\mathbb{E}\left[\sum_{i=0}^{n-1} H_{t_i}(W_{t_{i+1}} - W_{t_i})\right] &\underset{n\to\infty}{\longrightarrow} \mathbb{E}\left[\int_0^T H_t \mathrm{d}W_t\right] \\
\mathbb{V}\mathrm{ar}\left(\sum_{i=0}^{n-1} H_{t_i}(W_{t_{i+1}} - W_{t_i})\right) &\underset{n\to\infty}{\longrightarrow} \mathbb{V}\mathrm{ar}\left(\int_0^T H_t \mathrm{d}W_t\right).
\end{aligned}$$

Therefore, using the above discussion, we can conclude that

$$\begin{aligned}
\mathbb{E}\left[\int_0^T H_t \mathrm{d}W_t\right] &= \lim_{n\to\infty} 0 = 0 \\
\mathbb{V}\mathrm{ar}\left(\int_0^T H_t \mathrm{d}W_t\right) &= \lim_{n\to\infty} \sum_{i=0}^{n-1} \mathbb{E}[H_{t_i}^2](t_{i+1} - t_i) = \int_0^T \mathbb{E}[H_t^2]\, \mathrm{d}t,
\end{aligned}$$

where the last equality comes from the definition of the (deterministic) Riemann integral of the function $t \mapsto \mathbb{E}[H_t^2]$, as defined in (15.1.1).

Finally, if $H = \{H_t, 0 \leq t \leq T\}$ is a non-random function of time, then the H_{t_i}s are no longer *genuine* random variables: they are simple real numbers. Therefore, the n-th Riemann sum

$$\sum_{i=0}^{n-1} H_{t_i}(W_{t_{i+1}} - W_{t_i}),$$

is a linear combination of independent normal random variables given by the increments $W_{t_{i+1}} - W_{t_i}$. This means that the Riemann sum itself is normally distributed. We have already checked that the mean is equal to zero and now the variance becomes

$$\sum_{i=0}^{n-1} \mathbb{E}[H_{t_i}^2](t_{i+1} - t_i) = \sum_{i=0}^{n-1} H_{t_i}^2(t_{i+1} - t_i),$$

---

4 We will not provide rigorous proofs for these two statements.

because the expectation of a constant is equal to the constant itself. If we summarize, we have: if H is deterministic, then

$$\sum_{i=0}^{n-1} H_{t_i}\left(W_{t_{i+1}} - W_{t_i}\right) \sim \mathcal{N}\left(0, \sum_{i=0}^{n-1} H_{t_i}^2 (t_{i+1} - t_i)\right).$$

The *normality* of the Riemann sums is not affected by the limiting procedure, so if H is deterministic, then $\int_0^T H_t \mathrm{d}W_t$ is indeed normally distributed with mean 0 and variance given by:

$$\lim_{n\to\infty} \sum_{i=0}^{n-1} H_{t_i}^2 (t_{i+1} - t_i) = \int_0^T H_t^2 \mathrm{d}t.$$

Finally, since Riemann sums are clearly linear (they are sums) and since limits are themselves linear, it should be clear that stochastic integrals are linear with respect to their integrands. The details are left to the reader.

Example 15.2.9 *Stochastic integral of a Brownian motion*
In Example 15.2.8, we considered the stochastic integral of a standard Brownian motion, namely the random variable $\int_0^T W_t \mathrm{d}W_t$.

By the above properties, we have that $\mathbb{E}[\int_0^T W_t \mathrm{d}W_t] = 0$. Note that, in Example 15.2.3, we had already verified that the corresponding n-th Riemann sums were also zero-mean random variables. By Ito's isometry given in (15.2.7), the variance is given by

$$\mathbb{V}\mathrm{ar}\left(\int_0^T W_t \mathrm{d}W_t\right) = \int_0^T \mathbb{E}\left[W_t^2\right] \mathrm{d}t = \int_0^T t\,\mathrm{d}t = \frac{T^2}{2}.$$

Note that since the integrand is W, i.e. a stochastic process (as opposed to a function), we do not know the distribution of $\int_0^T W_t \mathrm{d}W_t$. ■

Example 15.2.10 *Linearity of stochastic integrals*
In example 15.2.8, we also considered the stochastic integral $\int_0^T \exp(W_t)\mathrm{d}W_t$.

Let us now consider the following stochastic integral:

$$\int_0^T \left(\frac{W_t}{\sqrt{\pi}} - \exp(W_t)\right) \mathrm{d}W_t.$$

We know that it has a zero mean and that its variance can be computed using Ito's isometry.

Moreover, by the linearity property of stochastic integrals, we can write

$$\int_0^T \left(\frac{W_t}{\sqrt{\pi}} - \exp(W_t)\right) \mathrm{d}W_t = \frac{1}{\sqrt{\pi}} \int_0^T W_t \mathrm{d}W_t - \int_0^T \exp(W_t)\mathrm{d}W_t,$$

where the two stochastic integrals on the right-hand-side have been defined in example 15.2.8. ■

15.3 Ito's lemma for Brownian motion

As we now know, a stochastic integral is a random variable with an expectation equal to 0 and an easy-to-compute variance (using Ito's isometry). We also discovered that in most cases its probability distribution is unknown.

On the other hand, because stochastic integrals are also integrals, we would like to be able to simplify their expressions, in the spirit of the *fundamental theorem of calculus* in classical calculus.

Before going any further, let us emphasize some differences between stochastic calculus and classical calculus. In classical calculus, we have that

$$\int_0^T t\mathrm{d}t = \frac{T^2}{2}$$

and, more generally,

$$\int_0^T f(t)\mathrm{d}f(t) = \int_0^T f(t)f'(t)\mathrm{d}t = \int_0^T \frac{1}{2}(f(t)^2)'\mathrm{d}t = \frac{f(T)^2}{2},$$

if f is differentiable and such that $f(0) = 0$. We used the fact that for a differentiable function, we have $\mathrm{d}f(t) = f'(t)\mathrm{d}t$.

The analog stochastic integral is given by

$$\int_0^T W_t \mathrm{d}W_t.$$

Unfortunately, we cannot write something like $\mathrm{d}W_t = W_t'\mathrm{d}t$ because the trajectories of Brownian motion are not differentiable functions. Even if we know that this stochastic integral has a mean equal to zero and a variance equal to $T^2/2$ (see Example 15.2.9), we can wonder whether a simpler expression can be obtained or not.

As alluded to above, in classical calculus, one of the main results is the *fundamental theorem of calculus*: if f is a differentiable function, then

$$f(b) - f(a) = \int_a^b f'(x)\mathrm{d}x. \tag{15.3.1}$$

The proof of this result is based on a first-order Taylor expansion of f.

There is a similar result in stochastic calculus whose simplest version is as follows: if f is a twice differentiable function, then

$$f(W_t) - f(W_0) = \int_0^t f'(W_s)\mathrm{d}W_s + \frac{1}{2}\int_0^t f''(W_s)\mathrm{d}s, \tag{15.3.2}$$

for all $0 \le t \le T$. This result is known as **Ito's lemma** and its proof is based on a second-order Taylor expansion of f.

Note that there are two types of integrals in Ito's lemma of equation (15.3.2):

1) the stochastic integral of $f'(W_s)$;
2) the stochastic Riemann integral of $f''(W_s)$.

Both types have been studied above.

Example 15.3.1 Let us go back to the following stochastic integral:

$$\int_0^T W_t \mathrm{d}W_t.$$

In order to use Ito's lemma, we must find a function $f(\cdot)$ such that $W_t = f'(W_t)$. If we choose $f(x) = x^2/2$, then $f'(x) = x$ and $f''(x) = 1$. By Ito's lemma in (15.3.2), we have

$$\frac{W_T^2}{2} - \frac{W_0^2}{2} = \int_0^T W_t \mathrm{d}W_t + \frac{1}{2}\int_0^T 1\mathrm{d}t = \int_0^T W_t \mathrm{d}W_t + \frac{T}{2},$$

where we used the linearity property of stochastic integrals.

In conclusion, if we reorganize the above equation and since $W_0 = 0$, then we can write

$$\int_0^T W_t \mathrm{d}W_t = \frac{W_T^2}{2} - \frac{T}{2}.$$

As $W_T \sim \mathcal{N}(0, T)$, we can obtain the probability density function of W_T^2 (by using the *change-of-variable formula* as seen in an elementary probability course) and then the distribution of this stochastic integral. ■

Example 15.3.2 Let us use Ito's lemma to simplify $\int_0^T \mathrm{e}^{W_t}\mathrm{d}W_t$.

For $f(x) = \mathrm{e}^x$, we have $f'(x) = f''(x) = \mathrm{e}^x$. Using Ito's lemma in (15.3.2), we can write

$$\mathrm{e}^{W_T} - \mathrm{e}^{W_0} = \int_0^T \mathrm{e}^{W_t}\mathrm{d}W_t + \frac{1}{2}\int_0^T \mathrm{e}^{W_t}\mathrm{d}t.$$

Since $\mathrm{e}^{W_0} = 1$, we can write

$$\int_0^T \mathrm{e}^{W_t}\mathrm{d}W_t = \mathrm{e}^{W_T} - 1 - \frac{1}{2}\int_0^T \mathrm{e}^{W_t}\mathrm{d}t. \tag{15.3.3}$$

Somehow, the expression on the right-hand-side is simpler:

- there is no stochastic integral left, only a stochastic Riemman integral;
- the distribution of e^{W_T} is known: it is a lognormal distribution. ■

A glimpse at the proof of Ito's lemma

Note the similarities between equation (15.3.1) and equation (15.3.2). To gain more intuition as to why they are similar but yet different, consider the following second-order Taylor expansion of f around x:

$$f(x+h) - f(x) \approx f'(x)h + \frac{1}{2}f''(x)h^2,$$

where the quality of this approximation is linked directly to the size of h. For standard Riemann integrals, the second-order term is negligible. However, this term does play an important role for stochastic integrals. The reason is that standard Brownian motion has a non-zero *quadratic variation*: for each $T > 0$, we have

$$\lim_{n\to\infty} \sum_{i=0}^{n-1} \left(W_{t_{i+1}} - W_{t_i}\right)^2 = T,$$

where $t_i = iT/n$, for each $i = 0, 1, \dots, n$.

Heuristically speaking, we could replace $x + h$ by W_{t+h} and replace x by W_t, while keeping in mind that $h = (x + h) - x$. Then, we can write symbolically

$$f(W_{t+h}) - f(W_t) \approx f'(W_t)(W_{t+h} - W_t) + \frac{1}{2}f''(W_t)(W_{t+h} - W_t)^2.$$

With $h = T/n$ and n large, if we sum up and take the limit:

$$\begin{aligned} & f(W_T) - f(W_0) \\ & \approx \sum_{i=0}^{n-1} f'(W_{t_i})(W_{t_{i+1}} - W_{t_i}) + \frac{1}{2}\sum_{i=0}^{n-1} f''(W_{t_i})(W_{t_{i+1}} - W_{t_i})^2 \\ & \longrightarrow \int_0^T f'(W_t)\mathrm{d}W_t + \frac{1}{2}\int_0^T f''(W_t)\mathrm{d}t. \end{aligned}$$

15.4 Diffusion processes

An important class of processes, applied in various areas including mathematical and actuarial finance, is known as **diffusion processes.** A diffusion process $X = \{X_t, 0 \leq t \leq T\}$ is a continuous-time stochastic process that can be written in the following form:

$$X_t - X_0 = \int_0^t a(X_s)\mathrm{d}s + \int_0^t b(X_s)\mathrm{d}W_s, \tag{15.4.1}$$

where $a(\cdot)$ and $b(\cdot)$ are deterministic functions[5] and where $X_0 = x_0 \in \mathbb{R}$ is the initial value of the process X. The function $a(x)$ is usually known as the **drift coefficient** whereas $b(x)$ is typically called the **diffusion** or **volatility coefficient.**

In some sense, diffusion processes are generalizations of linear Brownian motions. Indeed, recall from example 15.2.1 that a linear Brownian motion $X_t = X_0 + \mu t + \sigma W_t$ can also be written as

$$X_t = X_0 + \int_0^t \mu\,\mathrm{d}s + \int_0^t \sigma\,\mathrm{d}W_s.$$

In other words, such a linear Brownian motion is a diffusion process as defined in (15.4.1), with a constant drift coefficient $a(x) = \mu$ and a constant volatility coefficient $b(x) = \sigma$. Note also that the terminology is the same as the one used in Chapter 14. Diffusion processes also extend the commonly used GBM to create richer and more realistic dynamics for asset prices and interest rates.

15.4.1 Stochastic differential equations

In many actuarial and financial models, the dynamics of a financial variable over time (e.g. the stock price) is expressed with a **stochastic differential equation.** Such an equation specifies the desirable behavior of this quantity over the next infinitesimal period of time.

Symbolically, a stochastic differential equation (SDE) is a *stochastic equation*: for given functions $a(\cdot)$ and $b(\cdot)$, we are looking for a process X such that

$$\mathrm{d}X_t = a(X_t)\mathrm{d}t + b(X_t)\mathrm{d}W_t, \tag{15.4.2}$$

5 As for all integrals, these functions must be *sufficiently well-behaved* for the two integrals to exist.

with initial condition $X_0 = x_0$. More rigorously, we are looking for a diffusion process X such that

$$X_t - X_0 = \int_0^t a(X_s)\mathrm{d}s + \int_0^t b(X_s)\mathrm{d}W_s,$$

as already defined in (15.4.1). The stochastic differential equation in (15.4.2) is also called the *differential form* whereas the expression in (15.4.1) is known as the *integral form* of a diffusion process.

Intuitively, a SDE can be interpreted as follows: given that we know the value of X_t at time t, the future value of the process *one millisecond* after is explained by two components:

- a component $a(X_t)\mathrm{d}t$: a change of magnitude $a(X_t)$ in the direction of the infinitesimal increase in time $\mathrm{d}t$;
- a component $b(X_t)\mathrm{d}W_t$: a change of magnitude $b(X_t)$ in the direction of the *infinitesimal increment* of the Brownian motion $\mathrm{d}W_t$.

As a SDE describes the evolution of a quantity X_t as time passes, we are interested in finding an explicit expression for X_t: this is what we will mean by *solving the SDE.* But recall that a SDE of the form given in (15.4.2) has no meaning by itself: it is rather a symbolic representation for the integral equation given in (15.4.1). Hence, solving a SDE and studying the underlying stochastic process means analyzing its corresponding integral form and its existence. Most of the time, this is a difficult task.

Example 15.4.1 *Examples of diffusion processes*
Let us consider the following SDEs: for given numbers μ and σ,

1) $\mathrm{d}X_t = \mu\mathrm{d}t + \sigma\mathrm{d}W_t$, with initial value $X_0 = -1$;
2) $\mathrm{d}X_t = \mu X_t\mathrm{d}t + \sigma X_t\mathrm{d}W_t$, with initial value $X_0 = 2$.

For the first SDE, according to the general form given in (15.4.2), we have a constant drift coefficient $a(x) = \mu$ and a constant volatility coefficient $b(x) = \sigma$. Therefore, the integral form is given by

$$X_t = -1 + \int_0^t \mu\,\mathrm{d}s + \int_0^t \sigma\,\mathrm{d}W_s = -1 + \mu t + \sigma W_t,$$

and we have recovered a linear Brownian motion issued from -1. Note also that for this SDE, there is clearly a solution (we found it).

For the second SDE, according to the general form given in (15.4.2), we have a drift coefficient given by $a(x) = \mu x$ and a volatility coefficient given by $b(x) = \sigma x$. Therefore, the integral form is given by

$$X_t = 2 + \int_0^t \mu X_s\mathrm{d}s + \int_0^t \sigma X_s\mathrm{d}W_s.$$

For this SDE, it is not clear whether there exists a solution X that verifies this equality. ∎

15.4.2 Ito's lemma for diffusion processes

In the last example, we were not able to obtain an explicit expression for the diffusion process solving

$$\mathrm{d}X_t = \mu X_t\mathrm{d}t + \sigma X_t\mathrm{d}W_t,$$

with initial value $X_0 = 2$. In order to solve such SDEs, we will need a generalized version of Ito's lemma for diffusion processes. Indeed, in Section 15.3, we saw the simplest form of Ito's lemma; the most general form, still known as Ito's lemma, applies to deterministic transformations $f(t, X_t)$ of a diffusion process X.

Let $f(t,x)$ be a function of two variables, i.e. of *time* and *space* respectively, such that $\partial f/\partial t$ and $\partial^2 f/\partial x^2$, i.e. the first partial derivative in time and the second partial derivative in space,[6] are both continuous functions. If $X = \{X_t, 0 \leq t \leq T\}$ is a diffusion process solving the general SDE given in (15.4.2), then the process $Y_t = f(t, X_t)$ solves the following SDE:

$$df(t,X_t) = \frac{\partial f}{\partial t}(t,X_t)\mathrm{d}t + \frac{\partial f}{\partial x}(t,X_t)\mathrm{d}X_t + \frac{1}{2}\frac{\partial^2 f}{\partial x^2}(t,X_t)(\mathrm{d}X_t)^2. \tag{15.4.3}$$

Before going any further, we need to make two clarifications. First, the terms $\frac{\partial f}{\partial t}(t,X_t)$ and $\frac{\partial f}{\partial x}(t,X_t)$ must be understood as follows:

1) Compute the partial derivatives $\frac{\partial f}{\partial t}$, $\frac{\partial f}{\partial x}$ and $\frac{\partial^2 f}{\partial x^2}$ of the function $f(t,x)$.
2) Evaluate each partial derivative at (t, X_t).

Second, the expression $(\mathrm{d}X_t)^2$ must be understood as follows: it can be shown that

$$(\mathrm{d}X_t)^2 = (b(X_t))^2\mathrm{d}t.$$

This is an application of the **product rule**; see the box below for more details. Hence, substituting $\mathrm{d}X_t$ by its expression given in (15.4.2) and reorganizing equation (15.4.3), we get

$$\begin{aligned} df(t,X_t) = &\left(\frac{\partial f}{\partial t}(t,X_t) + a(X_t)\frac{\partial f}{\partial x}(t,X_t) + \frac{1}{2}(b(X_t))^2\frac{\partial^2 f}{\partial x^2}(t,X_t)\right)\mathrm{d}t \\ &+ \frac{\partial f}{\partial x}(t,X_t)b(X_t)\mathrm{d}W_t. \end{aligned} \tag{15.4.4}$$

This last SDE is equivalent to the one in (15.4.3). They are both referred to as **Ito's lemma for diffusion processes.**

While the last expression in (15.4.4) can be tedious, the one in (15.4.3) is more intuitive: it is more or less a Taylor expansion, of order one in time and order two in space. Then, one only needs to be familiar with the *product rule* to further simplify the expressions.

Product rule

Whenever we need to *multiply* two SDEs, there exists a convenient and symbolic tool (backed by rigorous mathematics) known as the **product rule.** It is described in the following table:

$\times$	$\mathrm{d}W_t$	$\mathrm{d}t$
$\mathrm{d}W_t$	$\mathrm{d}t$	0
$\mathrm{d}t$	0	0

More precisely, if we have two SDEs

$$\mathrm{d}X_t = a(X_t)\mathrm{d}t + b(X_t)\mathrm{d}W_t,$$
$$\mathrm{d}Y_t = c(Y_t)\mathrm{d}t + d(Y_t)\mathrm{d}W_t,$$

6 Remember that if $\partial^2 f/\partial x^2$ exists, then $\partial f/\partial x$ also exists and is continuous.

then

$$\begin{aligned} &\mathrm{d}X_t \times \mathrm{d}Y_t \\ &= [a(X_t)\mathrm{d}t + b(X_t)\mathrm{d}W_t] \times [c(Y_t)\mathrm{d}t + d(Y_t)\mathrm{d}W_t] \\ &= a(X_t)c(Y_t)(\mathrm{d}t)^2 + [a(X_t)d(Y_t) + b(X_t)c(Y_t)]\mathrm{d}t\mathrm{d}W_t + b(X_t)d(Y_t)(\mathrm{d}W_t)^2 \\ &= b(X_t)d(Y_t)\mathrm{d}t. \end{aligned}$$

In particular,

$$(\mathrm{d}X_t)^2 = (b(X_t))^2\mathrm{d}t,$$

as used in Ito's lemma (15.4.3) above.

Example 15.4.2 *Simple geometric Brownian motion*
Using Ito's lemma, let us verify that the following simple geometric Brownian motion, i.e. the process

$$S_t = \mathrm{e}^{W_t - \frac{t}{2}}$$

is a diffusion process. We must show that S satisfies an SDE as given in (15.4.2), that is we must find the SDE coefficients.

Clearly, we have $S_t = f(t, W_t)$, with $f(t, x) = \mathrm{e}^{x-t/2}$, and $S_0 = 1$. First, we compute the partial derivatives:

$$\frac{\partial f}{\partial t}(t, x) = -(1/2)f(t, x),$$

$$\frac{\partial f}{\partial x}(t, x) = f(t, x),$$

$$\frac{\partial^2 f}{\partial x^2}(t, x) = f(t, x).$$

Second, we replace x by W_t and we use Ito's lemma (15.4.3) to $f(t, W_t) = \exp(W_t - t/2)$, noticing that $\mathrm{d}X_t = \mathrm{d}W_t$. We get

$$\begin{aligned} \mathrm{d}f(t, W_t) &= \{-(1/2)f(t, W_t)\}\mathrm{d}t + f(t, W_t)\mathrm{d}W_t + \frac{1}{2}f(t, W_t)(\mathrm{d}W_t)^2 \\ &= f(t, W_t)\mathrm{d}W_t, \end{aligned}$$

since $(\mathrm{d}W_t)^2 = \mathrm{d}t$. Substituting $f(t, W_t)$ by S_t, we see that $\exp(W_t - t/2)$ is a solution to the SDE given by

$$\mathrm{d}S_t = S_t\mathrm{d}W_t,$$

with $S_0 = 1$. In conclusion, we have indeed verified that the simple geometric Brownian motion

$$S_t = \mathrm{e}^{W_t - \frac{t}{2}}$$

is a diffusion process with drift $a(S_t) = 0$ and diffusion $b(S_t) = S_t$. ■

Note that if we consider a transformation $f(X_t)$ in space only, all partial derivatives simplify to ordinary derivatives and then Ito's lemma of (15.4.3) and (15.4.4) simplify respectively to:

$$\mathrm{d}f(X_t) = f'(X_t)\mathrm{d}X_t + \frac{1}{2}f''(X_t)(\mathrm{d}X_t)^2 \tag{15.4.5}$$

and, equivalently,

$$df(X_t) = (a(X_t)f'(X_t) + \frac{1}{2}(b(X_t))^2 f''(X_t))dt + f'(X_t)b(X_t)dW_t. \tag{15.4.6}$$

15.4.3 Geometric Brownian motion

We saw in example 15.4.1 that an SDE of the form

$$dX_t = \mu dt + \sigma dW_t$$

is that of a linear Brownian motion with drift μ and diffusion/volatility σ. In other words, $X_t = x_0 + \mu t + \sigma W_t$ is a diffusion process.

In this section, we are interested in obtaining a similar characterization, as a diffusion process, for geometric Brownian motions. We want to extend what we did in example 15.4.2 and thus solve the second SDE in example 15.4.1.

We know from Chapter 14 that a geometric Brownian motion S is obtained as the exponential of a linear Brownian motion X. As this is a simple deterministic transformation of a diffusion process, then, using Ito's lemma, we should be able to find its corresponding SDE.

Let us apply Ito's lemma to $S_t = \exp(X_t)$, where $X_t = x_0 + \mu t + \sigma W_t$. Here, as $f(x) = e^x$, we will use the version of Ito's lemma in (15.4.5). First, we have

$$f(x) = f'(x) = f''(x) = e^x.$$

Consequently, as $S_t = f'(X_t) = f''(S_t)$, using Ito's lemma we can write

$$dS_t = S_t dX_t + \frac{1}{2} S_t (dX_t)^2$$

where $dX_t = \mu dt + \sigma dW_t$. Since, by the product rule,

$$(dX_t)^2 = (\mu dt + \sigma dW_t)^2 = \sigma^2 dt,$$

we further have

$$\begin{aligned} dS_t &= S_t(\mu dt + \sigma dW_t) + \frac{1}{2} S_t \sigma^2 dt \\ &= S_t(\mu + \sigma^2/2)dt + \sigma S_t dW_t, \end{aligned}$$

with $S_0 = e^{x_0}$. By analogy with Example 15.4.1, this is an SDE with drift coefficient given by $a(x) = (\mu + \sigma^2/2)x$ and volatility coefficient given by $b(x) = \sigma x$. In other words, a geometric Brownian motion is indeed a diffusion process.

It is important to realize that if we consider instead the GBM S given by

$$S_t = \exp(x_0 + (\mu - \sigma^2/2)t + \sigma W_t),$$

then it will be the solution to the following SDE:

$$dS_t = \mu S_t dt + \sigma S_t dW_t.$$

This representation corresponds to the SDE in Equation (17.1.8) of Chapter 17.

In summary, we have obtained Table 15.1, for converting a geometric Brownian motion from its SDE representation to its exponential representation, and vice versa.

Table 15.1 Equivalence of representations for geometric Brownian motions (with $S_0 = 1$)

SDE representation		Exponential representation
$dS_t = \mu S_t dt + \sigma S_t dW_t$	$\Leftrightarrow$	$S_t = \exp((\mu - \sigma^2/2)t + \sigma W_t)$
$dS_t = (\mu + \sigma^2/2)S_t dt + \sigma S_t dW_t$	$\Leftrightarrow$	$S_t = \exp(\mu t + \sigma W_t)$

15.4.4 Ornstein-Uhlenbeck process

In this section, we analyze the **Ornstein-Uhlenbeck (OU) process**, which is widely used in actuarial finance, especially to model mortality/longevity risk and interest rates in continuous-time short rate models.

First, let us look at a simple version of the Ornstein-Uhlenbeck SDE, namely

$$dX_t = -X_t dt + dW_t, \tag{15.4.7}$$

with an arbitrary initial condition $X_0 = x_0$. We know this SDE means that

$$X_t = x_0 - \int_0^t X_s ds + W_t,$$

for all $t \geq 0$. Unfortunately, this cannot be considered as an explicit expression: the process X itself appears on both sides of the equation.

To find the solution to the SDE in (15.4.7), we can use Ito's lemma along with the function $f(t,x) = xe^t$. The partial derivatives of this function are

$$\frac{\partial f}{\partial t}(t,x) = xe^t, \quad \frac{\partial f}{\partial x}(t,x) = e^t, \quad \frac{\partial^2 f}{\partial x^2}(t,x) = 0.$$

Thus, using the version of Ito's lemma in (15.4.3), we can write

$$d\left(X_t e^t\right) = \left(X_t e^t\right) dt + e^t dX_t = \left(X_t e^t - X_t e^t\right) dt + e^t dW_t = e^t dW_t,$$

with $f(0, x_0) = x_0$, which is equivalent to

$$X_t e^t - x_0 = \int_0^t e^s dW_s.$$

If we move things around, we get

$$X_t = e^{-t}\left(x_0 + \int_0^t e^s dW_s\right). \tag{15.4.8}$$

In other words, $e^{-t}(x_0 + \int_0^t e^s dW_s)$ is a solution to $dX_t = -X_t dt + dW_t$. This is now an explicit expression for the OU process X.

For a fixed time t, since the integrand is deterministic, the stochastic integral in (15.4.8) is normally distributed (see Section 15.2.4) with

$$\mathbb{E}[X_t] = e^{-t} x_0$$

and

$$\mathbb{V}\text{ar}(X_t) = e^{-2t}\mathbb{V}\text{ar}\left(\int_0^t e^s dW_s\right) = e^{-2t}\int_0^t e^{2s} ds = \frac{1 - e^{-2t}}{2}.$$

In conclusion, for each fixed time t, we have

$$X_t \sim \mathcal{N}\left(e^{-t}x_0, \frac{1-e^{-2t}}{2}\right).$$

More generally, the simple OU SDE in (15.4.7) can be extended to get the following general OU SDE:

$$dX_t = \alpha(\beta - X_t)dt + \sigma dW_t,$$

for parameters $\alpha, \beta, \sigma > 0$, and initial value $X_0 = x_0$.

This OU dynamic is said to be **mean reverting**:

- if $X_t < \beta$, then the drift is positive, *pushing* the process upward (toward β);
- if $X_t > \beta$, then the drift is negative, *pushing* the process downward (toward β).

Note that this is a desirable feature for interest rate modeling. We interpret β as an equilibrium level whereas α is the speed of mean reversion, thus amplifying or mitigating the effect of the gap between X_t and β. This version of the mean-reverting OU process is used in the Vasicek model for the short rate.

It is also possible to find an explicit expression for the general OU process. Again, using Ito's lemma with $f(t,x) = xe^{\alpha t}$, we obtain

$$X_t = e^{-\alpha t}\left(x_0 + \beta(e^{\alpha t} - 1) + \sigma\int_0^t e^{\alpha s}dW_s\right).$$

As a consequence, we have, for each fixed time $t > 0$, that

$$X_t \sim \mathcal{N}\left(e^{-\alpha t}x_0 + \beta(1 - e^{-\alpha t}), \frac{\sigma^2}{2\alpha}(1 - e^{-2\alpha t})\right).$$

15.4.5 Square-root diffusion process

A general OU process can take both positive and negative values. Indeed, since the random variable X_t is normally distributed, it has a non-zero probability of being negative. This can be a major impediment for some actuarial and financial modeling.

This is not the case for the *square-root diffusion process*. A simple version of this other process can be obtained as the square of a specific OU process. Let X be given by the SDE in (15.4.7), i.e.

$$dX_t = -X_t dt + dW_t.$$

Set $Y_t = X_t^2$. If we use Ito's lemma with the function $f(x) = x^2$, for which the first two derivatives are

$$f'(x) = 2x \qquad \text{and} \qquad f''(x) = 2,$$

then, after a few manipulations, we get

$$\begin{aligned} dY_t &= \left(1 - 2(X_t)^2\right)dt + 2X_t dW_t \\ &= (1 - 2Y_t)dt + 2\sqrt{Y_t}dW_t. \end{aligned}$$

The general **mean-reverting square-root process** is defined as the solution of

$$dX_t = \alpha(\beta - X_t)dt + \sigma\sqrt{X_t}dW_t,$$

for parameters $\alpha > 0, \beta > 0, \sigma > 0$, and initial condition $X_0 = x_0$. In general, there is no closed-form expression for the solution of this SDE.

This version of the square-root dynamic is used for interest rate modeling, e.g. in the Cox-Ingersoll-Ross (CIR) model, and for volatility modeling, e.g. in the Heston model.

15.5 Summary

Notation

- Stochastic processes:
  - $W = \{W_t, 0 \leq t \leq T\}$ is a standard Brownian motion;
  - $H = \{H_t, 0 \leq t \leq T\}$ is a continuous-time stochastic process.
- Partition of the time interval $[0, t]$: for each $i = 0, 1, \dots, n$, set $t_i = it/n$.

Stochastic Riemann integrals

- Definition:

$$\int_0^T H_t \mathrm{d}t = \lim_{n\to\infty} \sum_{i=0}^{n-1} H_{t_i}(t_{i+1} - t_i).$$

- Important property:

$$\mathbb{E}\left[\int_0^T H_t \mathrm{d}t\right] = \int_0^T \mathbb{E}[H_t]\mathrm{d}t.$$

Ito's stochastic integrals

- Definition:

$$\int_0^T H_t \mathrm{d}W_t = \lim_{n\to\infty} \sum_{i=0}^{n-1} H_{t_i}\left(W_{t_{i+1}} - W_{t_i}\right)$$

  where $\sum_{i=0}^{n-1} H_{t_i}(\omega)(W_{t_{i+1}}(\omega) - W_{t_i}(\omega))$ is the n-th Riemann sum.
- Elementary stochastic process:

$$H_t = \sum_{i=0}^{n-1} \phi_i \mathbb{1}_{(t_i, t_{i+1}]}(t),$$

  where the ϕ_is are random variables.
- Stochastic integral of an elementary process:

$$\int_0^T H_t \mathrm{d}W_t = \sum_{i=0}^{n-1} \phi_i\left(W_{t_{i+1}} - W_{t_i}\right).$$

- Important properties:
  1) Mean:

$$\mathbb{E}\left[\int_0^T H_t \mathrm{d}W_t\right] = 0.$$

  2) Variance (Iso's isometry):

$$\mathbb{V}\mathrm{ar}\left(\int_0^T H_t \mathrm{d}W_t\right) = \int_0^T \mathbb{E}\left[H_t^2\right] \mathrm{d}t.$$

3) If H is a non-random function of time, then

$$\int_0^T H_t \mathrm{d}W_t \sim \mathcal{N}\left(0, \int_0^T H_t^2 \mathrm{d}t\right).$$

4) Linearity:

$$\int_0^T (\alpha_1 H_t + \alpha_2 G_t)\mathrm{d}W_t = \alpha_1 \int_0^T H_t \mathrm{d}W_t + \alpha_2 \int_0^T G_t \mathrm{d}W_t.$$

Diffusion processes

- Diffusion process: a continuous-time stochastic process X verifying

$$X_t - X_0 = \int_0^t a(X_s)\mathrm{d}s + \int_0^t b(X_s)\mathrm{d}W_s.$$

- Drift coefficient: $a(X_s)$.
- Diffusion coeffcient: $b(X_s)$.
- Stochastic differential equation (SDE) representation:

$$\mathrm{d}X_t = a(X_t)\mathrm{d}t + b(X_t)\mathrm{d}W_t.$$

- Linear Brownian motion: $X_t = X_0 + \mu t + \sigma W_t$ solves

$$\mathrm{d}X_t = \mu \mathrm{d}t + \sigma \mathrm{d}W_t.$$

- Geometric Brownian motion: $X_t = S_0 \exp((\mu - \sigma^2/2)t + \sigma W_t)$ solves

$$\mathrm{d}X_t = \mu X_t \mathrm{d}t + \sigma X_t \mathrm{d}W_t.$$

- Ornstein-Uhlenbeck process: $X_t = \mathrm{e}^{-\alpha t}(x_0 + \beta(\mathrm{e}^{\alpha t} - 1) + \sigma \int_0^t \mathrm{e}^{\alpha s}\mathrm{d}W_s)$ solves

$$\mathrm{d}X_t = \alpha(\beta - X_t)\mathrm{d}t + \sigma \mathrm{d}W_t$$

  and is such that

$$X_t \sim \mathcal{N}(\mathrm{e}^{-\alpha t}x_0 + \beta(1 - \mathrm{e}^{-\alpha t}), \frac{\sigma^2}{2\alpha}(1 - \mathrm{e}^{-2\alpha t})).$$

- Square-root diffusion process:

$$\mathrm{d}X_t = \alpha(\beta - X_t)\mathrm{d}t + \sigma\sqrt{X_t}\mathrm{d}W_t.$$

Ito's lemma

- Ito's lemma for Brownian motion: if f is a twice differentiable function, then

$$f(W_t) - f(W_0) = \int_0^t f'(W_s)\mathrm{d}W_s + \frac{1}{2}\int_0^t f''(W_s)\mathrm{d}s.$$

- Ito's lemma for diffusion processes (simple version): if X is a diffusion process and if $f(x)$ is twice differentiable, then

$$\mathrm{d}f(X_t) = f'(X_t)\mathrm{d}X_t + \frac{1}{2}f''(X_t)(\mathrm{d}X_t)^2.$$

- Ito's lemma for diffusion processes (general version): if X is a diffusion process and if $f(t,x)$ is twice differentiable in x and differentiable in t, then

$$\mathrm{d}f(t,X_t) = \frac{\partial f}{\partial t}(t,X_t)\mathrm{d}t + \frac{\partial f}{\partial x}(t,X_t)\mathrm{d}X_t + \frac{1}{2}\frac{\partial^2 f}{\partial x^2}(t,X_t)(\mathrm{d}X_t)^2.$$

- From the product rule: $(\mathrm{d}X_t)^2 = (b(X_t))^2\mathrm{d}t$.

15.6 Exercises

15.1 Compute the expected value of the following stochastic Riemann integrals:

(a) $\int_0^T (W_t)^2 dt$;

(b) $\int_0^T (W_t + t)dt$;

(c) $\int_0^T e^{\sigma W_t} dt$.

15.2 Find the distribution of the following random variables:

(a) $\int_0^\pi (e^t + e^{2t})dW_t$;

(b) $\int_0^T W_t dt$ (hint: use the definition of the Riemann integral as a limit).

15.3 Let $S = \{S_t, t \geq 0\}$ be a geometric Brownian motion given by $dS_t = \mu S_t dt + \sigma S_t dW_t$. Let $Y_t = f(t, S_t)$ be a two-variable transformation of S. Derive the SDE of the following transformations and determine the SDE of which $Y = \{Y_t, t \geq 0\}$ is a solution:

(a) $Y_t = e^{-rt} S_t$;

(b) $Y_t = (S_t)^\alpha$.

15.4 Use Ito's lemma to show that $X_t = \frac{t+W_t}{\exp\{W_t + \frac{t}{2}\}}$ is a stochastic integral.

15.5 Compute $\mathbb{E}[X_T]$ for a mean-reverting square-root diffusion process X given by

$$dX_t = \alpha(\beta - X_t)dt + \sigma\sqrt{X_t}dW_t.$$

Hint: consider the process $Y_t = e^{\alpha t} X_t$.

15.6 Consider the stochastic process

$$X_t = (1+t)^{-2}(1 + t + W_t).$$

Verify that X_t is the solution to the following SDE:

$$dX_t = \left(\frac{1}{(1+t)^2} - \frac{2}{1+t}X_t\right) dt + \frac{1}{(1+t)^2} dW_t.$$

16

Introduction to the Black-Scholes-Merton model

The year 1973 shall be remembered as the year of breakthroughs in the history of options and derivatives. First of all, the largest U.S. options market, the Chicago Board Options Exchange (CBOE), was created that year. Moreover, the works from Fischer Black, Myron Scholes and Robert C. Merton, gave birth to modern financial mathematics and thus largely contributed to the shape of today's derivatives markets.

More precisely, in the paper *The Pricing of Options and Corporate Liabilities*, published in 1973, Black and Scholes pioneered rational option pricing by dynamically replicating the option payoff until maturity. During the same period, Robert C. Merton published *Theory of Rational Option Pricing*. He extended Black and Scholes' approach in several ways, e.g. by pricing options on dividend-paying stocks and down-and-out barrier options. See [19] and [20].

In his paper, Merton refers to Black and Scholes' framework as the "Black-Scholes' theory of option pricing" and even nowadays the model is widely known as the Black-Scholes' model. Their work was recognized by the Royal Swedish Academy of Sciences (Nobel Prizes) in 1997.[1] The model is still very popular today on the financial markets.

In this chapter, our main objective is to lay the foundations of the famous *Black-Scholes-Merton market model* and its pricing formula. We will provide a heuristic approach to this formula by linking as much as possible the derivations to the binomial model of Part I using a limiting argument. More specifically, the learning objectives are to:

- understand the main assumptions of the Black-Scholes-Merton model, including the dynamics of the risk-free and risky assets;
- connect the Black-Scholes-Merton model to the binomial model;
- differentiate real-world (actuarial) and risk-neutral probabilities;
- compute call and put options price with the Black-Scholes formula;
- price simple derivatives using risk-neutral probabilities;
- analyze the impact of various determinants of the call or put option price, namely the strike price, stock price, risk-free rate, volatility and time to maturity;
- derive the replicating portfolio for simple derivatives;
- apply the delta-hedging strategy over several periods to hedge simple derivatives.

1 Nearly 2 years after Black's death in 1995.

Actuarial Finance: Derivatives, Quantitative Models and Risk Management, First Edition.
Mathieu Boudreault and Jean-François Renaud.

16.1 Model

As in the binomial and the trinomial models, the **Black-Scholes-Merton model** (BSM model) is assumed to be a *frictionless* market composed of two assets:

- a risk-free asset (a bank account or a bond) which evolves according to the risk-free interest rate r;
- a risky asset (a stock or an index) with a known initial value and unknown future values.

Trades in this market can occur at any *continuous* time $t \geq 0$ and therefore the BSM framework is a continuous-time market model. We assume that time is always expressed in years (even if it is possible to do otherwise) so that parameters are written on an annual basis. We will mostly work on a time horizon $[0, T]$ where T is very often the maturity of a derivative.

16.1.1 Risk-free asset

In the BSM model, there is an asset $B = \{B_t, 0 \leq t \leq T\}$ from which investors can earn an interest rate of r. This asset is said to be risk-free because capital and interest is repaid with certainty: there is no default. Thus, $r > 0$ is known as the risk-free rate and it is assumed to be continuously compounded, i.e. for $0 \leq t \leq T$,

$$B_t = B_0 e^{rt}.$$

As before, we set $B_0 = 1$. Note that, we could also have fixed $B_T = 1$ instead of B_0, in which case B would be modeling a zero-coupon bond.

16.1.2 Risky asset

In the BSM model, the price of the risky asset (often an index or a stock) $S = \{S_t, 0 \leq t \leq T\}$ evolves according to a geometric Brownian motion

$$S_t = S_0 \exp\left\{\left(\mu - \frac{\sigma^2}{2}\right)t + \sigma W_t\right\}, \tag{16.1.1}$$

where $W = \{W_t, 0 \leq t \leq T\}$ is a standard Brownian motion (with respect to the probability measure $\mathbb{P}$) and S_0 is a known quantity (initial asset price).

It follows from Section 14.5 that

$$S_t \overset{\mathbb{P}}{\sim} \mathcal{LN}\left(\ln(S_0) + \left(\mu - \frac{\sigma^2}{2}\right)t, \sigma^2 t\right)$$

i.e. for each time $0 < t \leq T$, the random variable S_t follows a lognormal distribution with parameters $\ln(S_0) + (\mu - \frac{\sigma^2}{2})t$ and $\sigma^2 t$ (with respect to the actuarial probability $\mathbb{P}$ measure).

Finally, note that the BSM model can be fully specified by four parameters: S_0, μ, r and $\sigma > 0$.

We can also express the model for the risky asset in terms of log-returns. On any given year, the log-return is such that

$$\ln(S_{t+1}/S_t) \overset{\mathbb{P}}{\sim} \mathcal{N}\left(\mu - \frac{\sigma^2}{2}, \sigma^2\right)$$

i.e. the continuously compounded annual return is normally distributed with mean $\mu - \frac{\sigma^2}{2}$ and variance σ^2. As a result, σ is called the volatility of the log-return.

In many textbooks, μ is interpreted as the expected annual return on the asset. This is because the expected asset price in the BSM model is

$$\mathbb{E}[S_t] = S_0 e^{\mu t}.$$

However, the expected annual *log*-return is

$$\mathbb{E}[\ln(S_{t+1}/S_t)] = \mu - \frac{\sigma^2}{2}.$$

To avoid any confusion with other textbooks and unless stated otherwise, we will interpret μ as the mean (annual) *real* rate of return on the asset.

Usual description of the BSM model

One may wonder why the BSM model is based on a geometric Brownian motion with drift $\mu - \frac{\sigma^2}{2}$ instead of μ. This is because, in the literature, the dynamics of the risky asset is usually given by the following stochastic differential equation:

$$\mathrm{d}S_t = \mu S_t \mathrm{d}t + \sigma S_t \mathrm{d}W_t.$$

More details on stochastic differential equations can be found in Chapter 15. It turns out that the solution to this equation is the process defined in equation (16.1.1), which is a geometric Brownian motion with drift $\mu - \frac{\sigma^2}{2}$.

Moreover, it provides a financial interpretation to parameter μ: it is then the expected asset price is $S_0 e^{\mu t}$.

Recall from the binomial tree model that in order to price derivatives with risk-neutral formulas, we had to contrast between probabilities p and q for each up-move. More formally, we had to distinguish between the *real-world* or actuarial probability measure $\mathbb{P}$ and the artificial *risk-neutral* probability measure $\mathbb{Q}$ obtained as a by-product of an equation rearrangement. A similar situation will occur in the BSM model.

At this point, it is important to note that the notation $\overset{\mathbb{P}}{\sim}$ is used to emphasize that at each time $t > 0$, the risky asset price S_t *really* follows a lognormal distribution with parameters $\ln(S_0) + \left(\mu - \frac{\sigma^2}{2}\right)t$ and $\sigma^2 t$. Any computations involving the likelihood that the risky asset price S reaches a given level over the next year, or that it crashes, have to be done with this *real-world* distribution.

16.1.3 Derivatives

We now introduce a third asset in our market, namely a European derivative, whose value is given by the stochastic process $V = \{V_t, 0 \leq t \leq T\}$. The price of the derivative at *inception* is given by V_0 which is a constant that will be determined later. As the value of the derivative depends upon the underlying asset S, then, for each time $0 < t \leq T$, the value V_t is random. In particular, the derivative matures at time T, so the random variable V_T represents the payoff of this derivative.

One of our main objectives is to compute the no-arbitrage price of derivatives in the BSM model, i.e. to determine V_t, for each time $0 \leq t < T$, given a payoff V_T. In other words, we will compute the initial price V_0 and characterize the price V_t, at any time $0 < t < T$, using no-arbitrage arguments.

This chapter will focus on simple European derivatives, i.e. derivatives whose payoff only depends on the final asset price S_T:

$$V_T = g(S_T)$$

where $g(\cdot)$ is a function specified in the derivative's contract. As seen previously in this book, examples include a call option with strike price K, i.e.

$$g(x) = (x - K)_+ = \max(x - K, 0),$$

or a put option, i.e.

$$g(x) = (K - x)_+ = \max(K - x, 0).$$

Exotic path-dependent options, those whose payoff depends on the risky asset price at several times preceding maturity, will be treated later in Chapter 18.

As in discrete-time models, instead of using $V = \{V_t, 0 \leq t \leq T\}$, we will use $C = \{C_t, 0 \leq t \leq T\}$ and $P = \{P_t, 0 \leq t \leq T\}$ for the price process of a call option and a put option respectively, in which cases we clearly have

$$C_T = (S_T - K)_+ \qquad \text{and} \qquad P_T = (K - S_T)_+,$$

when their common strike price is K.

Example 16.1.1 *Probability of expiration in the money*
A 3-year call option, with a strike price of \$120, has been issued on a stock. If the initial stock price is \$100 and it is assumed the stock price evolves in a BSM model with mean return of 7% and volatility of 30%, calculate the probability that the call option expires in the money.

We have $K = 120$ and $T = 3$ years, so the payoff is given by

$$C_3 = (S_3 - 120)_+.$$

The probability that the option expires in the money is $\mathbb{P}((S_3 - 120)_+ > 0)$, which in this case is equal to $\mathbb{P}(S_3 > 120)$.

In this BSM model, we also have that $\mu = 0.07$, $\sigma = 0.3$ and $S_0 = 100$. Then,

$$S_3 \overset{\mathbb{P}}{\sim} \mathcal{LN}\left(\ln(100) + \left(0.07 - \frac{0.3^2}{2}\right) \times 3, 0.3^2 \times 3\right)$$

or, equivalently,

$$\ln(S_3/100) \overset{\mathbb{P}}{\sim} \mathcal{N}\left(\left(0.07 - \frac{0.3^2}{2}\right) \times 3, 0.3^2 \times 3\right) = \mathcal{N}(0.075, 0.27).$$

Finally, we can write

$$\begin{aligned}
\mathbb{P}((S_3 - 120)_+ > 0) &= \mathbb{P}(\ln(S_3/100) > \ln(120/100)) \\
&= \mathbb{P}\left(\frac{\ln(S_3/100) - 0.075}{\sqrt{0.27}} > \frac{\ln(120/100) - 0.075}{\sqrt{0.27}}\right) \\
&= 1 - N\left(\frac{\ln(120/100) - 0.075}{\sqrt{0.27}}\right) \\
&= 1 - N(0.2065404) \\
&= 0.418184405,
\end{aligned}$$

where we used the fact that

$$\frac{\ln(S_3/100) - 0.075}{\sqrt{0.27}} \overset{\mathbb{P}}{\sim} \mathcal{N}(0,1). \qquad \blacksquare$$

16.2 Relationship between the binomial and BSM models

In this section, we will see how closely related the BSM model is to the binomial model. In fact, the geometric Brownian motion used in the BSM model to represent the evolution of the asset price can be seen as the limit of random walks from binomial trees. Therefore, on many occasions we will be able to compare results from both models.

16.2.1 Second look at the binomial model

Suppose we have a given time horizon $[0, T]$ in mind, with 0 being the initial time and T the end of the investment period (expressed in years). In an n-period binomial tree model, we divide this time interval $[0, T]$ in n periods: $(0, T/n), (T/n, 2T/n), \dots, ((n-1)T/n, T)$. Then, there are $n+1$ time points: $0, T/n, 2T/n, \dots, (n-1)T/n, T$. In Chapter 11, we put a significant amount of effort into characterizing the corresponding stock price process

$$S^{(n)} = \left\{S_0, S^{(n)}_{T/n}, S^{(n)}_{2T/n}, \dots, S^{(n)}_{(n-1)T/n}, S^{(n)}_T\right\},$$

with $S^{(n)}_0 = S_0$, as the initial price is not affected by the partition of the interval $[0, T]$. Here, we have added the superscript (n) to emphasize that we are in an n-period binomial tree model.

For each $k = 1, 2, \dots, n$, we can write[2]

$$S^{(n)}_{\frac{kT}{n}} = S_0 \times u^{I_k} \times d^{k-I_k},$$

where the random variable I_k keeps track of the number of upward movements in the trajectory after k periods. In other words, I_k can be viewed as the number of successes in k trials. Therefore, I_k follows a binomial distribution with parameters k and p (w.r.t. to $\mathbb{P}$), and then, for a given $j = 0, 1, 2, \dots, k$, we have

$$\mathbb{P}\left(S^{(n)}_{kT/n} = S_0 u^j d^{k-j}\right) = \binom{k}{j} p^j (1-p)^{k-j}.$$

However, note that the random variable $S^{(n)}_{kT/n}$ does not follow a binomial distribution. It is a random variable whose probability mass function is linked to that of a binomial distribution, namely the distribution of I_k. Figure 16.1 illustrates an eight-period binomial tree (left panel) along with the probability mass function of the asset price at time $k = 6$ (right panel).

The collection of random variables $S = \left\{S_0, S^{(n)}_{T/n}, S^{(n)}_{2T/n}, \dots, S^{(n)}_{(n-1)T/n}, S^{(n)}_T\right\}$ is a discrete-time stochastic process that can also be written as

$$S^{(n)}_{kT/n} = S^{(n)}_{(k-1)T/n} U_k,$$

2 In Chapter 11, the time index is expressed as an integer $0, 1, 2, \dots, n$. But since we will be taking limits when n goes to infinity, the time index needs to be defined in years.

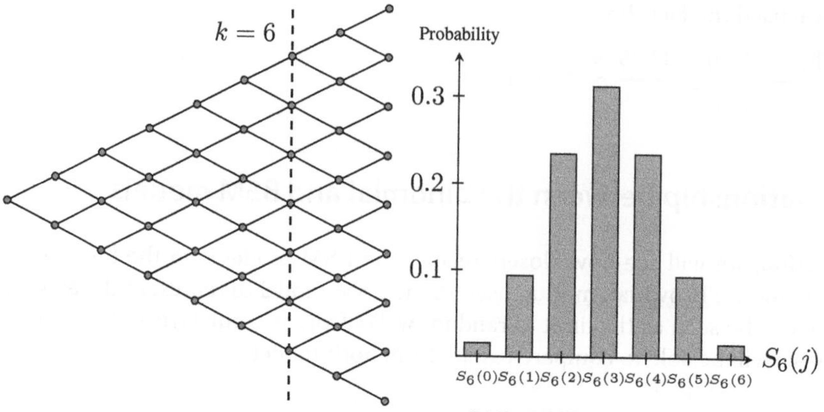

(a) Eight-period binomial tree (b) Probability mass function at $k = 6$

Figure 16.1 An eight-period binomial tree (left panel) along with the probability mass function of the asset price at time $k = 6$ (right panel)

for each $k = 1, 2, \dots, n$, where $U_k \in \{u, d\}$ (up and down factors) with probability p and $1 - p$, respectively. Note that we can also write

$$\frac{S^{(n)}_{kT/n}}{S^{(n)}_{(k-1)T/n}} = U_k,$$

for each $k = 1, 2, \dots, n$. These last two equations describe how the asset price evolves over time.

In conclusion, the risky asset price should be viewed as a *stochastic process*, with each possible path in the tree being a realization of this process. Figure 16.2 illustrates four different paths taken by the asset price in a 20-period binomial tree.

16.2.2 Convergence of the binomial model

Suppose we represent the evolution of the price of a financial index over 1 year, using a binomial model with monthly time steps. After each month, the price can take one of two possible values. Thus, after a year, there will be 13 possible values. More generally, after k months, there are $k + 1$ possible values for the random variable $S^{(12)}_{k/12}$, where $k = 0, 1, \dots, 12$. But what happens if we want to squeeze in additional up and down moves? In other words, what would happen if we were to have daily time steps or even hourly time steps?

Said differently, for a fixed investment horizon $[0, T]$, we will look at what happens if the number of time steps increases, i.e. if we let n go to infinity. If we increase the number of time steps in the binomial model, the *mesh* of realizations will increase.

For a given investment horizon, i.e. for a fixed T, the probability distribution of the asset price $S^{(n)}_{kT/n}$, which is a discrete distribution for each n, will slowly converge to a continuous distribution, as n goes to infinity. To what distribution exactly?

Since

$$S^{(n)}_{kT/n} = S^{(n)}_{(k-1)T/n} U_k,$$

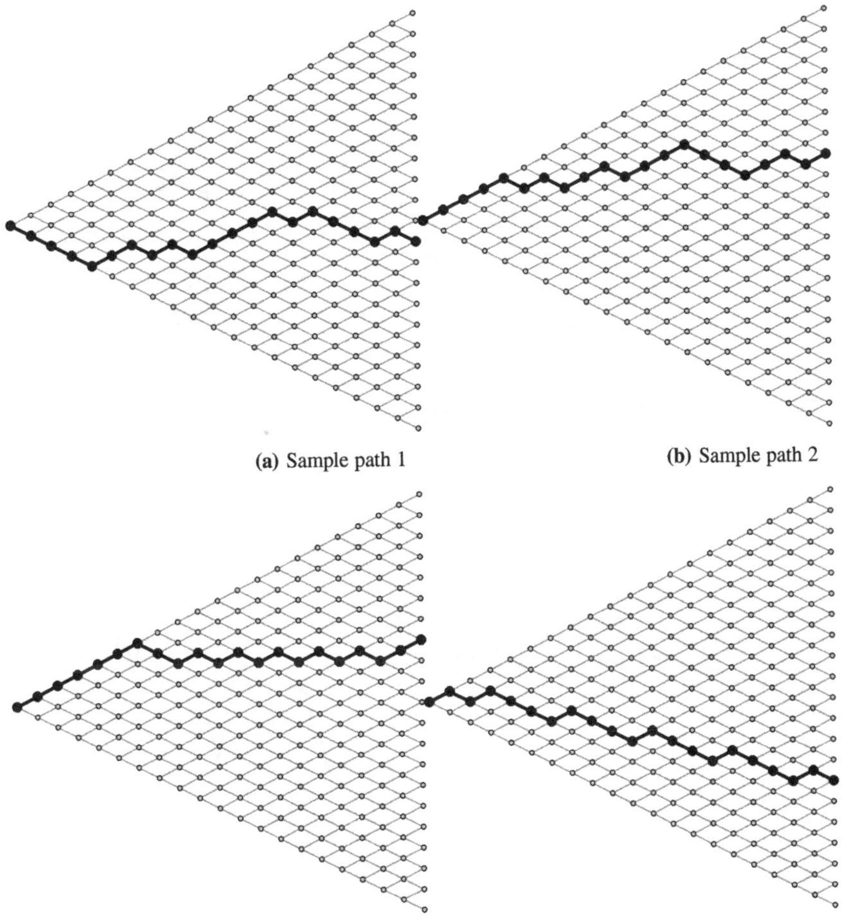

(a) Sample path 1 (b) Sample path 2

(c) Sample path 3 (d) Sample path 4

Figure 16.2 Four different sample paths of the asset price in a 20-period binomial tree

taking the log on both sides, we get

$$X_k^{(n)} = X_{k-1}^{(n)} + \ln(U_k),$$

where $X_k^{(n)} = \ln\left(S_{kT/n}^{(n)}\right)$, for each $k = 1, 2, \ldots, n$. In other words, $X^{(n)} = \left\{X_k^{(n)}, k = 0, 1, \ldots, n\right\}$ is given by $X_0^{(n)} = \ln(S_0)$ and, for each $k = 1, 2, \ldots, n$, by

$$X_k^{(n)} = \sum_{i=1}^{k} \ln(U_i) = \ln(U_1) + \ln(U_2) + \cdots + \ln(U_k).$$

The latter process is a random walk because its increments $\epsilon_i = \ln(U_i)$, where $i = 1, 2, \ldots, n$, are independent and identically distributed random variables taking their values in $\{\ln(u), \ln(d)\}$. However, it is a non-symmetric random walk.

We know from Section 14.2 that a standard Brownian motion is obtained as the continuous-time limit of symmetric random walks. Similarly, non-symmetric random walks will converge

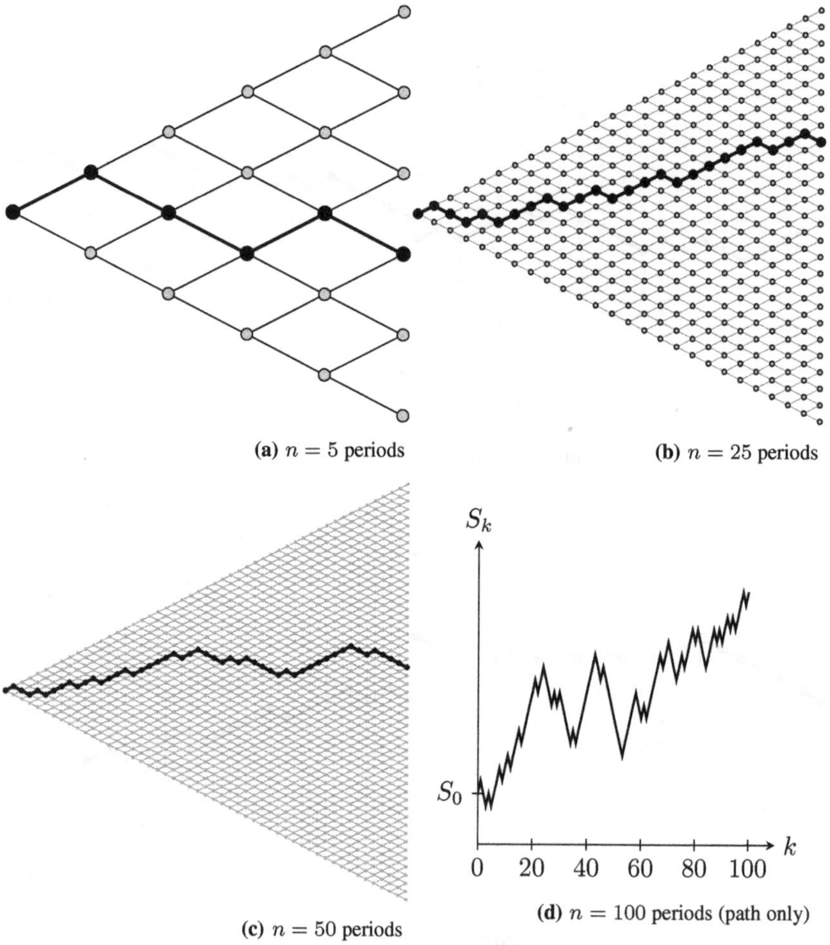

Figure 16.3 Illustration of a path in the binomial tree as the number of time steps increases. The lattice in the lower right plot was removed to highlight the path characteristics

to a *linear* Brownian motion. Consequently, the non-symmetric random walks given by $X^{(n)} = \{X_k^{(n)}, k = 0, 1, \ldots, n\}$ will converge to the linear Brownian motion

$$\left\{ \ln(S_t/S_0) = \left(\mu - \frac{\sigma^2}{2} \right) t + \sigma W_t, 0 \le t \le T \right\}$$

used in the BSM model.

Figure 16.3 illustrates how paths in the binomial model resemble paths of a geometric Brownian motion as the number of time steps increases, for a fixed time horizon $[0, T]$.

16.2.3 Formal proof

In this section, we show how the distribution of the risky asset price in the binomial model converges to a lognormal distribution as the number of time steps increases. Let us focus on the convergence of the final price $S_T^{(n)}$. The same work would be necessary to prove the convergence of each $S_{kT/n}^{(n)}$, for $k = 1, 2, \ldots, n-1$.

For the convergence to go through, we need to let the up and down factors depend also on n. It would not make sense to increase the number of time steps, for a given time horizon, without adapting the size of the jumps. We did the same thing when proving that some symmetric random walks were converging to a standard Brownian motion. Therefore, the up and down factors are denoted by u_n and d_n, emphasizing that they depend on the number of time steps.

Recall that we can write

$$S_T^{(n)} = S_0 u_n^{I_n} d_n^{n-I_n},$$

where the random variable I_n records the number of up moves in the full trajectory of the underlying asset's price. As discussed above, we can further write

$$\begin{aligned} \ln\left(S_T^{(n)}/S_0\right) &= \ln(u_n)I_n + \ln(d_n)(n - I_n) \\ &= \ln(u_n/d_n)I_n + n\ln(d_n). \end{aligned}$$

The random variable I_n follows a binomial distribution with parameters n and p_n, i.e. $I_n \overset{\mathbb{P}}{\sim} \text{Bin}(n, p_n)$, where

$$p_n = \frac{e^{\mu(T/n)} - d_n}{u_n - d_n}.$$

Hence, by the Central Limit Theorem,[3] we deduce that

$$\frac{I_n - np_n}{\sqrt{np_n(1-p_n)}} \underset{n\to\infty}{\longrightarrow} \mathcal{N}(0,1)$$

or, more precisely,

$$\mathbb{P}_n\left(a \le \frac{I_n - np_n}{\sqrt{np_n(1-p_n)}} \le b\right) \underset{n\to\infty}{\longrightarrow} N(b) - N(a),$$

for any real numbers $a < b$.

On the other hand, since $\ln\left(S_T^{(n)}/S_0\right) = \ln(u_n/d_n)I_n + n\ln(d_n)$, we have

$$\begin{aligned} \mathbb{E}\left[\ln\left(S_T^{(n)}/S_0\right)\right] &= \ln(u_n/d_n)np_n + n\ln(d_n), \\ \mathbb{V}\text{ar}\left(\ln\left(S_T^{(n)}/S_0\right)\right) &= (\ln(u_n/d_n))^2 np_n(1-p_n). \end{aligned}$$

We are interested in the convergence of $\ln\left(S_T^{(n)}/S_0\right)$, which can also be rewritten as

$$\ln\left(S_T^{(n)}/S_0\right) = a_n \frac{I_n - np_n}{\sqrt{np_n(1-p_n)}} + b_n,$$

where

$$\begin{aligned} a_n &= \sqrt{\mathbb{V}\text{ar}\left[\ln\left(S_T^{(n)}/S_0\right)\right]} \\ b_n &= \mathbb{E}\left[\ln\left(S_T^{(n)}/S_0\right)\right]. \end{aligned}$$

3 In fact, a more general version of the Central Limit Theorem than the standard one.

With the specifications of the Jarrow-Rudd tree (see Chapter 11), we have

$$u_n = \exp\left(\left(\mu - \sigma^2/2\right)\frac{T}{n} + \sigma\sqrt{\frac{T}{n}}\right),$$

$$d_n = \exp\left(\left(\mu - \sigma^2/2\right)\frac{T}{n} - \sigma\sqrt{\frac{T}{n}}\right),$$

where μ and σ are the BSM model parameters. Then it can be shown that

$$a_n \underset{n\to\infty}{\longrightarrow} \sigma\sqrt{T} \quad \text{and} \quad b_n \underset{n\to\infty}{\longrightarrow} \left(\mu - \sigma^2/2\right)T.$$

Consequently,

$$\ln\left(S_T^{(n)}/S_0\right) \underset{n\to\infty}{\longrightarrow} \ln(S_T/S_0) \overset{\mathbb{P}}{\sim} \mathcal{N}((\mu - \sigma^2/2)T, \sigma^2 T). \tag{16.2.1}$$

This means that, at the limit, the final asset price S_T follows a lognormal distribution with parameters $\ln(S_0) + (\mu - \sigma^2/2)T$ and $\sigma^2 T$ (with respect to the real-world or actuarial probability measure $\mathbb{P}$).

Note that we have obtained the limiting distribution for a single random variable S_T. We could repeat the above procedure for any time $0 < t \leq T$ and get

$$\ln\left(S_{\lfloor nt\rfloor/n}^{(n)}/S_0\right) \underset{n\to\infty}{\longrightarrow} \ln(S_t/S_0) \overset{\mathbb{P}}{\sim} \mathcal{N}((\mu - \sigma^2/2)t, \sigma^2 t).$$

Putting all these random variables together, we get the linear Brownian motion of the BSM model, namely the stochastic process

$$\left\{\ln(S_t/S_0) = \left(\mu - \frac{\sigma^2}{2}\right)t + \sigma W_t, 0 \leq t \leq T\right\},$$

or, equivalently, the geometric Brownian motion

$$\{S_t, 0 \leq t \leq T\},$$

defined in equation (16.1.1).[4]

16.2.4 Risk-neutral probabilities

Recall that in the binomial model, for any given simple derivative, we were able to find its replicating portfolio and then rewrite its value as an expectation using new probability weights q and $1 - q$ known as risk-neutral probabilities. It is important to emphasize that the risk-neutral (conditional) probability q had nothing to do with the real-world (conditional) probability p: it is only a convenient way to rewrite the value of the replicating portfolio as an expectation.

This approach tells us that whenever we need to *find the no-arbitrage price of a derivative*, and only then, we are allowed to use the artificial binomial distribution with parameters n and q_n in order to write the option price as a discounted (with r) expectation. We should insist that q_n has no other actuarial or financial meaning.

As the convergence of the binomial model to the BSM model does not depend on the specific value taken by p, we deduce that the underlying binomial distribution with parameter q_n for the

4 Note that it is a much bigger challenge to deduce the dependence structure between these random variables and describe the full dynamics obtained at the limit.

log asset price will also converge to a normal distribution, but with slightly different parameters. Following from equation (16.2.1), we summarize the results as follows.

The risk-neutral distributions of the random variables $\ln(S_T^{(n)}/S_0)$, which depend on the (conditional) probability q_n of an up-move, converge to the risk-neutral distribution of $\ln\left(S_T/S_0\right)$, as n tends to infinity. Therefore,

$$\ln(S_T/S_0) \overset{\mathbb{Q}}{\sim} \mathcal{N}((r-\sigma^2/2)T, \sigma^2 T)$$

where we use the notation $\overset{\mathbb{Q}}{\sim}$ to emphasize that this is a risk-neutral distribution (rather than a real-world distribution). This is the equivalent in the BSM model of using q rather than p in the binomial model.

We can repeat the above procedure for any time $0 < t \leq T$ and get at the limit

$$\ln(S_t/S_0) \overset{\mathbb{Q}}{\sim} \mathcal{N}\left(\left(r-\frac{\sigma^2}{2}\right)t, \sigma^2 t\right)$$

or, equivalently,

$$S_t \overset{\mathbb{Q}}{\sim} \mathcal{LN}\left(\ln(S_0) + \left(r-\frac{\sigma^2}{2}\right)t, \sigma^2 t\right),$$

at each time $0 < t \leq T$.

16.3 Black-Scholes formula

In the binomial model, we first approached the pricing problem with replication. This is the same approach as the one used by Black and Scholes in their pioneering article of 1973. However, now that the underlying asset price follows a geometric Brownian motion, the mathematical arguments needed to obtain option prices by replication are more complicated.[5]

Therefore, to find the no-arbitrage price of (simple) derivatives in the BSM model, we will take a different approach. Black and Scholes' original derivation of the replicating portfolio will be presented in detail in Chapter 17, a chapter that requires a stronger mathematical background.

The **Black-Scholes formula** is a mathematical expression for the no-arbitrage price of a call (or a put) option in the BSM model. We present two methods to obtain this formula: (1) as a limit of prices obtained in binomial models, and (2) as the stop-loss transform of a well-chosen lognormally distributed random variable.

16.3.1 Limit of binomial models

As above, assume the maturity of the call option is T and split the time interval $[0, T]$ into n subintervals of length T/n. We want to find the no-arbitrage price of a vanilla call option. Recall from Chapter 11 that in an n-period binomial tree, the initial price of a call option with payoff $(S_T^{(n)} - K)_+$ can be written as a binomial sum:

$$\begin{aligned} C_0^{(n)} &= \mathrm{e}^{-rT}\mathbb{E}^{\mathbb{Q}}\left[\left(S_T^{(n)} - K\right)_+\right] \\ &= \mathrm{e}^{-rT}\sum_{j=0}^{n}\left(S_0 u^j d^{n-j} - K\right)_+ \binom{n}{j} q_n^j (1-q_n)^{n-j} \end{aligned}$$

5 Results from stochastic calculus and partial differential equations are needed.

where

$$q_n = \frac{e^{rT/n} - d}{u - d}.$$

See equation (11.3.5) in Chapter 11. In many cases, i.e. when the realization of $S_T^{(n)}$ is not *big enough*, the first terms of this summation are all equal to zero. In other words, if we define

$$k_n^* = \min\left\{i \geq 1 : u^i d^{n-i} S_0 > K\right\},$$

then we can write

$$\begin{aligned} C_0^{(n)} &= e^{-rT} \sum_{j=k_n^*}^{n} \left(u^j d^{n-j} S_0 - K\right) \binom{n}{j} q_n^j (1 - q_n)^{n-j} \\ &= S_0 \sum_{j=k_n^*}^{n} \binom{n}{j} \left(u q_n e^{-r(T/n)}\right)^j \left(d(1 - q_n) e^{-r(T/n)}\right)^{n-j} \\ &\quad - e^{-rT} K \sum_{j=k_n^*}^{n} \binom{n}{j} q_n^j (1 - q_n)^{n-j}, \end{aligned}$$

where we have distributed the term $e^{-nr(T/n)} = e^{-jr(T/n)} e^{-(n-j)r(T/n)}$ inside the first summation.

Then, let us define the incomplete binomial summation, for integers $k \leq m$ and real numbers $0 < \theta < 1$, by

$$\Phi(k, m, \theta) = \sum_{j=k}^{m} \binom{m}{j} \theta^j (1 - \theta)^{m-j}.$$

Therefore, the call option price can be written as

$$C_0^{(n)} = e^{-rT} \mathbb{E}^{\mathbb{Q}_n}\left[\left(S_T^{(n)} - K\right)_+\right] = S_0 \Phi\left(k_n^*, n, q_n^*\right) - e^{-rT} K \Phi\left(k_n^*, n, q_n\right),$$

where $q_n^* = u q_n e^{-r(T/n)}$. It is easy to show that $1 - q_n^* = d(1 - q_n) e^{-r(T/n)}$ and $0 < q_n^* < 1$.

Intuitively, since the incomplete binomial summation Φ is a (survival) probability for binomial distributions, by the Central Limit Theorem, we expect it to converge to a similar probability for normal distributions.

Letting $C_0 = \lim_{n \to \infty} C_0^{(n)}$, then

$$\begin{aligned} C_0 &= \lim_{n \to \infty} e^{-rT} \mathbb{E}^{\mathbb{Q}_n}\left[\left(S_T^{(n)} - K\right)_+\right] \\ &= \lim_{n \to \infty} S_0 \Phi\left(k_n^*, n, q_n^*\right) - e^{-rT} K \Phi\left(k_n^*, n, q_n\right) \end{aligned}$$

and one can show that we obtain

$$C_0 = S_0 N(d_1) - e^{-rT} K N(d_2), \tag{16.3.1}$$

with

$$\begin{aligned} d_1 &= \frac{\ln(S_0/K) + (r + \sigma^2/2)T}{\sigma\sqrt{T}}, \\ d_2 &= \frac{\ln(S_0/K) + (r - \sigma^2/2)T}{\sigma\sqrt{T}} = d_1 - \sigma\sqrt{T}. \end{aligned}$$

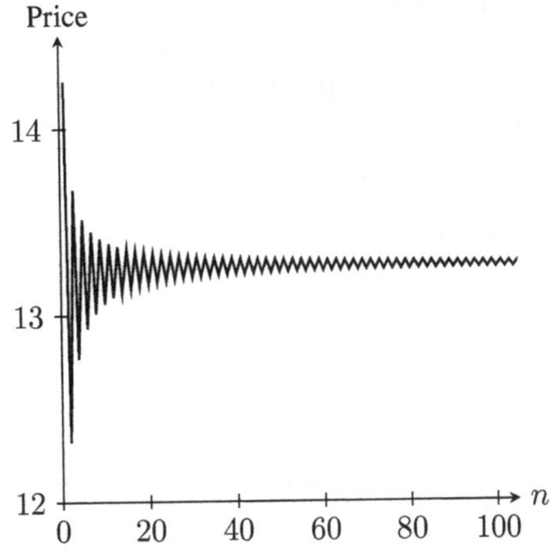

Figure 16.4 Convergence of a call option price in Cox-Ross-Rubinstein (CRR) binomial trees as a function of the number of time steps. The parameters are $\sigma = 0.2$, $T = 1$, $r = 0.1$ and $S_0 = K = 100$

This is the **classical Black-Scholes pricing formula** for the initial price C_0 of a European call option in the Black-Scholes-Merton model.

Figure 16.4 illustrates the convergence of the call option price in a Cox-Ross-Rubinstein (CRR) binomial tree as a function of the number of time steps. The plot shows an oscillatory pattern around $C_0 = 13.26967658$, where the wave amplitude goes to 0 as $n \to \infty$.

Example 16.3.1 *Price of a call option*
A share of stock currently trades for \$65. The mean annual log-return on this stock is 6% and the volatility of the log-return is 27%. Calculate the Black-Scholes price of a 6-month 62-strike call option if the risk-free rate is 3%.

We have $S_0 = 65$, $K = 62$, $T = 0.5$, $\sigma = 0.27$, $r = 0.03$. Then

$$d_1 = \frac{\ln(65/62) + (0.03 + 0.27^2/2) \times 0.5}{0.27\sqrt{0.5}} = 0.421529319$$

$$d_2 = d_1 - 0.27\sqrt{0.5} = 0.230610488$$

with corresponding probabilities

$$N(d_1) = N(0.421529319) = 0.663315697$$
$$N(d_2) = N(0.230610488) = 0.591191291.$$

Using the Black-Scholes formula in equation (16.3.1), the price of this call option is thus

$$\begin{aligned} C_0 &= S_0 N(d_1) - K e^{-rT} N(d_2) \\ &= 65 \times 0.663315697 - 62 e^{-0.03 \times 0.5} \times 0.591191291 \\ &= 7.007365155. \end{aligned}$$

■

We shall note that μ, the mean (annual) *real* rate of return on the risky asset, does not appear in the Black-Scholes formula. This is because the Black-Scholes formula can be obtained as the limit of *binomial prices* which were obtained by first replicating the call option payoff. No matter what is the expected return on the stock, the replicating strategy will replicate the option payoff in any scenario. This parameter is irrelevant for option pricing.

Example 16.3.2 *Probability of expiring in the money*
Your investment bank has issued the call option of example 16.3.1. Your boss asks about the probability that the option will end up in the money at maturity.

We are thus asked to compute

$$\mathbb{P}((S_{0.5} - 62)_+ > 0) = \mathbb{P}(S_{0.5} > 62) = \mathbb{P}(S_{0.5}/65 > 62/65),$$

where $S_0 = 65$. Using the fact that

$$\frac{S_T}{65} \overset{\mathbb{P}}{\sim} \mathcal{LN}((\mu - 0.5\sigma^2)T, \sigma^2 T)$$

we have

$$\begin{aligned}\mathbb{P}(S_{0.5} > 62) &= 1 - N\left(\frac{\ln(62/S_0) - (\mu - 0.5\sigma^2)T}{\sigma\sqrt{T}}\right)\\ &= 1 - N\left(\frac{\ln(62/65) - (0.06 - 0.5 \times 0.27^2) \times 0.5}{0.27\sqrt{0.5}}\right)\\ &= 1 - N(-0.309177909) = 0.621406901.\end{aligned}$$

We have to be careful here: we should not compute $\mathbb{Q}((S_{0.5} - 62)_+ > 0)$, i.e. using the distribution

$$\frac{S_T}{65} \overset{\mathbb{Q}}{\sim} \mathcal{LN}((r - 0.5\sigma^2)T, \sigma^2 T).$$

This risk-neutral distribution has no *real-world* meaning and should be used for option pricing purposes only. ■

Using the put-call parity in equation (6.2.4) of Chapter 6, we have

$$P_0 = C_0 - S_0 + Ke^{-rT}.$$

Substituting the Black-Scholes formula of equation (16.3.1), we get

$$\begin{aligned}P_0 &= \left(S_0 N(d_1) - Ke^{-rT} N(d_2)\right) - S_0 + Ke^{-rT}\\ &= S_0\left(N(d_1) - 1\right) - Ke^{-rT}(N(d_2) - 1).\end{aligned}$$

Since $N(x) - 1 = -N(-x)$ for all $x \in \mathbb{R}$, we get

$$\begin{aligned}P_0 &= S_0\left(N(d_1) - 1\right) - Ke^{-rT}(N(d_2) - 1)\\ &= -S_0 N(-d_1) + Ke^{-rT} N(-d_2).\end{aligned}$$

Reorganizing yields the desired result:

$$P_0 = Ke^{-rT} N(-d_2) - S_0 N(-d_1). \tag{16.3.2}$$

This is the **classical Black-Scholes pricing formula** for the initial price P_0 of a European put option.

We could also define $P_0^{(n)}$ as the time-0 price of a put option in an n-period binomial tree and verify that

$$P_0^{(n)} = e^{-rT}\mathbb{E}^{\mathbb{Q}}\left[(K - S_T^{(n)})_+\right]$$

converges to

$$P_0 = Ke^{-rT} N(-d_2) - S_0 N(-d_1),$$

as n tends to infinity.

Example 16.3.3 In a BSM model, you are given:

- $S_0 = 45$;
- $r = 0.05$;
- $\sigma = 0.25$.

Compute the price of a 3-month at-the-money put option.

We can find:

$$\begin{aligned} d_1 &= 0.1625; \\ d_2 &= 0.0375; \\ N(-d_1) &= 0.435456064; \\ N(-d_2) &= 0.48504317. \end{aligned}$$

Using the Black-Scholes formula in equation (16.3.2), the price of this put option is

$$\begin{aligned} P_0 &= K e^{-rT} N(-d_2) - S_0 N(-d_1) \\ &= 45 e^{-0.05 \times 0.25} \times 0.48504317 - 45 \times 0.435456064 = 1.960281131. \end{aligned}$$

■

16.3.2 Stop-loss transforms

The attentive reader will have noticed that the Black-Scholes formulas of equations (16.3.1) and (16.3.2) have a familiar look. Indeed, the call option price is related to the stop-loss transform of a lognormally distributed random variable. In equation (14.1.7) of Chapter 14, we obtained the following identity: for $X \overset{\mathbb{P}}{\sim} \mathcal{LN}(m, s^2)$ and a real number $a > 0$,

$$\mathbb{E}[(X - a)_+] = e^{m + \frac{s^2}{2}} N\left(-\frac{\ln(a) - m}{s} + s\right) - aN\left(-\frac{\ln(a) - m}{s}\right).$$

Moreover, we know from the binomial model that

$$C_0^{(n)} = e^{-rT} \mathbb{E}^{\mathbb{Q}}\left[\left(S_T^{(n)} - K\right)_+\right]$$

converges, as $n \to \infty$, to

$$C_0 = e^{-rT} \mathbb{E}^{\mathbb{Q}}[(S_T - K)_+].$$

In other words, the no-arbitrage price of a call option is the *discounted* stop-loss transform of S_T using its risk-neutral distribution:

$$S_T \overset{\mathbb{Q}}{\sim} \mathcal{LN}\left(\ln(S_0) + \left(r - \frac{\sigma^2}{2}\right) T, \sigma^2 T\right)$$

or, equivalently,

$$\ln(S_T / S_0) \overset{\mathbb{Q}}{\sim} \mathcal{N}((r - \sigma^2/2)T, \sigma^2 T).$$

Using this lognormal random variable with its risk-neutral distribution along with the above formula for the stop-loss transform yields the Black-Scholes formula for the call option, that is

$$C_0 = S_0 N(d_1) - K e^{-rT} N(d_2),$$

with

$$d_1 = \frac{\ln(S_0/K) + (r + \sigma^2/2)T}{\sigma\sqrt{T}}$$

$$d_2 = \frac{\ln(S_0/K) + (r - \sigma^2/2)T}{\sigma\sqrt{T}} = d_1 - \sigma\sqrt{T}.$$

The same can be done for a put option; see exercise 14.1 in Chapter 14.

16.3.3 Dynamic formula

As the price of the risky asset changes over time, the prices of call and put options also change. We intend to determine how to calculate those option prices when the underlying asset price moves from S_0 to S_t at time t.

In an n-period binomial tree, the random variable $V_{kT/n}$ is the no-arbitrage price after k steps of a derivative with payoff V_T. To find this price, we can compute recursively the risk-neutral conditional expectation

$$V_{kT/n} = \mathrm{e}^{-rT/n} \times \mathbb{E}^{\mathbb{Q}}[V_{(k+1)T/n} | S_{kT/n}]$$

for the corresponding nodes. This is equivalent to applying equation (11.3.1), i.e. it is as if we were issuing, at time kT/n, a derivative maturing one period later with random payoff $V_{(k+1)T/n}$, given that the "initial price" of the risky asset is $S_{kT/n}$. Let us emphasize once more that this risk-neutral expectation was obtained as a rewriting of the value of the replicating portfolio for this derivative. At that point, only elementary algebra was needed.

Furthermore, applying iterated expectations, we find that

$$V_{kT/n} = \mathrm{e}^{-r(n-k)\frac{T}{n}} \times \mathbb{E}^{\mathbb{Q}}[V_T | S_{kT/n}].$$

We can interpret this expectation as if we were issuing, at time kT/n, a derivative maturing $n-k$ periods later with payoff V_T, given that the "initial price" of the risky asset is $S_{kT/n}$. In other words, when we observe the value $S_{kT/n}$ after k periods, we can determine the value $V_{kT/n}$ by analyzing the sub-tree corresponding to the dates $kT/n, (k+1)T/n, \ldots, T$ with a root corresponding to the observed value of $S_{kT/n}$. Figure 16.5 illustrates a 20-period binomial tree

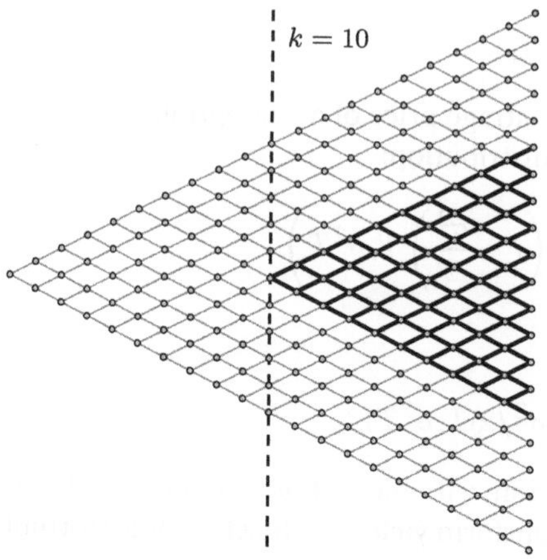

Figure 16.5 A highlighted sub-tree (in black) of 10 periods in a tree with 20 periods

conditional upon observing $S_{10T/n}$: a 10-period binomial sub-tree is highlighted in the figure for a given root.

In the BSM model, the argument is similar. Indeed, suppose now that at any time $0 \leq t < T$, we observe that the underlying asset price is equal to S_t. Then, the time interval $[t, T]$ can be split into n sub-periods and the evolution of the asset price is approximated by a binomial tree. We can determine what happens when n tends to infinity using the same arguments as in Section 16.2.2.

We deduce that, in the BSM model, given the value of S_t, we have the following conditional risk-neutral probability distributions

$$\ln(S_T/S_t) \mid S_t \overset{\mathbb{Q}}{\sim} \mathcal{N}((r - \sigma^2/2)(T-t), \sigma^2(T-t))$$

or, equivalently,

$$S_T \mid S_t \overset{\mathbb{Q}}{\sim} \mathcal{LN}\left(\ln(S_t) + \left(r - \frac{\sigma^2}{2}\right)(T-t), \sigma^2(T-t)\right).$$

Consequently, the time-t Black-Scholes formula, where $0 \leq t < T$, is thus

$$C_t = S_t N(d_1(t, S_t)) - e^{-r(T-t)} K N(d_2(t, S_t)), \tag{16.3.3}$$

where

$$d_1(t, S_t) = \frac{\ln(S_t/K) + (r + \sigma^2/2)(T-t)}{\sigma\sqrt{T-t}},$$

$$d_2(t, S_t) = d_1(t, S_t) - \sigma\sqrt{T-t}.$$

Note that $d_1(0, S_0) = d_1$ and $d_2(0, S_0) = d_2$ as used in formula (16.3.1).

Using the put-call parity relationship at time t, we deduce that the price of a put option at time t is

$$P_t = e^{-r(T-t)} K N(-d_2(t, S_t)) - S_t N(-d_1(t, S_t)). \tag{16.3.4}$$

Example 16.3.4 *Call and put prices at intermediate times*
For a stock currently trading at \$34, you have:

- $\sigma = 0.32$;
- $\mu = 0.09$;
- $r = 0.04$.

In this BSM model, calculate the initial price of 6-month 32-strike call and put options. Determine the price of these options 2 months later, in the scenario that the stock price drops to \$31.

Using $S_0 = 34, K = 32$ and $T = 0.5$ in equations (16.3.1) and (16.3.2), we get

$$C_0 = 4.444883108 \quad \text{and} \quad P_0 = 1.811240654.$$

Let us now consider the following scenario. Suppose that, 2 months later, the price of the stock goes down to $S_{2/12} = 31$. Using $t = 2/12$ and $T = 0.5$ in equations (16.3.3) and (16.3.4), we get

$$C_{2/12} = 2.026140357 \quad \text{and} \quad P_{2/12} = 2.602305535.$$

These last two option prices are realizations of the corresponding random variables given that $S_{2/12} = 31$. If we were to consider another scenario, i.e. another value for $S_{2/12}$, then the values of $C_{2/12}$ and $P_{2/12}$ would be different. ■

16.4 Pricing simple derivatives

Now, we will find the price of other simple derivatives, such as forwards, binary options and gap options, i.e. those with a payoff of the form $V_T = g(S_T)$.

Since, in an n-period binomial tree model, we would have

$$V_{kT/n} = \mathrm{e}^{-r(n-k)\frac{T}{n}} \times \mathbb{E}^{\mathbb{Q}}[g(S_T)|S_{kT/n}],$$

then at the limit, i.e. in the BSM model, we will have the following corresponding risk-neutral pricing formula:

$$V_t = \mathrm{e}^{-r(T-t)}\mathbb{E}^{\mathbb{Q}}[g(S_T)|S_t], \tag{16.4.1}$$

for all $0 \leq t \leq T$. Therefore, to find the no-arbitrage price of a simple derivative, we will use the fact that

$$S_T \mid S_t \overset{\mathbb{Q}}{\sim} \mathcal{LN}\left(\ln(S_t) + \left(r - \frac{\sigma^2}{2}\right)(T-t), \sigma^2(T-t)\right) \tag{16.4.2}$$

to compute the risk-neutral conditional expectation in equation (16.4.1).

Clearly, the Black-Scholes formulas for call and put options can be (re-)obtained using the risk-neutral formula in equation (16.4.1) with $g(S_T) = (S_T - K)_+$ and $g(S_T) = (K - S_T)_+$, respectively. In what follows, we will focus on forwards, binary options and gap options.

16.4.1 Forward contracts

We know from Chapter 3 that the payoff of a forward contract on a stock is $g(S_T) = S_T - K$. Applying the risk-neutral pricing formula in (16.4.1), we deduce that the value of this forward is given by

$$V_t = \mathrm{e}^{-r(T-t)}\mathbb{E}^{\mathbb{Q}}[S_T - K|S_t] = \mathrm{e}^{-r(T-t)}\mathbb{E}^{\mathbb{Q}}[S_T|S_t] - \mathrm{e}^{-r(T-t)}K,$$

at any time $0 \leq t \leq T$.

Using the fact that

$$S_T \mid S_t \overset{\mathbb{Q}}{\sim} \mathcal{LN}\left(\ln(S_t) + \left(r - \frac{\sigma^2}{2}\right)(T-t), \sigma^2(T-t)\right),$$

together with formula (14.1.5) for the mean of a lognormal distribution, we get

$$\begin{aligned}\mathbb{E}^{\mathbb{Q}}[S_T|S_t] &= \exp\left(\ln(S_t) + \left(r - \frac{\sigma^2}{2}\right)(T-t) + \frac{1}{2}\sigma^2(T-t)\right)\\ &= S_t\mathrm{e}^{r(T-t)}.\end{aligned}$$

In conclusion, the price process of a forward contract is given by

$$V_t = S_t - K\mathrm{e}^{-r(T-t)},$$

for every time $0 \leq t \leq T$. As expected, we have recovered the *model-free formula* of equation (3.2.2), which was obtained with replication arguments and without any assumption on the distribution of S_T.

16.4.2 Binary options

Recall from Chapter 6 that there are two classes of binary options: cash-or-nothing and asset-or-nothing binary options. Binary call options pay whenever $S_T \geq K$ whereas binary put options pay whenever $S_T < K$.

More precisely, binary call options with maturity date T admit the following payoffs (up to a multiplicative constant): for a strike price $K > 0$, we have

- cash-or-nothing call: $C_T^{\text{CoN}} = \mathbb{1}_{\{S_T \geq K\}}$;
- asset-or-nothing call: $C_T^{\text{AoN}} = S_T \mathbb{1}_{\{S_T \geq K\}}$.

Let the stochastic process $\{C_t^{\text{CoN}}, 0 \leq t \leq T\}$ represent the price process of a cash-or-nothing call option and $\left\{C_t^{\text{AoN}}, 0 \leq t \leq T\right\}$ that of an asset-or-nothing call option. Note that if we buy one unit of the asset-or-nothing call and (short-)sell K units of the cash-or-nothing call, then the payoff of our position is equal to

$$C_T^{\text{AoN}} - K\, C_T^{\text{CoN}} = S_T \mathbb{1}_{\{S_T \geq K\}} - K \mathbb{1}_{\{S_T \geq K\}} = (S_T - K)_+,$$

which is the payoff of a vanilla call option. By the no-arbitrage assumption, we have the same relationship regarding time-t prices, that is

$$C_t^{\text{AoN}} - K\, C_t^{\text{CoN}} = C_t,$$

for all $0 \leq t \leq T$. As a consequence, it suffices to obtain an expression for C_t^{CoN} and then, together with the Black-Scholes formula in (16.3.3), we will automatically obtain an expression for C_t^{AoN}.

For a cash-or-nothing call, using the risk-neutral pricing formula (16.4.1), we have

$$C_t^{\text{CoN}} = \mathrm{e}^{-r(T-t)} \mathbb{E}^{\mathbb{Q}}[\mathbb{1}_{\{S_T \geq K\}} | S_t] = \mathrm{e}^{-r(T-t)} \mathbb{Q}(S_T \geq K | S_t),$$

for all $0 \leq t \leq T$. Using the lognormal risk-neutral distribution of $S_T | S_t$, as given in (16.4.2), together with the c.d.f. of the lognormal distribution as seen in Section 14.1.2, we can write

$$\begin{aligned}
\mathbb{Q}(S_T \geq K | S_t) &= 1 - \mathbb{Q}(S_T < K | S_t) \\
&= 1 - N\left(\frac{\ln(K) - \left(\ln(S_t) + \left(r - \frac{\sigma^2}{2}\right)(T-t)\right)}{\sigma\sqrt{T-t}} \right) \\
&= 1 - N\left(-\frac{\ln(S_t/K) + \left(r - \frac{\sigma^2}{2}\right)(T-t)}{\sigma\sqrt{T-t}} \right) \\
&= 1 - N(-d_2(t, S_t)) \\
&= N(d_2(t, S_t)).
\end{aligned}$$

Consequently, the time-t value of this cash-or-nothing call is given by

$$C_t^{\text{CoN}} = \mathrm{e}^{-r(T-t)} N\left(\mathrm{d}_2(t, S_t)\right). \tag{16.4.3}$$

As mentioned above, since

$$C_t^{\text{AoN}} = K\, C_t^{\text{CoN}} + C_t,$$

using the Black-Scholes formula in (16.3.3) together with the pricing formula for a cash-or-nothing call in (16.4.3), we get the following pricing formula for an asset-or-nothing call: for all $0 \leq t \leq T$,

$$\begin{aligned} C_t^{\mathrm{AoN}} &= K\left\{\mathrm{e}^{-r(T-t)}N\left(\mathrm{d}_2(t,S_t)\right)\right\} \\ &\quad + \left\{S_tN(\mathrm{d}_1(t,S_t)) - \mathrm{e}^{-r(T-t)}KN(\mathrm{d}_2(t,S_t))\right\} \\ &= S_tN(\mathrm{d}_1(t,S_t)). \end{aligned} \tag{16.4.4}$$

Example 16.4.1 *Price of binary call options*
In the BSM model, you are given:

- $S_0 = K = 75$;
- $r = 0.05$;
- $\sigma = 0.25$.

Calculate the initial price of 3-month binary call options.
Since we have

$$\begin{aligned} d_1 &= 0.1625, \quad N(d_1) = 0.564543936, \quad N(-d_1) = 0.435456064, \\ d_2 &= 0.0375, \quad N(d_2) = 0.51495683, \quad N(-d_2) = 0.48504317, \end{aligned}$$

then, using the pricing formulas in (16.4.3) and in (16.4.4), we get

$$\begin{aligned} C_0^{\mathrm{CoN}} &= \mathrm{e}^{-rT}N(d_2) = 0.508559933, \\ C_0^{\mathrm{AoN}} &= S_0N(d_1) = 42.34079519. \end{aligned}$$

■

Recall that binary put options with maturity date T and strike price K admit the following payoffs:

- cash-or-nothing put: $P_T^{\mathrm{CoN}} = \mathbb{1}_{\{S_T<K\}}$;
- asset-or-nothing put: $P_T^{\mathrm{AoN}} = S_T\mathbb{1}_{\{S_T<K\}}$.

Using parity relationships, it is easy to obtain expressions for the time-t prices of these options (see exercise 16.5). More precisely, for each $0 \leq t \leq T$, we will obtain

$$\begin{aligned} P_t^{\mathrm{CoN}} &= \mathrm{e}^{-r(T-t)}N(-\mathrm{d}_2(t,S_t)), \\ P_t^{\mathrm{AoN}} &= S_tN(-\mathrm{d}_1(t,S_t)). \end{aligned}$$

Example 16.4.2 *Price of binary put options*
In the same BSM model as the previous example, calculate the initial price of 3-month binary put options.
Using the above formulas, we obtain

$$\begin{aligned} P_0^{\mathrm{CoN}} &= \mathrm{e}^{-rT}N(-d_2) = 0.479017867, \\ P_0^{\mathrm{AoN}} &= S_0N(-d_1) = 32.65920481. \end{aligned}$$

■

16.4.3 Gap options

Recall from Chapter 6 that gap call and put options are options to buy or sell an asset for a strike price of K only if the underlying asset price at maturity is greater or less than a trigger price H.

First, a gap call option has a payoff given by

$$C_T^{\text{gap}} = (S_T - K)\mathbb{1}_{\{S_T \geq H\}}.$$

We know this payoff can be decomposed as a long position in one unit of an asset-or-nothing call with strike price H and a short position in K units of a cash-or-nothing call with strike price H. Mathematically,

$$C_T^{\text{gap}} = S_T \mathbb{1}_{\{S_T \geq H\}} - K\,\mathbb{1}_{\{S_T \geq H\}}.$$

Therefore, by linearity, and using the pricing formulas in (16.4.4) and in (16.4.3) for the time-t prices of binary call options, we get

$$C_t^{\text{gap}} = S_t N(\mathrm{d}_1(t, S_t; H)) - K\,\mathrm{e}^{-r(T-t)} N(\mathrm{d}_2(t, S_t; H)),$$

where

$$d_1(t, S_t; H) = \frac{\ln(S_t/H) + (r + \sigma^2/2)(T-t)}{\sigma\sqrt{T-t}},$$

$$d_2(t, S_t; H) = d_1(t, S_t; H) - \sigma\sqrt{T-t}.$$

This formula is very similar to the Black-Scholes formula for a vanilla call option as given in (16.3.3). In fact, when $K = H$, they are equal.

Example 16.4.3 *Price of a gap call option*
A share of stock currently trades for \$35 and the volatility of the log-returns on this stock is 28%. If the risk-free rate is 3%, calculate the initial Black-Scholes price of a 6-month gap call option with strike price \$35 and trigger price \$37.

We have $\sigma = 0.28$, $r = 0.03$, $T = 0.5$, $S_0 = K = 35$ and $H = 37$. If we compute $d_1(0, S_0; H)$ and $d_2(0, S_0; H)$ with these quantities, then we get

$$d_1(0, 35; 37) = -0.105913743, \quad N(d_1(0, 35; 37)) = 0.457825395,$$

$$d_2(0, 35; 37) = -0.303903641, \quad N(d_2(0, 35; 37)) = 0.380600652.$$

Therefore, the initial price of this gap call option is

$$\begin{aligned} C_0^{\text{gap}} &= S_0 N(\mathrm{d}_1(0, S_0; H)) - K\,\mathrm{e}^{-rT} N(\mathrm{d}_2(0, S_0; H)) \\ &= 35 \times 0.457825395 - 35 \times 0.380600652\mathrm{e}^{-0.03\times 0.5} \\ &= 2.901190197. \end{aligned}$$

■

Second, a gap put option has a payoff given by:

$$P_T^{\text{gap}} = (K - S_T)\mathbb{1}_{\{S_T < H\}} = K\mathbb{1}_{\{S_T < H\}} - S_T \mathbb{1}_{\{S_T < H\}}.$$

In other words, it corresponds to a long position in K units of a cash-or-nothing put with strike price H and a short position in one unit of a cash-or-nothing put with strike price H.

Using this parity relationship together with the formulas for binary put options, the price at time t of a gap put option is

$$P_t^{\text{gap}} = K\,\mathrm{e}^{-r(T-t)} N(\mathrm{d}_2(t, S_t; H)) - S_t N(\mathrm{d}_1(t, S_t; H)).$$

Again, this pricing formula is very similar to the Black-Scholes formula for a vanilla put option as given in (16.3.4) and the two formulas coincide when $K = H$.

16.5 Determinants of call and put prices

To compute the Black-Scholes formulas for call and put options, we needed to know in advance the model parameters r and σ and the asset-related quantities K, S_0 and T. They are the main determinants of call and put option prices and, as such, they are inputs for the Black-Scholes formulas. If the value of one of these quantities changes, then the option price also changes accordingly. In this section, we analyze how such changes affect the value of call and put options in the Black-Scholes-Merton model.

First, recall the Black-Scholes formulas for call and put options, as given in (16.3.1) and (16.3.2) respectively:

$$C_0 = S_0 N(d_1(S_0, K, T; r, \sigma)) - e^{-rT} K N(d_2(S_0, K, T; r, \sigma))$$
$$P_0 = e^{-rT} K N(-d_2(S_0, K, T; r, \sigma)) - S_0 N(-d_1(S_0, K, T; r, \sigma)),$$

where

$$d_1(S_0, K, T; r, \sigma) = \frac{\ln(S_0/K) + (r + \sigma^2/2)T}{\sigma\sqrt{T}},$$
$$d_2(S_0, K, T; r, \sigma) = d_1(S_0, K, T; r, \sigma) - \sigma\sqrt{T}.$$

Note that the notation has been slightly modified to emphasize the dependence on the quantities and parameters of interest. In fact, both C_0 and P_0 are functions of $(S_0, K, T; r, \sigma)$.

We would like to determine whether C_0 increases or decreases if, for example, the value of r increases. If we consider the Black-Scholes formula as a function of r only, keeping all other quantities constant, then differentiating with respect to r will provide an answer to our question. Mathematically, we will compute the partial derivative $\partial C_0/\partial r$ of the option price with respect to r and determine whether it is positive or negative.

The behavior of each determinant is plotted in Figure 16.6. In words, we computed the Black-Scholes formulas for the call and the put keeping all but one parameter constant.

16.5.1 Stock price

From a financial point of view, we expect the value of a call option to increase if S_0 increases. Indeed, for a fixed strike price K, a greater stock price S_0 increases the intrinsic value of the call, which in turn increases the price. It should be the opposite for a put option.

Let us verify this mathematically. We can show that[6]

$$\frac{\partial C_0}{\partial S_0} = N(d_1(S_0, K, T; r, \sigma))$$

and then, using the put-call parity $P_0 = C_0 - S_0 + e^{-rT}K$, it is easy to deduce that

$$\frac{\partial P_0}{\partial S_0} = N(d_1(S_0, K, T; r, \sigma)) - 1.$$

6 It requires more work than it might appear at first: it is a tedious but elementary differentiation exercise. See exercise 16.10.

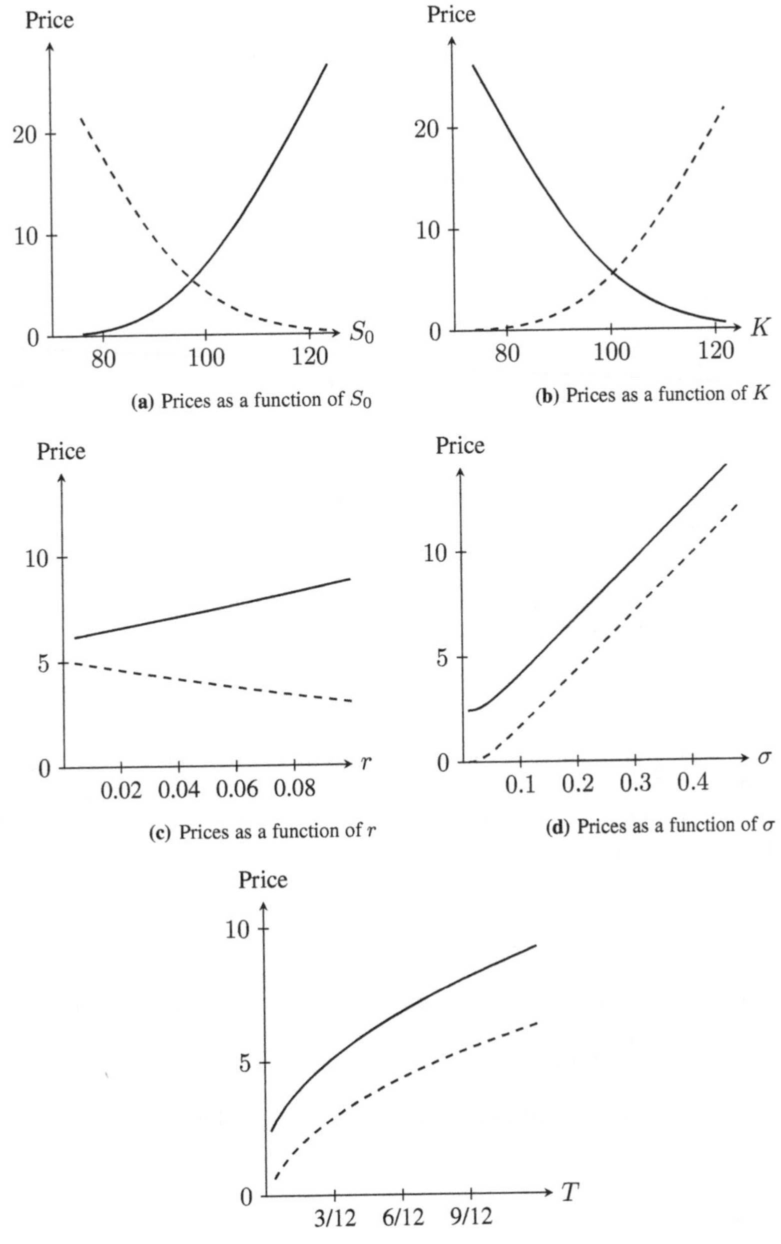

(a) Prices as a function of S_0

(b) Prices as a function of K

(c) Prices as a function of r

(d) Prices as a function of σ

(e) Prices as a function of T

Figure 16.6 Black-Scholes formulas (solid line: call, dashed line: put) as functions of S_0, K, r, σ and T. Otherwise, the values are $S_0 = 100, K = 98, r = 0.03, \sigma = 0.2$ and $T = 0.5$

Since the standard normal c.d.f. $N(\cdot)$ is a probability, its value lies between 0 and 1 and then we get

$$0 \leq \frac{\partial C_0}{\partial S_0} \leq 1$$

and

$$-1 \leq \frac{\partial P_0}{\partial S_0} \leq 0.$$

As anticipated, the call option value increases with the initial stock price because $\frac{\partial C_0}{\partial S_0} \geq 0$, i.e. C_0 is a non-decreasing function of S_0. Similarly, the put option value decreases with the initial stock price because $\frac{\partial P_0}{\partial S_0} \leq 0$.

16.5.2 Strike price

Because it is the price at which we buy or sell the underlying asset at maturity, the strike price K has the opposite effect, compared with the initial stock price, on the up-front premium. Indeed, if $K_1 \leq K_2$, then

$$(S_T - K_1)_+ \geq (S_T - K_2)_+ \quad \text{and} \quad (K_1 - S_T)_+ \leq (K_2 - S_T)_+,$$

in all possible scenarios, i.e. for any value of the random variable S_T. By the usual no-arbitrage arguments, it is easy to verify that this also holds true at any time t prior to maturity.

In conclusion, if we increase the strike price, then we reduce the initial value of the call option and increase that of the put option. Note that we did not use the Black-Scholes formulas, so this conclusion is model-free: it holds in any market model.

16.5.3 Risk-free rate

From a financial perspective, if a long call option ends up in the money at maturity, the holder will pay K (and she will receive a share of the stock). An increase in the risk-free interest rate reduces the present value of K and therefore increases the value of this call option. For a long put option, it is the opposite.

We can verify that

$$\frac{\partial C_0}{\partial r} = KT\mathrm{e}^{-rT}N(d_2) \quad \text{and} \quad \frac{\partial P_0}{\partial r} = -KT\mathrm{e}^{-rT}N(-d_2).$$

In both cases, as we are multiplying positive quantities, we can further deduce that

$$\frac{\partial C_0}{\partial r} \geq 0 \quad \text{and} \quad \frac{\partial P_0}{\partial r} \leq 0.$$

In words, an increase in the risk-free rate increases the value of the call and decreases the value of the put.

16.5.4 Volatility

Volatility in general is a measure of the range of outcomes of a random variable. Higher volatility means there is a wider spectrum of scenarios in which call and put options can end up in the money. Therefore, more volatility increases the values of call and put options.

We can also compute the partial derivatives of C_0 and P_0 with respect to σ to get:

$$\frac{\partial P_0}{\partial \sigma} = \frac{\partial C_0}{\partial \sigma} = S_0\phi(d_1)\sqrt{T} \geq 0,$$

where $\phi(x)$ is the p.d.f. of the standard normal distribution. As the partial derivatives are positive, it confirms that increasing the volatility increases both the call and put option prices.

16.5.5 Time to maturity

We want to determine how the time to maturity T affects an option price. Intuitively, the time to maturity impacts at least two elements:

1) A longer time to maturity increases the range of outcomes of asset prices. Indeed, it is intuitive to think that the distribution of stock prices after 1 year will be wider than the possible stock prices after 1 month.
2) A longer time to maturity decreases the present value of future cash flows. Indeed, one dollar paid in a year is worth less than one dollar paid in a month, or K paid in a year is better than K paid tomorrow.

In the case of call options, those effects add up, thus increasing the call option price. As for the put option, the effects go in opposite directions.

Differentiating C_0 with respect to T and then using the put-call parity, we get

$$\frac{\partial C_0}{\partial T} = \frac{S_0\phi(d_1)\sigma}{2\sqrt{T}} + rKe^{-rT}N(d_2),$$

$$\frac{\partial P_0}{\partial T} = \frac{S_0\phi(d_1)\sigma}{2\sqrt{T}} - rKe^{-rT}N(-d_2).$$

Since $\partial C_0/\partial T$ is the sum of two positive quantities, we can conclude that the call value increases with T. In other words, a greater time to maturity increases the call option price. For a put option, we cannot make such a definitive conclusion. However, Figure 16.6 suggests that it is also increasing.

16.6 Replication and hedging

In the binomial model, we derived a recursive algorithm to replicate the payoff of any simple derivative. As a consequence, the binomial tree model is a *complete* market model, meaning that every derivative admits a replicating portfolio (see Chapter 13). By no-arbitrage arguments, the cost of a replicating portfolio has to correspond to the price of the derivative it replicates. Replication is also known as *perfect hedging* because, at any time during the life of the option, the replicating portfolio and the derivative have the same value.

In the BSM model, now that the asset price follows a geometric Brownian motion, we cannot find an expression for the no-arbitrage price of a derivative using *only* replication (unless we use Black and Scholes' original approach through stochastic calculus and partial differential equations). Therefore, in this section, we will discuss replication of a given derivative *assuming we already know its no-arbitrage price* as studied in Sections 16.3 and 16.4.

In a continuous-time model such as the BSM model, replication of a derivative might require to trade/rebalance continuously. Of course, this is impossible in practice; we will come back to this issue later in the book.

16.6.1 Trading strategies/portfolios

In a continuous-time setup, a trading strategy (or portfolio) on the investment horizon $[0, T]$ is a pair of stochastic processes

$$\{(\Theta_t, \Delta_t), 0 \leq t \leq T\},$$

where

- Θ_t is the number of units of the risk-free asset B held in the portfolio at time t;
- Δ_t is the number of units of the risky asset S held in the portfolio at time t.

As in the discrete-time setup, (Θ_t, Δ_t) can be chosen with specific investment goals in mind. For the rest of this chapter, it will be for replication purposes.

For a given trading strategy $\{(\Theta_t, \Delta_t), 0 \leq t \leq T\}$, its value at time t is given by Π_t where

$$\Pi_t = \Delta_t S_t + \Theta_t B_t.$$

This clearly depends on the risky asset price S_t and on the (possibly) random number of units Δ_t and Θ_t. Therefore, $\Pi = \{\Pi_t, 0 \leq t \leq T\}$ is a stochastic process and it is called the **portfolio value process.**

In discrete-time financial models, we encountered several times the notion of a **self-financing strategy** and found many equivalent definitions:

- a strategy that does not involve deposits/withdrawals between inception and maturity;
- the portfolio value resulting from such a strategy is the same before and after rebalancing;
- the variation in the portfolio value comes only from variations in the risky asset and risk-free asset prices.

In a continuous-time model, a self-financing strategy can be defined heuristically as a pair (Θ_t, Δ_t) such that

$$\Theta_{t-} B_t + \Delta_{t-} S_t = \Theta_t B_t + \Delta_t S_t \,(= \Pi_t), \tag{16.6.1}$$

for all $t \geq 0$, i.e. the value of the strategy is the same just before rebalancing $(t-)$ and just after rebalancing (t). A rigorous definition of a self-financing strategy requires knowledge of stochastic calculus. The interested reader is referred to Section 17.1 for more details.

We seek to replicate the payoff of simple derivatives using a dynamically updated strategy $\{(\Theta_t, \Delta_t), 0 \leq t \leq T\}$. A strategy $\{(\Theta_t, \Delta_t), 0 \leq t \leq T\}$ is said to be a **replicating strategy for the payoff** V_T if:

- it is self-financing;
- their final values coincide, meaning that $\Pi_T = V_T$ in each possible scenario.

Since a replicating strategy is self-financing (by definition), which means it generates no cash flows except at time 0 and time T (just like the derivative it replicates), then by the no-arbitrage assumption we can conclude that the corresponding portfolio value process Π mimics the derivative price at each time point t and in each possible scenario. In other words, if $\{(\Theta_t, \Delta_t), 0 \leq t \leq T\}$ is a replicating strategy for V_T, then we have

$$V_t = \Pi_t = \Delta_t S_t + \Theta_t B_t$$

for all $0 \leq t \leq T$. In particular, the initial values are equal: $V_0 = \Pi_0 = \Delta_0 S_0 + \Theta_0 B_0$. In conclusion, the price of the derivative coincides with the value or cost of the replicating portfolio.

In a general continuous-time model, it is not clear whether each payoff admits a replicating portfolio or not. However, as the Black-Scholes-Merton model can be obtained as the limit of binomial trees, it is also a complete market model, i.e. all derivatives can be replicated.

16.6.2 Replication for call and put options

If we want to find the replicating portfolio of a call option, then we must find a self-financing portfolio $\{(\Theta_t, \Delta_t), 0 \le t \le T\}$ such that

$$\Delta_t S_t + \Theta_t B_t = C_t,$$

in each possible scenario, at any time $0 \le t \le T$.

The Black-Scholes formula is famous for many reasons, one of them being that it yields an explicit representation of the replicating portfolio within the expression for C_t. Indeed, since we have

$$C_t = S_t N(d_1(t, S_t)) - e^{-r(T-t)} K N(d_2(t, S_t)),$$

then, by identification, we directly obtain that

$$\begin{cases} \Delta_t &= N(d_1(t, S_t)), \\ \Theta_t &= -e^{-rT} K N(d_2(t, S_t)), \end{cases} \tag{16.6.2}$$

are the number of units of each asset needed to replicate a call option.[7]

Said differently, at any given time t, in order to replicate the value of the call, we must be

- *long* $N(\mathrm{d}_1(t, S_t))$ units of S;
- *short* $e^{-rT} K N(\mathrm{d}_2(t, S_t))$ units of B.

From the properties of the normal c.d.f., we further deduce that $0 \le \Delta_t \le 1$ and thus we should be holding a fraction of the underlying asset to replicate the call option. This is similar to what we obtained in the binomial tree model. Moreover, as the call option gets further into the money (or out of the money), then $\mathrm{d}_1(t, S_t)$ increases (or decreases) so that Δ_t approaches 1 (or 0).

This replicating portfolio for a call option, i.e.

$$\left\{\left(e^{-rT} K N(\mathrm{d}_2(t, S_t)), N(\mathrm{d}_1(t, S_t))\right), 0 \le t < T\right\},$$

requires to be able to trade continuously. Indeed, the number of units $\Delta_t = N(\mathrm{d}_1(t, S_t))$ and $\Theta_t = -e^{-rT} K N(\mathrm{d}_2(t, S_t))$ change *continuously* with t and S_t, because $d_1(t, S_t)$ and $d_2(t, S_t)$ change *continuously* with t and S_t. In fact, they are never constant for any time interval. Thus, at every instant, the replicating portfolio must be rebalanced to keep matching the call option price.

Example 16.6.1 *Replicating portfolio for a call option*
Assume a stock, currently trading for \$100, behaves according to a geometric Brownian motion with mean return of 6% and volatility of 30%. If the risk-free rate is 3%, find the replicating strategy of a 1-year at-the-money call option:

- at inception;
- after 6 months, in the scenario where $S_{0.5} = 115$;
- and *just before* maturity, in the scenario where $S_{1-} = 107$.

At inception, we have $d_1 = 0.25$, $d_2 = -0.05$ and thus, $N(d_1) = 0.598706326$ and $N(d_2) = 0.480061194$. Therefore, the Black-Scholes formula in equation (16.3.1) tells us that this call option price is 13.2833084. The corresponding replicating strategy is given by

$$\Delta_0 = N(d_1) = 0.5987 \quad \text{and} \quad \Theta_0 = -e^{-0.03} \times 100 \times N(d_2) = -46.5873,$$

at inception. This position is valid at time 0 only and will change *immediately after*.

7 The only missing justification here is that we should verify that it is indeed a self-financing portfolio. This will be verified in Chapter 17.

In the chosen scenario, the stock price is $S_{0.5} = 115$ after 6 months. Then we can compute $d_1(0.5, 115)$ and $d_2(0.5, 115)$, and find that

$$\Delta_{0.5} = 0.7983 \quad \text{and} \quad \Theta_{0.5} = -71.1840.$$

Again, this replicating portfolio is only valid at time $t = 0.5$ (after 6 months) and in this scenario. Note that they will change immediately after.

Moreover, it is interesting to realize that in this scenario, we have $\Delta_0 = 0.5987 < 0.7983 = \Delta_{0.5}$. This is because the call option is no longer at the money at time 0.5 in this scenario; it is then in the money because we have $S_0 = K = 100 < 115 = S_{0.5}$.

Suppose now that at (or one millisecond prior to) maturity, the stock price is equal to 107. We know that the payoff of the option will be $107 - 100 = 7$. Then, the replicating portfolio must have a balance of \$7.

To compute the positions slightly prior to maturity, we must look at the behavior of d_1 and d_2 when $t \to T$. Because $S_1 > K$, then $d_1 \to \infty$ and $d_2 \to \infty$ when $t \to T$. In this scenario, we get

$$\Delta_{1-} = 1 \quad \text{and} \quad \Theta_{1-} = -Ke^{-rT} = -97.0446$$

and the replicating portfolio value just before maturity is

$$S_1\Delta_{1-} + B_1\Theta_{1-} = 107 - e^{0.03} \times 97.0446 = 7,$$

as expected. ■

Given that the time-t Black-Scholes formula for a put is given by

$$P_t = Ke^{-r(T-t)}N(-d_2(t, S_t)) - S_tN(-d_1(t, S_t)),$$

then again by identification, we get that

$$\begin{cases} \Delta_t &= -N(-\mathrm{d}_1(t, S_t)), \\ \Theta_t &= e^{-rT}KN(-\mathrm{d}_2(t, S_t)), \end{cases} \tag{16.6.3}$$

are the number of units required in each asset to replicate a put option. In this case, at any given time t, in order to replicate the price of the put, we must be

- *short* $N(-\mathrm{d}_1(t, S_t))$ units of S;
- *long* $e^{-rT}KN(-\mathrm{d}_2(t, S_t))$ units of B.

As for the replicating portfolio of a call option, the replicating strategy for a put option requires *continuous rebalancing*.

Again, from the properties of the normal c.d.f., we deduce that $-1 \leq \Delta_t \leq 0$ and thus we should be selling a fraction of the underlying asset to replicate the put option. Moreover, as the put option gets further into the money (or out of the money), then $\mathrm{d}_1(t, S_t)$ decreases (or increases) so that Δ_t approaches -1 (or 0).

Example 16.6.2 *Replicating portfolio for a put option*
Consider a BSM model such that

- $S_0 = 75$;
- $\sigma = 0.25$;
- $r = 0.04$.

A 1-year at-the-money put option is issued. Find the replicating strategy for this put option:

- at inception;
- after 6 months, in the scenario where $S_{0.5} = 72$;
- and *just before* maturity, in the scenario where $S_{1-} = 77$.

At inception, using the Black-Scholes formula, we have that the put option price is equal to 5.93699. The corresponding replicating portfolio is

$$\Delta_0 = -0.3878 \quad \text{and} \quad \Theta_0 = 35.0236.$$

Again, this position is valid at time 0 only and will change *immediately after*.

In the chosen scenario, 6 months later, the stock price has gone down and the put option is then in the money. Using the formulas in (16.6.3) at time $t = 0.5$ and in the scenario where $S_{0.5} = 72$, we get

$$\Delta_{0.5} = -0.5117 \quad \text{and} \quad \Theta_{0.5} = 41.9149.$$

As the put option gets more into the money, we must short more stocks: from 0.3878 unit to 0.5117 unit.

Finally, *just before* maturity, the stock price is worth \$77. In this scenario, it is clear the put option expires unexercised as

$$S_{1-} = 77 > 75 = K.$$

Therefore, as t gets closer to $T = 1$, the values of d_1 and d_2 get closer to ∞, which means that

$$\Delta_{1-} = \Theta_{1-} = 0.$$

So, the put option price just before maturity is also equal to zero and in this scenario the option expires out of the money. ■

16.6.3 Replication for simple derivatives

We now focus on how to replicate simple derivatives, not just call and put options. Unfortunately, finding the replicating portfolio of general simple options will not be as easy as for calls and puts. To help us with this matter, we will relate once more with the binomial model.

Procedure

First of all, let us consider a simple option with payoff $V_T = g(S_T)$. We know from Section 16.4 that, for all $0 \leq t < T$, the time-t value of this option is given by $V_t = F(t, S_t)$, where

$$F(t, S_t) = e^{-r(T-t)}\mathbb{E}^{\mathbb{Q}}[g(S_T) \mid S_t]$$

with

$$S_T | S_t \overset{\mathbb{Q}}{\sim} \mathcal{LN}\left(\ln(S_t) + \left(r - \frac{\sigma^2}{2}\right)(T-t), \sigma^2(T-t)\right).$$

For call and put options, we will now use the notation $C_t = C(t, S_t)$ and $P_t = P(t, S_t)$ where

$$C(t, S_t) = \mathrm{e}^{-r(T-t)}\mathbb{E}^{\mathbb{Q}}[(S_T - K)_+ \mid S_t] \quad \text{and} \quad P(t, S_t) = \mathrm{e}^{-r(T-t)}\mathbb{E}^{\mathbb{Q}}[(K - S_T)_+ \mid S_t].$$

Second, we know from equation (11.2.5) in Chapter 11 that for a derivative with payoff V_T,

$$\Delta_k^{(n)}(j) = \frac{V_k^{(n)}(j+1) - V_k^{(n)}(j)}{S_k^{(n)}(j+1) - S_k^{(n)}(j)}$$

is the number of units of the risky asset that an investor should hold, at time step k and in node j, in order to replicate the derivative value. This formula is very specific to the binomial model because it involves discrete time steps. However, it tells us that the *delta of a derivative* is the ratio of the variation in the derivative value over the variation in the underlying asset value.

Therefore, we can expect that for a *very small* quantity ε, the *delta of a simple derivative* will be such that

$$\Delta_t \approx \frac{F(t, S_t + \varepsilon) - F(t, S_t)}{(S_t + \varepsilon) - S_t} = \frac{F(t, S_t + \varepsilon) - F(t, S_t)}{\varepsilon}.$$

If we take the limit when $\varepsilon \to 0$, formally we will obtain the partial derivative of F with respect to its second variable:

$$\Delta_t = \frac{\partial F}{\partial S}(t, S_t). \tag{16.6.4}$$

Here, the notation $\partial F/\partial S$ simply means taking the first-order partial derivative of the deterministic function F with respect to its second variable.[8]

The formula for Δ_t given in equation (16.6.4) is known as the **delta of the option** (see also Chapter 20). It is the number of units of the risky asset S to hold at time t in the replicating portfolio in order to match the value of the simple option with payoff $V_T = g(S_T)$.

Now that we have the option price V_t and the delta Δ_t at any time t, we can completely specify the replicating portfolio. Indeed, since we want the replicating portfolio to be such that

$$V_t = \Delta_t S_t + \Theta_t B_t,$$

then we must have

$$\Theta_t = \frac{1}{B_t}(V_t - \Delta_t S_t) = \mathrm{e}^{-rt}\left(F(t, S_t) - S_t \frac{\partial F}{\partial S}(t, S_t)\right).$$

This is the number of units of the risk-free asset B to hold at time t in the replicating portfolio in order to match the value of the simple option. In conclusion, we now have fully described the replicating portfolio[9] for a simple derivative with payoff $V_T = g(S_T)$.

To summarize, the replicating procedure for a simple derivative with payoff $V_T = g(S_T)$ is as follows:

1) Identify the function of two variables $F(\cdot, \cdot)$ such that $F(t, S_t) = \mathrm{e}^{-r(T-t)}\mathbb{E}^{\mathbb{Q}}[g(S_T) \mid S_t]$.
2) Compute $\partial F/\partial S$, i.e. the first-order partial derivative of $F(\cdot, \cdot)$ with respect to its second variable.
3) Set $\Delta_t = \frac{\partial F}{\partial S}(t, S_t)$.
4) Set

$$\Theta_t = \mathrm{e}^{-rt}\left(F(t, S_t) - S_t \frac{\partial F}{\partial S}(t, S_t)\right).$$

8 It would not make sense to take a derivative with respect to the stochastic process S.

9 Note that even if we have not verified the condition, this portfolio is self-financing.

It is important to emphasize that depending on the derivative, this replicating portfolio might need to be continuously rebalanced (or not).

Forward

A forward contract can be considered as a simple derivative with payoff of $g(S_T) = S_T - K$. The first step is to identify the function of two variables $F(\cdot, \cdot)$ such that $F(t, S_t) = \mathrm{e}^{-r(T-t)}\mathbb{E}^{\mathbb{Q}}[S_T - K \mid S_t]$. We have already found that

$$F(t, S_t) = S_t - K\mathrm{e}^{-r(T-t)}.$$

Then, the partial derivative $\partial F/\partial S$ with respect to its second variable is constant and equal to 1, i.e.

$$\frac{\partial F}{\partial S} \equiv 1.$$

Consequently,

$$\Delta_t = \frac{\partial F}{\partial S}(t, S_t) = 1.$$

In other words, the replicating portfolio of a forward contract should always be long one unit of the risky asset, at any time t and in any scenario.

Finally,

$$\begin{aligned}\Theta_t &= \mathrm{e}^{-rt}\left(F(t, S_t) - S_t\frac{\partial F}{\partial S}(t, S_t)\right) \\ &= \mathrm{e}^{-rt}\left(S_t - K\mathrm{e}^{-r(T-t)} - 1 \times S_t\right) \\ &= -K\mathrm{e}^{-rT},\end{aligned}$$

for all $0 \leq t < T$. Since $B_0 = 1$, then at time 0 we should borrow the present value of K (by investing in the risk-free asset) and hold on to this position.

Note that both expressions for Δ_t and Θ_t are constant, i.e. they do not depend on the value of t nor of S_t. In other words, the replicating strategy of a forward contract is a *static* replicating strategy. Such trading strategies are also called *buy-and-hold strategies* because no rebalancing is needed. This is consistent with the model-free results obtained in Chapter 3.

Call option

A call option is a simple derivative with payoff $g(S_T) = (S_T - K)_+$. The Black-Scholes formula for a call option gives us

$$C(t, S_t) = S_t N(d_1(t, S_t)) - \mathrm{e}^{-r(T-t)}KN(d_2(t, S_t))$$

with

$$\begin{aligned}d_1(t, S_t) &= \frac{\ln(S_t/K) + (r + \sigma^2/2)(T-t)}{\sigma\sqrt{T-t}}, \\ d_2(t, S_t) &= d_1(t, S_t) - \sigma\sqrt{T-t}.\end{aligned}$$

Following from Section 16.5.1, we directly get

$$\Delta_t = \frac{\partial F}{\partial S}(t, S_t) = N(d_1(t, S_t))$$

as obtained in equation (16.6.2).

Finally,

$$\begin{aligned}\Theta_t &= e^{-rt}\left(F(t,S_t) - S_t\frac{\partial F}{\partial S}(t,S_t)\right)\\ &= e^{-rt}(C_t - S_tN(d_1(t,S_t)))\\ &= -e^{-rT}KN(d_2(t,S_t)),\end{aligned}$$

as obtained in equation (16.6.2).

16.6.4 Delta-hedging strategy

The above *procedure* tells us how to perfectly hedge, i.e. how to replicate, a simple payoff in the BSM model. However, for many derivatives, replication requires continuous rebalancing to maintain the corresponding replicating portfolio, which is obviously impossible in practice. When trading can only occur *discretely*, i.e. daily, weekly or monthly, then attempting to replicate a payoff will inevitably yield errors for most derivatives. This is the case for vanilla call and put options. Few derivatives can be replicated using a static strategy, as for forward contracts, or a *discretely rebalanced* trading strategy.

Whenever the objective is to manage *as much as possible* the risk of a derivative, taking into account the practical constraints of trading, we say we are **hedging** (rather than replicating) the derivative. We will come back to hedging in Chapter 20.

The discrepancy between the price of the derivative and the value of its hedging portfolio is known as the **hedging error**. This is illustrated in the following example.

Example 16.6.3 *Hedging error over a month*
An investment bank sells an at-the-money call option with maturity $T = 1$ year and exercise price $K = \$1000$. Using a Black-Scholes-Merton model with volatility $\sigma = 0.2$ and risk-free rate $r = 0.04$ for the evolution of the underlying stock price, let us implement a monthly rebalanced hedging strategy for this call option by discretizing the Black-Scholes replicating strategy of equation (16.6.2). Let us also compute the hedging error made over 1 month in each of the following two scenarios:

1) if $S_{1/12} = \$975$;
2) if $S_{1/12} = \$1075$.

Note that $S_0 = \$1000$. At time 0, let us compute the number of units as given in (16.6.2). In this case, since

$$d_1(0,1000) = 0.3 \quad \text{and} \quad d_2(0,1000) = 0.1,$$

we have

$$\Delta_0 = 0.6179114 \quad \text{and} \quad \Theta_0 = -518.6609.$$

In other words, at time 0, we buy 0.6179114 shares of the underlying stock and short-sell 518.6609 units of the risk-free asset (this also means that we borrow \$518.6609 at time 0). Using the Black-Scholes formula, we see that $C_0 = 99.2505$, which coincides with the initial value of this hedging portfolio:

$$\begin{aligned}\Pi_0 &= \Theta_0B_0 + \Delta_0S_0\\ &= -518.6609 \times 1 + 0.6179114 \times 1000\\ &= 99.2505.\end{aligned}$$

Now, no matter what the value of $S_{1/12}$ will be, 1 month later our loan will be worth $518.6609 \times e^{0.04/12} = 520.3927$. Since no trading is allowed between time 0 and time 1/12, we should not expect the portfolio and the option to have the same value at time $t = 1/12$ (1 month later). On one side, the portfolio will be worth

$$\Theta_0 B_{1/12} + \Delta_0 S_{1/12} = -520.3927 + 0.6179114 \times S_{1/12}.$$

On the other side, using the Black-Scholes formula in (16.3.3), we know that the option will be worth

$$C(1/12, S_{1/12}) = S_{1/12} N(d_1(1/12, S_{1/12})) - e^{-0.04(1-1/12)} \times 1000 N(d_2(1/12, S_{1/12}))$$

where

$$d_1(1/12, S_{1/12}) = \frac{\ln(S_{1/12}/1000) + (0.04 + 0.2^2/2)(1 - 1/12)}{0.2\sqrt{1 - 1/12}},$$

$$d_2(1/12, S_{1/12}) = d_1(1/12, S_{1/12}) - 0.2\sqrt{1 - 1/12}.$$

Now, if we consider the scenario where $S_{1/12} = 975$, then our position in the risky asset would be worth $0.6179114 \times 975 = 602.4636$, the portfolio value would be

$$-520.3927 + 0.6179114 \times 975 = 82.07096,$$

while the option price would be $C(1/12, 975) = 79.57933$. In this specific scenario, the bank would be *lucky*: the hedging error will be negative, i.e. there will be an *overflow* of $82.07096 - 79.57933 = \$2.49$. The hedging portfolio is worth more than the call option on the market.

In the other scenario, that is if $S_{1/12} = 1075$, the portfolio value would be

$$-520.3927 + 0.6179114 \times 1075 = 143.8621,$$

while the option price would be $C(1/12, 1020) = 145.4774$. In this case, the hedging portfolio would be worth less than the option price, for an hedging error (loss) of \$1.62. ■

As in the previous example, we will call **delta-hedging strategy** the trading strategy obtained by discretizing the Black-Scholes replicating portfolio, i.e. whose positions are based upon the *procedure* of Section 16.6.3. It acknowledges the fact that, in practice, trading can only occur discretely.

The delta-hedging strategy for a call option is defined as follows. Assume that the portfolio can be rebalanced only at the (equidistant) time points $t_0, t_1, t_2, \dots, t_{n-1}$ such that $0 = t_0 < t_1 < t_2 < \dots < t_{n-1} < T$. Even if prices $\{S_t\}$ and $\{B_t\}$ are still following the continuous-time Black-Scholes-Merton dynamics, we want to set up a piecewise buy-and-hold trading portfolio. At each of these trading dates t_i, we will observe the realization of S_{t_i} and then choose $(\Delta_{t_i}, \Theta_{t_i})$ such that:

1) buy/sell $\Delta_{t_i} = N(d_1(S_{t_i}, K, T - t_i, r, \sigma))$ units of the risky asset S;
2) invest the remaining portfolio balance in the risk-free asset.

In other words, at the beginning of the $(i+1)$-th period we must make sure to hold Δ_{t_i} units of the risky asset. Then, the hedging portfolio's remaining balance is invested at the risk-free rate.

For each t_i, we hold on to the quantities $(\Delta_{t_i}, \Theta_{t_i})$ from time t_i (beginning of period) up to time t_{i+1} (end of period), and repeat for $i = 0, 1, 2, \ldots$ until $t_{i+1} = T$. Mathematically,

$$\Delta_t = \begin{cases} \Delta_0 & \text{for } 0 \leq t < t_1, \\ \Delta_{t_1} & \text{for } t_1 \leq t < t_2, \\ \ldots & \\ \Delta_{t_{n-1}} & \text{for } t_{n-1} \leq t \leq T. \end{cases}$$

Consequently, the corresponding portfolio value process evolves continuously and is given by

$$\Pi_t = \begin{cases} \Delta_0 S_t + \Theta_0 e^{rt} & \text{for } 0 \leq t < t_i, \\ \Delta_{t_1} S_t + \Theta_{t_1} e^{rt} & \text{for } t_1 \leq t < t_2, \\ \ldots & \\ \Delta_{t_{n-1}} S_t + \Theta_{t_{n-1}} e^{rt} & \text{for } t_{n-1} \leq t \leq T. \end{cases}$$

This is indeed a *discretization* of the Black-Scholes replicating strategy. To delta-hedge any other simple derivative, we only need to replace $\Delta_{t_i} = N(d_1(S_{t_i}, K, T - t_i, r, \sigma))$ by the corresponding $\Delta_{t_i} = \frac{\partial F}{\partial S}(t_i, S_{t_i})$, as given in (16.6.4).

In this case, the portfolio is self-financing by construction: no funds are injected or withdrawn from the portfolio at any of the trading dates. More precisely, at each of these trading dates t_i, we will observe the realization of S_{t_i} and then we will choose $(\Delta_{t_i}, \Theta_{t_i})$ such that

$$\Delta_{t_{i-1}} S_{t_i} + \Theta_{t_{i-1}} B_{t_i} = \Delta_{t_i} S_{t_i} + \Theta_{t_i} B_{t_i}.$$

Since we want $\Delta_{t_i} = \frac{\partial F}{\partial S}(t_i, S_{t_i})$, it implies that

$$\Theta_{t_i} \times B_{t_i} = (\Delta_{t_{i-1}} S_{t_i} + \Theta_{t_{i-1}} B_{t_i}) - \frac{\partial F}{\partial S}(t_i, S_{t_i}) S_{t_i}$$

or, equivalently,

$$\Theta_{t_i} = e^{-rt_i} \{ (\Delta_{t_{i-1}} S_{t_i} + \Theta_{t_{i-1}} e^{rt_i}) - \frac{\partial F}{\partial S}(t_i, S_{t_i}) S_{t_i} \},$$

for each $i = 0, 1, \ldots, n-1$.

The value of the portfolio and the derivative it is tracking are only compared at maturity. In a self-financing hedging portfolio, hedging errors are rolled over until maturity. This is illustrated in the following examples.

Example 16.6.4 *Rebalancing a delta-hedging portfolio*

Let us come back to example 16.6.3 and see how to rebalance the portfolio after 1 month in the scenario where $S_{1/12} = 975$.

We saw, in the previous example, that in this scenario the value of the portfolio after 1 month (before rebalancing) is \$82.07096, split into a loan of \$520.3927 and 0.6179114 units of the risky asset. This is the amount available for investing over the next month. Recall that this is more than the option price. However, since we are implementing a delta-hedging strategy, we do not withdraw this excess from the portfolio.

Given that the asset price went down to 975, we should trade to make sure we now own

$$N(d_1(975, 1000, 11/12, 0.04, 0.2)) = 0.5615934$$

units of the underlying asset. In this scenario, this means selling $0.6179114 - 0.5615934 = 0.056318$ unit of this asset for a net income of $0.056318 \times \$975 = \$54.91005$.

This amount is then used to partially reimburse the loan: we should now have a short position in the risk-free asset worth $520.3927 - 54.91005 = \$465.4826$. Since this risk-free asset is now worth

$$B_{1/12} = e^{0.04\times1/12} = 1.003339,$$

we should be short $465.4826/1.003339 = 463.9335$ units.

In conclusion, after rebalancing we have:

$$\Delta_{1/12} = 0.5615934 \quad \text{and} \quad \Theta_{1/12} = -463.9335.$$

We can verify that the portfolio value before and after rebalancing is the same:

$$\begin{aligned} 82.07096 &= \Delta_{1/12}S_{1/12} + \Theta_{1/12}B_{1/12} \\ &= 0.5615934 \times 975 - 463.9335 \times e^{0.04\times1/12}. \end{aligned}$$

■

It is important to remember that for most derivatives, a delta-hedging strategy is not a replicating strategy: hedging errors will occur. These errors are solely due to *discretization* because we cannot trade continuously. However, in principle, if we rebalance the portfolio more often, the hedging errors should decrease. We will come back to this issue in Chapter 20.

Example 16.6.5 *Delta-hedging until maturity*

A 4-month at-the-money European call option is issued. You are using a BSM model with parameters $\mu = 0.07$, $r = 0.04$ and $\sigma = 0.2$. The underlying asset price is $S_0 = 25$ and the call option price is given by the Black-Scholes formula. Calculate the profit or loss resulting from delta-hedging this option monthly until maturity in the following scenario:

$$S_{1/12} = 26, S_{2/12} = 28, S_{3/12} = 27, S_{4/12} = 29.$$

At inception

We first need to find the option price at inception. Using the Black-Scholes formula with a maturity of $T = 4/12$, we get $d_1 = 0.17$ (rounded to two decimals), $N(d_1) = 0.5675$ and thus $C_0 = 1.2635$.

Consequently, over the time interval $[0, \frac{1}{12}[$, we must hold 0.5675 share of stock for a total value of $25 \times 0.5675 = 14.1875$. But since we received only \$1.2635 from selling the call option, we will borrow the difference, i.e. $14.1875 - 1.2635 = 12.9240$ (recall that $B_0 = 1$). The delta-hedging portfolio over the time interval $[0, \frac{1}{12}[$ is

$$\Theta_0 = -12.9240 \quad \text{and} \quad \Delta_0 = 0.5675.$$

After 1 month

In our scenario, after a month, the stock price will increase to $S_{1/12} = \$26$ and hence the portfolio value (before rebalancing) will be

$$-12.9240 \times e^{0.04\frac{1}{12}} + 0.5675 \times 26 = 1.7878.$$

Now, we should hold $N(d_1(1/12, 26)) = 0.7054$ share of stocks over the time interval $[\frac{1}{12}, \frac{2}{12}[$. After rebalancing, the portfolio value must still be worth 1.7878 since it must be self-financing. Therefore, solving for $\Theta_{1/12}$ in

$$1.7878 = \Theta_{1/12} \times e^{0.04\frac{1}{12}} + 0.7054 \times 26,$$

we find that

$$\Theta_{1/12} = -16.4975 \quad \text{and} \quad \Delta_{1/12} = 0.7054,$$

in this scenario.

After 2 months

After 2 months, the stock price in this scenario is now \$28 and hence the portfolio value (before rebalancing) is

$$-16.4975 \times e^{0.04\frac{2}{12}} + 0.7054 \times 28 = 3.1433.$$

Now, we should hold $N(d_1(2/12, S_{2/12})) = 0.9345$ share of stocks over the time interval $[\frac{2}{12}, \frac{3}{12}[$. After rebalancing, since the portfolio value must still be 3.1433 (self-financing condition), solving for $\Theta_{2/12}$ in

$$3.1433 = \Theta_{2/12} \times e^{0.04\frac{2}{12}} + 0.9345 \times 28,$$

we get

$$\Theta_{2/12} = -22.8697 \quad \text{and} \quad \Delta_{2/12} = 0.9345.$$

At maturity

Repeating the procedure at times 3/12 and 4/12, we get that the delta-hedging portfolio is worth 3.9004 at the option maturity whereas the payoff is 29 − 25 = 4. In this scenario, there is a hedging error (loss) of 0.1.

The following table summarizes the computations:

t	$T-t$	S_t	B_t	$\Delta_{t+\frac{1}{12}}$	$\Theta_{t+\frac{1}{12}}$	Π_t
0	4/12	25	1.0000	0.5675	−12.9240	1.2635
1/12	3/12	26	1.0033	0.7054	−16.4975	1.7878
2/12	2/12	28	1.0067	0.9345	−22.8697	3.1434
3/12	1/12	27	1.0101	0.9222	−22.5409	2.1320
4/12	0	29	1.0134			**3.9004**

■

16.7 Summary

Model

- Risk-free asset price: $B_t = B_0 e^{rt}$ with $B_0 = 1$.
- Risky asset price:
  - it can be obtained as the continuous-time limit of the risky asset price in a binomial tree model;
  - it is a geometric Brownian motion:

$$S_t = S_0 \exp\left\{\left(\mu - \frac{\sigma^2}{2}\right)t + \sigma W_t\right\};$$

  - its time-t value is lognormally distributed (w.r.t. the real-world or actuarial probability $\mathbb{P}$):

$$S_t \overset{\mathbb{P}}{\sim} \mathcal{LN}\left(\ln(S_0) + \left(\mu - \frac{\sigma^2}{2}\right)t, \sigma^2 t\right);$$

- the corresponding risk-neutral distribution is also lognormal (w.r.t. the risk-neutral probability $\mathbb{Q}$):

$$S_t \overset{\mathbb{Q}}{\sim} \mathcal{LN}\left(\ln(S_0) + \left(r - \frac{\sigma^2}{2}\right) t, \sigma^2 t\right).$$

Black-Scholes formula

- Derivation: it can be obtained as the continuous-time limit of the corresponding pricing formula in a binomial tree model or as the stop-loss transform of a lognormally distributed random variable.
- Call option: $C_t = S_t N(d_1(t, S_t)) - e^{-r(T-t)} K N(d_2(t, S_t))$ with

$$d_1(t, S_t) = \frac{\ln(S_t/K) + (r + \sigma^2/2)(T-t)}{\sigma\sqrt{T-t}},$$

$$d_2(t, S_t) = d_1(t, S_t) - \sigma\sqrt{T-t}.$$

- Put option: $P_t = e^{-r(T-t)} K N(-d_2(t, S_t)) - S_t N(-d_1(t, S_t))$.

Pricing simple derivatives

- Payoff: $V_T = g(S_T)$.
- Risk-neutral pricing formula: $V_t = e^{-r(T-t)} \mathbb{E}^{\mathbb{Q}}[g(S_T)|S_t]$, where

$$S_T \mid S_t \overset{\mathbb{Q}}{\sim} \mathcal{LN}\left(\ln(S_t) + \left(r - \frac{\sigma^2}{2}\right)(T-t), \sigma^2(T-t)\right).$$

- Binary call options:
  - asset-or-nothing: $C_t^{\text{AoN}} = S_t N(d_1(t, S_t))$;
  - cash-or-nothing: $C_t^{\text{CoN}} = e^{-r(T-t)} N\left(d_2(t, S_t)\right)$.
- Binary put options:
  - asset-or-nothing: $P_t^{\text{AoN}} = S_t N(-d_1(t, S_t))$;
  - cash-or-nothing: $P_t^{\text{CoN}} = e^{-r(T-t)} N(-d_2(t, S_t))$.
- Gap call option: $C_t^{\text{gap}} = S_t N(d_1(t, S_t; H)) - K e^{-r(T-t)} N(d_2(t, S_t; H))$, where

$$d_1(t, S_t; H) = \frac{\ln(S_t/H) + (r + \sigma^2/2)(T-t)}{\sigma\sqrt{T-t}},$$

$$d_2(t, S_t; H) = d_1(t, S_t; H) - \sigma\sqrt{T-t}.$$

- Gap put option: $P_t^{\text{gap}} = K e^{-r(T-t)} N(d_2(t, S_t; H)) - S_t N(d_1(t, S_t; H))$.

Determinants of call and put prices

- Option price increases (+) or decreases (−) when this quantity increases:
  - stock price S_0: call (+), put (−);
  - strike price K: call (−), put (+);
  - risk-free rate r: call (+), put (−);
  - volatility σ: call (+), put (+);
  - time to maturity T: call (+), put (cannot be determined).

Replication and hedging

- Portfolio: a pair (Θ, Δ), where $\Theta = \{\Theta_t, 0 \le t \le T\}$ and $\Delta = \{\Delta_t, 0 \le t \le T\}$ are such that:
  - number of units of B held at time t: Θ_t;
  - number of units of S held at time t: Δ_t.

- Portfolio value process: $\Pi = \{\Pi_t, 0 \le t \le T\}$, where

$$\Pi_t = \Delta_t S_t + \Theta_t B_t.$$

- Self-financing condition: $\Theta_{t-} B_t + \Delta_{t-} S_t = \Theta_t B_t + \Delta_t S_t$.
- Replicating strategy: self-financing and such that $\Pi_T = V_T$.
- Replication of a call option:
  - *long* $N(\mathrm{d}_1(t, S_t))$ units of S;
  - *short* $\mathrm{e}^{-rT} K N(\mathrm{d}_2(t, S_t))$ units of B.
- Replication of a put option:
  - *short* $N(-\mathrm{d}_1(t, S_t))$ units of S;
  - *long* $\mathrm{e}^{-rT} K N(-\mathrm{d}_2(t, S_t))$ units of B.
- Replication procedure for an option with payoff $V_T = g(S_T)$:
  1) identify F such that $F(t, S_t) = \mathrm{e}^{-r(T-t)} \mathbb{E}^{\mathbb{Q}}[g(S_T) \mid S_t]$;
  2) compute $\partial F / \partial S$;
  3) set $\Delta_t = \frac{\partial F}{\partial S}(t, S_t)$;
  4) set

$$\Theta_t = \mathrm{e}^{-rt} \left(F(t, S_t) - S_t \frac{\partial F}{\partial S}(t, S_t) \right).$$

- In practice, replication cannot be implemented as most replicating portfolios must be continuously rebalanced.
- Hedging strategy: portfolio set up to manage partially the risks of a given derivative.
- Hedging error: discrepancy between the price of the derivative and the value of the hedging portfolio.
- Delta-hedging strategy: discretization of the Black-Scholes replicating strategy.

16.8 Exercises

In the following problems, assume the market is a Black-Scholes-Merton model.

16.1 Consider a Black-Scholes-Merton model with $\mu = 0.07, r = 0.04, \sigma = 0.25$ and $S_0 = 100$.
(a) What is the probability that the stock price will be greater than \$120 after 2 years?
(b) A call option with strike price \$110 and a maturity of 2 years is available. What is the probability that this option will mature in the money?

16.2 For each of the next payoffs, find the initial price and determine the replicating strategy.
(a) $V_T = K$;
(b) $V_T = S_T$;
(c) $V_T = 100 \times \mathbb{1}_{\{S_T \ge K\}}$;
(d) $V_T = S_T \mathbb{1}_{\{S_T \ge K\}}$.
For each strategy, determine whether it is static or dynamic.

16.3 A derivative is introduced whose payoff is \$5 whenever the stock price is below 100 or above 115. Otherwise, it pays nothing.
(a) Express the payoff of this derivative as a function of S_T.
(b) Write the payoff as a combination of binary option payoffs.

(c) Compute the initial price of this derivative if $\mu = 0.12$, $r = 0.04$, $\sigma = 0.27$, $T = 1$ and $S_0 = 100$.
(d) Using the parameters given in (c), what is the probability that the derivative will mature out of the money?

16.4 A share of stock currently trades for \$35 and the volatility of the log-return of this stock is 28%. If the risk-free rate is 3%, calculate the initial Black-Scholes price of a 6-month gap put option with strike \$35 and trigger price \$34.

16.5 Using parity relationships between standard and binary options:
(a) Verify that the price of a cash-or-nothing put option is $P_t^{\mathrm{CoN}} = \mathrm{e}^{-r(T-t)}N(-\mathrm{d}_2(t, S_t))$.
(b) Verify that the price of an asset-or-nothing put option is $P_t^{\mathrm{AoN}} = S_t N(-\mathrm{d}_1(t, S_t))$.

16.6 Find the price and the replicating portfolio of the following payoff:

$$V_T = S_T^n,$$

where $n \geq 1$ is a given integer.

16.7 Explain briefly (and intuitively) what happens to Δ_0 of a call option when S_0 is *very* big/small.

16.8 You signed an agreement such that you owe 100 if the stock price of ABC inc. ends up above \$45 4 months later and 0 otherwise. Note that the current stock price is \$50. Assume $r = 0.03$ and $\sigma = 0.18$. You intend to replicate that liability with a portfolio made of shares of ABC inc. and the risk-free asset.
(a) At time 0, what is the delta-hedging portfolio for this agreement?
(b) In the scenario that, a month later, the stock price is up to \$52, determine the delta-hedging portfolio such that the portfolio remains self-financing.
(c) In the scenario that, 1 month prior to maturity, the stock price is down to \$48, determine the delta-hedging portfolio such that the portfolio remains self-financing.
(d) In the scenario that, at maturity, the stock price is \$47, calculate the hedging error of your preceding strategy.
(e) Suppose that you approach maturity and the current stock price is close to \$45. What happens to Δ_t whenever the stock price moves a little?

16.9 Find the initial price of a derivative having payoff:

$$((S_1 S_2 S_3)^{1/3} - K)_+,$$

where the maturity is $T = 3$.

16.10 Using the chain rule of calculus, verify that, for a call option,

$$\frac{\partial C_0}{\partial S_0} = N(d_1(S_0, K, T; r, \sigma)).$$

16.11 A chooser option gives its holder the choice, at time T, between entering a call or a put option (both maturing later). It has the following payoff

$$V_T = \max(C(T, S_T), P(T, S_T)),$$

where the call and the put options have a common maturity $T^\star > T$, have the same strike price and are both written on the same underlying asset.

(a) Verify that the chooser option payoff can be rewritten as follows:

$$V_T = C(T, S_T) + \left(e^{-r(T^\star - T)}K - S_T\right)^+.$$

(b) Compute the initial price of this chooser option.

17

Rigorous derivations of the Black-Scholes formula***

In Chapter 16 we provided a derivation of the Black-Scholes formula based on limiting arguments. Indeed, we have shown that when the time step in a binomial tree becomes increasingly small, the geometric random walk followed by the stock price gradually approaches that of a geometric Brownian motion. Although it provides a lot of the intuition and explains where many results come from, the approach lacks some rigor.

This chapter intends to fill these gaps by providing a more advanced treatment of the BSM model based upon advanced tools such as stochastic calculus (see Chapter 15), partial differential equations and changes of probability measures. Our main objective is to provide two *rigorous* derivations of the Black-Scholes formula using either partial differential equations or changes of probability measures. This chapter is therefore targeted at readers with a stronger mathematical background and who have read Chapter 15. This chapter is not mandatory to understand the upcoming chapters.

More specifically, the learning objectives are to:

- distinguish an ordinary differential equation (ODE) from a partial differential equation (PDE);
- understand the link between PDEs and diffusion processes as given by the Feynman-Kač formula;
- derive and solve the Black-Scholes PDE for simple payoffs;
- understand how to price and replicate simple derivatives with the Black-Scholes PDE;
- apply a change of probability measure to random variables and to Brownian motions;
- understand the role of the Fundamental Theorem of Asset Pricing to determine the no-arbitrage price of a derivative;
- compute the price of simple and exotic derivatives using the risk-neutral probability measure;
- using either the Black-Scholes PDE or the change of measure techniques, derive the Black-Scholes formula with the stop-loss transform.

17.1 PDE approach to option pricing and hedging

In their seminal paper [19], Fischer Black and Myron Scholes derived the dynamics of the self-financing replicating portfolio of a call option using partial differential equations. This has led to an increasing use of PDEs in mathematical finance and to an approach known as the *PDE approach to option pricing.*

Our objective is not to provide an introduction to the theory of PDEs; our only objective here is to lay down sufficient background material to understand the role of the Black-Scholes PDE in the replication and pricing of simple derivatives.

Actuarial Finance: Derivatives, Quantitative Models and Risk Management, First Edition.
Mathieu Boudreault and Jean-François Renaud.

17.1.1 Partial differential equations

First, let us look at the following quadratic equation:

$$ax^2 + bx + c = 0. \tag{17.1.1}$$

For given constants a, b and c, the goal is to obtain the value(s) of x such that $ax^2 + bx + c = 0$ is verified. We know that, depending on the values of a, b and c, the value of the discriminant $b^2 - 4ac$ will decide whether the quadratic equation in (17.1.1) has one or two solutions, or even no (real) solution.

In physics and biology, it is common for a function $F(x)$ describing a physical/biological quantity to be the solution of an *equation* of the form

$$aF''(x) + bF'(x) + cF(x) = 0, \tag{17.1.2}$$

with a, b and c being known constants. An equation like the one in (17.1.2) is called a *differential equation* and solving it means finding an analytical expression for $F(x)$.

One one hand, solving a quadratic equation as (17.1.1) means looking for a *number* x whereas on the other hand, solving a differential equation as (17.1.2) means looking for a *function* $F(x)$. In both cases, the existence of a solution depends on certain conditions. In what follows we will consider *ordinary differential equations* and then *partial differential equations.*

An **ordinary differential equation** (ODE) is a differential equation whose solution (if it exists) is a function of only one variable, say x, and which gives a relationship between the function $F(x)$ itself and its derivatives $F'(x)$, $F''(x)$, etc.

Example 17.1.1 *A simple ODE*
Let b be a known constant and let us find a function $F(x)$ such that:

$$F(x) = \frac{1}{b}F'(x),$$

for all $x \in \mathbb{R}$. It is easy to verify that, for this ODE,

$$F(x) = a + e^{bx}$$

is one possible solution, for any value of a. ■

A **partial differential equation** (PDE) is a differential equation whose function we are looking for has two variables, say t and x. The PDE is thus expressed in terms of the partial derivatives of F, namely $\frac{\partial F}{\partial t}(t,x)$, $\frac{\partial F}{\partial x}(t,x)$, $\frac{\partial^2 F}{\partial x^2}(t,x)$, etc.

Example 17.1.2 *A simple PDE*
Let us find a function $F(t, x)$ such that:

$$\frac{\partial F}{\partial t}(t,x) + \frac{\partial^2 F}{\partial x^2}(t,x) + F(t,x) = 0, \tag{17.1.3}$$

for all $(t, x) \in (0, T) \times \mathbb{R}$. For this PDE, we can easily verify that the function

$$F(t,x) = xe^{-t}$$

is such that its partial derivatives $\partial F/\partial t$ and $\partial^2 F/\partial x^2$ satisfy the relationship in (17.1.3). Indeed,

$$\frac{\partial F}{\partial t}(t,x) = -xe^{-t} \quad \text{and} \quad \frac{\partial^2 F}{\partial x^2}(t,x) = 0.$$

Note that the function $F(t, x) = 0$, for all t and x, is also a solution to this PDE. ■

In the previous examples, we found more than one solution to the same ODE and PDE, respectively. In order to guarantee the uniqueness of a solution, ODEs and PDEs require *initial or final conditions*, known as **boundary conditions**.

For the ODE in example 17.1.1, if we add an *initial condition* such as $F(0) = 1$, then

$$F(x) = 1 + e^{bx}$$

becomes the only solution.

Without getting into the details, in what follows we will specify *final conditions* to obtain unique solutions to our PDEs. More details below.

In most cases, it is difficult or even impossible to find an explicit solution to a PDE. In those cases, numerical methods are used to obtain approximations of the solution. However, for one class of PDEs, the one most commonly used in finance, there is a way to obtain the solution using stochastic calculus, as presented in Chapter 15; this is what we will describe next.

17.1.2 Feynman-Kač formula

The **Feynman-Kač formula** provides a solution to PDEs in the family of *parabolic PDEs*. This solution is expressed as the conditional expectation of a diffusion process sampled at a fixed time. Here is a simple version of the Feynman-Kač formula, which is adapted to the broader objectives of this chapter.

As always, assume that $T > 0$ is fixed. For a given parameter $r > 0$ and given functions $a(\cdot)$, $b(\cdot)$, $h(\cdot)$, we seek to find a solution to the PDE

$$\frac{\partial F}{\partial t}(t,x) + a(x)\frac{\partial F}{\partial x}(t,x) + \frac{1}{2}b^2(x)\frac{\partial^2 F}{\partial x^2}(t,x) - rF(t,x) = 0 \tag{17.1.4}$$

for all $(t,x) \in (0,T) \times \mathbb{R}$, with final condition $F(T,x) = h(x)$, for all $x \in \mathbb{R}$.

According to the Feynman-Kač formula, the solution to this PDE can be written as follows: for each (t,x) in the domain, we have

$$F(t,x) = e^{-r(T-t)}\mathbb{E}[h(X_T) \mid X_t = x], \tag{17.1.5}$$

where $\{X_s, s \geq 0\}$ is the diffusion process determined by the following SDE:

$$dX_s = a(X_s)ds + b(X_s)dW_s. \tag{17.1.6}$$

When trying to solve a PDE of the form given in (17.1.4), we can use the Feynman-Kač formula by following these steps:

INPUTS: parameter $r > 0$ and functions $a(\cdot), b(\cdot), h(\cdot)$, as given by the PDE.

1) Find the diffusion process $\{X_s, s \geq 0\}$ given by equation (17.1.6), using stochastic calculus, as seen in Chapter 15.
2) For each t and x in the domain, determine the conditional probability distribution of X_T given that $X_t = x$.
3) Compute the conditional expectation in equation (17.1.5).

OUTPUT: function $F(t,x)$.

The following example illustrates this methodology.

Example 17.1.3 *Solving a simple PDE*
We would like to find the solution to the PDE

$$\frac{\partial F}{\partial t}(t,x) + \frac{1}{2}\sigma^2\frac{\partial^2 F}{\partial x^2}(t,x) = 0$$

with boundary condition $F(T,x) = x^2$, where σ is a given constant.

Here, our inputs are $a(x) = 0$ and $b(x) = \sigma$, $h(x) = x^2$ and $r = 0$. Consequently, looking at (17.1.6), we consider the following SDE:

$$dX_s = \sigma dW_s,$$

whose solution is simply $X_s = \sigma W_s$. Now, using $h(x) = x^2$ and $r = 0$ in (17.1.5), the solution is

$$F(t, x) = \mathbb{E}[(X_T)^2 | X_t = x] = \mathbb{E}[(\sigma W_T)^2 | \sigma W_t = x].$$

As already computed in Chapter 14 (see equation (14.3.2)), we can write

$$\begin{aligned}
\mathbb{E}[(\sigma W_T)^2 | \sigma W_t = x] &= \mathbb{E}[(\sigma W_T - \sigma W_t + \sigma W_t)^2 | \sigma W_t = x] \\
&= \mathbb{E}[(\sigma(W_T - W_t) + x)^2] \\
&= x^2 + 2x\sigma\mathbb{E}[W_T - W_t] + \sigma^2\mathbb{E}[(W_T - W_t)^2] \\
&= x^2 + \sigma^2(T - t).
\end{aligned}$$

In conclusion, the function $F(t, x) = x^2 + \sigma^2(T - t)$ is the solution to the above PDE.

For validation purposes, let us check this. First, we can verify that the final condition $F(T, x) = x^2 + \sigma^2(T - T) = x^2$ is indeed equal to $h(x) = x^2$. Second, we have

$$\frac{\partial F}{\partial t}(t, x) = -\sigma^2, \quad \frac{\partial F}{\partial x}(t, x) = 2x, \quad \frac{\partial^2 F}{\partial x^2}(t, x) = 2.$$

Therefore,

$$\frac{\partial F}{\partial t}(t, x) + \frac{1}{2}\sigma^2 \frac{\partial^2 F}{\partial x^2}(t, x) = -\sigma^2 + \frac{1}{2}\sigma^2 \times 2 = 0,$$

as expected. ∎

17.1.3 Deriving the Black-Scholes PDE

Let us recall that the Black-Scholes model can be summarized as a continuous-time financial market model composed of a risk-free asset $\{B_t, t \geq 0\}$ and a risky asset $\{S_t, t \geq 0\}$. The dynamics of the price of the risk-free asset can be determined by the following ODE:

$$dB_t = rB_t dt, \tag{17.1.7}$$

whose solution is simply $B_t = B_0 e^{rt}$. The dynamics of the price of the risky asset can be determined by the following SDE:

$$dS_t = \mu S_t dt + \sigma S_t dW_t, \tag{17.1.8}$$

whose solution is a geometric Brownian motion: $S_t = S_0 e^{(\mu - \sigma^2/2)t + \sigma W_t}$.

Let us consider a simple option with payoff $V_T = h(S_T)$. We want to compute the no-arbitrage price of this derivative and find its replicating portfolio, as defined in Chapter 16.

We make the following assumptions: the time-t value V_t of this option can be written as $F(t, S_t)$, where $F(t, x)$ is a *sufficiently smooth* function; assume also that there exists a replicating portfolio $\{(\Theta_t, \Delta_t), 0 \leq t \leq T\}$ for this payoff, i.e. $V_T = \Pi_T$, where

$$\Pi_T = \Delta_T S_T + \Theta_T B_T,$$

and where the only cash flows of this portfolio are at time 0 and time T. Then, by the no-arbitrage principle, we have that $V_t = \Pi_t$, for all $0 \leq t \leq T$, where

$$\Pi_t = \Delta_t S_t + \Theta_t B_t$$

is the value at time t of this replicating portfolio. By the above assumptions, we then have that $F(t, S_t) = \Pi_t$, for all $0 \leq t \leq T$. At this stage, we do not have an expression for the function F nor for the replicating portfolio $\{(\Theta_t, \Delta_t), 0 \leq t \leq T\}$: we only assume that they exist. Our goal is to find explicit expressions.

To meet this objective, we need to find a trading strategy that is both replicating and self-financing. This is similar to what we did in the binomial model, except that we are now in a continuous-time model.

In a continuous-time model, the **self-financing condition** is given by the following *stochastic equation*:

$$d\Pi_t = \Theta_t dB_t + \Delta_t dS_t. \tag{17.1.9}$$

It can be obtained as the limit of equation (11.2.3) in Chapter 11. In words, the variation in the portfolio value comes from changes in values from both assets within the portfolio.

Moreover, since $F(t, S_t) = \Pi_t$, for all $0 \leq t \leq T$, we must also have that

$$d\Pi_t = dF(t, S_t)$$

meaning that, at each instant, the variation in the portfolio value must match the variation in the derivative's price. This is also known as the *replicating condition.*

We are now ready to derive the Black-Scholes PDE. First of all, $F(t, S_t)$ is a transformation of a diffusion process. Therefore, using Ito's lemma, we can write

$$dF(t, S_t) = \frac{\partial F}{\partial t}(t, S_t)dt + \frac{\partial F}{\partial x}(t, S_t)dS_t + \frac{1}{2}\sigma^2 S_t^2 \frac{\partial^2 F}{\partial x^2}(t, S_t)dt.$$

Substituting dS_t, as given by equation (17.1.8), we further have

$$\begin{aligned} dF(t, S_t) = &\left(\frac{\partial F}{\partial t}(t, S_t) + \frac{\partial F}{\partial x}(t, S_t)\mu S_t + \frac{1}{2}\sigma^2 S_t^2 \frac{\partial^2 F}{\partial x^2}(t, S_t)\right) dt \\ &+ \frac{\partial F}{\partial x}(t, S_t)\sigma S_t dW_t. \end{aligned} \tag{17.1.10}$$

On the other hand, since $\Pi_t = F(t, S_t)$, the self-financing condition of (17.1.9) can be rewritten as

$$\begin{aligned} dF(t, S_t) &= \Theta_t dB_t + \Delta_t dS_t \\ &= (\Theta_t r B_t + \Delta_t \mu S_t)dt + \Delta_t \sigma S_t dW_t \\ &= rF(t, S_t)dt + (\mu - r)\Delta_t S_t dt + \sigma \Delta_t S_t dW_t, \end{aligned} \tag{17.1.11}$$

where we used the fact that $\Theta_t B_t = F(t, S_t) - \Delta_t S_t$ (from the definition of the portfolio's value) and where we have substituted the expressions for dB_t and dS_t as given in equations (17.1.7) and (17.1.8), respectively.

We have obtained two equivalent *stochastic dynamics* for $F(t, S_t)$: one in equation (17.1.10) and one in equation (17.1.11). This means that for these two SDEs, the terms in front of dt must be equal and the terms in front of dW_t must be equal. Consequently, we have the following system of equations:

$$\begin{cases} rF(t, S_t) + (\mu - r)\Delta_t S_t = \frac{\partial F}{\partial t}(t, S_t) + \mu S_t \frac{\partial F}{\partial x}(t, S_t) + \frac{1}{2}\sigma^2 S_t^2 \frac{\partial^2 F}{\partial x^2}(t, S_t), \\ \sigma S_t \Delta_t = \sigma S_t \frac{\partial F}{\partial x}(t, S_t). \end{cases}$$

From the second equation, we easily find that

$$\Delta_t = \frac{\partial F}{\partial x}(t, S_t).$$

Then, using this in the first equation, we deduce the so-called **Black-Scholes PDE** or **Black-Scholes equation**: if the function $F(t,x)$ is such that

$$\frac{\partial F}{\partial t}(t,x) + rx\frac{\partial F}{\partial x}(t,x) + \frac{1}{2}\sigma^2 x^2 \frac{\partial^2 F}{\partial x^2}(t,x) - rF(t,x) = 0, \tag{17.1.12}$$

with boundary (final) condition $F(T,x) = h(x)$, where h is the payoff function, then the above system of equations is verified.

To summarize, we have obtained that the time-t value of a simple derivative with payoff $V_T = h(S_T)$ is given by $F(t, S_t)$, where the function $F(t,x)$ is the solution to the Black-Scholes PDE. The PDE comes from the model assumptions (dynamics of each asset) whereas the payoff of the option provides the terminal condition that the PDE should abide to. This boundary condition assures that we have a unique solution to this PDE.

Once we have found the price $F(t, S_t)$ for the payoff $V_T = h(S_T)$, then the replicating portfolio $\{(\Theta_t, \Delta_t), 0 \leq t \leq T\}$ is given by:

$$\Delta_t = \frac{\partial F}{\partial x}(t, S_t) \quad \text{and} \quad \Theta_t = \mathrm{e}^{-rt}(F(t,S_t) - \Delta_t S_t), \tag{17.1.13}$$

at every time $t \in [0, T)$.

It is also important to notice from the Black-Scholes equation (17.1.12) that the parameter μ has completely disappeared. Consequently, the option price (through function F) will not/does not depend on this parameter. This is similar to the result obtained in the binomial tree: the expected stock return under the real-world probability measure, which in that case was determined by the parameter p, is not relevant to determine the no-arbitrage price of a derivative.

In conclusion, if we can find a solution $F(t,x)$ to the Black-Scholes PDE in (17.1.12), we will have a full solution to our *pricing and hedging problem*: at each time t, we have

- the price of the derivative $V_t = F(t, S_t)$;
- the replicating portfolio (Θ_t, Δ_t) given by

$$\Delta_t = \frac{\partial F}{\partial x}(t, S_t),$$
$$\Theta_t = \mathrm{e}^{-rt}(F(t,S_t) - \Delta_t S_t).$$

17.1.4 Solving the Black-Scholes PDE

Luckily enough, under the assumptions of the BSM model, the Black-Scholes PDE (17.1.12) can be solved using the Feynman-Kač formula. Indeed, since $F(t,x)$ is a solution of (17.1.12), then, from equation (17.1.5), it can be written in the following form:

$$F(t,x) = \mathrm{e}^{-r(T-t)}\mathbb{E}[h(X_T) \mid X_t = x],$$

where $\{X_s, s \geq 0\}$ is determined by the SDE

$$\mathrm{d}X_s = rX_s\mathrm{d}s + \sigma X_s \mathrm{d}W_s.$$

This is simply a geometric Brownian motion with drift r whose solution we already found earlier: $X_s = X_0 \exp\{(r - \sigma^2/2)s + \sigma W_s\}$.

Using the standard arguments, such as formula (14.3.2), we get:

$$\begin{aligned} F(t,x) &= e^{-r(T-t)}\mathbb{E}[h(X_T) \mid X_t = x] \\ &= e^{-r(T-t)}\mathbb{E}\left[h\left(x\frac{X_T}{X_t}\right)\right] \\ &= e^{-r(T-t)}\mathbb{E}\left[h\left(x\exp\left\{\left(r-\frac{\sigma^2}{2}\right)(T-t)+\sigma(W_T-W_t)\right\}\right)\right]. \end{aligned} \tag{17.1.14}$$

This is the pricing formula for simple derivatives with payoff $V_T = h(S_T)$ obtained with the PDE approach. Note that we then have an explicit expression for the replicating portfolio as given in (17.1.13).

Let us illustrate what we have just obtained with a forward contract.

Example 17.1.4 *Forward contract*
The payoff of a (long) forward contract with delivery price K is $V_T = S_T - K$. Then, using the pricing formula in (17.1.14) with $h(x) = x - K$, we get

$$\begin{aligned} &F(t,x) \\ &= e^{-r(T-t)}\mathbb{E}\left[x\exp\left\{\left(r-\frac{\sigma^2}{2}\right)(T-t)+\sigma(W_T-W_t)\right\}\right] - e^{-r(T-t)}K \\ &= x\mathbb{E}\left[\exp\left\{\left(-\frac{\sigma^2}{2}\right)(T-t)+\sigma(W_T-W_t)\right\}\right] - e^{-r(T-t)}K. \end{aligned}$$

From the formula in (14.1.2) for the moment generating function of a normal distribution, we know that

$$\mathbb{E}\left[\exp\left\{\left(-\frac{\sigma^2}{2}\right)(T-t)+\sigma(W_T-W_t)\right\}\right] = 1$$

and, consequently,

$$F(t,x) = x - e^{-r(T-t)}K.$$

This means that the time-t value of a forward contract is given by $F(t,S_t) = S_t - e^{-r(T-t)}K$, as we already knew from Chapter 3. In fact, this was a model-free result, so it had to be true in any model, including the Black-Scholes-Merton model.

Since we have

$$\frac{\partial F}{\partial x}(t,x) = 1,$$

the replicating portfolio is given by

$$\Delta_t = \frac{\partial F}{\partial x}(t,S_t) = 1$$

and

$$\begin{aligned} \Theta_t &= e^{-rt}(F(t,S_t) - \Delta_t S_t) \\ &= e^{-rt}((S_t - e^{-r(T-t)}K) - S_t) = -e^{-rT}K. \end{aligned}$$

Note that it is a *static* strategy, i.e. it is constant with respect to time. In other words, to replicate a long forward contract, we must hold (at any time t) one unit of the underlying asset and be short $e^{-rT}K$ units of the risk-free asset. This is also a result we had obtained in Chapter 3. ∎

Note that we have found our first solution to the Black-Scholes PDE (17.1.12): it is given by the function $F(t,x) = x - e^{-r(T-t)}K$. We will soon find other solutions.

17.1.5 Black-Scholes formula

Let us now specialize the above results to the case of a call option. Similar to what we did before, we will substitute the notation $F(t,x)$ by $C(t,x)$. In this case, the final condition become $C(T,x) = (x-K)_+$ and by (17.1.14) we have

$$C(t,x) = e^{-r(T-t)}\mathbb{E}\left[\left(x\exp\left\{\left(r-\frac{\sigma^2}{2}\right)(T-t)+\sigma(W_T-W_t)\right\}-K\right)_+\right].$$

Note that this expectation is the stop-loss transform of a lognormally distributed random variable. Indeed, we have

$$x\exp\left\{\left(r-\frac{\sigma^2}{2}\right)(T-t)+\sigma(W_T-W_t)\right\}$$
$$\sim \mathcal{LN}\left(\ln(x)+\left(r-\frac{\sigma^2}{2}\right)(T-t), \sigma^2(T-t)\right).$$

Consequently, using once again the stop-loss formula in (14.1.7) of Chapter 14, we have

$$C(t,x) = xN(d_1(t,x)) - e^{-r(T-t)}KN(d_2(t,x))$$

where the functions $d_1(t,x)$ and $d_2(t,x)$ have been defined in (16.3.3) of Chapter 16.

Again, we can verify (with tedious calculations) that $C(t,x)$ is indeed a solution to the Black-Scholes PDE (see the Greek letters and their relationship to the Black-Scholes equation (PDE) in equation (20.5.7)). We could also show that as we approach maturity, then

$$\lim_{t\to T} C(t,x) = (x-K)_+.$$

As a byproduct of the above verification, we get

$$\frac{\partial C}{\partial x}(t,x) = N(d_1(t,x)),$$

from which we deduce the replicating portfolio of a call option:

$$\Delta_t = \frac{\partial C}{\partial x}(t,S_t) = N(d_1(t,S_t)),$$
$$\Theta_t = e^{-rt}(C(t,S_t) - \Delta_t S_t) = -e^{-rT}KN(d_2(t,S_t)).$$

This is consistent with the results previously derived in Chapter 16.

Finally, using the put-call parity, we can deduce the analytic expression of the put option price $P(t,S_t)$. This can also be achieved by exploiting the linearity of the Black-Scholes PDE (see the exercises).

17.2 Risk-neutral approach to option pricing

In the binomial tree model, we obtained the price of a derivative as the value of its replicating portfolio and then we reorganized the expression of this price using weights q and $1-q$. In other words, the price of the derivative was also given by a *discounted risk-neutral expectation*, i.e. an expectation taken with respect to another probability measure $\mathbb{Q}$ as opposed to the actuarial/real-world probability measure $\mathbb{P}$. See, for example, the expectation in (9.3.3).

Thanks to the FTAP, as discussed in Section 13.3.2 of Chapter 13, we can hope to find such a risk-neutral probability $\mathbb{Q}$ in the Black-Scholes-Merton model since this model is assumed to be free of arbitrage opportunities. This is the foundation of a probabilistic approach[1] called the *risk-neutral approach* or the *martingale approach* to option pricing, as opposed to the PDE approach.

Therefore, the objective of this section is to obtain directly the value of both a call and a put option in the Black-Scholes-Merton model, without using the binomial limiting argument. Our goal is to formalize some of the statements made in Section 16.2.4 about risk-neutral probabilities. In that section, we said that there exists a risk-neutral probability $\mathbb{Q}$ such that, for any time $0 < t \leq T$, we have

$$\ln(S_t/S_0) \overset{\mathbb{Q}}{\sim} \mathcal{N}\left(\left(r - \frac{\sigma^2}{2}\right)t, \sigma^2 t\right)$$

or, equivalently,

$$S_t \overset{\mathbb{Q}}{\sim} \mathcal{LN}\left(\ln(S_0) + \left(r - \frac{\sigma^2}{2}\right)t, \sigma^2 t\right).$$

We will see how to rigorously obtain this *artificial* probability measure $\mathbb{Q}$ and the corresponding risk-neutral dynamics of the risky asset price.

Before going any further, we will recall the definition of a probability measure, as seen in an introductory probability course, and then see how to introduce another probability measure in the same probability model using a technique known as a *change of probability measure.*

17.2.1 Probability measure

In elementary and more advanced probability theory, for a given sample space Ω, we define a probability measure $\mathbb{P}$ as a real-valued mapping defined on the set of all events, i.e. subsets of Ω, with the following properties:

1) for any event $E \subseteq \Omega$, we have $0 \leq \mathbb{P}(E) \leq 1$;
2) $\mathbb{P}(\Omega) = 1$;
3) for any sequence of disjoint events $E_1, E_2, \ldots \subseteq \Omega$, we have

$$\mathbb{P}\left(\biguplus_{k=1}^{\infty} E_k\right) = \sum_{k=1}^{\infty} \mathbb{P}(E_k).$$

This last property is known as *infinite additivity* and it implies *finite additivity*: for any n disjoint events $E_1, E_2, \ldots, E_n \subseteq \Omega$, we also have

$$\mathbb{P}\left(\biguplus_{k=1}^{n} E_k\right) = \sum_{k=1}^{n} \mathbb{P}(E_k).$$

Example 17.2.1 *Throwing a regular die*
We throw a well-balanced die, as we did in Chapter 1. Therefore, the sample space Ω is

$$\Omega = \{⚀, ⚁, ⚂, ⚃, ⚄, ⚅\}.$$

1 Usually credited to a series of papers by J.M. Harrison, D.M. Kreps and S.R. Pliska, namely [21,22,23].

For this experiment, we should define $\mathbb{P}$ by:

$$\mathbb{P}(\text{⚀}) = \frac{1}{6}, \quad \mathbb{P}(\text{⚁}) = \frac{1}{6}, \ldots, \quad \mathbb{P}(\text{⚅}) = \frac{1}{6}.$$

The mapping $\mathbb{P}$ assigns the value $1/6$ to each event. It is easy to see that this $\mathbb{P}$ meets the three criteria necessary to be called a probability measure. ■

In an introductory probability course, the modelling is usually done with a fundamental space Ω and only one probability measure $\mathbb{P}$ (rarely defined in an explicit way). Unfortunately, it creates the false impression that $\mathbb{P}$ is just a symbol with no particular meaning and/or specific definition.

Example 17.2.2 *Probability measures for economic states*
Assume the sample space of economic states for the upcoming year is

$$\Omega = \{\text{Economic boom, Standard economic growth, Recession, Economic crisis}\}.$$

According to actuary A, the probability of observing these events is:

$$\begin{aligned} \mathbb{P}_A(\text{Economic boom}) &= 0.1 \\ \mathbb{P}_A(\text{Standard economic growth}) &= 0.45 \\ \mathbb{P}_A(\text{Recession}) &= 0.4 \\ \mathbb{P}_A(\text{Economic crisis}) &= 0.05. \end{aligned}$$

But actuary B is more pessimistic. According to her, the probability of observing these events is:

$$\begin{aligned} \mathbb{P}_B(\text{Economic boom}) &= 0.05 \\ \mathbb{P}_B(\text{Standard economic growth}) &= 0.3 \\ \mathbb{P}_B(\text{Recession}) &= 0.50 \\ \mathbb{P}_B(\text{Economic crisis}) &= 0.15. \end{aligned}$$

Verifying the three conditions in the definition of a probability measure presented above, we have thus defined two probability measures $\mathbb{P}_A$ and $\mathbb{P}_B$ for the same experiment. ■

We say that $\mathbb{P}$ and $\tilde{\mathbb{P}}$ are **equivalent probability measures** if, for any event E, we have

$$\mathbb{P}(E) = 0 \quad \text{if and only if} \quad \tilde{\mathbb{P}}(E) = 0,$$

or, equivalently,

$$\mathbb{P}(E) = 1 \quad \text{if and only if} \quad \tilde{\mathbb{P}}(E) = 1.$$

This is often written as $\mathbb{P} \sim \tilde{\mathbb{P}}$. More or less, two probability measures are equivalent if they agree on all *impossible* events, which means then that they also agree on all events of probability equal to one. For any other event, they can differ.

Example 17.2.3 *Probability measures for economic states (continued)*
Because actuaries A and B agree that all four events have a non-zero probability, we have that $\mathbb{P}_A \sim \mathbb{P}_B$.

A third actuary (C) assesses the likelihood of each event of Ω and concludes that an economic boom is impossible. Mathematically, $\mathbb{P}_C(\text{Economic boom}) = 0$. Therefore $\mathbb{P}_C$ is not equivalent to $\mathbb{P}_A$ and is not equivalent to $\mathbb{P}_B$. ■

17.2.2 Changes of probability measure

Let us now present a methodology leading to a **change of probability measure** in a given probability model.

Let $\mathbb{P}$ be a probability measure on a sample space Ω, both designed for a given experiment. Now, choose a random variable Z such that:

1) $Z \geq 0$;
2) $\mathbb{E}[Z] = 1$.

For such a given random variable Z, we can define another probability measure $\mathbb{P}_Z$ as follows: for any event $E \subseteq \Omega$, set

$$\mathbb{P}_Z(E) = \mathbb{E}[\mathbb{1}_E Z] \tag{17.2.1}$$

where we should stress that the expectation $\mathbb{E}$ is taken with respect to $\mathbb{P}$. The notation $\mathbb{P}_Z$ emphasizes that for each such random variable Z, there is a new probability measure attached to it. In probability theory, Z is known as the **Radon-Nikodym derivative** of $\mathbb{P}_Z$ with respect to $\mathbb{P}$. This is just terminology as no derivatives are really involved.

We need to make sure that $\mathbb{P}_Z$ meets the three conditions to be a probability measure. It is rather easy to verify that for any event $E \subseteq \Omega$, we have $0 \leq \mathbb{P}_Z(E) \leq 1$ and $\mathbb{P}_Z(\Omega) = 1$. A little bit more work is needed to verify that for any sequence of disjoint events $E_1, E_2, \ldots \subseteq \Omega$, we have

$$\mathbb{P}_Z\left(\biguplus_{k=1}^{\infty} E_k\right) = \sum_{k=1}^{\infty} \mathbb{P}_Z(E_k).$$

In conclusion, for each such random variable Z, we now have a methodology to construct a new probability measure $\mathbb{P}_Z$, usually different from the existing $\mathbb{P}$.

Let us now compute expectations of the form $\mathbb{E}_Z[X]$ with respect to the probability $\mathbb{P}_Z$, i.e. compute *weighted averages* using the *weights* given by $\mathbb{P}_Z$. First, by definition (17.2.1), we have

$$\mathbb{E}_Z[\mathbb{1}_E] = \mathbb{E}[\mathbb{1}_E Z].$$

Indeed, since the random variable $X = \mathbb{1}_E$ is a Bernoulli random variable, i.e. it takes only the values 0 and 1, then

$$\mathbb{E}_Z[\mathbb{1}_E] = 1 \times \mathbb{P}_Z(E) + 0 \times \mathbb{P}_Z(E^c) = \mathbb{P}_Z(E).$$

The result follows by the definition in (17.2.1).

More generally, one can show that[2] for any random variable X, we have a similar relationship:

$$\mathbb{E}_Z[X] = \mathbb{E}[XZ]. \tag{17.2.2}$$

The expectation of an arbitrary random variable X computed with the probability measure $\mathbb{P}_Z$ is equal to the expectation with respect to $\mathbb{P}$ where X is *distorted* by the random variable Z.

To successfully use the above methodology, we first need to choose a positive random variable Z with unit mean. Then, we can determine the distribution of another random variable X with respect to $\mathbb{P}_Z$ or, in other words, the impact of the change of measure on the distribution of X.

---

2 This is a rather difficult task.

Example 17.2.4 *Change of measure for a normal distribution*

Let us consider the case of a normally distributed random variable $X \overset{\mathbb{P}}{\sim} \mathcal{N}(0,1)$ where the notation $\overset{\mathbb{P}}{\sim}$ emphasizes that the distribution is affected by the choice of the probability measure ($\mathbb{P}$ in this case). This statement means that: for any $a \in \mathbb{R}$, we have

$$\mathbb{P}(X \leq a) = \int_{-\infty}^{a} \frac{1}{\sqrt{2\pi}} \mathrm{e}^{-\frac{x^2}{2}} \mathrm{d}x.$$

An equivalent characterization of the normal distribution is given by its m.g.f.: for any $\lambda \in \mathbb{R}$, we have

$$\mathbb{E}[\mathrm{e}^{\lambda X}] = \mathrm{e}^{\lambda^2/2}.$$

See Chapter 14.

Fix a real number α. Let us now find the distribution of X with respect to a probability measure $\mathbb{P}_\alpha$ defined with the random variable

$$Z_\alpha = \mathrm{e}^{\alpha X - (1/2)\alpha^2}.$$

For simplicity, we have chosen to write $\mathbb{P}_\alpha$ instead of $\mathbb{P}_{Z_\alpha}$.

First of all, it is easy to verify (see exercise 17.4) that Z_α is indeed such that

1) $Z_\alpha \geq 0$;
2) $\mathbb{E}[Z_\alpha] = 1$.

Now, let us show that $X \overset{\mathbb{P}_\alpha}{\sim} \mathcal{N}(\alpha, 1)$. Using the characterization of the normal distribution by its m.g.f., it suffices to verify that

$$\mathbb{E}_\alpha[\mathrm{e}^{\lambda X}] = \mathrm{e}^{\lambda\alpha + \frac{\lambda^2}{2}},$$

for all $\lambda \in \mathbb{R}$, where $\mathbb{E}_\alpha$ stands for the expectation with respect to $\mathbb{P}_\alpha$. Indeed, using (17.2.2), we have

$$\begin{aligned}
\mathbb{E}_\alpha[\mathrm{e}^{\lambda X}] &= \mathbb{E}[\mathrm{e}^{\lambda X} Z_\alpha] \\
&= \mathbb{E}[\mathrm{e}^{\lambda X} \mathrm{e}^{\alpha X - (1/2)\alpha^2}] \\
&= \mathrm{e}^{-(1/2)\alpha^2} \mathbb{E}[\mathrm{e}^{(\lambda+\alpha)X}] \\
&= \mathrm{e}^{-(1/2)\alpha^2} \mathrm{e}^{(1/2)(\lambda+\alpha)^2},
\end{aligned}$$

where, in the last step, we used the fact that $X \overset{\mathbb{P}}{\sim} \mathcal{N}(0,1)$ or, equivalently, that $\mathbb{E}[\mathrm{e}^{(\lambda+\alpha)X}] = \mathrm{e}^{(\lambda+\alpha)^2/2}$. The result follows. ■

This last example shows that it is possible to apply a well-chosen change of measure to *shift the mean* of a normally distributed random variable, i.e. to move from a mean of 0 under $\mathbb{P}$ to a mean of α under $\mathbb{P}_\alpha$. Note that the variance is not affected by this change of measure. This will be of paramount importance for linear Brownian motions and the Black-Scholes-Merton model.

Example 17.2.5 *Change of measure for an exponential distribution*

Let us now consider an exponentially distributed random variable $X \overset{\mathbb{P}}{\sim} \exp(\alpha)$, with $\alpha > 0$. This means that, with respect to $\mathbb{P}$, the random variable X admits the following probability density function:

$$f_X(x) = \alpha \mathrm{e}^{-\alpha x} \mathbb{1}_{(0,\infty)}(x).$$

Again, an equivalent characterization of the exponential distribution is given by its m.g.f. Indeed, for any $\lambda < \alpha$, we have

$$\mathbb{E}[e^{\lambda X}] = \frac{\alpha}{\alpha - \lambda}. \tag{17.2.3}$$

It is possible to change the probability measure from $\mathbb{P}$ to $\mathbb{P}_\beta$ to make X an exponentially distributed random variable with a larger parameter $\beta > \alpha$ with respect to $\mathbb{P}_\beta$. Indeed, it suffices to use

$$Z_\beta = \frac{\beta}{\alpha} e^{(\alpha-\beta)X}$$

to define $\mathbb{P}_\beta$.

Again, it is easy to verify (see exercise 17.5) that Z_β is such that

1) $Z_\beta \geq 0$;
2) $\mathbb{E}[Z_\beta] = 1$.

Let us now compute the m.g.f. of X with respect to $\mathbb{P}_\beta$:

$$\begin{aligned}
\mathbb{E}_\beta[e^{\lambda X}] &= \mathbb{E}[e^{\lambda X} Z_\beta] \\
&= \mathbb{E}\left[e^{\lambda X} \frac{\beta}{\alpha} e^{(\alpha-\beta)X}\right] \\
&= \frac{\beta}{\alpha} \mathbb{E}[e^{(\alpha-\beta+\lambda)X}] \\
&= \frac{\beta}{\alpha} \times \frac{\alpha}{\alpha - (\alpha - \beta + \lambda)} \\
&= \frac{\beta}{\beta - \lambda},
\end{aligned}$$

where in the second last step we used the formula in (17.2.3). Since

$$\mathbb{E}_\beta[e^{\lambda X}] = \frac{\beta}{\beta - \lambda},$$

for all $\lambda < \beta$, we can conclude that $X \overset{\mathbb{P}_\beta}{\sim} \exp(\beta)$. ■

Likelihood ratio

One might wonder how to choose the random variable Z in order to change the probability distribution of X (with respect to the original probability $\mathbb{P}$) to a prescribed distribution with respect to the probability measure $\mathbb{P}_Z$.

For a continuous random variable, this is given by a ratio of probability density functions (p.d.f.). Assume that X has a p.d.f. $f_X(x)$ with respect to $\mathbb{P}$ and that we are looking for the change of probability measure such that we would obtain the p.d.f. $f_X^Z(x)$ with respect to $\mathbb{P}_Z$. In particular, after the change of measure, we should have the following:

$$\mathbb{E}_Z[X] = \int x f_X^Z(x) \mathrm{d}x.$$

This is the expectation of X with respect to $\mathbb{P}_Z$.

First, we know that the $\mathbb{P}$-expectation of the random variable given by

$$X \times \frac{f_X^Z(X)}{f_X(X)}$$

is given by

$$\mathbb{E}\left[X \times \frac{f_X^Z(X)}{f_X(X)}\right] = \int_{-\infty}^{\infty} \left(u \times \frac{f_X^Z(u)}{f_X(u)}\right) f_X(u)\mathrm{d}u = \int_{-\infty}^{\infty} u f_X^Z(u)\mathrm{d}u = \mathbb{E}^Z[X].$$

In other words, we must choose:

$$Z = \frac{f_X^Z(X)}{f_X(X)}.$$

We easily verify that $Z \geq 0$, as the p.d.f.s are positive functions and

$$\mathbb{E}[Z] = \int \frac{f_X^Z(u)}{f_X(u)} f_X(u)\mathrm{d}u = 1.$$

Similarly, for discrete random variables, a similar change of probability measures, given by the ratio of probability mass functions, will allow a prescribed distribution to be obtained.

17.2.3 Girsanov theorem

The last two examples have shown the effect of a given change of measure on the distribution of a random variable X. In the Black-Scholes-Merton model, we will need to change the distribution of a whole stochastic process, namely that of a linear Brownian motion.

This situation will be taken care of by the so-called Girsanov theorem, which is essentially the analog for Brownian motions of the change of measure discussed in example 17.2.4 for normal random variables. The Girsanov theorem is the last building block needed before we can finally determine the no-arbitrage price of a derivative and obtain risk-neutral pricing formulas. In this section we present a simple version of the latter.

Consider a $\mathbb{P}$-standard Brownian motion $W = \{W_t, 0 \leq t \leq T\}$, i.e. a standard Brownian motion under $\mathbb{P}$, and for a fixed number θ define the process $W^\theta = \{W_t^\theta, 0 \leq t \leq T\}$ by

$$W_t^\theta = \theta t + W_t.$$

The Girsanov theorem states that if we use the random variable

$$Z_\theta = \mathrm{e}^{-\theta W_T - \frac{1}{2}\theta^2 T}$$

to obtain the probability measure $\mathbb{P}^\theta$, then $W^\theta = \{W_t^\theta, 0 \leq t \leq T\}$ will be a $\mathbb{P}^\theta$-standard Brownian motion.

At first, this result can be rather confusing. Indeed, by definition W is a $\mathbb{P}$-SBM and since

$$W_t^\theta = \theta t + W_t, \text{ for all } t \geq 0,$$

then W^θ is a linear Brownian motion with drift θ with respect to $\mathbb{P}$. On the other hand, since by the Girsanov theorem W^θ is a $\mathbb{P}^\theta$-SBM and

$$W_t = -\theta t + W_t^\theta, \text{ for all } t \geq 0,$$

then W is a linear Brownian motion with drift of $-\theta$ with respect to $\mathbb{P}^\theta$. We have summarized this discussion in Table 17.1.

Whereas example 17.2.4 showed how we can change the probability measure *to shift the mean* of a normally distributed random variable, the Girsanov theorem tells us how to change the probability measure *to shift the drift* of a linear Brownian motion.

Table 17.1 Impact of the Girsanov theorem on (linear) Brownian motions

with respect to	$\mathbb{P}$	$\mathbb{P}^\theta$
W is a	standard BM	linear BM
W^θ is a	linear BM	standard BM

17.2.4 Risk-neutral probability measures

As in the general n-period binomial model and in the trinomial tree model (see e.g. Section 13.3.2), the concept of *risk-neutral probability measure* is fundamental in the theory of continuous-time derivatives pricing. In *any* continuous-time model with a stock price process $S = \{S_t, 0 \le t \le T\}$ and a risk-free asset price process $B = \{B_t, 0 \le t \le T\}$, a probability measure $\mathbb{Q}$ is called a **risk-neutral probability measure** if it is such that:

1) $\mathbb{Q}$ is equivalent to $\mathbb{P}$;
2) the stochastic process $\{\frac{S_t}{B_t}, 0 \le t \le T\}$ is a $\mathbb{Q}$-martingale, i.e. a martingale with respect to the probability measure $\mathbb{Q}$, which means that, for all $0 \le t_1 < t_2 \le T$,

$$\frac{S_{t_1}}{B_{t_1}} = \mathbb{E}^{\mathbb{Q}}\left[\left.\frac{S_{t_2}}{B_{t_2}}\right| S_u : 0 \le u \le t_1\right].$$

The probability measure $\mathbb{Q}$ is also known as an **equivalent martingale measure (EMM)**.

As discussed earlier in this chapter, the *equivalence* condition means that $\mathbb{P}$ and $\mathbb{Q}$ have the same zero-probability events or, equivalently, the same probability-one events. Intuitively, if an event is *impossible* or *certain* with respect to $\mathbb{P}$, then it is still the case under the risk-neutral probability measure $\mathbb{Q}$.

The *martingale condition* says that if the risky asset price S is discounted with the risk-free asset B, then, with respect to a risk-neutral probability $\mathbb{Q}$, this discounted price does not have any remaining *trend* as it behaves as a martingale.

Using now a more formal definition of probability measure, recall from Section 13.3.2 that the FTAP states that:

- a market model is free of arbitrage opportunities if and only if there exists at least one risk-neutral probability measure $\mathbb{Q}$;
- an arbitrage-free market model is complete if and only if there exists a unique risk-neutral probability measure $\mathbb{Q}$. Otherwise the market is said to be incomplete.

17.2.5 Risk-neutral dynamics

Let us now put all the pieces together and determine how to price derivatives in the BSM model using the risk-neutral approach. In the BSM model, the risky asset S is a geometric Brownian motion of the form

$$S_t = S_0 \exp\left\{\left(\mu - \frac{\sigma^2}{2}\right)t + \sigma W_t\right\}$$

and the risk-free asset is such that

$$B_t = B_0 e^{rt}$$

with $B_0 = 1$. The resulting financial market is free of arbitrage opportunities.[3] Thus, according to the FTAP, if the market is arbitrage-free then there exists at least one risk-neutral probability measure $\mathbb{Q}$ under which the process given by $\{\frac{S_t}{B_t}, 0 \leq t \leq T\}$ is a martingale.

Dividing S_t by B_t we get

$$\frac{S_t}{B_t} = S_0 \exp\left\{\left(\mu - r - \frac{\sigma^2}{2}\right)t + \sigma W_t\right\}.$$

In other words, the discounted stock price process S/B is still a geometric Brownian motion. We know that this GBM is a $\mathbb{P}$-martingale if and only if $\mu = r$ (see Section 14.5.3). Therefore, if we need to find an EMM, the probability measure $\mathbb{P}$ is not going to be our candidate. That is why we called $\mathbb{P}$ the real-world probability measure to mark the difference.

In our quest for an EMM $\mathbb{Q}$, let us define (using a very suggestive notation) $W^{\mathbb{Q}} = \{W_t^{\mathbb{Q}}, 0 \leq t \leq T\}$ by

$$W_t^{\mathbb{Q}} = \left(\frac{\mu - r}{\sigma}\right)t + W_t, \tag{17.2.4}$$

and $S^{\mathbb{Q}} = \{S_t^{\mathbb{Q}}, 0 \leq t \leq T\}$ by

$$S_t^{\mathbb{Q}} = S_0^{\mathbb{Q}} \exp\left\{-\frac{\sigma^2}{2}t + \sigma W_t^{\mathbb{Q}}\right\}. \tag{17.2.5}$$

One has to be careful: recall from Table 17.1 that, with respect to $\mathbb{P}$, the new process $W^{\mathbb{Q}} = \{W_t^{\mathbb{Q}}, 0 \leq t \leq T\}$ is not a standard Brownian motion, it is a linear Brownian motion.

Using the Girsanov theorem, if we set $\mathbb{Q} = \mathbb{P}^{\theta}$ with $\theta = \dfrac{\mu - r}{\sigma}$, i.e. if we use the following change of measure

$$Z^{\mathbb{Q}} = \mathrm{e}^{-\left(\frac{\mu-r}{\sigma}\right)W_T - \frac{1}{2}\left(\frac{\mu-r}{\sigma}\right)^2 T}, \tag{17.2.6}$$

then $W^{\mathbb{Q}} = \{W^{\mathbb{Q}}, 0 \leq t \leq T\}$ is a $\mathbb{Q}$-standard Brownian motion and, consequently, $S^{\mathbb{Q}} = \{S_t^{\mathbb{Q}}, 0 \leq t \leq T\}$ is a $\mathbb{Q}$-martingale. This last statement about the martingale property of $S^{\mathbb{Q}}$ is backed up by the work we did in Section 14.5.3.

So, the probability measure $\mathbb{Q}$ defined with the random variable in (17.2.6) is the (unique) risk-neutral probability measure in the BSM model. Indeed, it is now easy to verify that $\{\frac{S_t}{B_t}, 0 \leq t \leq T\}$ is a $\mathbb{Q}$-martingale. By definition of the geometric Brownian motion S in the BSM model, we have

$$\begin{aligned}
\frac{S_t}{B_t} &= S_0 \exp\left\{\left(\mu - r - \frac{\sigma^2}{2}\right)t + \sigma W_t\right\} \\
&= S_0 \exp\left\{-\frac{\sigma^2}{2}t + \sigma\left(W_t + \left(\frac{\mu - r}{\sigma}\right)t\right)\right\} \\
&= S_0 \exp\left\{-\frac{\sigma^2}{2}t + \sigma W_t^{\mathbb{Q}}\right\}, \\
&= S_t^{\mathbb{Q}},
\end{aligned}$$

where in the second last step we used the definition of $W^{\mathbb{Q}}$ given in (17.2.4) and in the last step we used the definition of $S^{\mathbb{Q}}$ given in (17.2.5).

3 As long as $\sigma > 0$.

Since $S_t = S_t^{\mathbb{Q}} \times B_t$, then we can summarize our findings as: for all $0 \leq t \leq T$,

$$S_t \overset{\mathbb{P}}{\sim} \mathcal{LN}\left(\ln(S_0) + \left(\mu - \frac{\sigma^2}{2}\right)t, \sigma^2 t\right),$$

$$S_t \overset{\mathbb{Q}}{\sim} \mathcal{LN}\left(\ln(S_0) + \left(r - \frac{\sigma^2}{2}\right)t, \sigma^2 t\right).$$

We can also say that the continuous-time stochastic process $\{S_t, t \geq 0\}$ is

- a GBM with drift coefficient $\mu - \frac{\sigma^2}{2}$ under the real-world/actuarial probability measure $\mathbb{P}$;
- a GBM with drift coefficient $r - \frac{\sigma^2}{2}$ under the risk-neutral probability measure $\mathbb{Q}$.

Note that the volatility coefficient is σ with respect to both probability measures.

17.2.6 Risk-neutral pricing formulas

Imagine a financial derivative with payoff $V_T = S_T$, that is a derivative providing one unit of the underlying risky asset at maturity T. To avoid arbitrage opportunities, this derivative must be worth exactly S_t, i.e. we must have $V_t = S_t$, at any time $0 \leq t \leq T$. From the definition of a risk-neutral probability measure, we already have that

$$\frac{S_t}{B_t} = \mathbb{E}^{\mathbb{Q}}\left[\left.\frac{S_T}{B_T}\right| S_u : 0 \leq u \leq t\right].$$

In other words, for this specific derivative, we have

$$\frac{V_t}{B_t} = \mathbb{E}^{\mathbb{Q}}\left[\left.\frac{V_T}{B_T}\right| S_u : 0 \leq u \leq t\right],$$

for all $0 \leq t \leq T$. This should be reminiscent of the risk-neutral pricing formulas obtained in the general binomial tree model (see e.g. equation (11.3.1)) and in the BSM model (see e.g. Section 16.4).

It turns out that[4] for any European payoff V_T, simple or exotic, we have a similar risk-neutral pricing formula: for all $0 \leq t \leq T$,

$$\frac{V_t}{B_t} = \mathbb{E}^{\mathbb{Q}}\left[\left.\frac{V_T}{B_T}\right| S_u : 0 \leq u \leq t\right]$$

or, written differently,

$$V_t = \mathrm{e}^{-r(T-t)}\mathbb{E}^{\mathbb{Q}}[V_T | S_u : 0 \leq u \leq t]. \tag{17.2.7}$$

Note that as a particular case of this pricing formula, the initial no-arbitrage price is given by

$$V_0 = \mathrm{e}^{-rT}\mathbb{E}^{\mathbb{Q}}[V_T].$$

This is why the risk-neutral probability $\mathbb{Q}$, under which we can find the no-arbitrage price of any derivative, is also called the *pricing* probability measure.

Note that, for a simple payoff $V_T = h(S_T)$, the pricing formula in (17.2.7) coincides with the pricing formula in (17.1.14), obtained with the PDE approach

⚠ Again, let us emphasize that $\mathbb{Q}$ is an *artificial* probability measure meant to find the price of a derivative. In other words, it is a byproduct of no-arbitrage pricing. The measure $\mathbb{Q}$ should not be used for risk management purposes such as computing likelihoods of events and scenarios.

Let us put the risk-neutral pricing formula of (17.2.7) to the test.

4 It is beyond the scope of this book to provide a rigorous justification of this result.

Example 17.2.6 *Forward contract*
The payoff of a forward contract with fixed delivery price K is given by $V_T = S_T - K$. Then by equation (17.2.7) we have

$$V_t = e^{-r(T-t)}\mathbb{E}^{\mathbb{Q}}[S_T - K|S_u : 0 \leq u \leq t]$$
$$= e^{rt}\mathbb{E}^{\mathbb{Q}}[e^{-rT}S_T|S_u : 0 \leq u \leq t] - e^{-r(T-t)}K,$$

for any $0 \leq t \leq T$. Since $\{e^{-rt}S_t, 0 \leq t \leq T\}$ is a $\mathbb{Q}$-martingale (by definition of $\mathbb{Q}$), we have

$$\mathbb{E}^{\mathbb{Q}}[e^{-rT}S_T|S_u : 0 \leq u \leq t] = e^{-rt}S_t$$

and then

$$V_t = S_t - e^{-r(T-t)}K,$$

as obtained a few times already in this book. ∎

17.2.6.1 Black-Scholes formula

Now, let us consider the no-arbitrage price of a call and a put option. In particular, if we consider $C_T = (S_T - K)_+$, then from (17.2.7) we have

$$C_t = e^{-r(T-t)}\mathbb{E}^{\mathbb{Q}}[(S_T - K)_+|S_u : 0 \leq u \leq t],$$

for any $0 \leq t \leq T$. Since S is a $\mathbb{Q}$-geometric Brownian motion, it possesses the Markovian property as discussed in Chapter 14. Consequently, as we did in Chapter 16, we can further write

$$C_t = e^{-r(T-t)}\mathbb{E}^{\mathbb{Q}}[(S_T - K)_+ \mid S_t] = \mathbb{E}^{\mathbb{Q}}\left[\left(a\frac{S_T}{S_t} - K\right)_+\right]\Bigg|_{a=S_t},$$

where

$$\frac{S_T}{S_t} = \exp\left\{\left(r - \frac{\sigma^2}{2}\right)(T-t) + \sigma\left(W_T^{\mathbb{Q}} - W_t^{\mathbb{Q}}\right)\right\}$$
$$\overset{\mathbb{Q}}{\sim} \mathcal{LN}\left(\left(r - \frac{\sigma^2}{2}\right)(T-t), \sigma^2(T-t)\right).$$

Applying once more the stop-loss formula in equation (14.1.7), we obtain a dynamic version of the **Black-Scholes formula**: for $0 \leq t < T$,

$$C_t = S_t N(d_1(t, S_t)) - e^{-r(T-t)}KN(d_2(t, S_t)),$$

where

$$d_1(t, S_t) = \frac{\ln(S_t/K) + (r + \sigma^2/2)(T-t)}{\sigma\sqrt{T-t}},$$
$$d_2(t, S_t) = d_1(t, S_t) - \sigma\sqrt{T-t}.$$

By the put-call parity, we have

$$C_t - P_t = S_t - Ke^{-r(T-t)}$$

and we can recover the Black-Scholes formula for the price of a put option: for all $0 \leq t \leq T$,

$$P_t = e^{-r(T-t)}KN(-d_2(t, S_t)) - S_t N(-d_1(t, S_t)).$$

It is important to emphasize that the risk-neutral pricing formula in equation (17.2.7) can be applied to any European payoff V_T, not only simple payoffs of the form $V_T = h(S_T)$. Consequently, we now have a powerful methodology to price exotic options such as Asian, barrier and lookback options, as seen in Chapter 7, using the risk-neutral dynamics of $\{S_t, t \geq 0\}$ and the corresponding path-dependent payoff functional.

17.3 Summary

Black-Scholes-Merton model

- Risk-free asset: $\{B_t, t \geq 0\}$.
- Risky asset: $\{S_t, t \geq 0\}$.
- BSM model:

$$\begin{aligned} dB_t &= rB_t dt \\ dS_t &= \mu S_t dt + \sigma S_t dW_t \end{aligned}$$

or, equivalently,

$$\begin{aligned} B_t &= B_0 e^{rt} \\ S_t &= S_0 e^{(\mu-\sigma^2/2)t+\sigma W_t}. \end{aligned}$$

PDE approach to option pricing and hedging

- Assumptions: for a payoff $V_T = h(S_T)$,
  - $V_t = F(t, S_t)$, where $F(t,x)$ is a *sufficiently smooth* function;
  - exists $\{(\Theta_t, \Delta_t), 0 \leq t \leq T\}$, with time-$t$ value $\Pi_t = \Delta_t S_t + \Theta_t B_t$ such that:

$$\begin{aligned} d\Pi_t &= \Theta_t dB_t + \Delta_t dS_t \quad \text{(self-financing)} \\ \Pi_t &= F(t, S_t) \quad \text{(replicating)}. \end{aligned}$$

- Objective: find explicit expressions for $F(t,x)$, Θ_t and Δ_t.
- First step toward a solution:
  - $F(t,x)$ is the solution of the Black-Scholes PDE:

$$\frac{\partial F}{\partial t}(t,x) + rx\frac{\partial F}{\partial x}(t,x) + \frac{1}{2}\sigma^2 x^2 \frac{\partial^2 F}{\partial x^2}(t,x) - rF(t,x) = 0,$$

  with final condition $F(T,x) = h(x)$;
  - the replicating portfolio is given by:

$$\begin{aligned} \Delta_t &= \frac{\partial F}{\partial x}(t, S_t), \\ \Theta_t &= e^{-rt}(F(t, S_t) - \Delta_t S_t). \end{aligned}$$

- Using the Feynman-Kač formula, we can solve the Black-Scholes PDE:

$$F(t,x) = e^{-r(T-t)}\mathbb{E}\left[h\left(x \exp\left\{\left(r - \frac{\sigma^2}{2}\right)(T-t) + \sigma\left(W_T - W_t\right)\right\}\right)\right].$$

Risk-neutral approach to option pricing

- Change of probability measure: for a positive random variable Z with unit mean, define

$$\mathbb{P}_Z(E) = \mathbb{E}\left[\mathbb{1}_E Z\right].$$

- Girsanov theorem: if $W_t^\theta = \theta t + W_t$, then the probability measure $\mathbb{P}^\theta$ defined with

$$Z_\theta = e^{-\theta W_T - \frac{1}{2}\theta^2 T}$$

is such that $\{W_t^\theta, 0 \leq t \leq T\}$ is a $\mathbb{P}^\theta$-standard Brownian motion.

with respect to	$\mathbb{P}$	$\mathbb{P}^\theta$
W is a	standard BM	linear BM
W^θ is a	linear BM	standard BM

- In the BSM model, if we choose $\theta = (\mu - r)/\sigma$, then $\mathbb{Q} = \mathbb{P}^\theta$ is an EMM.
- With respect to $\mathbb{Q}$:
  - $\{S_t/B_t, 0 \leq t \leq T\}$ is a martingale;
  - the (risk-neutral) distribution of S_t is given by

$$S_t \overset{\mathbb{Q}}{\sim} \mathcal{LN}\left(\ln(S_0) + \left(r - \frac{\sigma^2}{2}\right)t, \sigma^2 t\right).$$

- Risk-neutral pricing formula: for any European payoff V_T, simple or exotic, we have:

$$V_t = \mathrm{e}^{-r(T-t)}\mathbb{E}^{\mathbb{Q}}\left[V_T | S_u : 0 \leq u \leq t\right]$$

17.4 Exercises

17.1 Verify that the following functions are solutions to the Black-Scholes PDE:
(a) $V(t, x) = 0$;
(b) $V(t, x) = x$;
(c) $V(t, x) = \mathrm{e}^{rt}$;
(d) $V(t, x) = x - K\mathrm{e}^{-r(T-t)}$;
(e) $P(t, x) = \mathrm{e}^{-r(T-t)}KN(-\mathrm{d}_2(t, x)) - xN(-\mathrm{d}_1(t, x))$.
In each case, identify the corresponding financial derivative.

17.2 Use the Feynman-Kač formula to find the solution to the following PDE problem:

$$\begin{cases} \frac{\partial F}{\partial t}(t, x) + \mu x \frac{\partial F}{\partial x}(t, x) + \frac{1}{2}\sigma^2 x^2 \frac{\partial^2 F}{\partial x^2}(t, x) = 0, \\ F(T, x) = \ln\left(x^2\right), \end{cases}$$

where μ, σ are given constants.

17.3 Let us consider a financial market such that $S_t = W_t$ (a SBM) and $B_t = 1$ for $t \geq 0$. Determine whether the trading strategy (Δ_t, Θ_t) given by $\Delta_t = 2W_t$ and $\Theta_t = -t - W_t^2$ is self-financing.

17.4 Verify that Z_α in example 17.2.4 is such that $Z_\alpha \geq 0$ and $\mathbb{E}[Z_\alpha] = 1$.

17.5 Verify that Z_β in example 17.2.5 is such that $Z_\beta \geq 0$ and $\mathbb{E}[Z_\beta] = 1$.

17.6 Verify that if $X \overset{\mathbb{P}}{\sim} \mathcal{N}(\alpha, 1)$ and if the change of measure is given by

$$Z_{-\alpha} = \mathrm{e}^{-\alpha X + (1/2)\alpha^2} = \mathrm{e}^{-\alpha(X-\alpha) - (1/2)\alpha^2},$$

then $X \overset{\mathbb{P}_{-\alpha}}{\sim} \mathcal{N}(0, 1)$.

17.7 Find the explicit change of measure behind the conditional probability associated with a fixed event F such that $\mathbb{P}(F) > 0$: $E \mapsto \mathbb{P}(E \mid F)$.

17.8 Suppose that in a financial market, three assets are traded: a risk-free asset such that

$$dB_t = rB_t dt$$

and two risky assets, such that

$$\begin{aligned} dS_t^{(1)} &= \mu_1 S_t^{(1)} dt + \sigma_1 S_t^{(1)} dW_t \\ dS_t^{(2)} &= \mu_2 S_t^{(2)} dt + \sigma_2 S_t^{(2)} dW_t. \end{aligned}$$

Note that both risky assets depend on the same Brownian motion $W = \{W_t, t \geq 0\}$. We would like to determine the conditions such that this market is free of arbitrage opportunities. You will find that

$$\frac{\mu_1 - r}{\sigma_1} = \frac{\mu_2 - r}{\sigma_2}$$

must hold to prevent arbitrage opportunities. This is known as the *Sharpe ratio* or the *market price of risk*. There are two approaches to reach such a conclusion:

(a) Build a self-financing strategy with both risky assets that replicates the risk-free asset. Show that the Sharpe ratio must hold to prevent arbitrage opportunities.

(b) Invoke the Fundamental Theorem of Asset Pricing and let $W_t^{\mathbb{P}} = W_t^{\mathbb{Q}} - \lambda t$ to show that λ must correspond to the Sharpe ratio to prevent arbitrage opportunities.

18

Applications and extensions of the Black-Scholes formula

In Chapter 16 we derived the so-called Black-Scholes formula, which gives compact and instructive expressions for the price of a call and of a put option in the Black-Scholes-Merton model. It turns out it is also possible to further apply and extend the Black-Scholes formula in various contexts, including financial products sold by insurance companies.

The aim of this chapter is mostly to analyze the pricing of options and other derivatives such as options on dividend-paying assets, currency options and futures options, but also insurance products such as investment guarantees, equity-indexed annuities and variable annuities, as well as exotic options (Asian, lookback and barrier options). In most of these cases, we can easily find a Black-Scholes-like formula for the no-arbitrage price of these contracts. The specific objectives of this chapter are to:

- use the Black-Scholes formula to price options on dividend-paying assets, currency options, futures options and exchange options;
- use financial engineering arguments and the Black-Scholes formula to obtain pricing formulas for investment guarantees, equity-indexed annuities and variable annuities;
- understand how to compute the break-even participation rate (annual fee) for common equity-linked insurance and annuities;
- derive the distribution of the discrete geometric mean of lognormally distributed random variables;
- understand why it might be difficult for various path-dependent options to obtain closed-form solutions;
- compute the price of average price Asian options (with a geometric mean), floating-strike lookback options and barrier options.

18.1 Options on other assets

In this section, we will derive analytic expressions for the price of options on dividend-paying stocks, currencies and futures in the BSM framework. We then conclude by analyzing exchange options when both assets are represented by a (joint, correlated) BSM model.

18.1.1 Options on dividend-paying stocks

We want to determine the no-arbitrage price of call and put options whenever a continuous dividend yield γ is paid by the underlying stock. Following Chapter 16, we will begin with a binomial tree defined over a time interval $[0, T]$ that we divide into n sub-intervals of length T/n. The resulting tree is a n-step binomial tree in which the risky asset price, at any intermediate time k, is given by $S^{(n)}_{kT/n}$.

Actuarial Finance: Derivatives, Quantitative Models and Risk Management, First Edition.
Mathieu Boudreault and Jean-François Renaud.

The main difference with Chapter 16 is that the tree now represents the evolution of the *ex-dividend* price, i.e. the price at which we can buy the asset on the market and upon which the option payoff is computed.

As usual, at maturity time T, i.e. when $k = n$, the (ex-dividend) asset price can be written as

$$S_T^{(n)} = S_0 u_n^{I_n} d_n^{n-I_n},$$

where the number of up-moves $I_n \overset{\mathbb{P}}{\sim} \text{Bin}(n, p_n)$ with

$$p_n = \frac{e^{(\mu-\gamma)(T/n)} - d_n}{u_n - d_n},$$

with respect to the actuarial probability measure $\mathbb{P}$. Using the specifications of the Jarrow-Rudd tree, we have

$$u_n = \exp\left((\mu - \gamma - \sigma^2/2)\frac{T}{n} + \sigma\sqrt{\frac{T}{n}} \right),$$

$$d_n = \exp\left((\mu - \gamma - \sigma^2/2)\frac{T}{n} - \sigma\sqrt{\frac{T}{n}} \right).$$

In contrast to Chapter 16, the mean rate of return on the stock is now $\mu - \gamma$ rather than just μ. Indeed, for $k = 0, 1, \ldots, n-1$,

$$\mathbb{E}\left[S_{(k+1)T/n}^{(n)} \middle| S_{kT/n}^{(n)} \right] = S_{kT/n}^{(n)}(u_n p_n + d_n(1 - p_n)),$$

so simplifications yield

$$\mathbb{E}\left[S_{(k+1)T/n}^{(n)} \middle| S_{kT/n}^{(n)} \right] = S_{kT/n}^{(n)} e^{(\mu-\gamma)T/n}.$$

A dividend is a cash outflow for the corporation and it slows down the growth of the ex-dividend stock price. However, a shareholder being entitled to these dividends, an investment of $S_{kT/n}^{(n)}$ at time kT/n yields an expected return of $\mu - \gamma$ (from capital gains) plus a dividend yield of γ. Overall, the total expected return is μ.

As in Section 16.2.2, if the number of time steps goes to infinity ($n \to \infty$), then

$$\ln\left(S_T^{(n)}/S_0 \right) \underset{n\to\infty}{\longrightarrow} \ln(S_T/S_0) \overset{\mathbb{P}}{\sim} \mathcal{N}((\mu - \gamma - \sigma^2/2)T, \sigma^2 T). \tag{18.1.1}$$

Note that this distribution is essentially the same as the one obtained in Section 16.2.2, i.e. when we had $\gamma = 0$, except that now, since dividends are paid, the parameter μ is replaced by $\mu - \gamma$.

If we want to find the no-arbitrage price of a derivative, we have to determine the *risk-neutral* dynamics of the tree. The risk-neutral version of the Jarrow-Rudd tree, subject to dividend payments, is then based on the following parameters:

$$q_n = \frac{e^{(r-\gamma)(T/n)} - d_n}{u_n - d_n},$$

$$u_n = \exp\left((r - \gamma - \sigma^2/2)\frac{T}{n} + \sigma\sqrt{\frac{T}{n}} \right),$$

$$d_n = \exp\left((r - \gamma - \sigma^2/2)\frac{T}{n} - \sigma\sqrt{\frac{T}{n}} \right).$$

Thus, if $n \to \infty$, at the limit we obtain the risk-neutral version of the distribution in (18.1.1):

$$\ln(S_T/S_0) \overset{\mathbb{Q}}{\sim} \mathcal{N}((r-\gamma-\sigma^2/2)T, \sigma^2 T).$$

To find the Black-Scholes formula with a continuous dividend yield γ, it suffices to compute, for $0 \leq t < T$,

$$C_t = \mathrm{e}^{-r(T-t)}\mathbb{E}^{\mathbb{Q}}[(S_T-K)_+|S_t],$$

which is again the discounted stop-loss transform of a lognormal distribution. In this case, we must use the risk-neutral conditional distribution of S_T (given S_t) which is

$$S_T \mid S_t \overset{\mathbb{Q}}{\sim} \mathcal{LN}(\ln(S_t)+(r-\gamma-\sigma^2/2)(T-t), \sigma^2(T-t)).$$

In conclusion, the Black-Scholes formula for a call option on a dividend-paying stock is, for all $0 \leq t < T$,

$$C_t = S_t\mathrm{e}^{-\gamma(T-t)}N(d_1(t,S_t;\gamma)) - \mathrm{e}^{-r(T-t)}KN(d_2(t,S_t;\gamma)) \tag{18.1.2}$$

where

$$d_1(t,S_t;\gamma) = \frac{\ln(S_t/K)+(r-\gamma+\sigma^2/2)(T-t)}{\sigma\sqrt{T-t}},$$

$$d_2(t,S_t;\gamma) = d_1(t,S_t;\gamma) - \sigma\sqrt{T-t}.$$

Note that if $\gamma = 0$, then the *classical* Black-Scholes formula for a call option, as given in equation (16.3.3), is recovered.

Example 18.1.1 *Call option price with a dividend yield*
The initial price of a share of a dividend-paying stock is \$34 with a continuous dividend yield of 1.25% per year. You know the stock price has a volatility of 27% and the risk-free rate is 4%. Compute the price of a 32-strike 6-month call option.

First, we compute $d_1(0,S_0;\gamma)$ and $d_2(0,S_0;\gamma)$ as given above: since $\gamma = 0.0125$, we have

$$d_1 = \frac{\ln(34/32)+(0.04-0.0125+(0.27)^2/2)\times 0.5}{0.027\sqrt{0.5}} = 0.485020893,$$

$$d_2 = d_1 - 0.027\sqrt{0.5} = 0.294102062.$$

Therefore, the formula for the call option price in (18.1.2) yields

$$\begin{aligned} C_0 &= 34\mathrm{e}^{-0.0125\times 0.5}\times N(0.485020893) - \mathrm{e}^{-0.04\times 0.5}32\times N(0.294102062) \\ &= 3.873384805. \end{aligned}$$

■

From the put-call parity relationship (when the underlying asset is paying continuous dividends) given in equation (6.4.6), we can deduce that the time-t value of the corresponding put option is

$$P_t = C_t - S_t\mathrm{e}^{-\gamma(T-t)} + K\mathrm{e}^{-r(T-t)}.$$

Using the Black-Scholes formula in (18.1.2), we get

$$P_t = \mathrm{e}^{-r(T-t)}KN(-d_2(t,S_t;\gamma)) - S_t\mathrm{e}^{-\gamma(T-t)}N(-d_1(t,S_t;\gamma)). \tag{18.1.3}$$

Example 18.1.2 *Put option price with a dividend yield*
In the context of example 18.1.1, compute the value of the corresponding put option.
Using the Black-Scholes formula in (18.1.3), we find that $P_0 = 1.45157967$. ■

As in Section 16.6.2, the replicating portfolio of a call option written on a dividend-paying stock is already embedded in the Black-Scholes formula. Indeed, by inspection of the Black-Scholes formula given in (18.1.2), we see that the replicating portfolio is given by

$$\begin{cases} \Delta_t = e^{-\gamma(T-t)} N(d_1(t, S_t; \gamma)), \\ \Theta_t = -e^{-rT} K N(d_2(t, S_t; \gamma)), \end{cases} \tag{18.1.4}$$

where Δ_t is the number of units of the stock and where Θ_t is the number of units of the risk-free asset, both held at time t.

This is very similar to the replicating portfolio of a call option on a non-dividend-paying stock. The only difference with the case $\gamma = 0$ is the magnitude of the number of shares we should hold at time t: it is now $e^{-\gamma(T-t)} N(d_1(t, S_t; \gamma))$ shares instead of $N(d_1(t, S_t; 0))$ shares. We can see that the number of shares of the stock needed in the replicating portfolio is smaller than in the case $\gamma = 0$. This is because dividends are reinvested in the stock. It also ensures the replicating strategy is self-financing; otherwise there would be a cash outflow.

Finally, it is important to note that in the BSM model, replication of options written on dividend-paying stocks requires continuous trading.

Example 18.1.3 *Replicating a call on a dividend-paying stock*
Consider again the setting of example 18.1.1. Initially, the replicating portfolio, as given in equation (18.1.4), is

$$\Delta_0 = e^{-0.0125 \times 0.5} \times 0.686169237 = 0.681894053,$$
$$\Theta_0 = -e^{-0.04 \times 0.5} 32 \times 0.615660042 = -19.31101301.$$

In other words, at time 0, we buy 0.681894053 share of the stock at a (ex-dividend) price of \$34 and we borrow \$19.31101301.

We hold on to these quantities for the next 3 months, except for the reinvestment of dividends. This means that in 3 months we will hold $0.681894053 \times e^{0.0125 \times 0.25}$ shares of the stock.

Now, consider the following scenario: 3 months later, the stock price has increased to \$36. Given that dividends have been reinvested in the stock, the investment in the stock has grown from $0.681894053 \times 34 = 23.1843978$ units to

$$0.681894053 \times e^{0.0125 \times 0.25} \times 36 = 24.62501898.$$

The loan has grown with interest to

$$19.31101301 \times e^{0.04 \times 0.25} = 19.50509192.$$

The portfolio value at time $t = 0.25$, i.e. 3 months later, but prior to rebalancing, is thus

$$24.62501898 - 19.50509192 = 5.11992706.$$

This is in contrast with the call option price, which is in this scenario given by $C_{0.25} = 4.643364432$. Recall that the difference is the *hedging error*, which is a profit in this scenario. ■

The replicating portfolio of a put option can also be obtained by identification. Inspecting the Black-Scholes formula in (18.1.3), we deduce that the replicating portfolio is, at any time $0 \leq t < T$, given by

$$\begin{cases} \Delta_t = -\mathrm{e}^{-\gamma(T-t)} N(-d_1(t, S_t; \gamma)), \\ \Theta_t = \mathrm{e}^{-rT} K N(-d_2(t, S_t; \gamma)). \end{cases} \tag{18.1.5}$$

18.1.2 Currency options

We now analyze currency options which are options to buy or sell a foreign currency at a predetermined price. Recall from Section 12.3 that the notation is such that S_t corresponds to the price (expressed in the local currency) of one unit of the foreign currency.

Holding one unit of the foreign currency earns interest at the foreign risk-free rate r_f. As seen previously in Section 12.3, a currency option can thus be viewed as an option on a dividend-paying stock with a dividend yield of r_f. Therefore, the Black-Scholes formula for currency options is equivalent to equation (18.1.2) with $\gamma = r_f$, i.e.

$$C_t = S_t \mathrm{e}^{-r_f(T-t)} N(d_1(t, S_t; r_f)) - \mathrm{e}^{-r(T-t)} K N(d_2(t, S_t; r_f)),$$
$$P_t = \mathrm{e}^{-r(T-t)} K N(-d_2(t, S_t; r_f)) - S_t \mathrm{e}^{-r_f(T-t)} N(-d_1(t, S_t; r_f)),$$

where $d_1(t, S_t; r_f)$ and $d_2(t, S_t; r_f)$ are as in equation (18.1.2) with r_f taking the place of γ in the expression.

Example 18.1.4 *Currency call option*
You are a U.S. investor wishing to buy 1,000 Euros 3 months from now. Therefore, you enter into 1,000 3-month currency call options with a strike of 1.1 USD/EUR. You know that the risk-free rate in the U.S. is 2% whereas it is 3% in Europe. If the volatility on the exchange rate is 20% and the current exchange rate is 1.05 USD/EUR, calculate the amount required to buy those 1,000 call options.

We have $r = 0.02, r_f = 0.03, \sigma = 0.2, S_0 = 1.05, K = 1.1$. As of right now, buying one Euro costs $S_0 = 1.05$ USD and we want to fix the buying price at $K = 1.10$ USD. Using the above Black-Scholes formula, we find that $C_0 = 0.02143608$ per option and therefore the 1,000 options cost 21.43608. ■

Replicating currency call options is similar to replicating options on dividend-paying stocks. Because the foreign currency earns interest at the foreign risk-free rate r_f, the quantity of foreign currencies needed to replicate the payoff has to be smaller to account for this additional income.

18.1.3 Futures options and Black's formula

Recall from Chapter 12 that a futures option is an option to buy or sell a futures contract at a predetermined price (which is different from the delivery/futures/forward price). In this section we will analyze options on futures (or futures options) in the BSM model, i.e. if the underlying asset price S_t follows the BSM dynamic. Note that in a binomial environment, we have modeled the futures price with a tree instead of the underlying asset price.

Assume that $T^\star$ is the maturity of a futures contract and T is the maturity of the futures option, with $T^\star \geq T$. We know that, at time t, the $T^\star$-forward price of an asset is simply

$$F_t^{T^\star} = S_t \mathrm{e}^{r(T^\star - t)}.$$

First of all, note that

$$\frac{F_T^{T\star}}{F_t^{T\star}} = \frac{S_T e^{r(T^\star - T)}}{S_t e^{r(T^\star - t)}} = \frac{S_T}{S_t} e^{-r(T-t)}$$

regardless of how S_T and S_t are distributed. Since

$$\ln(S_T/S_t) \mid S_t \overset{\mathbb{Q}}{\sim} \mathcal{N}((r - \sigma^2/2)(T-t), \sigma^2(T-t))$$

for a non-dividend-paying stock, then

$$\ln\left(\frac{F_T^{T\star}}{F_t^{T\star}}\right) = \ln(S_T/S_t) - r(T-t)$$

follows also a normal distribution but with different parameters:

$$\ln\left(\frac{F_T^{T\star}}{F_t^{T\star}}\right) \overset{\mathbb{Q}}{\sim} \mathcal{N}\left(-\frac{\sigma^2}{2}(T-t), \sigma^2(T-t)\right). \tag{18.1.6}$$

The price of a futures option is then obtained by computing the *usual* conditional expectation

$$C_t = e^{-r(T-t)} \mathbb{E}^{\mathbb{Q}}\left[\left(F_T^{T\star} - K\right)_+ \middle| F_t^{T\star}\right]$$

using the risk-neutral distribution of $\ln(F_T^{T\star}/F_t^{T\star})$ given in equation (18.1.6). We can either evaluate the expectation using the stop-loss transform or simply let $\gamma = r$ in equation (18.1.2) to obtain

$$C_t = e^{-r(T-t)} \left(F_t^{T\star} N(d_1) - KN(d_2)\right) \tag{18.1.7}$$

where

$$d_1 = \frac{\ln\left(F_t^{T\star}/K\right) + \frac{1}{2}\sigma^2(T-t)}{\sigma\sqrt{T-t}},$$

$$d_2 = d_1 - \sigma\sqrt{T-t}.$$

Equation (18.1.7) is also known as **Black's formula** for a futures call option. A similar formula exists for a futures put option, i.e.

$$P_t = e^{-r(T-t)} \left(KN(-d_2) - F_t^{T\star} N(-d_1)\right).$$

Example 18.1.5 *Futures options*
Futures on the S&P 500 index currently trade for 1200 whereas the volatility on the latter index is 18%. If the risk-free rate is 3%, find the price of 1-year at-the-money call and put options on that futures contract expiring after the options.

We have $F_0^{T\star} = 1200 = K, \sigma = 0.18, r = 0.03$ and $T = 1$. Therefore, we obtain

$$d_1 = \frac{\ln(1) + 0.5 \times 0.18^2}{0.18} = 0.5 \times 0.18 = 0.09$$

$$d_2 = d_1 - \sigma = 0.09 - 0.18 = -0.09$$

$$N(d_1) = 0.535856393$$

$$N(d_2) = N(-d_1) = 0.464143607.$$

Because the options are at the money, it is easy to remark that call and put options prices are exactly the same. Using Black's formula, we see that

$$C_0 = P_0 = e^{-0.03} \times 1200 \times (0.535856393 - 0.464143607) = 83.51202345. \qquad \blacksquare$$

18.1.4 Exchange options

Recall from Section 7.1.4, that an exchange option gives its holder the right to exchange an asset for another asset. Again, let us denote by $S^{(1)}$ and $S^{(2)}$ two risky assets whose price dynamics $\{S_t^{(1)}, t \geq 0\}$ and $\{S_t^{(2)}, t \geq 0\}$ are given by two *dependent* geometric Brownian motions. More precisely, the (real-world, actuarial) distributions of these prices are such that

$$\ln\left(\frac{S_T^{(1)}}{S_t^{(1)}}\right) \overset{\mathbb{P}}{\sim} \mathcal{N}\left(\left(\mu_1 - \gamma_1 - \frac{1}{2}\sigma_1^2\right)(T-t), \sigma_1^2(T-t)\right)$$

and

$$\ln\left(\frac{S_T^{(2)}}{S_t^{(2)}}\right) \overset{\mathbb{P}}{\sim} \mathcal{N}\left(\left(\mu_2 - \gamma_2 - \frac{1}{2}\sigma_2^2\right)(T-t), \sigma_2^2(T-t)\right),$$

where, for the i-th asset, μ_i and σ_i are the drift and diffusion coefficients and γ_i is the dividend yield, with $i = 1, 2$.

In a BSM environment, we assume that the random vector $\left(\ln\left(\frac{S_T^{(1)}}{S_t^{(1)}}\right), \ln\left(\frac{S_T^{(2)}}{S_t^{(2)}}\right)\right)$ has a bivariate normal distribution. The dependence structure (between these asset prices) is given through the correlation between the corresponding (normally distributed) log-returns, i.e.

$$\rho = \mathrm{Corr}\left(\ln\left(\frac{S_T^{(1)}}{S_t^{(1)}}\right), \ln\left(\frac{S_T^{(2)}}{S_t^{(2)}}\right)\right).$$

To compute the no-arbitrage price of an exchange option, we know we will need to find the *risk-neutral* dynamics of these assets, that is, the risk-neutral distribution of the random vector $\left(\ln\left(\frac{S_T^{(1)}}{S_t^{(1)}}\right), \ln\left(\frac{S_T^{(2)}}{S_t^{(2)}}\right)\right)$. It turns out that the log-returns have very similar real-world and risk-neutral distributions (and correlation), the only difference being that μ_1 and μ_2 are replaced by r in the latter. More precisely, the risk-neutral distributions of $S^{(1)}$ and $S^{(2)}$ are such that

$$\ln\left(\frac{S_T^{(i)}}{S_t^{(i)}}\right) \overset{\mathbb{Q}}{\sim} \mathcal{N}\left(\left(r - \gamma_i - \frac{1}{2}\sigma_i^2\right)(T-t), \sigma_i^2(T-t)\right),$$

where $i = 1, 2$, with correlation ρ as well.

From Section 7.1.4, the payoff of an option to obtain a share of $S^{(1)}$ in exchange for a share of $S^{(2)}$ at maturity T, is given by

$$\left(S_T^{(1)} - S_T^{(2)}\right)_+ = \max\left(S_T^{(1)} - S_T^{(2)}, 0\right).$$

Therefore, to derive the initial price of this exchange option, we need to compute

$$C_0^{\mathrm{Ex}} = e^{-rT}\mathbb{E}^{\mathbb{Q}}\left[\left(S_T^{(1)} - S_T^{(2)}\right)_+\right]$$

and the price at any time t is

$$C_t^{\text{Ex}} = e^{-r(T-t)}\mathbb{E}^{\mathbb{Q}}\left[\left(S_T^{(1)} - S_T^{(2)}\right)_+ \middle| S_t^{(1)}, S_t^{(2)}\right].$$

However, the risk-neutral distribution of $S_T^{(1)} - S_T^{(2)}$, being the difference of two dependent and lognormally distributed random variables, is not easy to find. But if we condition on $S_T^{(2)}$, then we can write

$$C_0^{\text{Ex}} = e^{-rT}\mathbb{E}^{\mathbb{Q}}\left[\mathbb{E}^{\mathbb{Q}}\left[\left(S_T^{(1)} - S_T^{(2)}\right)_+ \middle| S_T^{(2)}\right]\right],$$

where the inside conditional expectation is easy to compute. Indeed, if we consider $S_T^{(2)}$ or equivalently $\ln(S_T^{(2)}/S_0)$ as a *known value*, then using a property of the bivariate normal distribution,[1] the risk-neutral conditional expectation

$$\mathbb{E}^{\mathbb{Q}}\left[\left(S_T^{(1)} - S_T^{(2)}\right)_+ \middle| S_T^{(2)}\right]$$

can be computed using the Black-Scholes formula (or the stop-loss formula for lognormally distributed random variables), with the strike price replaced by $S_T^{(2)}$. The last (and tedious) step consists of taking the risk-neutral expectation of this *modified Black-Scholes formula*.

Finally, the price of this exchange option at any time t is given by

$$C_t^{\text{Ex}} = S_t^{(1)}e^{-\gamma_1(T-t)}N\left(d_1\left(S_t^{(1)}, S_t^{(2)}\right)\right) - S_t^{(2)}e^{-\gamma_2(T-t)}N\left(d_2\left(S_t^{(1)}, S_t^{(2)}\right)\right) \tag{18.1.8}$$

where

$$d_1\left(S_t^{(1)}, S_t^{(2)}\right) = \frac{\ln\left(S_t^{(1)}/S_t^{(2)}\right) + \left(\gamma_2 - \gamma_1 + \frac{\sigma^2}{2}\right)(T-t)}{\sigma\sqrt{T-t}},$$

$$d_2\left(S_t^{(1)}, S_t^{(2)}\right) = d_1\left(S_t^{(1)}, S_t^{(2)}\right) - \sigma\sqrt{T-t},$$

and where

$$\sigma^2 = \frac{\sigma_1^2 + \sigma_2^2 - 2\rho\sigma_1\sigma_2}{T-t}.$$

It is interesting to note that the interest rate r does not appear in this formula.

Example 18.1.6 Consider a 1-year exchange option allowing to trade one share of ABC inc. for one share of XYZ inc. Both stocks currently sell for $S_0^{(1)} = S_0^{(2)} = 100$. Let us find the initial price of this exchange option in a BSM environment with the following parameters: $\mu_1 = 0.07$, $\mu_2 = 0.06$, $\sigma_1 = \sigma_2 = 0.2$, $\gamma_1 = \gamma_2 = 0$ and $\rho = -0.1$.

In this case, the formula in (18.1.8) becomes

$$C_0^{\text{Ex}} = S_0^{(1)}N(d_1) - S_0^{(2)}N(d_2)$$

with

$$\sigma^2 = \frac{(0.2)^2 + (0.2)^2 - 2(-0.1)(0.2)^2}{1} = 0.088$$

1 If (X, Y) follows a bivariate normal distribution, then $X \mid Y = y$ also follows a normal distribution with known parameters.

and

$$d_1 = \frac{\ln(S_0^{(1)}/S_0^{(2)}) + \left(\frac{\sigma^2}{2}\right)T}{\sigma\sqrt{T}} = \frac{0 + \frac{(0.088)^2}{2}}{0.088} = 0.044,$$

$$d_2 = d_1 - \sigma\sqrt{T} = 0.044 - 0.088 = -0.044.$$

In conclusion,

$$\begin{aligned} C_0^{\text{Ex}} &= 100(N(0.044) - N(-0.044)) \\ &= 100(0.517547798 - (1 - 0.517547798)) = 3.509559613. \end{aligned}$$

Using the parity relationship seen in Section 7.1.4, we can also compute the exchange option allowing to get one share of $S^{(2)}$ in exchange for one share of $S^{(1)}$:

$$P_0^{\text{Ex}} = C_0^{\text{Ex}} + S_0^{(2)} - S_0^{(1)} = 3.509559613 + 100 - 100 = 3.509559613.$$ ■

Suppose now that $S_T^{(2)}$ is an asset paying K with certainty at maturity T. It is easy to show that the no-arbitrage price of the exchange option to buy $S^{(1)}$ (with a dividend yield of $\gamma_1 = \gamma$) in exchange for $S^{(2)}$ is the Black-Scholes formula for a call option on a dividend-paying asset.

If $S_T^{(2)} = K$ (in all scenarios), then $\sigma_2 = 0$, $\gamma_2 = 0$ and $S_t^{(2)} = Ke^{-r(T-t)}$ at any time t. Using the pricing formula in equation (18.1.8), we find that

$$C_t^{\text{Ex}} = S_t e^{-\gamma(T-t)} N(d_1) - Ke^{-r(T-t)} N(d_2)$$

with

$$d_1 = \frac{\ln(S_t/K) + \left(r - \gamma + \frac{1}{2}\sigma_1^2\right)(T-t)}{\sigma_1\sqrt{T-t}},$$

$$d_2 = d_1 - \sigma_1\sqrt{T-t},$$

which is the classical Black-Scholes formula for a call option.

18.2 Equity-linked insurance and annuities

We would like to find the price of various equity-linked insurance and annuities (ELIAs), as introduced in Chapter 8, when the underlying follows a BSM dynamic. It is important to remember that finding the no-arbitrage price of a derivative assumes implicitly that agents are *rational*, i.e. they trade to optimize their gains and thus are ready to exploit any arbitrage opportunity if any. Does this apply to ELIAs typically sold to common people? Not so much.

However, no-arbitrage pricing matters as long as ELIAs are sold by many insurance companies in a competitive market. Indeed, from economic arguments, competition will drive prices down to the marginal cost. For ELIAs, the marginal cost represents the cost of the replicating portfolio and hence the corresponding no-arbitrage price.

Therefore, we will use Black-Scholes formulas for the pricing of ELIAs, when applicable. Note that when this is not possible, one can use simulation methods (see Chapter 19 for more details). We begin by analyzing investment guarantees, then we move on to equity-indexed annuities (EIAs) and variable annuities (VAs).

18.2.1 Investment guarantees

As seen in Chapter 8, an investment guarantee with payoff $V_T^{IG} = \max(S_T, K)$ plays an important role in life insurance as it forms the basis of many ELIA policies. We know from Chapter 6 that the payoff of an investment guarantee can be easily replicated with simple options:

1) a call option and a risk-free investment, i.e.

$$V_T^{IG} = \max(S_T, K) = K + (S_T - K)_+;$$

2) a put option and a share of the underlying asset, i.e.

$$V_T^{IG} = \max(S_T, K) = S_T + (K - S_T)_+.$$

Consequently, we can obtain the time-t value of an investment guarantee in two equivalent ways. Indeed:

1) First, note that the time-t value of K *guaranteed at time* T is equal to $e^{-r(T-t)}K$. Then, if we use the Black-Scholes formula for a call as given in (18.1.2), we obtain

$$\begin{aligned} V_t^{IG} &= e^{-r(T-t)}K + C_t \\ &= e^{-r(T-t)}K + \left[e^{-\gamma(T-t)}S_t N(d_1(t, S_t; \gamma)) - e^{-r(T-t)}KN(d_2(t, S_t; \gamma))\right] \\ &= e^{-\gamma(T-t)}S_t N(d_1(t, S_t; \gamma)) + e^{-r(T-t)}K(1 - N(d_2(t, S_t; \gamma))) \\ &= e^{-\gamma(T-t)}S_t N(d_1(t, S_t; \gamma)) + e^{-r(T-t)}KN(-d_2(t, S_t; \gamma)). \end{aligned} \tag{18.2.1}$$

2) Second, note that the time-t value of a derivative paying S_T is equal to $e^{-\gamma(T-t)}S_t$. Indeed, buying $e^{-\gamma(T-t)}$ unit of the asset at time t, for a price of S_t, and then reinvesting the dividends in the asset until time T yields a final value of S_T. Then, if we use the Black-Scholes formula for a put as given in (18.1.3), we obtain

$$\begin{aligned} V_t^{IG} &= e^{-\gamma(T-t)}S_t + P_t \\ &= e^{-\gamma(T-t)}S_t + \left[e^{-r(T-t)}KN(-d_2(t, S_t; \gamma)) - e^{-\gamma(T-t)}S_t N(-d_1(t, S_t; \gamma))\right] \\ &= e^{-r(T-t)}KN(-d_2(t, S_t; \gamma)) + S_t e^{-\gamma(T-t)}(1 - N(-d_1(t, S_t; \gamma))) \\ &= e^{-r(T-t)}KN(-d_2(t, S_t; \gamma) + S_t e^{-\gamma(T-t)}N(d_1(t, S_t; \gamma)) \\ &= e^{-\gamma(T-t)}S_t N(d_1(t, S_t; \gamma)) + e^{-r(T-t)}KN(-d_2(t, S_t; \gamma)). \end{aligned} \tag{18.2.2}$$

As expected, the pricing formulas in (18.2.1) and (18.2.2) are equal.

Example 18.2.1 *Investment guarantee in a BSM environment*
In the BSM model, you are given:

- $S_0 = 10$;
- $\sigma = 0.25$;
- $r = 0.03$;
- $\gamma = 0$.

How much should you pay at time 0 for a 5-year investment guarantee where 80% of the initial stock price is guaranteed? Also, analyze this product from the point of view of the insurance company.

In this case, since we have $T = 5$ and $K = 0.8 \times S_0 = 8$, we get

$$\begin{aligned} d_1 &= 0.947007974 \\ d_2 &= 0.38799098 \\ N(d_1) &= 0.828182642 \\ N(-d_2) &= 0.349011354. \end{aligned}$$

Using the pricing formula in (18.2.1), we find that

$$V_0^{\mathrm{IG}} = 10 \times 0.828182642 + 8 \times \mathrm{e}^{-5\times 0.03} \times 0.349011354 = 10.68500127.$$

This means that an insured who invests (at inception) \$10,685 in this product will receive (at maturity) the largest value between \$8,000, the guaranteed amount and the accumulated value of \$10,000 invested at time 0 in the underlying asset, i.e. 1,000 shares of S bought at time 0. Despite the name *investment guarantee*, there is still a risk of losing some money, i.e. the final payout could be less than the initial investment (not even taking into account the time value of money). To be entitled to the (upside) benefits of the risky investment, the insured/investor has to bear some (downside) risk; there is no arbitrage in this market.

On the other side of the transaction, the insurer will receive \$10,685 (at inception), which corresponds to the cost for setting up a replicating portfolio, i.e. buying 100 shares of the stock and 100 put options (with strike price 8 and maturity 5 years). If these put options are not available on the market, the insurer can replicate each option with the Black-Scholes replicating strategy seen in Chapter 16. The insurer can also take this \$10,685 to buy call options and lend at the risk-free rate, which is the other replicating strategy for an investment guarantee. ■

18.2.2 Equity-indexed annuities

A variety of indexing schemes for EIAs has been presented in Chapter 8. We will analyze two of them in a BSM framework:

1) point-to-point (PTP);
2) compound periodic ratchet (CPR).

For these two schemes, we will obtain closed-form expressions for the no-arbitrage value of their benefits. However, for most indexing schemes, simulation methods are needed (see Chapter 19).

18.2.2.1 Point-to-point

In Chapter 8, we saw that the maturity benefit (payoff) of a PTP EIA is given by

$$\max(I(1 + \beta R_T), G_T)$$

where I is the initial investment, where $0 < \beta < 1$ is the participation rate in the cumulative return over $[0, T]$, namely $R_T = S_T/S_0 - 1$, and where G_T is the guaranteed amount at maturity (usually a fraction of the initial investment I). As the initial investment I can be factorized in the above expression, to simplify the presentation, we will now work *per dollar invested*, i.e. we will assume that $I = 1$. Therefore, the *normalized* maturity benefit, or per-dollar benefit, is now given by

$$V_T^{\mathrm{PTP}} = \max\left(1 + \beta\left(\frac{S_T}{S_0} - 1\right), G_T\right).$$

Clearly, we can write

$$V_T^{\text{PTP}} = \max\left((1-\beta) + \beta\frac{S_T}{S_0}, G_T\right).$$

Thus, the payoff function is of the form $\max(a + bx, c)$ and we can use the following identity

$$\max(a + bx, c) = b \max\left(x, \frac{c-a}{b}\right) + a,$$

as seen in Chapter 6. For the PTP indexing scheme, since we have $a = 1 - \beta$, $b = \beta/S_0$ and $c = G_T$, we can also write

$$V_T^{\text{PTP}} = \frac{\beta}{S_0}\max(S_T, K) + (1-\beta),$$

if we set $K = S_0\frac{G_T-(1-\beta)}{\beta}$.

Clearly, this EIA is equivalent to a position in β/S_0 units of an investment guarantee, with guaranteed amount K just given, and a risk-free investment (bond with face value $1-\beta$). Consequently, using the pricing formula in equation (18.2.1), we deduce that the time-t value of the PTP EIA is given by

$$\begin{aligned} V_t^{\text{PTP}} &= \frac{\beta}{S_0}V_t^{\text{IG}} + (1-\beta)\mathrm{e}^{-r(T-t)} \\ &= \mathrm{e}^{-r(T-t)}(G_T - (1-\beta))N(-\mathrm{d}_2(t, S_t; \gamma)) + \frac{\beta S_t}{S_0}N(\mathrm{d}_1(t, S_t; \gamma)) + (1-\beta)\mathrm{e}^{-r(T-t)} \\ &= \mathrm{e}^{-r(T-t)}G_T N(-\mathrm{d}_2(t, S_t; \gamma)) + \frac{\beta S_t}{S_0}N(\mathrm{d}_1(t, S_t; \gamma)) + (1-\beta)\mathrm{e}^{-r(T-t)}N(\mathrm{d}_2(t, S_t; \gamma)), \end{aligned}$$

where $K = S_0\frac{G_T-(1-\beta)}{\beta}$ and where

$$\begin{aligned} d_1(t, S_t; \gamma) &= \frac{\ln(S_t/K) + (r - \gamma + \sigma^2/2)(T-t)}{\sigma\sqrt{T-t}}, \\ d_2(t, S_t; \gamma) &= d_1(t, S_t; \gamma) - \sigma\sqrt{T-t}. \end{aligned}$$

Example 18.2.2 *Point-to-point indexing method (continued)*
In Example 8.2.1, we had a PTP indexing scheme with a participation rate of 80% in the cumulative return of the S&P 500. We know that the guarantee applies to 100% of the initial investment. If the S&P 500 index is modeled by a geometric Brownian motion, as in the BSM model, with $S_0 = 2200$, $\sigma = 0.215$, $\gamma = 0.01$ and if the risk-free rate is $r = 0.03$, then let us find the initial value of this PTP EIA for an initial investment of \$1.

We have $\beta = 0.8$, $G_T = 1$ and $T = 8$. First of all, we compute

$$K = S_0\frac{G_T - (1-\beta)}{\beta} = 2200\frac{1-(1-\beta)}{\beta} = 2200.$$

Second, the initial value V_0^{IG} is given by

$$V_0^{\text{IG}} = 2345.003165.$$

Finally,

$$\begin{aligned} V_0^{\text{PTP}} &= \frac{\beta}{S_0} V_0^{\text{IG}} + (1-\beta)\mathrm{e}^{-rT} \\ &= \frac{0.8}{2200} \times 2345.003165 + 0.2 \times \mathrm{e}^{-0.03\times 8} \\ &= 1.010053996. \end{aligned}$$

Therefore, it costs \$1.01 to provide the largest value between \$1 at maturity and an investment of \$1 growing at 80% of the cumulative return on the S&P 500 (per dollar invested by the insured). Unless the insurance company charges a separate premium of \$0.01, this product is not viable, from the insurance company's point of view. The insurance company could, for example, lower the participation rate to *break even* at least. We will come back to this issue later. ■

18.2.2.2 Compound periodic ratchet

For the CPR indexing scheme, the maturity benefit is

$$\max\left(I \times \prod_{k=1}^{n} (1 + \beta \max(y_k, 0)), G_T\right),$$

where

$$y_k = \frac{S_k}{S_{k-1}} - 1, \quad k = 1, 2, \ldots, n.$$

To obtain compact formulas and to simplify the presentation, we will assume that:

- $I = 1$, to obtain normalized values or prices per dollar invested;
- $T = n$;
- $G_T \leq I$, so that the outer maximum operator disappears;
- compounding of the return y_k is annual.

Therefore, the benefit of a **compound annual ratchet (CAR)** scheme becomes

$$V_T^{\text{CAR}} = \prod_{k=1}^{n} (1 + \beta \max(y_k, 0)).$$

We are ready to derive the time-0 value of this benefit. As we are in a BSM model, the random variables y_{k_1} and y_{k_2} (for $k_1 \neq k_2$) are log-increments of a GBM. We know from Chapter 14 that relative increments of a GBM over non-overlapping intervals are independent and lognormally distributed. Therefore, the value at time 0 of the benefit from this CAR scheme is given by

$$V_0^{\text{CAR}} = \mathrm{e}^{-rT} \mathbb{E}^{\mathbb{Q}} \left[\prod_{k=1}^{n} (1 + \beta \max(y_k, 0)) \right].$$

Therefore, using the independence between the y_ks, we have

$$\begin{aligned} V_0^{\text{CAR}} &= \mathrm{e}^{-rT} \mathbb{E}^{\mathbb{Q}} \left[\prod_{k=1}^{n} (1 + \beta \max(y_k, 0)) \right] \\ &= \prod_{k=1}^{n} \mathbb{E}^{\mathbb{Q}} [\mathrm{e}^{-r} (1 + \beta \max(y_k, 0))] \\ &= \left(\mathbb{E}^{\mathbb{Q}} [\mathrm{e}^{-r} (1 + \beta \max(y_1, 0))] \right)^n, \end{aligned}$$

where in the last equality we used the fact that the y_ks have the same distribution (e.g. the distribution of y_1).

Inside the expectation, we have

$$1 + \beta \max(y_1, 0) = 1 + \beta \max\left(\frac{S_1}{S_0} - 1, 0\right) = 1 + \beta\left(\frac{S_1}{S_0} - 1\right)_+$$

and then

$$\mathbb{E}^{\mathbb{Q}}[(1 + \beta \max(y_1, 0))] = 1 + \frac{\beta}{S_0}\mathbb{E}^{\mathbb{Q}}[(S_1 - S_0)_+].$$

Defining

$$C_0^{\text{atm}} = \mathrm{e}^{-r}\mathbb{E}^{\mathbb{Q}}[(S_1 - S_0)_+],$$

which is the initial value of a one-year at-the-money call option on S, we obtain[2]

$$V_0^{\text{CAR}} = \left(\mathrm{e}^{-r} + \beta C_0^{\text{atm}}\right)^n.$$

Finally, in the BSM model, an explicit expression for C_0^{atm} is given by the Black-Scholes formula:

$$C_0^{\text{atm}} = \mathrm{e}^{-\gamma}N(d_1) - \mathrm{e}^{-r}N(d_2),$$

where

$$d_1 = \frac{r - \gamma + \sigma^2/2}{\sigma},$$
$$d_2 = d_1 - \sigma.$$

Example 18.2.3 *Value of a CAR scheme*

Find the initial value of a 10-year CAR scheme where the participation rate is 75%. Assume the CAR is written on a non-dividend paying asset whose volatility is 20%. Use a BSM model with a risk-free rate of 4% and suppose $G_T \leq I$.

We compute

$$d_1 = \frac{0.04 - 0 + 0.2^2/2}{0.2} = 0.3,$$
$$d_2 = 0.3 - 0.2 = 0.1,$$
$$N(d_1) = 0.617911422,$$
$$N(d_2) = 0.539827837,$$
$$C_0^{\text{atm}} = 0.617911422 - 0.539827837\mathrm{e}^{-0.04} = 0.099250537.$$

Therefore,

$$V_0^{\text{CAR}} = (\mathrm{e}^{-0.04} + 0.75 \times 0.099250537)^{10} = 1.035227342^{10} = 1.413700262.$$

To provide the greatest between \$10,000 (if the annual return on the stock is negative for each of the 10 years) and \$10,000 accumulated with the latter ratcheting scheme costs \$14,137. It is the replication value for the insurer. If the policyholder invests \$10,000, the insurance company is likely to suffer an important loss.

This product seems too expensive: will an investor be willing to pay \$14,137 and be guaranteed to receive at least \$10,000? The company could lower the participation rate to bring these two quantities (initial value and guaranteed amount) closer to each other, as we will see below. ■

2 No matter what the value of S_0 is, it will not appear in the final expression.

18.2.2.3 Break-even participation rate

For the products described in examples 18.2.2 and 18.2.3, the cost of replicating their maturity benefits was larger than the initial investment made by the policyholder. Unless the policyholder is willing to pay an extra premium at inception, upon which no guarantee applies (0.01 per dollar invested in example 18.2.2, or 0.41 per dollar invested in example 18.2.3), these products are not very attractive for an investor, at least at first sight.

Instead of requiring an extra upfront premium, insurance companies can adjust the participation rate[3] β so that the initial investment $I = 1$ covers the replication cost of the indexing scheme. In other words, we need to find β such that

$$V_0(\beta) = 1,$$

where V_0 is the initial value of the contract, e.g. $V_0 = V_0^{\mathrm{PTP}}$ or $V_0 = V_0^{\mathrm{CAR}}$. Note that we have emphasized the fact that V_0 depends on the value of the parameter β by writing $V_0(\beta)$. This notation means that all other quantities involved in the pricing formula stay fixed when β varies, especially the guaranteed amount. This value of β is known as the **break-even participation rate.** Most of the time, the solution to such an equation (the break-even participation rate) can only be found numerically using a *root-finding algorithm.*

However, for the CAR scheme described above, it is possible to solve this equation analytically. We want to find the value of β such that

$$V_0^{\mathrm{CAR}} = \left(\mathrm{e}^{-r} + \beta C_0^{\mathrm{atm}}\right)^n = 1.$$

We easily find that the break-even participation rate is given by

$$\beta = \frac{1 - \mathrm{e}^{-r}}{C_0^{\mathrm{atm}}}.$$

Example 18.2.4 *(High) Value of a CAR scheme (continued)*
Let us find the break-even participation rate in the CAR scheme of example 18.2.3. We easily find

$$\beta = \frac{1 - \mathrm{e}^{-r}}{C_0^{\mathrm{atm}}} = \frac{1 - \mathrm{e}^{-0.04}}{0.099250537} = 0.395066486.$$

The break-even participation rate is about 40%, which is much lower than the initial 75%.

Let us further analyze this product and its risk-return characteristics from the perspective of the policyholder. With a participation rate of 40%, the annual return on the underlying index should be more than 10% to compete with the risk-free rate of 4%. In other words, over this 10-year period, the CAR indexing scheme will generate more value than the guaranteed amount if the index constantly earns more (or much more) than 10%. In fact, the likelihood of accumulating more than $10000\mathrm{e}^{rT} = 14918.25$ by investing in this CAR scheme is about 55% (since $\mu = 0.08$).[4] This analysis highlights the fact that there is a high *hidden* cost to the minimum annual guaranteed return of 0%. ■

---

3 They could also adjust the guaranteed amount.

4 Remember the annual log-return is normally distributed with mean $\mu - 0.5\sigma^2 = 0.06$.

18.2.3 Variable annuities

It is also possible to find explicit no-arbitrage values for variable annuities with a GMMB or a GMDB rider in a BSM framework. Just like the PTP indexing scheme, we will be able to rewrite the benefits in terms of investment guarantees. As for the valuation of a GMWB or a GLWB, one would need to use simulations (see Chapter 19) or other numerical schemes.

18.2.3.1 GMMB

In a GMMB, the sub-account balance evolves with credited returns whereas fees are deducted constantly. The policyholder is typically not allowed to make withdrawals and we will assume there are no deaths occurring prior to maturity.

In the BSM framework, we assume that the sub-account balance is adjusted continuously and thus

$$A_t = A_0 \frac{S_t}{S_0} e^{-\alpha t}$$

where the annual fee α is withdrawn continuously. Over the time interval $[t, T]$, the sub-account accumulates according to

$$\frac{A_T}{A_t} = \frac{S_T}{S_t} e^{-\alpha(T-t)}.$$

Given that the distribution of the accumulation factor S_T/S_t is

$$\frac{S_T}{S_t} \overset{\mathbb{P}}{\sim} \mathcal{LN}((\mu - \sigma^2/2)(T-t), \sigma^2(T-t)),$$

we deduce that the sub-account balance is such that

$$\frac{A_T}{A_t} \overset{\mathbb{P}}{\sim} \mathcal{LN}((\mu - \alpha - \sigma^2/2)(T-t), \sigma^2(T-t))$$

(under the real-world/actuarial probability measure $\mathbb{P}$). For no-arbitrage pricing purposes, we also need to find the *risk-neutral distribution* of the sub-account balance. With similar steps, we obtain

$$\frac{A_T}{A_t} \overset{\mathbb{Q}}{\sim} \mathcal{LN}((r - \alpha - \sigma^2/2)(T-t), \sigma^2(T-t))$$

(under the risk-neutral/pricing probability measure $\mathbb{Q}$).

In other words, it is possible to view the dynamics of the sub-account balance as that of a dividend-paying asset with a continuous dividend yield of α. The fees paid by the insured act as a sort of dividend for the insurer. Indeed, from the policyholder's perspective, the sub-account balance is like the ex-dividend price of an asset. Since the insurance company holds the sub-account in the name of the insured and therefore collects the fees, this is similar to holding a dividend-paying asset (which entitles its holder to capital gain and dividends).

In Chapter 8, we have seen that the maturity benefit of a GMMB is

$$V_T^{\text{GMMB}} = \max(A_T, G_T),$$

with G_T being defined similarly as in the context of EIAs. The maturity benefit is thus equivalent to an investment guarantee with the difference that the underlying asset is the sub-account balance. However, in this case, the sub-account is not a traded asset. Note that the dynamics of A_t is that of a GBM with a dividend yield α. Therefore, the GMMB is an investment guarantee on a *dividend-paying asset* whose time-t value is given by equation (18.2.1). Consequently, we have

$$V_t^{\mathrm{GMMB}} = A_t \mathrm{e}^{-\alpha(T-t)} N(\mathrm{d}_1(t, A_t; \alpha)) + \mathrm{e}^{-r(T-t)} G_T N(-\mathrm{d}_2(t, A_t; \alpha)), \tag{18.2.3}$$

with K replaced by G_T, and with γ replaced by α. More precisely, we have

$$d_1(t, A_t; \alpha) = \frac{\ln(A_t/G_T) + (r - \alpha + \sigma^2/2)(T-t)}{\sigma\sqrt{T-t}},$$
$$d_2(t, A_t; \alpha) = d_1(t, A_t; \alpha) - \sigma\sqrt{T-t}.$$

It is important to note that V_0^{GMMB} is the initial cost for the insurance company to replicate this product and this replicating procedure relies on the inflow of fee payments (they are immediately reinvested in the replicating portfolio). Said differently, the initial value V_0^{GMMB} takes into account the amount of accumulated fees collected by the insurance company during the life of this product.

This is significantly different from the pricing of standard financial derivatives where a single premium is collected at inception by the seller. For variable annuities, in particular for GMMB riders, the insurance company collects the initial investment I as well as fee payments at rate α during the life of the contract.

Example 18.2.5 *Initial value of a GMMB*
Suppose that a policyholder enters into a 10-year GMMB such that the guaranteed amount at maturity corresponds to the initial investment. The sub-account credits the returns of an asset following a BSM dynamic with parameters $\mu = 0.07$, $\sigma = 0.25$ and $r = 0.02$ and an annual fee rate of $\alpha = 2.5\%$ is deducted continuously. Compute the initial value of the GMMB if the initial investment is \$1.

In this case, we have $T = 10$ and $A_0 = I = 1 = G_{10}$. To use the Black-Scholes formula of equation (18.2.3), we need:

$$d_1(0, 1; 0.025) = \frac{\ln(1/1) + (0.02 - 0.025 + (0.25)^2/2)10}{0.25\sqrt{10}} = 0.332039154$$
$$d_2(0, 1; 0.025) = -0.458530261$$
$$N(d_1) = 0.630070153$$
$$N(-d_2) = 0.676714236.$$

Therefore, the value of the GMMB at inception is

$$V_0^{\mathrm{GMMB}} = \mathrm{e}^{-0.025\times 10} \times 0.630070153 + \mathrm{e}^{-0.02\times 10} \times 0.676714236 = 1.044745885.$$

To provide this protection on the sub-account balance, the insurance company needs \$1.044745885 per dollar invested by the policyholder. With an initial investment of \$1 and a fee rate of 2.5% per annum, this GMMB is thus worth \$1.044745885 to the policyholder. This means it is sold at a loss by the insurance company. In conclusion, the annual fee charged by the company is too low or the guaranteed amount is too high. ■

18.2.3.2 GMDB

When a guarantee applies upon death of the policyholder, we know from Chapter 8 that this is a GMDB rider. Therefore, to protect an invested capital upon death or maturity, whichever comes first, one needs to combine both protections.[5]

If we knew in advance when the policyholder will die, we could treat the GMDB as a GMMB with a maturity corresponding to the time of death. This is (fortunately) not possible in practice. In this case, the insurance company can sell GMDBs to a *very large* number of *independent and identically distributed* (iid) policyholders so that the insurer can predict with a high probability the number of policyholders surviving each year. Consequently, we can treat a very large portfolio of GMDBs with each policy having a random effective maturity as a portfolio of GMMBs with deterministic maturities.

Suppose the insurance company holds a portfolio of M iid policyholders (e.g. all aged x) and let us denote M_k as the *expected* number of deaths from the pool in the k-th year. In practice, the real number of deaths should be different from its expectation, but with M large enough, the relative variability (M_k/M) should be small enough to be ignored.

Then, we know that selling T-year GMDBs to these M policyholders (assume T is an integer for the sake of the illustration) is roughly equal to holding a portfolio of:

- M_1 1-year GMMBs whose maturity benefit is replaced by the death benefit that would apply at the end of year 1;
- M_2 2-year GMMBs whose maturity benefit is replaced by the death benefit that would apply at the end of year 2;
- $\ldots$;
- M_T T-year GMMBs whose maturity benefit is replaced by the death benefit that would apply at the end of year T.

Let $V_0^{\text{GMMB}}(k, G_k)$ be the initial value of a GMMB as given in (18.2.3) with an emphasis on the fact that it is maturing in k years with maturity guarantee G_k. Therefore, the *total* initial value of the GMDBs sold to the pool of policyholders is

$$\sum_{k=1}^{T} M_k V_0^{\text{GMMB}}(k, G_k). \tag{18.2.4}$$

The initial value of such a GMDB for a single policyholder is thus given by the total amount in (18.2.4) divided by the number of participants M in the pool, i.e.

$$V_0^{\text{GMDB}} = \sum_{k=1}^{T} \frac{M_k}{M} V_0^{\text{GMMB}}(k, G_k).$$

Note that $\{M_1/M, M_2/M, \ldots, M_T/M\}$ is a discrete probability distribution (w.r.t. the real-world/actuarial probability $\mathbb{P}$) over $\{1, 2, \ldots, T\}$ linked to the mortality distribution of these insureds. Consequently, V_0^{GMDB} is given by the average of the values $V_0^{\text{GMMB}}(1, G_1)$, $V_0^{\text{GMMB}}(2, G_2), \ldots, V_0^{\text{GMMB}}(T, G_T)$ weighted by this mortality distribution.

More generally, if we let τ be the random variable representing the time of death of a given policyholder, we would have the following general relationship:

$$V_0^{\text{GMDB}} = \mathbb{E}\left[V_0^{\text{GMMB}}(\tau, G_\tau)\mathbb{1}_{\{\tau \leq T\}}\right].$$

---

5 Recall that this is a convention taken in this book. From one insurance company to another, the protections may be named differently.

It is very important to note that the expectation is taken with respect to the actuarial probability distribution of τ, which can be computed for example with a mortality table corresponding to the insureds common profile.

18.2.3.3 Break-even fee rate

In example 18.2.5, we found that the cost of replicating the maturity benefit was greater than what was received by the insurance company from the policyholder, at inception (initial investment) and from the sub-account (fees). Just like for EIAs, for which we found a break-even participation rate, we can find a **break-even fee rate** for variable annuities, given a guaranteed amount G_T. As before, the idea is to determine the value of α such that

$$V_0(\alpha) = A_0,$$

where V_0 is the initial value of the contract, e.g. $V_0 = V_0^{\text{GMMB}}$ or $V_0 = V_0^{\text{GMDB}}$.

Again, in most cases, the solution to such an equation (the break-even fee rate) can only be found numerically using a *root-finding algorithm.*

Example 18.2.6 *Annual fee in a GMMB*
Let us find the break-even annual fee rate α for the GMMB in example 18.2.5. We need to determine the value of α such that

$$V_0^{\text{GMMB}}(\alpha) = 1,$$

since $A_0 = 1$. Using for example the *Goal Seek* tool from Excel™, we find $\alpha = 0.034954116$, i.e. the fee rate needs to be at least 3.5%, which is higher than the rate of 2.5% used in example 18.2.5.

Instead, if we assume the policyholder is willing to bear a 25% loss over 10 years, meaning that we now set $G_T = 0.75 \times A_0$, then the break-even annual fee rate drops to $\alpha = 1.2575606\%$. ■

18.3 Exotic options

Recall from Chapter 7 that a European payoff V_T is said to be exotic or path-dependent if it depends on the underlying asset price at more than one date during the life of the derivative. There are two types of path-dependent payoffs determined by the *monitoring frequency* of the option:

1) **discretely monitored**: for $n \geq 2$ and for a set of dates $0 \leq t_1 < t_2 < \cdots < t_n \leq T$, the payoff is given by

$$V_T = g\left(S_{t_1}, S_{t_2}, \ldots, S_{t_n}\right);$$

2) **continuously monitored**: the payoff is given by

$$V_T = g(S_t, 0 \leq t \leq T).$$

These options are also said to be *exotic* to contrast with *vanilla*/simple options such as standard call and put options.

The main objective of this section is to derive formulas for the initial price of two types of exotic options: average price Asian options (with geometric mean) and floating-strike lookback options. By focusing on these options, we will be able to obtain Black-Scholes-like expressions that are easy to manipulate and to compute. As for barrier options, to derive expressions for their initial prices, more advanced mathematical techniques are needed; we will simply state the pricing formulas.

18.3.1 Asian options

We know from Chapter 7 that Asian options are derivatives whose payoff depends on the *average* asset price observed between inception and maturity. There are many types of Asian options depending on:

- how often the underlying asset price is monitored (discrete or continuous monitoring);
- how the average is computed (arithmetic or geometric average);
- how the average is used in the calculation of the payoff (average price or average strike).

In general, there are no simple formulas for the initial value of *average strike* Asian options. This is because we can buy or sell the underlying asset for an average strike of $\overline{S}_T$ and hence we need to determine the (risk-neutral) joint distribution of $(S_T, \overline{S}_T)$. Moreover, *average price* Asian options computed with an *arithmetic* average are also difficult to evaluate because we need to find the distribution of the sum of asset prices. Hence, for many Asian options, simulation methods are needed (see Chapter 19).

There is one case where we can obtain closed-form expressions for the initial price of an *average price* Asian option in the BSM model: when the average is *geometric*, computed either discretely or continuously. Recall that the payoff of these Asian options is given by

$$V_T = (\overline{S}_T - K)_+ \quad \text{or} \quad V_T = (K - \overline{S}_T)_+,$$

where the geometric average price $\overline{S}_T$ can be computed continuously or discretely. More precisely, the geometric average price of S is given by

$$\overline{S}_T = \exp\left(\frac{1}{T}\int_0^T \ln(S_t)\mathrm{d}t\right),$$

in the continuous monitoring case, and by

$$\overline{S}_T = \left(\prod_{i=1}^{n} S_{t_i}\right)^{\frac{1}{n}} = \left(S_{t_1} \times S_{t_2} \times \cdots \times S_{t_n}\right)^{\frac{1}{n}},$$

for given dates $0 \leq t_1 < t_2 < \cdots < t_n \leq T$, with $n \geq 1$, in the discrete monitoring case.

In a BSM setting, Asian options based on geometric averages are more tractable because the lognormal distribution is *stable* under multiplication (the product of independent lognormal random variables is also lognormally distributed). This is not the case when arithmetic averages are used.

Therefore, in what follows, we first analyze the distribution of the geometric average $\overline{S}_T$. We will find that both the real-world/actuarial distribution and the risk-neutral distribution of the average are lognormally distributed. Second, we will use the *stop-loss transform formula for lognormal random variables* to obtain explicit formulas similar to the Black-Scholes formula.

18.3.1.1 Discrete geometric average

Let us begin by considering a simple case: the discrete geometric average $\bar{S}_T = \sqrt{S_{T/2} \times S_T}$, i.e. the discrete geometric average of a geometric Brownian motion sampled at only two evenly-spaced dates. We know that at any time t, we have

$$S_t = S_0 \exp\left\{ \left(\mu - \frac{\sigma^2}{2}\right) t + \sigma W_t \right\}.$$

Therefore, the product of $S_{T/2}$ and S_T is equal to

$$S_{T/2} \times S_T = S_0^2 \exp\left\{ \left(\mu - \frac{\sigma^2}{2}\right) 3T/2 + \sigma(W_{T/2} + W_T) \right\}.$$

However, the random variables $S_{T/2}$ and S_T are not independent because $W_{T/2}$ and W_T have a non-zero covariance. Indeed, recall from Chapter 14 that the covariance between $W_{T/2}$ and W_T is given by $\mathbb{C}\text{ov}(W_{T/2}, W_T) = \min(T/2, T) = T/2$.

On the other hand, the random variables $W_T - W_{T/2}$ and $W_{T/2}$ are independent because they are Brownian increments over non-overlapping time intervals. Consequently,

$$W_{T/2} + W_T = (W_T - W_{T/2}) + 2W_{T/2}$$

follows a normal distribution with mean 0 and variance $5T/2$ (with respect to $\mathbb{P}$).

Taking the square root of the product, we obtain

$$\begin{aligned}
\sqrt{S_{T/2} \times S_T} &= S_0 \exp\left\{ \frac{1}{2}\left(\mu - \frac{\sigma^2}{2}\right) 3T/2 + \frac{\sigma}{2}(W_{T/2} + W_T) \right\} \\
&= S_0 \exp\left\{ \left(\mu - \frac{\sigma^2}{2}\right) 3T/4 + \frac{\sigma}{2}((W_T - W_{T/2}) + 2W_{T/2}) \right\} \\
&= S_0 \exp\left\{ \left(\mu - \frac{\sigma^2}{2}\right) 3T/4 \right\} e^{\sigma/2(W_T - W_{T/2})} e^{\sigma W_{T/2}}
\end{aligned}$$

and we get

$$\sqrt{S_{T/2} \times S_T} \overset{\mathbb{P}}{\sim} \mathcal{LN}\left(\ln(S_0) + \left(\mu - \frac{\sigma^2}{2}\right) 3T/4, 5\sigma^2 T/8 \right).$$

More generally, if the average is based on n given dates $0 \leq t_1 < t_2 < \cdots < t_n \leq T$, then we can write

$$\left(\prod_{i=1}^{n} S_{t_i} \right)^{\frac{1}{n}} = S_0 \exp\left(\left(\mu - \frac{\sigma^2}{2}\right) \frac{1}{n} \sum_{i=1}^{n} t_i + \frac{\sigma}{n} \sum_{i=1}^{n} W_{t_i} \right),$$

where the random vector $(W_{t_1}, W_{t_2}, \ldots, W_{t_n})$ follows a multivariate normal distribution. Consequently, $\sum_{i=1}^{n} W_{t_i}$ follows a (univariate) normal distribution with mean

$$\mathbb{E}\left[\sum_{i=1}^{n} W_{t_i} \right] = 0$$

and variance

$$\mathbb{V}\text{ar}\left(\sum_{i=1}^{n} W_{t_i} \right) = \sum_{i=1}^{n} \sum_{j=1}^{n} \mathbb{C}\text{ov}\left(W_{t_i}, W_{t_j}\right) = \sum_{i=1}^{n} \sum_{j=1}^{n} \min(t_i, t_j) = (2i-1)t_{n+1-i}.$$

Consequently, the geometric average follows a lognormal distribution:

$$\overline{S}_T = \left(\prod_{i=1}^{n} S_{t_i}\right)^{\frac{1}{n}} \overset{\mathbb{P}}{\sim} \mathcal{LN}\left(\ln(S_0) + \left(\mu - \frac{\sigma^2}{2}\right)\frac{1}{n}\sum_{i=1}^{n} t_i, \frac{\sigma^2}{n^2}\sum_{i=1}^{n}(2i-1)t_{n+1-i}\right). \quad (18.3.1)$$

Example 18.3.1 *Probability of expiring in the money*
A share of stock currently trades for \$50 and a 1-year 51-strike average price Asian call option with geometric average is issued. If the average is computed over monthly prices, then calculate the probability the option matures in the money in a BSM model with $\mu = 0.08, r = 0.02, \sigma = 0.25$.

We have $t_i = i/12$, for each $i = 1, 2, \ldots, 12$, and we have $t_{n+1-i} = (12 + 1 - i)/12 = (13 - i)/12$. Therefore,

$$\sum_{i=1}^{n} t_i = \sum_{i=1}^{12} \frac{i}{12} = \frac{1}{12} \times \frac{12 \times 13}{2} = 6.5$$

and

$$\sum_{i=1}^{n}(2i-1)t_{n+1-i} = \sum_{i=1}^{12}(2i-1)\frac{13-i}{12} = 54.16667.$$

The probability of the option maturing in the money, i.e. the probability the payoff $(\overline{S}_1 - 51)_+$ is positive, is computed using the actuarial distribution of equation (18.3.1) (i.e. with the real-world probability $\mathbb{P}$). We have

$$\ln(S_0) + \left(\mu - \frac{\sigma^2}{2}\right)\frac{1}{n}\sum_{i=1}^{n} t_i = \ln(50) + \left(0.08 - \frac{1}{2}0.25^2\right)\frac{78}{12^2} = 3.938429$$

and

$$\frac{\sigma^2}{n^2}\sum_{i=1}^{n}(2i-1)t_{n+1-i} = \frac{0.25^2}{12^2} \times \frac{650}{12} = 0.02350984.$$

Finally,

$$\begin{aligned}\mathbb{P}(\overline{S}_1 > 51) &= 1 - \mathbb{P}(\ln(\overline{S}_1) \leq \ln(51)) \\ &= 1 - N\left(\frac{\ln(51) - 3.938429}{\sqrt{0.02350984}}\right) \\ &= 1 - N(-0.0430666) \\ &= 0.51717\end{aligned}$$

is the probability for this Asian option of expiring in the money. ■

18.3.1.2 Continuous geometric average

The continuous geometric average of S over the time interval $[0, T]$ is defined by

$$\overline{S}_T = \exp\left(\frac{1}{T}\int_0^T \ln(S_t)\mathrm{d}t\right).$$

Let $\{X_t, t \geq 0\}$ be the linear Brownian motion with drift $\mu - \frac{1}{2}\sigma^2$ and diffusion σ, i.e.

$$X_t = \left(\mu - \frac{\sigma^2}{2}\right) t + \sigma W_t,$$

such that we can write $S_t = S_0 e^{X_t}$ and $\ln(S_t) = \ln(S_0) + X_t$. Hence, the continuous geometric average can be written as follows:

$$\overline{S}_T = \exp\left(\frac{1}{T}\int_0^T (\ln(S_0) + X_t)\mathrm{d}t\right) = S_0 \exp\left(\frac{1}{T}\int_0^T X_t \mathrm{d}t\right).$$

Since

$$\int_0^T X_t \mathrm{d}t = \int_0^T \left\{\left(\mu - \frac{\sigma^2}{2}\right) t + \sigma W_t\right\} \mathrm{d}t = \left(\mu - \frac{\sigma^2}{2}\right)\frac{T^2}{2} + \sigma \int_0^T W_t \mathrm{d}t,$$

using tools from stochastic calculus (see Chapter 15), we could verify that

$$\int_0^T W_t \mathrm{d}t \overset{\mathbb{P}}{\sim} \mathcal{N}\left(0, \frac{T^3}{3}\right)$$

and therefore

$$\int_0^T X_t \mathrm{d}t \overset{\mathbb{P}}{\sim} \mathcal{N}\left(\left(\mu - \frac{\sigma^2}{2}\right)\frac{T^2}{2}, \sigma^2 \frac{T^3}{3}\right).$$

In conclusion, the continuous geometric average of a geometric Brownian motion also follows a lognormal distribution (with respect to $\mathbb{P}$). More specifically, we have

$$\overline{S}_T = \exp\left(\frac{1}{T}\int_0^T \ln(S_t)\mathrm{d}t\right) \overset{\mathbb{P}}{\sim} \mathcal{LN}\left(\ln(S_0) + \left(\mu - \frac{\sigma^2}{2}\right)\frac{T}{2}, \sigma^2\frac{T}{3}\right). \quad (18.3.2)$$

Example 18.3.2 *Probability of expiring in the money (continued)*
Let us assume now that the average is computed continuously. From equation (18.3.2), $\overline{S}_T$ follows a lognormal distribution with parameters

$$\ln(S_0) + \left(\mu - \frac{\sigma^2}{2}\right)\frac{T}{2} = \ln(50) + \left(0.08 - \frac{0.25^2}{2}\right)\frac{1}{2} = 3.936398005$$

and

$$\sigma^2 \frac{T}{3} = 0.25^2 \frac{1}{3} = 0.020833333.$$

Consequently, the probability of the option maturing in the money is

$$\begin{aligned}
\mathbb{P}(\overline{S}_T > 51) &= 1 - \mathbb{P}(\ln(\overline{S}_T) \leq \ln(51)) \\
&= 1 - N\left(\frac{\ln(51) - 3.936398005}{\sqrt{0.020833333}}\right) \\
&= 1 - N(-0.031678327) \\
&= 0.512635711.
\end{aligned}$$

■

18.3.1.3 Black-Scholes formulas

We just showed that the geometric average of a geometric Brownian motion follows a lognormal distribution whose parameters depend on the monitoring frequency (discrete or continuous); see equation (18.3.1) and equation (18.3.2).

Now, we are interested in finding the no-arbitrage value of an average price (fixed-strike) Asian call option. This requires the computation of

$$C_0^A = e^{-rT}\mathbb{E}^{\mathbb{Q}}[(\overline{S}_T - K)_+],$$

which is a discounted stop-loss premium whose expectation is computed with the risk-neutral distribution (risk-neutral pricing probability $\mathbb{Q}$). Therefore, we cannot directly use the results from equations (18.3.1) and (18.3.2)) as they are based on the actuarial distribution of asset prices.

Relying on the mathematical tools developed in Section 17.2 of Chapter 17,[6] it turns out that the risk-neutral distribution of the geometric average is also lognormal:

$$\overline{S}_T = \left(\prod_{i=1}^{n} S_{t_i}\right)^{\frac{1}{n}} \overset{\mathbb{Q}}{\sim} \mathcal{LN}\left(\ln(S_0) + \left(r - \frac{\sigma^2}{2}\right)\frac{1}{n}\sum_{i=1}^{n} t_i, \frac{\sigma^2}{n^2}\sum_{i=1}^{n}(2i-1)t_{n+1-i}\right), \tag{18.3.3}$$

in the discrete case, and

$$\overline{S}_T = \exp\left(\frac{1}{T}\int_0^T \ln(S_t)dt\right) \overset{\mathbb{Q}}{\sim} \mathcal{LN}\left(\ln(S_0) + \left(r - \frac{\sigma^2}{2}\right)\frac{T}{2}, \sigma^2\frac{T}{3}\right), \tag{18.3.4}$$

in the continuous case. The risk-neutral distributions in (18.3.3) and (18.3.4) are equivalent to those in (18.3.1) and (18.3.2) with μ replaced by r.

Therefore, we can say that $\overline{S}_T$ follows a risk-neutral lognormal distribution, i.e.

$$\overline{S}_T \overset{\mathbb{Q}}{\sim} \mathcal{LN}(m, s^2),$$

where m and s^2 are given by either the expressions in (18.3.3) or in (18.3.4), depending on the monitoring frequency. It follows from formula (14.1.7) in Chapter 14 that

$$C_0^A = e^{m-rT+\frac{s^2}{2}}N\left(-\frac{\ln(K) - m}{s} + s\right) - e^{-rT}KN\left(-\frac{\ln(K) - m}{s}\right). \tag{18.3.5}$$

It is the Black-Scholes formula for the initial value of an average price Asian call option.

Example 18.3.3 *Average price Asian call option*
Your company sells an average price Asian call option with a strike price of \$34 maturing in 4 months. The behavior of the stock price is such that its log-return volatility is 30%, its current price is \$36 and the dynamics is that of the BSM model. Compute the price of the latter Asian call option if the average is computed continuously and the risk-free rate is 2%.

We have $S_0 = 36$, $K = 34$, $T = 0.25$, $\sigma = 0.3$ and $r = 0.02$. We compute the parameters m and s^2 of the risk-neutral lognormal distribution for the continuous geometric mean $\overline{S}_T$

6 Namely, the application of the Fundamental Theorem of Asset Pricing and the Girsanov theorem.

given in (18.3.4): we get

$$m = \ln(36) + \left(0.02 - \frac{0.3^2}{2}\right)\frac{0.25}{2} = 3.580393939$$

$$s^2 = 0.3^2 \frac{0.25}{3} = 0.0075.$$

The price of the option is thus given by the formula in (18.3.5):

$$C_0^A = 35.84284405 \times N(0.710526667) - 34e^{-0.02\times 0.25} \times N(0.623924127) = 2.467485793.$$

■

We know that

$$(\overline{S}_T - K)_+ + K = (K - \overline{S}_T)_+ + \overline{S}_T.$$

Therefore, we can deduce that the initial value of an average price Asian put option is such that

$$P_0^A = C_0^A + e^{-rT}\left(K - \mathbb{E}^{\mathbb{Q}}[\overline{S}_T]\right),$$

where

$$\mathbb{E}^{\mathbb{Q}}[\overline{S}_T] = e^{m+\frac{s^2}{2}}.$$

Again, m and s^2 are given by the expressions in (18.3.3) or in (18.3.4), depending on the monitoring frequency. This is the parity relationship for average price Asian options.

Example 18.3.4 *Average price Asian put option*
Using the context of the previous example, find the price of the corresponding average price Asian put option.

Since

$$\mathbb{E}^{\mathbb{Q}}[\overline{S}_T] = e^{m+\frac{s^2}{2}} = 36.02250705,$$

we get

$$P_0^A = 2.467485793 + e^{-0.02\times 0.25}(34 - 36.02250705) = 0.455066039.$$

■

18.3.2 Lookback options

We know from Chapter 7 that lookback options are derivatives whose payoff depends on the minimum or maximum price of the underlying asset observed during the life of the option. The types of lookback options are based upon:

- how often the underlying asset price is monitored (discrete or continuous monitoring);
- how the maximum/minimum is used in the calculation of the payoff (as the underlying asset price or as the strike price).

There are closed-form expressions for the no-arbitrage price of lookback options when the minimum or maximum is monitored continuously. For simplicity, in what follows, we will discuss only floating-strike lookback options (see below).

18.3.2.1 Distribution of the maximum and the minimum

First, let us analyze the distribution of the maximal and minimal values taken by S in a BSM setting, in which case S is a geometric Brownian motion. Recall from Chapter 7 the definitions of the following random variables:

$$M_T^S = \max_{0\leq t\leq T} S_t \qquad \text{and} \qquad m_T^S = \min_{0\leq t\leq T} S_t.$$

They are respectively the maximal value and the minimal value taken by the underlying asset price S over the time interval $[0, T]$.

As the asset price is given by $S_t = \exp(X_t)$, where

$$X_t = \ln(S_0) + \left(\mu - \frac{\sigma^2}{2}\right) t + \sigma W_t,$$

i.e. where X is a linear Brownian motion with drift $\mu - \frac{\sigma^2}{2}$ and diffusion σ, if we define

$$M_T^X = \max_{0\leq t\leq T} X_t \qquad \text{and} \qquad m_T^X = \min_{0\leq t\leq T} X_t,$$

then, since the exponential function $x \mapsto \exp(x)$ is an increasing function, we have that

$$M_T^S = S_0 \exp\left(M_T^X\right) \qquad \text{and} \qquad m_T^S = S_0 \exp\left(m_T^X\right).$$

In words, the maximum (or minimum) value taken by the geometric Brownian motion S can be rewritten in terms of the maximum (or minimum) taken by the underlying Brownian motion with drift X. Consequently, the distribution of M_T^S (resp. m_T^S) is equivalent to finding the distribution of M_T^X (resp. m_T^X).

It is known[7] that, for $x > 0$,

$$\mathbb{P}\left(M_T^X \leq x\right) = N\left(\frac{x - aT}{\sigma\sqrt{T}}\right) - e^{2ax/\sigma^2} N\left(\frac{-x - aT}{\sigma\sqrt{T}}\right) \tag{18.3.6}$$

and, for $x < 0$,

$$\mathbb{P}\left(m_T^X \leq x\right) = 1 - \left(N\left(\frac{-x + aT}{\sigma\sqrt{T}}\right) - e^{2ax/\sigma^2} N\left(\frac{x + aT}{\sigma\sqrt{T}}\right)\right) \tag{18.3.7}$$

where $a = \mu - \sigma^2/2$. Consequently, using the relationships presented above, we can write

$$\mathbb{P}\left(M_T^S \leq x\right) = \mathbb{P}\left(S_0 \exp\left(M_T^X\right) \leq x\right) = \mathbb{P}\left(M_T^X \leq \ln\left(\frac{x}{S_0}\right)\right)$$

and

$$\mathbb{P}\left(m_T^S \leq x\right) = \mathbb{P}\left(S_0 \exp\left(m_T^X\right) \leq x\right) = \mathbb{P}\left(m_T^X \leq \ln\left(\frac{x}{S_0}\right)\right).$$

18.3.2.2 Black-Scholes formulas

In a BSM model, the payoff of a floating-strike lookback put option is given by $P_T^{fS} = M_T^S - S_T$, so its no-arbitrage initial price can be written as

$$P_0^{fS} = e^{-rT} \mathbb{E}^{\mathbb{Q}}\left[M_T^S - S_T\right] = e^{-rT} \mathbb{E}^{\mathbb{Q}}\left[M_T^S\right] - S_0.$$

7 An accessible proof is available in [24].

The problem amounts to finding the expected value of the maximum using the *risk-neutral* distribution of the stock price (i.e. with the pricing probability $\mathbb{Q}$).

Since M_T^S is a positive random variable, its expectation is equal to the integral of its survival function, i.e.

$$\mathbb{E}^{\mathbb{Q}}\left[M_T^S\right] = \int_0^\infty \mathbb{Q}(M_T^S > x)\mathrm{d}x = \int_0^\infty \mathbb{Q}\left(M_T^X > \ln\left(\frac{x}{S_0}\right)\right)\mathrm{d}x.$$

The risk-neutral distribution of M_T^S is slightly different from the real-world distribution in (18.3.6), as it is based on the risk-neutral dynamic of the linear Brownian motion X. It can be shown that, for $x > 0$,

$$\mathbb{Q}(M_T^X \leq x) = N\left(\frac{x - aT}{\sigma\sqrt{T}}\right) - \mathrm{e}^{2ax/\sigma^2} N\left(\frac{-x - aT}{\sigma\sqrt{T}}\right) \tag{18.3.8}$$

where $a = r - \sigma^2/2$ is used instead of $\mu - \sigma^2/2$. Finally, the price of the floating-strike lookback put option is found by integrating the risk-neutral survival function of M_T^S, i.e. the function $x \mapsto \mathbb{Q}\left(M_T^X > \ln\left(\frac{x}{S_0}\right)\right)$ over the interval $(0, \infty)$, with the help of the expression given in (18.3.8).

Similarly, for a floating-strike lookback call option with payoff $C_T^{\mathrm{f}S} = S_T - m_T^S$, we have

$$C_0^{\mathrm{f}S} = \mathrm{e}^{-rT}\mathbb{E}^{\mathbb{Q}}\left[S_T - m_T^S\right] = S_0 - \mathrm{e}^{-rT}\mathbb{E}^{\mathbb{Q}}\left[m_T^S\right],$$

where

$$\mathbb{E}^{\mathbb{Q}}\left[m_T^S\right] = \int_0^\infty \mathbb{Q}\left(m_T^S > x\right)\mathrm{d}x = \int_0^\infty \mathbb{Q}\left(m_T^X > \ln\left(\frac{x}{S_0}\right)\right)\mathrm{d}x.$$

Again, it can be shown that, for $x < 0$,

$$\mathbb{Q}\left(m_T^X \leq x\right) = 1 - \left(N\left(\frac{-x + aT}{\sigma\sqrt{T}}\right) - \mathrm{e}^{2ax/\sigma^2} N\left(\frac{x + aT}{\sigma\sqrt{T}}\right)\right). \tag{18.3.9}$$

The expression in (18.3.9) can be used directly to derive the price of the call option.

As the final computations are tedious, for both lookback options we provide the pricing formulas without proof. First, let us define

$$\begin{aligned} d_1 &= \frac{(r + 0.5\sigma^2)\sqrt{T}}{\sigma}, \\ d_2 &= d_1 - \sigma\sqrt{T}, \\ d_3 &= \frac{(-r + 0.5\sigma^2)\sqrt{T}}{\sigma}. \end{aligned}$$

The initial value of the lookback call option is

$$C_0^{\mathrm{f}S} = S_0 \times \left(N(d_1) - \mathrm{e}^{-rT}N(d_2) - \frac{\sigma^2}{2r}N(-d_1) + \mathrm{e}^{-rT}\frac{\sigma^2}{2r}N(-d_3)\right)$$

whereas the initial value of the lookback put option is

$$P_0^{\mathrm{f}S} = S_0 \times \left(-N(-d_1) + \mathrm{e}^{-rT}N(-d_2) + \frac{\sigma^2}{2r}N(d_1) - \mathrm{e}^{-rT}\frac{\sigma^2}{2r}N(d_3)\right).$$

18.3.3 Barrier options

We know from Chapter 7 that barrier options are derivatives whose payoff is activated or deactivated if the asset price crosses a predetermined barrier before maturity. Finding the no-arbitrage price of barrier options in the BSM framework is not easy: it requires the knowledge of the joint distribution of (S_T, M_T^S) or (S_T, m_T^S). In what follows, we will provide Black-Scholes-like pricing formulas for barrier options without proof.

Recall from Chapter 7 that there are four types of barrier options: up-and-in, up-and-out, down-and-in and down-and-out barrier options. Pricing formulas also depend on the relationship between the barrier level H and the strike price K, so that overall there are 16 possible combinations. To lighten the presentation, we will use the fact that a knock-in barrier option combined with the corresponding knock-out barrier option has the same payoff as a regular option.

18.3.3.1 Call options

Using the same acronyms as in Chapter 7 to name the different barrier call options, let us denote their initial prices by UIC_0, UOC_0, DOC_0 and DIC_0. We do not recall the payoffs here, as they can be found in Chapter 7.

Let us begin with the case where the initial price S_0 is above the trigger level H, i.e. $S_0 \geq H$, meaning the asset price must go under H at some point for the barrier calls to be knocked-in or knocked-out.

If we further assume $S_0 \geq H \geq K$, we can show that

$$\text{DOC}_0 = S_0 N(d_1^+) - K\mathrm{e}^{-rT} N(d_2^+) - S_0 (H/S_0)^{2\lambda} N(d_1^-) + K\mathrm{e}^{-rT} (H/S_0)^{2\lambda-2} N(d_2^-)$$

where

$$\begin{aligned}
d_1^+ &= \frac{\ln(S_0/H)}{\sigma\sqrt{T}} + \lambda\sigma\sqrt{T}, \\
d_2^+ &= d_1^+ - \sigma\sqrt{T}, \\
d_1^- &= \frac{-\ln(S_0/H)}{\sigma\sqrt{T}} + \lambda\sigma\sqrt{T}, \\
d_2^- &= d_1^- - \sigma\sqrt{T},
\end{aligned}$$

and where

$$\lambda = \frac{r}{\sigma^2} + \frac{1}{2}.$$

Recalling from Chapter 7 that

$$\text{knock-in option} + \text{knock-out option} = \text{vanilla option},$$

we can conclude that $\text{DIC}_0 = C_0 - \text{DOC}_0$, where C_0 is the initial price of a vanilla call option, as given by the Black-Scholes formula.

Instead, if we further assume $H \leq K$, which means now that S_0 can be either greater than or less than the strike price K, then we can show that

$$\text{DIC}_0 = S_0 (H/S_0)^{2\lambda} N(d_1) - K\mathrm{e}^{-rT} (H/S_0)^{2\lambda-2} N(d_2)$$

where

$$d_1 = \frac{2\ln(H/(S_0K))}{\sigma\sqrt{T}} + \lambda\sigma\sqrt{T},$$
$$d_2 = d_1 - \sigma\sqrt{T},$$

and the corresponding down-and-out call option is such that $\mathrm{DOC}_0 = C_0 - \mathrm{DIC}_0$.

Now, let us consider the case where the initial price S_0 is below the trigger level H, i.e. $S_0 \leq H$, meaning the asset price must go above H at some point for the barrier options to be knocked-in or knocked-out.

If we further assume $H \geq K$, i.e. if $S_0 \leq K \leq H$ or if $K \leq S_0 \leq H$, then we can show that

$$\begin{aligned}\mathrm{UIC}_0 = {} & S_0N(d_1^+) - Ke^{-rT}N(d_2^+) - S_0(H/S_0)^{2\lambda}(N(-d_1) - N(-d_1^-)) \\ & + Ke^{-rT}(H/S_0)^{2\lambda-2}(N(-d_2) - N(-d_2^-))\end{aligned}$$

and the corresponding up-and-out call option is such that $\mathrm{UOC}_0 = C_0 - \mathrm{UIC}_0$.

Finally, if we further assume $S_0 \leq H \leq K$, it is easy to see that $\mathrm{UOC}_0 = 0$ as its payoff will be equal to zero in all possible scenarios. Indeed, for the barrier call to mature in the money, i.e. to have $S_T > K$, the underlying asset price must go above H during the life of the option and, by doing so, will be knocked-out. Therefore, in this case, we have $\mathrm{UIC}_0 = C_0$.

18.3.3.2 Put options

We follow similar steps to present the various formulas for barrier put options. Again, using the same acronyms as in Chapter 7 to name the different barrier put options, let us denote their initial prices by UIP_0, UOP_0, DOP_0 and DIP_0.

Let us begin with the case where the initial price S_0 is above the trigger level H, i.e. $S_0 \geq H$, meaning the asset price must go under H at some point for the barrier puts to be activated or deactivated.

If we further assume $S_0 \geq H \geq K$, it is easy to see that $\mathrm{DOP}_0 = 0$ as its payoff will be equal to zero in all possible scenarios. Therefore, in this case, we have $\mathrm{DIP}_0 = P_0$, where P_0 is the initial price of a vanilla put option, as given by the Black-Scholes formula.

If we further assume $H \leq K$, i.e. if $H \leq K \leq S_0$ or if $H \leq S_0 \leq K$, then we have

$$\begin{aligned}\mathrm{DIP}_0 = {} & -S_0N(-d_1^+) + Ke^{-rT}N(-d_2^+) + S_0(H/S_0)^{2\lambda}(N(d_1) - N(d_1^-)) \\ & - Ke^{-rT}(H/S_0)^{2\lambda-2}(N(d_2) - N(d_2^-))\end{aligned}$$

and then $\mathrm{DOP}_0 = P_0 - \mathrm{DIP}_0$.

Now, let us consider the case where the initial price S_0 is below the trigger level H, i.e. $S_0 \leq H$, meaning the asset price must go above H at some point for the barrier options to be knocked-in or knocked-out.

If we further assume $H \geq K$, i.e. if $S_0 \leq K \leq H$ or if $K \leq S_0 \leq H$, then we have

$$\mathrm{UIP}_0 = -S_0(H/S_0)^{2\lambda}N(-d_1) + Ke^{-rT}(H/S_0)^{2\lambda-2}N(-d_2)$$

and then $\mathrm{UOP}_0 = P_0 - \mathrm{UIP}_0$.

Finally, if we further assume $S_0 \leq H \leq K$, then

$$\mathrm{UOP}_0 = -S_0N(-d_1^+) + Ke^{-rT}N(-d_2^+) + S_0(H/S_0)^{2\lambda}N(-d_1^-) - Ke^{-rT}(H/S_0)^{2\lambda-2}N(-d_2^-)$$

and then $\mathrm{UIP}_0 = P_0 - \mathrm{UOP}_0$.

18.4 Summary

Options on other assets

- Black-Scholes formula for a call option on an asset paying continuous dividends with yield γ:

$$C_t = S_t e^{-\gamma(T-t)} N(d_1(t, S_t; \gamma)) - e^{-r(T-t)} K N(d_2(t, S_t; \gamma))$$

with

$$d_1(t, S_t; \gamma) = \frac{\ln(S_t/K) + (r - \gamma + \sigma^2/2)(T-t)}{\sigma\sqrt{T-t}},$$
$$d_2(t, S_t; \gamma) = d_1(t, S_t; \gamma) - \sigma\sqrt{T-t}.$$

- Replication portfolio for a call option on an asset paying continuous dividends with yield γ:

$$\begin{cases} \Delta_t = e^{-\gamma(T-t)} N(d_1(t, S_t; \gamma)), \\ \Theta_t = -e^{-rT} K N(d_2(t, S_t; \gamma)). \end{cases}$$

- Currency options: in the above formulas, interpret S_t as the price (expressed in the local currency) of one unit of the foreign currency and set $\gamma = r_f$.
- Futures options: in the above formulas, assume the maturity of the futures $T^\star \geq T$, replace S_t by $F_t^{T^\star}$ and set $\gamma = r$. The pricing formula of a futures call option is known as Black's formula.
- Using put-call parity, we obtain the corresponding put option prices.
- Exchange option: the Black-Scholes formula of an option to buy $S^{(2)}$ in exchange for $S^{(1)}$ is:

$$C_t^{\text{Ex}} = S_t^{(1)} e^{-\gamma_1(T-t)} N\left(d_1\left(S_t^{(1)}, S_t^{(2)}\right)\right) - S_t^{(2)} e^{-\gamma_2(T-t)} N\left(d_2\left(S_t^{(1)}, S_t^{(2)}\right)\right)$$

where

$$d_1\left(S_t^{(1)}, S_t^{(2)}\right) = \frac{\ln\left(S_t^{(1)}/S_t^{(2)}\right) + \left(\gamma_2 - \gamma_1 + \frac{\sigma^2}{2}\right)(T-t)}{\sigma\sqrt{T-t}},$$
$$d_2\left(S_t^{(1)}, S_t^{(2)}\right) = d_1\left(S_t^{(1)}, S_t^{(2)}\right) - \sigma\sqrt{T-t},$$
$$\sigma^2 = \frac{\sigma_1^2 + \sigma_2^2 - 2\rho\sigma_1\sigma_2}{T-t}.$$

Equity-linked insurance and annuities

- The Black-Scholes formula for an investment guarantee with payoff $\max(S_T, K)$ is:

$$V_t^{\text{IG}} = e^{-\gamma(T-t)} S_t N(\mathrm{d}_1(t, S_t; \gamma)) + e^{-r(T-t)} K N(-\mathrm{d}_2(t, S_t; \gamma)).$$

- The Black-Scholes formula for a PTP EIA is:

$$V_t^{\text{PTP}} = e^{-r(T-t)} G_T N(-\mathrm{d}_2(t, S_t; \gamma)) + \frac{\beta S_t}{S_0} N(\mathrm{d}_1(t, S_t; \gamma)) + (1-\beta) e^{-r(T-t)} N(\mathrm{d}_2(t, S_t; \gamma)).$$

- CAR: if we assume $G_T \leq I$ and $T = n$ is an integer, then the price of a CAR is:

$$V_0^{\text{CAR}} = \left(e^{-r} + \beta C_0^{\text{atm}}\right)^n$$

with

$$C_0^{\text{atm}} = e^{-\gamma}N(d_1) - e^{-r}N(d_2),$$
$$d_1 = \frac{r - \gamma + \sigma^2/2}{\sigma},$$
$$d_2 = d_1 - \sigma.$$

- The Black-Scholes formula for a GMMB is:

$$V_t^{\text{GMMB}} = A_t e^{-\alpha(T-t)} N(\mathrm{d}_1(t, A_t; \alpha)) + e^{-r(T-t)} G_T N(-\mathrm{d}_2(t, A_t; \alpha))$$

where

$$d_1(t, A_t; \alpha) = \frac{\ln(A_t/G_T) + (r - \alpha + \sigma^2/2)(T-t)}{\sigma\sqrt{T-t}},$$
$$d_2(t, A_t; \alpha) = d_1(t, A_t; \alpha) - \sigma\sqrt{T-t}.$$

- GMDB: if we have a large portfolio of iid policyholders, we can view a GMDB as a portfolio of GMMBs with varying maturities given by the expected number of deaths.

Exotic options
- Asian options: when average is geometric, we have closed-form formulas.
- Lookback options: for floating-strike lookback options, we have closed-form formulas.
- Barrier options: in all cases, we have Black-Scholes-like formulas.

18.5 Exercises

In the following problems, assume the market is a Black-Scholes-Merton model.

18.1 Working for a U.S. insurance company, you are holding 1,000 Euros (EUR) and you will have to sell them 6 months from now. The current exchange rate is quoted at 1.05 USD/EUR and you are afraid the exchange rate will attain parity or worse be lower. Therefore, you seek to buy 1,000 put options at a strike of 1 USD (per Euro). Compute the amount required to acquire such a protection assuming the volatility of the exchange rate is 0.15, whereas the risk-free rate in the U.S. is 3% and in Europe 2.5%.

18.2 A 6-month futures on a stock index currently trades for 1,000. The volatility on the underlying index is 0.2 and the risk-free rate is 4%. Compute the price of a 3-month futures call option with a strike price of 1,050.

18.3 Assume that a unique dollar dividend of D is paid at time $T/2$ and that it is reinvested at the risk-free rate. Determine and explain how to modify the Black-Scholes formula for a call option in this situation.

18.4 Assume $r = 0.07$, $\sigma = 0.3$, $T = 0.5$ and $S_0 = K = 100$. Compute the prices of the corresponding call and put options in each of the following situations:
(a) the (continuously) dividend rate paid by the underlying asset is 3%;
(b) a unique dividend of \$3 will be paid in 3 months (use the result in the preceding exercise).

18.5 Consider a derivative with payoff $\max(S_{T/2}, S_T)$. Using the independence between relative increments of a GBM, derive the formula for the initial price of this derivative. Assume there is no dividend on the stock.

18.6 For a stock paying a continuous dividend yield γ (whose dividends are reinvested in the stock), derive the time-t value of a forward contract on that stock maturing at time T with delivery price K. Compare your answer with what was obtained in Chapter 3.

18.7 Your company has issued a fixed-strike lookback put option with payoff $(70 - m_T^S)_+$. If $\mu = 0.07$, $r = 0.03$, $\sigma = 0.25$ and if the initial stock price is \$100, determine the probability the derivative ends in the money if it matures in a year. Again, assume there is no dividend on the stock.

18.8 Consider a 10-year investment guarantee protecting 75% of the invested capital and written on an index whose volatility is 0.16. Assume that $\mu = 0.08$ and $r = 0.02$.
(a) If the capital you intend to invest is \$1, how much more should you pay in extra (premium) to get such a protection?
(b) What is the probability of losing money 10 years from now (where a loss occurs whenever the payoff is below the total amount paid at inception, i.e. capital and premium)?

18.9 The initial investment of \$1,000 you made in a life insurance policy is credited with the returns of a major stock index whereas 80% of the invested capital is protected from market downturns after 15 years. Suppose the stock index has a volatility of 22%, the risk-free rate is 3%.
(a) Find the break-even annual fee on that contract using a numerical tool.
(b) Two years have passed and the sub-account value is now \$1,150. In this scenario, using the fee rate computed in (a), compute the value of the contract. Explain how the insurance company should manage that contract.

18.10 You enter into a 10-year CAR equity-linked life insurance. The underlying investment has a volatility $\sigma = 0.2$ whereas the risk-free rate is 2%. There are no dividends on the underlying investment.
(a) Find the break-even participation rate.
(b) Two years have passed. The annual return on the investment in the first 2 years has been −4% and +12%. In this scenario, using the participation rate computed in (a), compute the value of the contract. Explain how the insurance company should manage that contract.

18.11 You are holding a 3-month up-and-in call option that is knocked-in whenever the asset price reaches \$73. Assuming $\mu = 0.08$, $r = 0.02$, $\sigma = 0.2$ and an initial stock price of \$70, compute the probability the barrier option is knocked-in prior to maturity. Use the distribution of the maximum of the underlying stock price.

19

Simulation methods

As we have seen in Chapter 18, it can be a real challenge to obtain explicit expressions for the price of complex derivatives. In practice, contracts are even more complex than what we have studied so far and the market models used are usually more sophisticated than the Black-Scholes-Merton model. Therefore, deriving analytical expressions for the price of a financial contract is often cumbersome or even impossible. Then, we need to resort to numerical methods.

This chapter focuses on a popular approach to price complex derivatives, namely *simulation*. **Simulation** in general refers to a set of techniques meant to artificially imitate a complex phenomenon. For example, pilots commonly use flight simulators to train in situations that would be too costly or dangerous to perform otherwise. An oft-cited example in mathematics is *Monte Carlo integration* in which random numbers are used to approximate the area under a curve when a closed-form expression is not available.

In statistics and actuarial science, simulation is used to generate artificial random scenarios from a specified model, e.g. a random sample from a given distribution. Once the sample is generated, it can be used to approximate quantities that could not be computed otherwise. Typical applications of simulation in actuarial finance include the pricing of advanced derivatives such as ELIAs and path-dependent exotic options, and the analysis of sensitivities (of profits and solvency) with respect to various economic scenarios, as required by regulators for capital adequacy requirements.

The main objective of this chapter is to apply simulation techniques to compute approximations of the no-arbitrage price of derivatives when the underlying market model is a BSM model. The specific objectives are to:

- understand that the price of a complex derivative does not have a closed-form expression and that simulation can be helpful;
- estimate the price of simple and path-dependent derivatives with crude Monte Carlo methods;
- apply variance reduction techniques, such as stratified sampling, antithetic and control variates, to accelerate convergence of the price estimator.

The next section lays the foundations by showing how to simulate random variables in general and from the normal distribution in particular. This is instrumental for the following sections on Monte Carlo simulations and variance reduction techniques as we will generate sample paths of geometric Brownian motions.

Actuarial Finance: Derivatives, Quantitative Models and Risk Management, First Edition.
Mathieu Boudreault and Jean-François Renaud.

19.1 Primer on random numbers

Say you want to generate 1,000 outcomes from the discrete uniform distribution on {1, 2, 3, 4, 5, 6}. Naively, you could roll a die, mark the result and repeat this 999 other times. This is clearly a random experiment as it is impossible to predict the outcome on each throw, even if you try to recreate the same throw. However, this is clearly not a cost-efficient way to generate *random numbers* from this probability distribution. Instead of *generating randomness* by hand, we can use algorithms and computers.

19.1.1 Uniform random numbers

Creating randomness (or apparence thereof) with a computer is usually achieved with what is known as a **pseudo-random number generator** (PRNG). A PRNG is an algorithm that generates a sequence of numbers that appear to be random. It is known as *pseudo*-random because it will generate the same set of numbers once we know the PRNG's parameters and the first number of the sequence, known as the **seed**.

Let $U \sim \mathcal{U}(0, 1)$ be a random variable uniformly distributed over the interval $[0, 1]$. PRNGs from common statistical softwares will *generate* independent and identically distributed realizations of U, namely $u_1, u_2, \dots, u_n$, also known as **uniform random numbers**. The following list shows the function names in common softwares and programming languages generating uniform random numbers:

- Excel™: `RAND`;
- R : `runif`;
- SAS : `RAND`, specifying the distribution as `'UNIFORM'`;
- Matlab : `rand`;
- Python : `random` function from the `random` module.

Generating uniform random numbers, also known as sampling from the uniform distribution on $[0, 1]$, is at the core of all the simulation methods we are about to present.

19.1.2 Inverse transform method

Once we can generate uniform random numbers, we can simulate/sample from another probability distribution using its *quantile function*. This method is known as the **inverse transform method** (technique) and is described below. But first, let us recall the definition of the quantile function of a probability distribution.

Let X be a random variable with cumulative distribution function (c.d.f.) F_X. The quantile function of this probability distribution (of this random variable) is denoted by F_X^{-1}. Mathematically, the **quantile function** is defined, for any $0 \leq u \leq 1$, by

$$F_X^{-1}(u) = \inf\{x \in \mathbb{R} : \mathbb{P}(X \leq x) \geq u\}. \tag{19.1.1}$$

It can be shown that, for any real number x, we have $F_X^{-1}(u) \leq x$ if and only if $u \leq F_X(x)$.

In particular, if F_X is continuous and strictly increasing, which is the case if for example its probability density function (p.d.f.) f_X is strictly positive, then the quantile function F_X^{-1} is the *real inverse* function of F_X and

$$F_X\left(F_X^{-1}(u)\right) = u.$$

In that case, $x = F_X^{-1}(u)$ is the unique value such that $F_X(x) = u$.

Example 19.1.1 *Exponential distribution*
It is well known that the c.d.f. of an exponential distribution with mean $\lambda^{-1} > 0$ is

$$F_X(x) = 1 - e^{-\lambda x}, \ x > 0.$$

To identify its quantile function, we need to solve the following equation: for a fixed value of $0 \leq u \leq 1$, find x such that

$$1 - e^{-\lambda x} = u.$$

Consequently, we obtain

$$F_X^{-1}(u) = -\frac{1}{\lambda} \ln(1 - u).$$

■

For a given random variable X and a uniformly distributed random variable $U \sim \mathcal{U}(0, 1)$, it can be shown that the random variable $F_X^{-1}(U)$, which is the random variable U transformed by the function F_X^{-1}, has the same distribution as X, i.e.

$$X \sim F_X^{-1}(U). \tag{19.1.2}$$

Indeed, let us assume for simplicity that F_X is strictly increasing.[1] Then, for any $x \in \mathbb{R}$, we have

$$\begin{aligned} \mathbb{P}\left(F_X^{-1}(U) \leq x\right) &= \mathbb{P}(U \leq F_X(x)) \\ &= F_U(F_X(x)) \\ &= F_X(x), \end{aligned}$$

since the c.d.f. of a uniform distribution on $[0, 1]$ is given by $F_U(u) = u$ when $0 \leq u \leq 1$.

The inverse transform method builds on this relationship as follows:

1) Simulate a uniform random number u from $\mathcal{U}(0, 1)$.
2) Using u, compute $x = F_X^{-1}(u)$.

Then, we say that x is a random number from the probability distribution given by F_X.

Example 19.1.2 *Sampling from an exponential distribution*
We found in the previous example that if X is exponentially distributed with mean $\lambda^{-1} > 0$, then

$$F_X^{-1}(u) = -\frac{1}{\lambda} \ln(1 - u).$$

If we generate a uniform random number u, then its corresponding exponential number x is

$$x = -\frac{1}{\lambda} \ln(1 - u).$$

Suppose that, with your favorite software, you have generated the following two uniform random numbers: $u_1 = 0.158$ and $u_2 = 0.496$. The corresponding realizations of the exponential distribution with mean 100 are given by:

$$\begin{aligned} x_1 &= F_X^{-1}(0.158) = -100 \ln(1 - 0.158) = 17.19752647, \\ x_2 &= F_X^{-1}(0.496) = -100 \ln(1 - 0.158) = 68.51790109. \end{aligned}$$

■

1 As mentioned above, this assumption is not needed.

19.1.3 Normal random numbers

Not all random variables have explicit and invertible cumulative distribution functions. Recall from Chapter 14 that the c.d.f. of a standard normal distribution, i.e. the function

$$N(x) = \int_{-\infty}^{x} \frac{1}{\sqrt{2\pi}} e^{-\frac{y^2}{2}} dy,$$

cannot be written in an explicit fashion in terms of easy-to-compute functions. We need to use a *normal table* or a computer to evaluate $N(x)$. This is also true for its quantile function N^{-1}. Therefore, sampling from a normal distribution using the inverse transform method is conceptually easy but not always numerically efficient. For simple applications, we can still count on the inverse transform method to simulate from a standard normal distribution.

Example 19.1.3 *Sampling from a normal distribution*
Suppose that you have generated the following two uniform random numbers: $u_1 = 0.158$ and $u_2 = 0.496$.

Mapping these uniform random numbers with the standard normal quantile function N^{-1}, we get

$$N^{-1}(0.158) = -1.002711665$$
$$N^{-1}(0.496) = -0.010026681.$$

■

To generate standard *normal random numbers* from common statistical softwares, we can use the following functions:

- Excel: `NORM.S.INV` with `RAND` (through the inefficient inverse transform method);
- R : `rnorm`;
- SAS : `RAND`, specifying the distribution as `'NORMAL'`;
- Matlab : `randn`;
- Python : `normalvariate` or `gauss`, both from the `random` module.

Instead of relying on the inverse transform method, a commonly used technique to sample from a normal distribution is known as the **Box-Muller method** (or transform). It is based on the following fact: if U_1 and U_2 are independent $\mathcal{U}(0,1)$-distributed random variables and if we set $R = -2\ln(U_1)$ and $B = 2\pi U_2$, then $Z_1 = \sqrt{R}\cos(B)$ and $Z_2 = \sqrt{R}\sin(B)$ are independent and each follow a standard normal distribution $\mathcal{N}(0,1)$.

To implement the Box-Muller method and generate two normal random numbers z_1 and z_2, it suffices to follow these steps:

1) Generate independent uniform random numbers u_1 and u_2.
2) Set $r = -2\ln(u_1)$ and $b = 2\pi u_2$.
3) Set $z_1 = \sqrt{r}\cos(b)$ and $z_2 = \sqrt{r}\sin(b)$.

19.2 Monte Carlo simulations for option pricing

The main objective of Monte Carlo simulations is to compute approximative values of expectations such as

$$\theta = \mathbb{E}[X],$$

where X is a random variable. Whenever the distribution of X is complicated, obtaining a simple expression for θ can be difficult or even impossible. Fortunately, if we can sample from the probability distribution of X, then we can use simulation methods to approximate θ.

The **crude Monte Carlo estimator** of θ is defined as

$$\hat{\theta} = \frac{1}{N} \sum_{i=1}^{N} X_i, \tag{19.2.1}$$

where $X_1, X_2, \dots, X_N$ are independent random variables having the same distribution as X, i.e. all with the same c.d.f. F_X, and where N is the sample size. In other words, if we consider $X_1, X_2, \dots, X_N$ as a random sample from the distribution given by F_X, then $\hat{\theta}$ is the sample mean $\overline{X}$ of that distribution.

Example 19.2.1 *Monte Carlo estimator*
You have generated the following sample from the distribution of a random variable X, for example using the inverse transform method:

$$1178.53, 695.09, 235.82, 316.73, 219.66, 827.2, 339.62.$$

Then, the corresponding value of $\hat{\theta}$, as given in equation (19.2.1), is

$$\hat{\theta} = \frac{1}{7} \sum_{i=1}^{N} X_i = \frac{1}{7}(1178.53 + 695.09 + \cdots + 339.62) = 544.6642857. \qquad \blacksquare$$

The Monte Carlo estimator $\hat{\theta}$ is unbiased, i.e. $\mathbb{E}[\hat{\theta}] = \theta$, and its variance goes to zero as the sample size N increases. Indeed, using the independence between the X_is, we can write

$$\begin{aligned}
\mathbb{V}\text{ar}(\hat{\theta}) &= \mathbb{V}\text{ar}\left(\frac{1}{N}\sum_{i=1}^{N} X_i\right) \\
&= \frac{1}{N^2}\sum_{i=1}^{N} \mathbb{V}\text{ar}(X_i) \\
&= \frac{1}{N}\mathbb{V}\text{ar}(X) \longrightarrow 0, \quad \text{as } N \to \infty.
\end{aligned}$$

The precision of a Monte Carlo estimator is tied to its uncertainty which is often measured by its variance $\mathbb{V}\text{ar}(\hat{\theta})$. Therefore, the preceding derivation shows that as the sample size N increases, $\hat{\theta}$ gets closer to θ and hence the estimator becomes more precise.

19.2.1 Notation

In this chapter we will simulate random variables and stochastic processes and therefore the reader needs to understand how notation will change and be interpreted in these contexts.

For the distribution of a random variable X, we will use the notation X_i to denote the i-th independent realization of X, i.e. from its distribution F_X. For example:

- $U_i, i = 1, 2, \dots, N$ denotes a sample of size N coming from the distribution of $U \sim \mathcal{U}(0, 1)$;
- $Z_i, i = 1, 2, \dots, N$ denotes a sample of size N coming from the distribution of $Z \sim \mathcal{N}(0, 1)$.

Also, as we did in Chapter 14, we will simulate stochastic processes. In particular, we will generate samples for the stock price process $S = \{S_t, t \geq 0\}$ following the Black-Scholes-Merton dynamic. In this case, to avoid confusion, the index will still designate *time*. To identify the i-th realization, we will add the notation (i) next to the random variable. More precisely, $S_T(i), i = 1, 2, \dots, N$ will denote a random sample of size N coming from the distribution of S_T (which is a lognormal distribution in the BSM model).

On the other hand, we will generate values of S at other times t than the final time T. Again, as we did in Chapter 14, we will simulate *discretized trajectories* of S. Recall that for n time points $0 < t_1 < t_2 < \cdots < t_n = T$, the random vector

$$\left(S_0, S_{t_1}(i), S_{t_2}(i), \dots, S_{t_n}(i)\right)$$

will denote the i-th discretized trajectory of $S = \{S_t, 0 \le t \le T\}$, where $i = 1, 2, \dots, N$.

Note that there are n time points and N simulated (discretized) trajectories. Those are represented in the next table where each row is a discretized path and each column is a sample for the corresponding stock price at the corresponding time point:

Path/Time	0	1	2	...	n
1	S_0	$S_{t_1}(1)$	$S_{t_2}(1)$	...	$S_{t_n}(1)$
2	S_0	$S_{t_1}(2)$	$S_{t_2}(2)$	...	$S_{t_n}(2)$
$\vdots$	$\vdots$	$\vdots$	$\vdots$	$\ddots$	$\vdots$
N	S_0	$S_{t_1}(N)$	$S_{t_2}(N)$	...	$S_{t_n}(N)$

19.2.2 Application to option pricing

We will now see how to apply Monte Carlo estimations to approximate the initial price of a European derivative in the BSM model. We know that the initial value can be written as

$$V_0 = e^{-rT}\mathbb{E}^{\mathbb{Q}}[V_T]. \tag{19.2.2}$$

Therefore, if we can sample from the risk-neutral distribution of the random variable V_T, then we will be able to approximate $\mathbb{E}^{\mathbb{Q}}[V_T]$ with its Monte Carlo estimator.

19.2.2.1 Simple derivatives

First, recall that simple derivatives have payoffs of the form $V_T = g(S_T)$ where $g(\cdot)$ is the payoff function. In this case, to generate a sample from the risk-neutral distribution of the random variable V_T, it suffices to generate a sample from the (much simpler) risk-neutral distribution of S_T which is

$$S_T \overset{\mathbb{Q}}{\sim} \mathcal{LN}\left(\ln(S_0) + \left(r - \frac{\sigma^2}{2}\right)T, \sigma^2 T\right).$$

Since the lognormal distribution is an exponential function away from the normal distribution, instead of directly sampling from the risk-neutral distribution of S_T, we can use the following relationship:

$$S_T \overset{\mathbb{Q}}{\sim} S_0 \exp\left(\left(r - \frac{\sigma^2}{2}\right)T + \sigma\sqrt{T}Z\right) \tag{19.2.3}$$

where Z is a standard normal random variable.

Therefore, the Monte Carlo pricing algorithm becomes: choose $N \geq 1$ and then

1) generate a sample Z_i, $i = 1, 2, \ldots, N$ from the standard normal distribution;
2) for each $i = 1, 2, \ldots, N$, set

$$S_T(i) = S_0 \exp\left(\left(r - \frac{\sigma^2}{2}\right)T + \sigma\sqrt{T}Z_i\right);$$

3) for each $i = 1, 2, \ldots, N$, set

$$V_T(i) = g(S_T(i)).$$

Finally, the Monte Carlo estimator of this option's price is given by

$$\hat{V}_0 = e^{-rT} \times \frac{1}{N}\sum_{i=1}^{N} V_T(i).$$

This is an approximation for the no-arbitrage price in equation (19.2.2).

Example 19.2.2 *Call option price*
A share of stock trades for \$50 on the market. You also know that the risk-free rate is 4% and the volatility of the log-return is 30%. Compute the Monte Carlo estimator for the price of a 6-month 48-strike call option in the BSM model using the following normal random numbers:

$$-0.14, -1.19, 0.21, -1.57, -0.5.$$

Using the above algorithm, the second step is to compute the corresponding five realizations of the stock price $S_{0.5}$ with respect to the risk-neutral probability distribution. Note that $S_0 = 50$, $T = 0.5$, $r = 0.04$ and $\sigma = 0.3$. Then, for the first normal random number $Z_1 = -0.14$, we set

$$S_{0.5}(1) = 50\exp\left(\left(0.04 - \frac{0.3^2}{2}\right)0.5 + 0.3\sqrt{0.5} \times -0.14\right) = 48.41571837.$$

We compute $S_{0.5}(2)$, $S_{0.5}(3)$, $S_{0.5}(4)$ and $S_{0.5}(5)$ similarly.
The next step consists in computing $C_{0.5}(i) = (S_{0.5}(i) - 48)_+$, for each $i = 1, \ldots, 5$.
The next table summarizes these computations:

i	Z_i	$S_{0.5}(i)$	$C_{0.5}(i)$
1	−0.14	48.4157	0.41571837
2	−1.19	38.7483	0
3	0.21	52.1472	4.1472
4	−1.57	35.7473	0
5	−0.50	44.8560	0

For example, if $S_{0.5} = 52.1472$, the call option pays off $\max(52.1472 - 48, 0) = 4.1472$ at maturity.
Finally, an estimate for the initial price of this call option is given by

$$\hat{C}_0 = e^{-0.5\times0.04} \times \frac{1}{5}(0.4157 + 0 + 4.1472 + 0 + 0) = 0.894509705.$$

■

The purpose of the last example was to illustrate the Monte Carlo algorithm for simple derivatives in the BSM model. Of course, the price C_0 can be obtained with the Black-Scholes formula: in this case, we have $C_0 = 5.740763479$. We see that the Monte Carlo estimate is far from being accurate. This was to be expected given the very small size of the sample used, i.e. $N = 5$. To benefit from Monte Carlo simulations, one has to generate much larger samples.

19.2.2.2 Discretely-monitored exotic derivatives

Recall that a discretely-monitored exotic derivative has a payoff of the form

$$V_T = g(S_{t_1}, S_{t_2}, \dots, S_{t_n})$$

where the time points $t_0 < t_1 < \cdots < t_n$ are the monitoring times. In this case, the simulation algorithm is not very different from the one we just presented for simple derivatives.

Consider for example a fixed-strike Asian put option with arithmetic average, i.e. an option with payoff

$$V_T = \left(K - \frac{1}{n}\sum_{j=1}^{n} S_{t_j}\right)_+ .$$

Again, as the risk-neutral distribution of V_T is not known, we will:

1) generate realizations of the random vector

$$(S_0, S_{t_1}, S_{t_2}, \dots, S_{t_n}),$$

using the algorithm given in Chapter 14 for a geometric Brownian motion;
2) for each realization, compute the corresponding average (of the Asian option);
3) for each realization, compute the value of the payoff.

For simplicity, let $h = t_j - t_{j-1} = T/n$, for each $j = 1, 2, \dots, n$. Equivalently, we have $t_j = jT/n = jh$. Similarly as in equation (19.2.3), we will use the following equality of distributions: for each $j = 1, 2, \dots, n$, we have

$$\frac{S_{t_j}}{S_{t_{j-1}}} \overset{\mathbb{Q}}{\sim} \exp\left(\left(r - \frac{\sigma^2}{2}\right)h + \sigma\sqrt{h}Z\right)$$

where Z is a standard normal random variable.

Here is the pricing algorithm for discretely-monitored exotic derivatives. First, choose the number of simulations $N \geq 1$. Second, perform the following steps:

1) Generate a sample $Z_{i,j}$, $i = 1, 2, \dots, N, j = 1, 2, \dots, n$, of size $N \times n$, from the standard normal distribution.
2) For each $i = 1, 2, \dots, N$, set recursively for each $j = 1, 2, \dots, n$

$$S_{t_j}(i) = S_{t_{j-1}}(i) \exp\left(\left(r - \frac{\sigma^2}{2}\right)h + \sigma\sqrt{h}Z_{i,j}\right).$$

3) For each $i = 1, 2, \dots, N$, set

$$V_T(i) = g(S_{t_1}(i), S_{t_2}(i), \dots, S_{t_n}(i)).$$

Finally, the Monte Carlo estimator of the option price is given by

$$\widehat{V}_0 = e^{-rT} \times \frac{1}{N}\sum_{i=1}^{N} V_T(i).$$

Example 19.2.3 *Asian put option price*
In the BSM model, a share of stock trades for \$34, the volatility of its log-return is 25% and the risk-free rate is 3%. Let us calculate an estimate of the price of a 3-month fixed-strike Asian put option with a strike price of \$35 whose average is computed at the end of each month.

Since the option matures in 3 months ($T = 3/12$), the arithmetic average is that of $S_{1/12}$, $S_{2/12}$ and $S_{3/12}$. Therefore, we need to generate values of the random vector $(S_{1/12}, S_{2/12}, S_{3/12})$ according to its risk-neutral distribution. If we choose $N = 5$, then we need to generate 5×3 normal random numbers. Assume we have generated the following 15 normal random numbers $Z_{i,j}$:

−1	−0.79	0.85
−1.25	−1.23	0.41
1.21	−1.4	1.62
0.83	−0.29	1.06
0.26	0.21	−0.31

Let us use each row as a sample for each random vector. For example, let us set

$$(Z_{4,1}, Z_{4,2}, Z_{4,3}) = (0.83, -0.29, 1.06).$$

Using the above algorithm, with $S_0 = 34$, $\sigma = 0.25$ and $r = 0.03$, we get the following table:

i	S_0	$S_{1/12}(i)$	$S_{2/12}(i)$	$S_{3/12}(i)$	Average	$V_T(i)$
1	34	31.6294	29.8734	31.7601	31.0876	3.9124
2	34	31.0639	28.4223	29.2728	29.5863	5.4137
3	34	37.0987	33.5300	37.6845	36.1044	0.0000
4	34	36.0951	35.3438	38.1497	36.5295	0.0000
5	34	34.6404	35.1657	34.3841	34.7301	0.2699

For example, we have computed $S_{1/12}(4)$ and $S_{2/12}(4)$ as follows:

$$\begin{aligned} S_{1/12}(4) &= S_0 \exp\left(\left(r - \frac{\sigma^2}{2}\right) h + \sigma\sqrt{h} Z_{4,1}\right) \\ &= 34 \times \exp\left(\left(0.03 - \frac{0.25^2}{2}\right)(1/12) + 0.25\sqrt{1/12} \times 0.83\right) \\ &= 36.0951 \end{aligned}$$

and

$$\begin{aligned} S_{2/12}(4) &= S_{1/12}(4) \exp\left(\left(r - \frac{\sigma^2}{2}\right) h + \sigma\sqrt{h} Z_{4,2}\right) \\ &= 36.0951 \times \exp\left(\left(0.03 - \frac{0.25^2}{2}\right)(1/12) + 0.25\sqrt{1/12} \times -0.29\right) \\ &= 35.3438. \end{aligned}$$

Finally, an estimate of this Asian option price is given by

$$\hat{V}_0 = e^{-0.03\times0.25} \times \frac{1}{5}(3.9124 + 5.4137 + 0 + 0 + 0.2699) = 1.904853358.$$

■

19.2.2.3 Continuously-monitored exotic derivatives

Recall that the payoff of a continuously-monitored exotic derivative is written as some function g of the entire path, i.e.

$$V_T = g(S_t, 0 \leq t \leq T).$$

For example, the payoff of a continuously-monitored version of a fixed-strike Asian put option with arithmetic average is

$$V_T = \left(K - \frac{1}{T}\int_0^T S_t \mathrm{d}t\right)_+.$$

This is definitely more challenging because it is not possible to simulate a full trajectory $\{S_t, 0 \leq t \leq T\}$ for the geometric Brownian motion S in order to compute the above integral. In this case, we need to add an extra layer of approximation, namely for the computation of the payoff itself.

For these derivatives, one will need to generate discretized trajectories of the geometric Brownian motion $S = \{S_t, 0 \leq t \leq T\}$. In the case of the Asian put option, we need to choose n large enough and dates $0 = t_0 < t_1 < \cdots < t_n = T$ to approximate the average:

$$\frac{1}{n}\sum_{j=1}^{n} S_{t_j} \approx \frac{1}{T}\int_0^T S_t \mathrm{d}t.$$

Then, we apply the algorithm for discretely-monitored exotic derivatives.

In general, the idea is that for large n, the payoff can be approximated:

$$V_T = g(S_t, 0 \leq t \leq T) \approx g_n(S_0, S_{t_1}, S_{t_2}, \ldots, S_{t_{n-1}}, S_T), \tag{19.2.4}$$

where g is a *functional* of the whole trajectory $(S_t, 0 \leq t \leq T)$, while g_n is a function of the random vector $(S_0, S_{t_1}, S_{t_2}, \ldots, S_{t_{n-1}}, S_T)$.

In the previous example, we had

$$g(S_t, 0 \leq t \leq T) = \left(K - \frac{1}{T}\int_0^T S_t \mathrm{d}t\right)_+$$

and

$$g_n\left(S_0, S_{t_1}, S_{t_2}, \ldots, S_{t_{n-1}}, S_T\right) = \left(K - \frac{1}{n}\sum_{j=1}^{n} S_{t_j}\right)_+.$$

We are ready to formalize our pricing algorithm for continuously-monitored exotic derivatives. First, choose the number of simulations $N \geq 1$. Then, choose the number of discretization time-points $n \geq 1$ and an approximated version $g_n(S_0, S_{t_1}, S_{t_2}, \ldots, S_{t_{n-1}}, S_T)$ of the payoff $V_T = g(S_t, 0 \leq t \leq T)$ for the approximation in Equation (19.2.4) to be *reasonable*. Finally, perform the following steps:

1) Generate a sample $Z_{i,j}$, $i = 1, 2, \ldots, N, j = 1, 2, \ldots, n$, of size $N \times n$, from the standard normal distribution.
2) For each $i = 1, 2, \ldots, N$, set recursively for each $j = 1, 2, \ldots, n$

$$S_{t_j}(i) = S_{t_{j-1}}(i) \exp\left(\left(r - \frac{\sigma^2}{2}\right)h + \sigma\sqrt{h}Z_{i,j}\right).$$

3) For each $i = 1, 2, \dots, N$, set

$$V_T(i) = g_n\left(S_0, S_{t_1}(i), S_{t_2}(i), \dots, S_{t_{n-1}}(i), S_T(i)\right).$$

Finally, the Monte Carlo estimator of the option price is given by

$$\widehat{V}_0 = \mathrm{e}^{-rT} \times \frac{1}{N} \sum_{i=1}^{N} V_T(i).$$

19.3 Variance reduction techniques

In practice, *crude* Monte Carlo simulations can be demanding, in terms of computer power, and time consuming. *Variance reduction techniques* aim at improving the *performance* of Monte Carlo estimators by reducing their variability.

Therefore, we will seek an alternate estimator $\widehat{\theta}^{\text{alt}}$ which is still unbiased for $\theta = \mathbb{E}[X]$ while having a *lower variance* (i.e. being more precise) than the crude Monte Carlo estimator $\widehat{\theta}$. Mathematically, we want the following two conditions to hold:

$$\mathbb{E}[\widehat{\theta}^{\text{alt}}] = \mathbb{E}[\widehat{\theta}] = \theta,$$
$$\mathbb{V}\text{ar}(\widehat{\theta}^{\text{alt}}) \leq \mathbb{V}\text{ar}(\widehat{\theta}).$$

This is the goal of **variance reduction techniques.**

In what follows, we will present three variance reduction techniques: *stratified sampling, antithetic variates* and *control variates.*

But first, here is a motivational example for variance reduction techniques. In financial engineering and actuarial finance, many situations are such that crude Monte Carlo simulations are inefficient because a significant proportion of simulated payoffs are equal to zero. Consider, for example, the case of a fixed-strike Asian put option with payoff

$$V_T = \left(K - \frac{1}{n}\sum_{j=1}^{n} S_{t_j}\right)_+ .$$

We will generate several realizations of the random vector $(S_0, S_{t_1}, S_{t_2}, \dots, S_{t_n})$ for which the corresponding mean is greater than K, resulting in a payoff equal to 0. In some sense, this is a waste of computer time. Of course, the probability of simulating such values, and thus the number of *wasted* simulations, depends on the parameters. More importantly, this has an impact on the variance of the Monte Carlo estimator.

19.3.1 Stratified sampling

Suppose you have generated 100 uniform random numbers. You should expect to have 10 numbers/realizations *on average* to lie within the interval $[0, 0.1]$. It should also be the case for each interval of the form $[\frac{k-1}{10}, \frac{k}{10}]$, where $k = 1, 2, \dots, 10$.

The left panel of Figure 19.1 shows a histogram counting the number of realizations in each bin, i.e. within each interval $[\frac{k-1}{10}, \frac{k}{10}]$, for $k = 1, 2, \dots, 10$. It is hard to see this is actually coming from a uniform distribution. This is unfortunately a typical situation whenever we simulate a small sample from a given probability distribution.

The right panel of Figure 19.1 shows the probability distribution (a binomial distribution) for the number of realizations in each $[\frac{k-1}{10}, \frac{k}{10}]$. Even if, *on average*, we should have 10 realizations

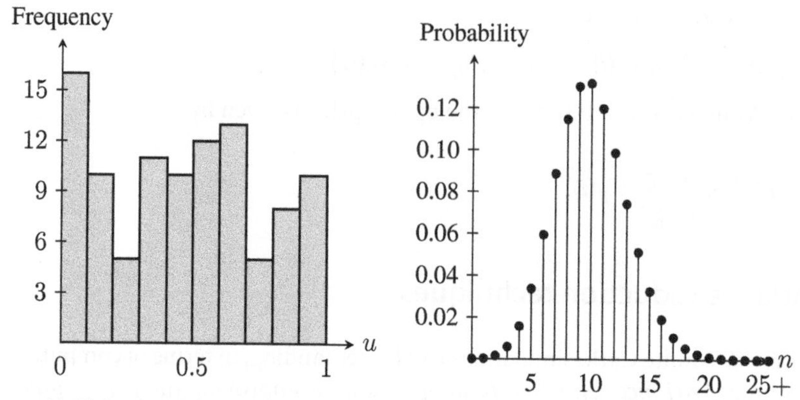

(a) Histogram of 100 simulated uniform random numbers

(b) Probability mass function (for each bin)

Figure 19.1 Histogram of 100 uniform random numbers (left panel) and the probability mass function of a binomial distribution Bin($n = 100$; $p = 0.1$) (right panel)

per bin, the probability of observing from 2 to 18 realizations (out of 100) of the sample within each interval is 99.5%. In other words, with a small sample of size 100, it is likely to deviate significantly from the expected "10 elements per bin."

An easy solution to thwart this variability is simply to sample more. Figure 19.2 shows the histogram of a sample of 10,000 uniform random numbers together with the probability distribution for the number of random numbers in each interval, for a sample of 10,000. In this case, the histogram looks much more *uniform* than what we had before in Figure 19.1. In this second experiment, we expect to have 1,000 realizations per bin. The sampling variability is also much smaller. The probability of having anything between 915 to 1085 elements per bin is 99.5%. In other words, we should have 9–11% of the sample to lie within each interval with almost certainty.

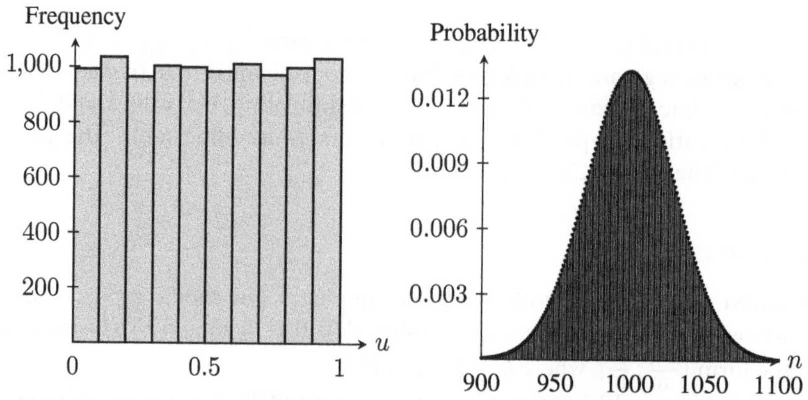

(a) Histogram of 10,000 simulated uniform random numbers

(b) Probability mass function (for each bin)

Figure 19.2 Histogram of 10,000 uniform random numbers (left panel) and the probability mass function of a binomial distribution Bin($n = 10,000$; $p = 0.1$) (right panel). For the latter, only the interval [900, 1100] is displayed

Sampling more is not always a cost-effective solution. To make sure we have exactly 10% of all uniform random numbers in each of the 10 bins is the idea of *stratified sampling*.

19.3.1.1 Method

Fix $m \geq 1$ and partition the interval $(0, 1)$ into m sub-intervals, called **strata**, i.e. let us consider each of the following sub-intervals:

$$\left(0, \frac{1}{m}\right), \left(\frac{1}{m}, \frac{2}{m}\right), \ldots, \left(\frac{m-1}{m}, 1\right).$$

For each $k = 1, 2, \ldots, m$, define

$$U^{(k)} = \frac{k-1}{m} + \frac{U}{m}, \tag{19.3.1}$$

where $U \sim \mathcal{U}(0, 1)$. It is easy to verify that $U^{(k)} \sim \mathcal{U}(\frac{k-1}{m}, \frac{k}{m})$, i.e. $U^{(k)}$ is uniformly distributed over the k-th stratum.

Instead of naively sampling N uniform random numbers from the uniform distribution $\mathcal{U}(0, 1)$, the idea of *stratified sampling* consists in sampling n times from $\mathcal{U}(0, \frac{1}{m})$, n times from $\mathcal{U}(\frac{1}{m}, \frac{2}{m})$, etc., until we have sampled n realizations from each stratum, for a total of $n \times m = N$ random numbers.

In other words, for each $k = 1, 2, \ldots, m$, we generate n realizations of $U^{(k)}$, as defined in (19.3.1), to get $U_j^{(k)}, j = 1, 2, \ldots, n$. The output is a set of uniform random numbers given by

$$U_j^{(k)}, j = 1, 2, \ldots, n \quad \text{and} \quad k = 1, 2, \ldots, m,$$

all taking values in $(0, 1)$ but with an equal number of realizations in each stratum. In other words, stratified sampling forces each stratum to have exactly $1/m$-th of the entire sample of random numbers.

Example 19.3.1 *Simulation from strata*

You sampled the following four uniform random numbers:

$$u_1 = 0.22, u_2 = 0.65, u_3 = 0.89, u_4 = 0.12.$$

First of all, we notice there are two numbers in the first quartile $(0, 1/4]$, none was sampled from the second quartile $(1/4, 1/2]$, while there is one for each of the last two quartiles $(1/2, 3/4]$ and $(3/4, 1]$. Using $m = 4$ strata, we can generate one number per stratum.

With $m = 4$ and $n = 1$, if we set

$$u_1^{(k)} = \frac{k-1}{4} + \frac{u_k}{4},$$

then the corresponding realizations are:

$$u_1^{(1)} = 0 + \frac{0.22}{4} = 0.055$$

$$u_1^{(2)} = \frac{1}{4} + \frac{0.65}{4} = 0.4125$$

$$u_1^{(3)} = \frac{2}{4} + \frac{0.89}{4} = 0.7225$$

$$u_1^{(4)} = \frac{3}{4} + \frac{0.12}{4} = 0.78.$$

Hence, there is a realization coming from each quartile. ■

The implementation of stratified sampling is simple and based on (19.3.1). The algorithm is as follows. First, choose $m \geq 1$, the number of strata, then choose $n \geq 1$, the number of realizations per stratum, so that $N = n \times m$. Then:

1) Simulate N uniform random numbers $U_i, i = 1, 2, \ldots, N$.
2) For each stratum $(\frac{k-1}{m}, \frac{k}{m})$, i.e. for each $k = 1, 2, \ldots, m$, and for each $j = 1, 2, \ldots, n$, set:

$$U_j^{(k)} = \frac{k-1}{m} + \frac{U_{(k-1)n+j}}{m}$$

or, alternatively, set

$$U_j^{(k)} = \frac{k-1}{m} + \frac{U_{(j-1)m+k}}{m}.$$

The uniform random numbers $U_1, U_2, \ldots, U_N$ are allocated to each stratum to simulate realizations of $U_j^{(k)}, j = 1, 2, \ldots, n$ and $k = 1, 2, \ldots, m$. There are two equivalent ways to allocate the U_is, as shown in the following tables.

- Here is the first way to use the uniform random numbers:

k/j	**1**	**2**	**...**	**n**
1	$U_1 \to U_1^{(1)}$	$U_2 \to U_2^{(1)}$	$\cdots$	$U_n \to U_n^{(1)}$
2	$U_{n+1} \to U_1^{(2)}$	$U_{n+2} \to U_2^{(2)}$	$\cdots$	$U_{2n} \to U_n^{(2)}$
$\vdots$	$\vdots$	$\vdots$	$\ddots$	$\vdots$
m	$U_{(m-1)n+1} \to U_1^{(m)}$	$U_{(m-1)n+2} \to U_2^{(m)}$	$\cdots$	$U_{mn} \to U_n^{(m)}$

  In this table, the first n realizations from $\mathcal{U}(0,1)$ are used to generate a sample of n realizations of $U^{(1)}$, then the next n realizations from $\mathcal{U}(0,1)$ are used to simulate n realizations of $U^{(2)}$, etc.
- Here is the second allocation:

k/j	**1**	**2**	**...**	**n**
1	$U_1 \to U_1^{(1)}$	$U_{m+1} \to U_2^{(1)}$	$\cdots$	$U_{(n-1)m+1} \to U_n^{(1)}$
2	$U_2 \to U_1^{(2)}$	$U_{m+2} \to U_2^{(2)}$	$\cdots$	$U_{(n-1)m+2} \to U_n^{(2)}$
$\vdots$	$\vdots$	$\vdots$	$\ddots$	$\vdots$
m	$U_m \to U_1^{(m)}$	$U_{2m} \to U_2^{(m)}$	$\cdots$	$U_{mn} \to U_n^{(m)}$

  In this table, the first m realizations from $\mathcal{U}(0,1)$ are used to generate one realization from each stratum, then the m following realizations from $\mathcal{U}(0,1)$ are used to generate the second realization from each stratum, etc.

To simulate a random variable X using stratified sampling, we simply use the inverse transform method together with the $U_j^{(k)}$s rather than the U_is. More precisely, for each $k = 1, 2, \ldots, m$ and $j = 1, 2, \ldots, n$, using the inverse transform method, set

$$X_j^{(k)} = F_X^{-1}(U_j^{(k)}).$$

Then, the corresponding **stratified sampling estimator** for $\theta = \mathbb{E}[X]$ is given by

$$\widehat{\theta}^{\text{strat}} = \frac{1}{N}\sum_{k=1}^{m}\sum_{j=1}^{n} X_j^{(k)}.$$

It is easy to see that this estimator is also unbiased for θ:

$$\mathbb{E}[\widehat{\theta}^{\text{strat}}] = \frac{1}{N}\sum_{k=1}^{m}\sum_{j=1}^{n} \mathbb{E}[X_j^{(k)}] = \frac{1}{N}\sum_{k=1}^{m}\sum_{j=1}^{n} \theta = \theta,$$

since $N = n \times m$.

It can be verified that the variance of $\widehat{\theta}^{\text{strat}}$ is less than the variance of $\widehat{\theta}$.

Example 19.3.2 *Naive sampling and stratified sampling from the standard normal distribution*
With the uniform random numbers of example 19.3.1, i.e.

$$0.22, 0.65, 0.89, 0.12,$$

we can simulate a standard normal distribution using both naive sampling and stratified sampling.

Naive simulation of a normal distribution is achieved by applying $Z_i = N^{-1}(U_i)$ whereas stratified sampling is obtained using $Z_j^{(k)} = N^{-1}(U_j^{(k)})$ with $U_j^{(k)}$ computed in example 19.3.1. With $n = 1$, the results are shown in the next table:

i, k	U_i	$U_1^{(k)}$	$N^{-1}(U_i)$	$N^{-1}(U_1^{(k)})$
1	0.22	0.055	−0.7722	−1.5982
2	0.65	0.4125	0.3853	−0.2211
3	0.89	0.7225	1.2265	0.5903
4	0.12	0.78	−1.1750	0.7722

Then, we can compute the corresponding stratified sampling estimate of $\theta = 0$:

$$\widehat{\theta}^{\text{strat}} = \frac{1}{4} \times (-1.5982 + (-0.2211) + 0.5903 + 0.7722) = -0.1142.$$

Stratified sampling can also be applied together with the Box-Muller method to generate normal random numbers. This is more efficient than inverting the standard normal cumulative distribution function $N(\cdot)$. ■

19.3.1.2 Application to option pricing

Pricing simple derivatives using stratified sampling is similar to using the standard Monte Carlo estimator of $\widehat{V}_0$, with the only difference that uniform random numbers are generated from stratified sampling.

Example 19.3.3 *Call option price*
For a share of stock trading at \$75, let us use stratified sampling to price a 1-year at-the-money call option in the BSM model with volatility parameter $\sigma = 0.25$ and risk-free rate parameter $r = 0.03$.

We will use the stratified sample of example 19.3.2, so $N = 4$ and

$$z_1 = -1.5982, z_2 = -0.2211, z_3 = 0.5903, z_4 = 0.7722.$$

Then, for each $i = 1, 2, 3, 4$, set

$$S_1(i) = S_0 \exp\left(\left(0.03 - \frac{(0.25)^2}{2}\right) \times 1 + 0.25 \times \sqrt{1} \times z_i\right)$$

and

$$V_1(i) = (S_1(i) - 75)_+.$$

For example, we have

$$S_1(1) = 75 \exp\left(\left(0.03 - \frac{0.25^2}{2}\right) + 0.25 \times -1.5982\right) = 50.23.$$

The other results are shown in the table:

i/k	$N^{-1}(u_1^{(k)})$	$S_1(i)$	$V_1(i)$
1	−1.5982	50.2338865	0
2	−0.2211	70.8778828	0
3	0.5903	86.8175866	11.8175866
4	0.7722	90.8569614	15.8569614

Consequently, the Monte Carlo estimate with stratified sampling is equal to

$$\hat{V}_0 = e^{-0.03} \times \frac{1}{4} \times (0 + 0 + 11.8176 + 15.8569) = 6.714160382.$$

■

Unfortunately, stratified sampling estimators for the pricing of path-dependent options are not easy to implement. We have to realize that a discretized path of a geometric Brownian motion is an n-dimensional vector $(S_{t_1}, S_{t_2}, \ldots, S_{t_n})$, which means that stratified sampling of N such vectors involves using n-dimensional strata. It can become cumbersome fairly quickly.

19.3.2 Antithetic variates

Recall that the variance of the Monte Carlo estimator $\hat{\theta}$ goes to 0 as the sample size N increases. If someone has time and computer power, there is a more efficient way to decrease the variance than simply simulating more values.

19.3.2.1 Method

Whenever $U \sim \mathcal{U}(0, 1)$, it is easy to see that $1 - U$ is also uniformly distributed over $(0, 1)$. Now, let $F(\cdot)$ be the c.d.f. of some distribution with finite mean θ and finite variance s^2. From the inverse transform method, we deduce that

$$X = F^{-1}(U) \quad \text{and} \quad X' = F^{-1}(1 - U) \tag{19.3.2}$$

have the same distribution. In other words,

$$F_X(x) = F_{X'}(x) = F(x),$$

for all $x \in \mathbb{R}$.

It results that

$$\mathbb{E}[X + X'] = 2\theta$$

and

$$\mathbb{V}\text{ar}(X + X') = \mathbb{V}\text{ar}(X) + \mathbb{V}\text{ar}(X') + 2\mathbb{C}\text{ov}(X, X') = 2s^2 + 2\mathbb{C}\text{ov}(X, X'). \tag{19.3.3}$$

Of course, the random variables X and X' are not independent because they are tranformations of the same random variable U.

Example 19.3.4 *Standard normal distribution*
Let $U \sim \mathcal{U}(0, 1)$. Because U and $1 - U$ are both uniformly distributed over $(0, 1)$, we deduce from the inverse transform method that

$$Z = N^{-1}(U) \quad \text{and} \quad Z' = N^{-1}(1 - U)$$

both have a standard normal distribution.

Since the c.d.f. of the standard normal distribution is strictly increasing, we can write $N(Z) = U$. Moreover, from the properties of the normal distribution, we know that $1 - N(x) = N(-x)$, for all $x \in \mathbb{R}$, so we can write

$$N(-Z) = 1 - N(Z) = 1 - U.$$

Finally, we deduce a very interesting property of the normal distribution for antithetic variates:

$$Z' = -Z,$$

so that Z' is the *true reflection* of Z. ■

The idea of the **antithetic variates technique** is to generate a single uniform random number u from which two realizations of X are obtained: $x = F_X^{-1}(u)$ and its *reflection* $x' = F_X^{-1}(1 - u)$. From the simulation of N uniform random numbers, we obtain $2N$ realizations from the distribution of X.

Algorithmically, we choose $N \geq 1$ and then we:

1) generate a sample $U_i, i = 1, 2, \ldots, N$ of size N coming from the distribution of $U \sim \mathcal{U}(0, 1)$;
2) for each $i = 1, 2, \ldots, N$, set $X_i = F_X^{-1}(U_i)$ and $X_i' = F_X^{-1}(1 - U_i)$.

The random couples $(X_1, X_1'), (X_2, X_2'), \ldots, (X_N, X_N')$ are independent and have the same bivariate distribution. But for each $i = 1, 2, \ldots, N$, the random variables X_i and X_i' are not independent.

Example 19.3.5 *Antithetic variates*
Assume you have generated the following five uniform random numbers:

$$u_1 = 0.05, u_2 = 0.74, u_3 = 0.41, u_4 = 0.23, u_5 = 0.88.$$

We want to generate standard normal random numbers.

Using the antithetic variates technique together with the inverse transform method, we can generate 10 standard normal random numbers from the above five uniform random numbers. Indeed, for $i = 1, 2, \ldots, 5$, if we set $z_i = N^{-1}(u_i)$, then we get

$$z_1 = N^{-1}(u_1) = N^{-1}(0.05) = -1.64485$$
$$z_2 = N^{-1}(0.74) = 0.64334$$

and

$$z_3 = -0.22754, \quad z_4 = -0.73884, \quad z_5 = 1.17498.$$

Since we are sampling from the standard normal distribution, we have seen in example 19.3.4 that the antithetic counterparts are such that $z_i' = -z_i$, for each $i = 1, 2, \ldots, 5$.

Consequently,

$$z_1' = 1.64485, \quad z_2' = -0.64334, \quad z_3' = 0.22754, \quad z_4' = 0.73884, \quad z_5' = -1.17498$$

are the other five normal random numbers. ■

In the last example, since we were dealing with the standard normal distribution, we could have used the antithetic variates technique together with the Box-Muller approach (for the sampling of the first five normal random numbers) rather than the inverse transform method. When the sample size is large, the former approach is numerically much more efficient.

Finally, the **antithetic variates estimator** of θ is defined as

$$\hat{\theta}^{\text{anti}} = \frac{1}{2N} \sum_{i=1}^{N} (X_i + X_i'). \tag{19.3.4}$$

In general, the antithetic variates estimator $\hat{\theta}^{\text{anti}}$ is an unbiased estimator for θ:

$$\mathbb{E}[\hat{\theta}^{\text{anti}}] = \frac{1}{2N}\mathbb{E}\left[\sum_{i=1}^{N} (X_i + X_i')\right] = \frac{N}{2N}(\mathbb{E}[X] + \mathbb{E}[X']) = \theta.$$

Given that the antithetic variates estimator is based on a sample twice as large as usual, we should at least make sure that $\hat{\theta}^{\text{anti}}$ is more efficient than simply using the crude Monte Carlo estimator $\hat{\theta}$ with a sample of $2N$. Hence, we need to compute the variance of $\hat{\theta}^{\text{anti}}$.

To do this, we shall note first that when $i \neq j$, the random variables $X_i + X_i'$ and $X_j + X_j'$ are independent and follow the same univariate distribution. Hence, using equation (19.3.3), we obtain

$$\begin{aligned}
\mathbb{V}\text{ar}(\hat{\theta}^{\text{anti}}) &= \mathbb{V}\text{ar}\left(\frac{1}{2N}\sum_{i=1}^{N}(X_i + X_i')\right) \\
&= \frac{N}{4N^2}\mathbb{V}\text{ar}(X + X') \\
&= \frac{1}{4N}(2s^2 + 2\mathbb{C}\text{ov}(X, X')) \\
&= \frac{1}{2N}(s^2 + \mathbb{C}\text{ov}(X, X')).
\end{aligned}$$

The variance of the crude Monte Carlo estimator, obtained with a sample of size $2N$, is given by $\frac{1}{2N}s^2$, as seen previously. Therefore, for the antithetic variates estimator to be more efficient than the crude one, we must have

$$\mathbb{C}\text{ov}(X, X') \leq 0.$$

Clearly, for the standard normal distribution, since $Z' = -Z$, then Z and Z' are negatively correlated. Indeed, we have

$$\mathbb{C}\text{ov}(Z, Z') = \mathbb{C}\text{ov}(Z, -Z) = -\mathbb{V}\text{ar}(Z) = -1 \leq 0.$$

Consequently, for any increasing transformation f, the random variables $f(Z)$ and $f(Z')$ will also be negatively correlated. Otherwise, the latter condition is not always met and needs to be verified.

19.3.2.2 Application to option pricing

Let us now apply the antithetic variates method to option pricing in the BSM model. The idea is to capitalize on the fact that for a standard normal random variable Z, its antithetic counterpart is its *true reflection* $-Z$.

First, we consider a simple derivative with payoff $V_T = g(S_T)$ where $g(\cdot)$ is the payoff function. As for crude Monte Carlo simulations, we will use the relationship of equation (19.2.3), i.e.

$$S_T \overset{\mathbb{Q}}{\sim} S_0 \exp\left(\left(r - \frac{\sigma^2}{2}\right) T + \sigma\sqrt{T}Z\right) \tag{19.3.5}$$

where Z is a standard normal random variable. In this case, since the antithetic counterpart of Z is $-Z$, we also have

$$S'_T \overset{\mathbb{Q}}{\sim} S_0 \exp\left(\left(r - \frac{\sigma^2}{2}\right) T - \sigma\sqrt{T}Z\right). \tag{19.3.6}$$

Therefore, the corresponding antithetic variates algorithm, based on (19.3.5) and (19.3.6), is: choose $N \geq 1$ and then

1) generate a sample Z_i, $i = 1, 2, \ldots, N$ from the standard normal distribution;
2) for each $i = 1, 2, \ldots, N$, set

$$S_T(i) = S_0 \exp\left(\left(r - \frac{\sigma^2}{2}\right) T + \sigma\sqrt{T}Z_i\right)$$

and

$$S'_T(i) = S_0 \exp\left(\left(r - \frac{\sigma^2}{2}\right) T - \sigma\sqrt{T}Z_i\right);$$

3) for each $i = 1, 2, \ldots, N$, set

$$V_T(i) = g(S_T(i)) \quad \text{and} \quad V'_T(i) = g\left(S'_T(i)\right).$$

Finally, the antithetic variates estimator of this option price is given by

$$\hat{V}_0^{\text{anti}} = \mathrm{e}^{-rT} \frac{1}{2N} \sum_{i=1}^{N} \left(V_T(i) + V'_T(i)\right).$$

To ensure that the antithetic variates method works for option pricing, we have to make sure that V_T and V'_T are negatively correlated. One quick approach to verify this is to compute the sample correlation between V_T and V'_T. The closer to -1, the more efficient the method.

There is one important case where the method always works efficiently and this is when V_T is an increasing function of S_T, as for example a call option. In this case, the relationship between V_T and V'_T being monotone, the covariance $\mathbb{C}\text{ov}(V_T, V'_T)$ is negative.

Example 19.3.6 *Antithetic variates estimator*
The current stock price is \$75 and the volatility of its log-return is 20%. If the risk-free rate is 4%, calculate the antithetic variates estimator for the price of an at-the-money put option maturing in 1 year. Use the following three normal variates:

$$z_1 = -1.45, z_2 = 0.22, z_3 = -0.07.$$

With three normal random numbers, we will get six realizations of the payoff. Since $S_0 = 75$, $\sigma = 0.2$ and $r = 0.04$, we get:

i	Z_i	Z'_i	$S_1(i)$	$S'_1(i)$	$V_1(i)$	$V'_1(i)$
1	−1.45	1.45	57.2535	102.2569	17.7465	0
2	0.22	−0.22	79.9569	73.2214	0	1.7786
3	−0.07	0.07	75.4514	77.5938	0	0

For example, with $z_1 = -1.45$, we obtain

$$S_1(1) = 75 \exp\left(\left(0.04 - \frac{0.2^2}{2}\right) + 0.2 \times -1.45\right) = 57.2535,$$

$$S'_1(1) = 75 \exp\left(\left(0.04 - \frac{0.2^2}{2}\right) + 0.2 \times 1.45\right) = 102.2569,$$

and the corresponding payoffs are

$$V_1(1) = \max(75 - 57.2535, 0) = 17.7465,$$
$$V'_1(1) = \max(75 - 102.2569, 0) = 0.$$

Then, the antithetic variates estimate of this option's initial value is

$$\hat{V}_0^{\text{anti}} = e^{-0.04}\frac{1}{6}(17.7465 + 1.7786) = 3.12658498.$$

■

Now, for path-dependent derivatives, both discretely- and continuously-monitored, the idea is similar: we apply the antithetic method for all time steps and trajectories. In the end, we get a final sample of normal random numbers which is twice the original size and hence we get twice the number of discretized trajectories of the geometric Brownian motion.

More precisely, consider dates $0 = t_0 < t_1 < \cdots < t_n = T$ and let us generate realizations of the random vector (discretized trajectory)

$$\left(S_0, S_{t_1}, S_{t_2}, \ldots, S_T\right)$$

using the following risk-neutral distributions:

$$\frac{S_{t_j}}{S_{t_{j-1}}} \overset{\mathbb{Q}}{\sim} \exp\left(\left(r - \frac{\sigma^2}{2}\right)(t_j - t_{j-1}) \pm \sigma\sqrt{t_j - t_{j-1}}Z\right)$$

where Z is a standard normal random variable. In other words, at each time step, we get an increment from Z and one from its antithetic counterpart $-Z$.

The antithetic variates pricing algorithm for path-dependent derivatives goes as follows. First, choose the number of sample paths $N \geq 1$. Then, choose the number of time steps $n \geq 1$ and set the time-points $t_j = jh, j = 1, 2, \ldots, n$ with $h = T/n$:

1) discrete monitoring: according to the option payoff $V_T = g(S_0, S_{t_1}, S_{t_2}, \ldots, S_{t_{n-1}}, S_T)$;
2) continuous monitoring: choose a good approximation

$$g_n(S_0, S_{t_1}, S_{t_2}, \ldots, S_{t_{n-1}}, S_T)$$

of the payoff $V_T = g(S_t, 0 \leq t \leq T)$, as in equation (19.2.4).

Then, perform the following steps:

1) Generate a sample $Z_{i,j}, i = 1, 2, \ldots, N, j = 1, 2, \ldots, n$, of size $N \times n$, from the standard normal distribution.
2) For each $i = 1, 2, \ldots, N$, set recursively for each $j = 1, 2, \ldots, n$

$$S_{t_j}(i) = S_{t_{j-1}}(i) \exp\left(\left(r - \frac{\sigma^2}{2}\right) h + \sigma\sqrt{h} Z_{i,j}\right),$$

$$S'_{t_j}(i) = S'_{t_{j-1}}(i) \exp\left(\left(r - \frac{\sigma^2}{2}\right) h - \sigma\sqrt{h} Z_{i,j}\right).$$

3) For each $i = 1, 2, \ldots, N$, set

$$V_T(i) = g_n\left(S_0, S_{t_1}(i), S_{t_2}(i), \ldots, S_{t_{n-1}}(i), S_T(i)\right),$$
$$V'_T(i) = g_n\left(S_0, S'_{t_1}(i), S'_{t_2}(i), \ldots, S'_{t_{n-1}}(i), S'_T(i)\right).$$

A path from a geometric Brownian motion along with its antithetic version is shown in Figure 19.3. Finally, the antithetic variates estimator of the option price with payoff V_T is given by

$$\hat{V}_0^{\text{anti}} = e^{-rT} \times \frac{1}{2N} \sum_{i=1}^{N} (V_T(i) + V'_T(i)).$$

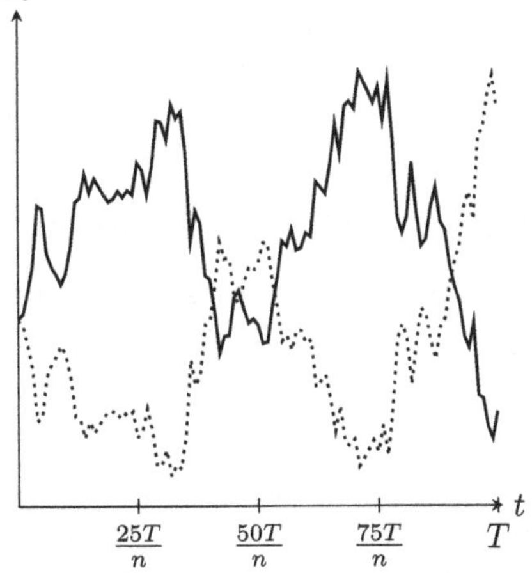

Figure 19.3 A sample path of a geometric Brownian motion (solid line) and its corresponding antithetic version (dotted line)

In the above algorithm, we use the sample of $Z_{i,j}$s to simulate N discretized trajectories and we compute the corresponding values of the payoff, i.e. the $V_T(i)$s, as we did in the crude Monte Carlo approach. Again, we capitalize on the antithetic variates method and the fact that for a standard normal random variable Z its antithetic counterpart is its *true reflection* $-Z$. Therefore, without generating any new random numbers, we use the $-Z_{i,j}$s and obtain an extra N discretized trajectories and N payoff values.

Example 19.3.7 *Floating-strike lookback call option*
Let us simulate values of the payoff of a 3-month floating-strike lookback call option using the antithetic variates method. Assume the underlying asset price is monitored at the end of each month. Consider a BSM model with $S_0 = 100$, $r = 0.02$, $\sigma = 0.3$. We will simulate paths of a GBM over a 3-month period with the following standard normal variates: $1.11, -1.69, 0.86$.

In this example, we have $N = 1$. Since $r - \sigma^2/2 = 0.02 - 0.5 \times 0.3^2 = -0.025$, applying the above algorithm, we obtain:

$$S_0 = 100$$
$$S_{1/12}(1) = S_0 \times \exp(-0.025 \times 1/12 + 0.3\sqrt{1/12} \times 1.11) = 109.8609717$$
$$S_{2/12}(1) = S_{1/12}(1) \times \exp(-0.025 \times 1/12 + 0.3\sqrt{1/12} \times -1.69) = 86.20503265$$
$$S_{3/12}(1) = S_{2/12}(1) \times \exp(-0.025 \times 1/12 + 0.3\sqrt{1/12} \times 0.86) = 107.5079757$$

and

$$S_0 = 100$$
$$S'_{1/12}(1) = S_0 \times \exp(-0.025 \times 1/12 + 0.3\sqrt{1/12} \times -1.11) = 90.64565757$$
$$S'_{2/12}(1) = S'_{1/12}(1) \times \exp(-0.025 \times 1/12 + 0.3\sqrt{1/12} \times 1.69) = 115.5201699$$
$$S'_{3/12}(1) = S'_{2/12}(1) \times \exp(-0.025 \times 1/12 + 0.3\sqrt{1/12} \times -0.86) = 92.62959286.$$

The payoff of this 3-month floating-strike lookback call option being

$$V_{3/12} = S_{3/12} - \min(S_{1/12}, S_{2/12}, S_{3/12}),$$

we have $V_{3/12}(1) = 107.5079757 - \min(109.86, 86.21, 107.51) = 21.30294305$ and $V'_{3/12}(1) = 92.62959286 - \min(90.65, 115.52, 92.63) = 1.983935288$, yielding an antithetic estimate of

$$\hat{V}_0^{\text{anti}} = e^{-0.02 \times 3/12} \times \frac{1}{2} \times (21.30294305 + 1.983935288) = 11.58537.$$

Of course, since $N = 1$, this is (probably) a very inaccurate estimate. ■

19.3.3 Control variates

The **control variates method** is a technique that seeks to improve the crude Monte Carlo estimator for $\theta = \mathbb{E}[X]$ using information about another expectation $\varphi = \mathbb{E}[Y]$, where Y is a random variable *related to* X and known as the **control variate**. For the method to be efficient, we need to know how to (analytically) compute φ quickly and, ideally, we should know the relationship between X and Y.

19.3.3.1 Method

The method is as follows. First, we choose another random variable Y. Then, we generate N independent random couples

$$(X_1, Y_1), (X_2, Y_2), \dots, (X_N, Y_N),$$

all having the same distribution as (X, Y). For a constant β, we define a new estimator for θ as

$$\widehat{\theta}^\beta = \frac{1}{N}\sum_{i=1}^{N}(X_i - \beta(Y_i - \varphi))$$

which can be rewritten as

$$\widehat{\theta}^\beta = \overline{X} - \beta(\overline{Y} - \varphi),$$

with $\overline{X} = (1/N)\sum_{i=1}^{N} X_i$ and $\overline{Y} = (1/N)\sum_{i=1}^{N} Y_i$.

The idea of the new estimator is simple. Whenever X and Y are closely related, the approximation error of estimating φ by

$$\widehat{\varphi} = \overline{Y}$$

should be tied to the approximation error we get by estimating θ with $\widehat{\theta}$. We expect some proportionality between the difference $\widehat{\varphi} - \varphi$ and $\widehat{\theta}^\beta - \theta$. Since we have knowledge about $\widehat{\varphi} - \varphi$, we will use this information to improve the Monte Carlo estimator $\widehat{\theta}^\beta$.

This new estimator for θ is clearly an unbiased estimator because

$$\mathbb{E}[\widehat{\theta}^\beta] = \frac{1}{N}\sum_{i=1}^{N}(\mathbb{E}[X_i] - \beta(\mathbb{E}[Y_i] - \varphi)) = \frac{1}{N}\sum_{i=1}^{N}(\theta - \beta \times 0) = \theta.$$

We should also analyze the variance of $\widehat{\theta}^\beta$. It is obtained by making use of the independence between the (X_i, Y_i)s. Recall that the couples are independent, but for each couple, i.e. for each i, the random variables X_i and Y_i are dependent. We get

$$\begin{aligned}
\mathbb{V}\mathrm{ar}(\widehat{\theta}^\beta) &= \frac{1}{N^2}\sum_{i=1}^{N}\mathbb{V}\mathrm{ar}(X_i - \beta(Y_i - \varphi)) \\
&= \frac{1}{N}(\mathbb{V}\mathrm{ar}(X) + \beta^2\mathbb{V}\mathrm{ar}(Y) - 2\beta\mathbb{C}\mathrm{ov}(X, Y)) \\
&= \mathbb{V}\mathrm{ar}(\widehat{\theta}) + \frac{1}{N}\left\{\beta^2\mathbb{V}\mathrm{ar}(Y) - 2\beta\mathbb{C}\mathrm{ov}(X, Y)\right\}.
\end{aligned}$$

In order for this estimator to *perform better* than the usual Monte Carlo estimator, we want to choose β to get

$$\mathbb{V}\mathrm{ar}(\widehat{\theta}^\beta) \leq \mathbb{V}\mathrm{ar}(\widehat{\theta}).$$

More than that, we will choose β to minimize this variance, i.e. to minimize the function

$$\beta \mapsto \mathbb{V}\mathrm{ar}(\widehat{\theta}^\beta).$$

Using elementary calculus, we see that it is minimal if we choose

$$\beta^* = \frac{\mathbb{C}\mathrm{ov}(X, Y)}{\mathbb{V}\mathrm{ar}(Y)}. \tag{19.3.7}$$

We can then verify that

$$\begin{aligned}\mathbb{V}\text{ar}(\widehat{\theta}^{\beta^*}) &= \mathbb{V}\text{ar}(\widehat{\theta}) + \frac{1}{N}\left\{\frac{(\mathbb{C}\text{ov}(X,Y))^2}{\mathbb{V}\text{ar}(Y)} - 2\frac{(\mathbb{C}\text{ov}(X,Y))^2}{\mathbb{V}\text{ar}(Y)}\right\}\\ &= \mathbb{V}\text{ar}(\widehat{\theta}) - \frac{1}{N}\left\{\frac{(\mathbb{C}\text{ov}(X,Y))^2}{\mathbb{V}\text{ar}(Y)}\right\}\\ &= \mathbb{V}\text{ar}(\widehat{\theta}) - \frac{1}{N}\left\{(\mathbb{C}\text{orr}(X,Y))^2\mathbb{V}\text{ar}(X)\right\}\\ &\leq \mathbb{V}\text{ar}(\widehat{\theta}),\end{aligned}$$

as desired. From the above computations, we see that only the magnitude of the correlation between X and Y has an impact, not its direction.

In conclusion, the **exact control variates estimator** is given by

$$\widehat{\theta}^{\text{cont}} = \frac{1}{N}\sum_{i=1}^{N}\left(X_i - \frac{\mathbb{C}\text{ov}(X,Y)}{\mathbb{V}\text{ar}(Y)}(Y_i - \varphi)\right) = \widehat{\theta} - \beta^*(\overline{Y} - \varphi). \tag{19.3.8}$$

Unfortunately, in most cases, $\mathbb{C}\text{ov}(X,Y)$ and/or $\mathbb{V}\text{ar}(Y)$ are unknown (or are too complicated to be computed), and so is β^*. Then, from a computational point of view, we will need to approximate β^* using an estimator that we denote by $\widehat{\beta}$. First, recall that the sample covariance is given by

$$\widehat{\mathbb{C}\text{ov}}(X,Y) = \frac{1}{N}\sum_{i=1}^{N}(X_i - \overline{X})(Y_i - \overline{Y}).$$

Consequently, we set

$$\widehat{\beta} = \frac{\sum_{i=1}^{N}(X_i - \overline{X})(Y_i - \overline{Y})}{\sum_{i=1}^{N}(Y_i - \overline{Y})^2} \tag{19.3.9}$$

and then we replace the estimator in (19.3.8) by the following **control variates estimator**:

$$\widehat{\theta}^{\text{cont}} = \widehat{\theta} - \widehat{\beta}(\overline{Y} - \varphi). \tag{19.3.10}$$

For simplicity, we used the same notation in (19.3.10) as in (19.3.8). Note that the estimator $\widehat{\beta}$ can be viewed as the slope of the regression of X on Y. This suggests that we can add more than one control variate to further reduce the variance of $\widehat{\theta}^{\text{cont}}$.

It is important to realize that replacing β^* by $\widehat{\beta}$ will introduce dependence and a bias. First of all, the simulated values of $X_i - \widehat{\beta}(Y_i - \varphi)$, for $i = 1, 2, \dots, N$, will be dependent (because of $\widehat{\beta}$). Moreover, $\widehat{\beta} \to \beta^*$ as $N \to \infty$ so there can be a bias in small samples. The combined effect is usually small and ignored in practice.

19.3.3.2 Application to option pricing

The application of the control variates method to approximate option prices implies that $X = V_T$ and that we want to estimate $\theta = \mathbb{E}^{\mathbb{Q}}[V_T]$. For the method to be successful, we must choose the control variate wisely. Once this is done, the control variates method is easy to implement.

Assume we have discretely-monitored payoffs of the form

$$V_T = g(S_0, S_h, S_{2h}, \dots, S_{(n-1)h}, S_T)$$

and

$$\widetilde{V}_T = \tilde{g}(S_0, S_h, S_{2h}, \dots, S_{(n-1)h}, S_T),$$

where g and $\tilde{g}$ are the payoff functions, respectively, of the targeted option and the control option. Of course, if $n = 1$, they are simple derivatives. The case of continously-monitored exotic payoffs V_T is treated as before: we approximate the continuously-monitored exotic payoff with a similar discretely-monitored payoff.

The control variates pricing algorithm is as follows:

1) Choose an appropriate control variate $Y = \widetilde{V}_T$ and compute analytically the value of $\varphi = \mathbb{E}^{\mathbb{Q}}[\widetilde{V}_T]$.
2) Choose the number of sample paths $N \geq 1$ and the number of discretization time-points $n \geq 1$.
3) For each $i = 1, 2, \dots, N$:
   (a) Generate a discretized path of the geometric Brownian motion using its risk-neutral distribution (i.e. with respect to $\mathbb{Q}$), i.e.

   $$(S_0, S_h(i), S_{2h}(i), \dots, S_{(n-1)h}(i), S_T(i)).$$

   (b) Compute the corresponding value of the payoff:

   $$X_i = V_T(i) = g(S_0, S_h(i), S_{2h}(i), \dots, S_{(n-1)h}(i), S_T(i)).$$

   (c) Compute the corresponding value of the control payoff:

   $$Y_i = \widetilde{V}_T(i) = \tilde{g}(S_0, S_h(i), S_{2h}(i), \dots, S_{(n-1)h}(i), S_T(i)).$$

4) Compute $\overline{X}$, $\overline{Y}$ and $\widehat{\beta}$, using the above simulations.

Finally, the control variates estimator of the option price is

$$\widehat{V}_0^{\text{cont}} = \mathrm{e}^{-rT}(\overline{X} - \widehat{\beta}(\overline{Y} - \varphi)).$$

One natural choice of control variate for simple payoffs of the form $X = g(S_T)$ is the underlying asset itself, i.e. $Y = S_T$. In principle, $Y = S_T$ should be strongly correlated with $X = g(S_T)$. The next example illustrates how to price a simple option using such a control variate.

Example 19.3.8 *Application of the control variates method*
Suppose we want to price a 6-month 70-strike call option, using the control variates method, in a BSM model with the following parameters: $S_0 = 80$, $\sigma = 0.2$ and $r = 0.02$.

In this case, we have $X = (S_T - K)_+$. First, we want to find a suitable control variate Y: it needs to be strongly correlated to $X = (S_T - K)_+$ and φ should be easily computable. As mentioned above, $Y = S_T$ can be an interesting choice of control variate for a simple payoff. Indeed, when the value of $Y = S_T$ increases, the option payoff $X = (S_T - K)_+$ increases as well. Moreover, we know that

$$\varphi = \mathbb{E}^{\mathbb{Q}}[S_T] = \mathrm{e}^{rT} S_0.$$

Next, we generate standard normal numbers. Assume we have obtained the following values:

$$2.5464, 0.8383, -1.0859, -0.8621, 1.4083.$$

This means $N = 5$. The next table shows the resulting values for the stock price along with their corresponding payoffs. Note that we have $T = 0.5$ and $K = 70$.

i	Z_i	$Y_i = S_{0.5}(i)$	$X_i = (S_{0.5}(i) - 70)_+$
1	2.5464	114.6796	44.67958
2	0.8383	90.06937	20.06937
3	−1.0859	68.6113	0
4	−0.8621	70.81758	0.817581
5	1.4083	97.63054	27.63054

Then, we compute the following quantities:

$$\overline{X} = \frac{1}{5}(44.68 + 20.07 + \cdots + 27.63) = 18.36167289$$

$$\overline{Y} = \frac{1}{5}(114.68 + 90.07 + \cdots + 97.63) = 88.36167289$$

$$\widehat{\text{Var}}(Y) = \frac{1}{5}\sum_{i=1}^{5}(Y_i - \overline{Y})^2 = 295.8665746$$

$$\widehat{\text{Cov}}(X, Y) = \frac{1}{5}\sum_{i=1}^{5}(X_i - \overline{X})(Y_i - \overline{Y}) = 290.3810885$$

$$\hat{\beta} = \frac{290.3810885}{295.8665746} = 0.981459595,$$

where, to compute $\hat{\beta}$, we used the expression in (19.3.9). Note that, in this case, we could have computed the exact value of β^* as given in equation (19.3.7).

Finally, using the above quantities in (19.3.10), we obtain

$$\begin{aligned}\hat{\theta}^{\text{cont}} &= \overline{X} - \hat{\beta}(\overline{Y} - \varphi)\\ &= 18.36167289 - 0.981459595 \times (88.36167289 - 80 \times e^{0.02\times0.5})\\ &= 10.94413543\end{aligned}$$

and the control variate estimate of the call option price is

$$\hat{V}_0 = e^{-rT}\hat{\theta}^{\text{cont}} = e^{-0.02\times0.5} \times 10.94413543 = 10.41038365.$$

To better understand how the control variates method works and how it improves on the crude Monte Carlo estimator $\overline{X}$, let us look at the approximation error made when estimating φ with $\overline{Y}$. In this example, we had

$$\varphi = 80e^{0.02\times0.5} = 80.80401337 < \overline{Y} = 88.36167289$$

which means $\overline{Y}$ overestimates φ. Since X and Y are positively correlated, we also expect $\overline{X}$ to be overestimating θ. This is why we correct the crude Monte Carlo estimator $\overline{X}$ by adding the extra term $-\hat{\beta}(\overline{Y} - \varphi)$. ■

A typical situation where the control variates method is particularly useful is for Asian options whose payoff is based on a discrete *arithmetic* average. More specifically, let us consider the following fixed-strike Asian call payoff:

$$X = \left(\frac{1}{n}(S_{t_1} + S_{t_2} + \cdots + S_{t_n}) - K\right)_+.$$

As a control variate, we can use the payoff of a similar Asian option based on a discrete *geometric* average instead, i.e. choose

$$Y = \left(\left(S_{t_1} \times S_{t_2} \times \cdots \times S_{t_n} \right)^{\frac{1}{n}} - K \right)_+ .$$

We expect the correlation between X and Y to be strong and, as we saw in Chapter 18, there exists a closed-form expression for the price of options with payoff Y in the BSM model. The control variates method can then be applied directly in this case.

Other simulation methods for option pricing

If we seek to increase the efficiency of crude Monte Carlo simulations, there are two popular approaches that we have not looked at in this chapter due to their complexity.

Importance sampling is a variance reduction technique that consists in finding an alternate distribution for X that puts more emphasis where it matters the most. For example, when pricing deep out-of-the-money options, a large proportion of the sample will have a zero payoff. Importance sampling will shift the distribution such that positive payoffs occur more often while making sure that $\widehat{\theta}$ remains unbiased. The method essentially relies on a change of probability measure (see Chapter 17) and requires a deep knowledge of the problem to make sure the variance of $\widehat{\theta}$ is always reduced.

Quasi Monte Carlo methods consist in generating uniform numbers by using a deterministic sequence (known as low-discrepancy sequences) rather than a PRNG. As a result, any quantity computed from quasi Monte Carlo is not random anymore. Because an expectation can be written as an integral, quasi Monte Carlo is a method that is in between pure Monte Carlo (as described in this chapter) and numerical integration (trapezoid rule, Newton-Cotes, etc.). To address the lack of randomness of quasi Monte Carlo methods, one can randomize the latter by combining it with a PRNG.

19.4 Summary

Random numbers and Monte Carlo simulations

- Pseudo-random number generator (PRNG): algorithm generating numbers that appear to be random.
- Uniform random numbers: independent realizations $u_1, u_2, \ldots, u_n$ of $U \sim \mathcal{U}(0,1)$ generated with a PRNG.
- Quantile function: inverse of F_X defined by

$$F_X^{-1}(u) = \inf\{x \in \mathbb{R} : \mathbb{P}(X \leq x) \geq u\}, \quad u \in [0,1].$$

- Inverse transform method: basic simulation method to generate random numbers $x_1, x_2, \ldots, x_n$ from F_X, i.e. generate independent realizations of X: for each i,
  1) simulate a uniform random number u_i from $\mathcal{U}(0,1)$;
  2) set $x_i = F_X^{-1}(u_i)$.
- Normal random numbers: independent realizations $z_1, z_2, \ldots, z_n$ of $Z \sim \mathcal{N}(0,1)$ generated with the inverse transform method (or preferably with the Box-Muller technique).
- Monte Carlo estimator of $\theta = \mathbb{E}[X]$:

$$\widehat{\theta} = \frac{1}{N} \sum_{i=1}^{N} X_i,$$

where $X_1, X_2, \ldots, X_N$ are independent and all following the distribution F_X.

- Monte Carlo estimator $\hat{\theta}$:
  1) is unbiased;
  2) its has a variance that goes to 0 as the sample size N increases.

Monte Carlo simulations for option pricing

- Monte Carlo pricing algorithm for the payoff $V_T = g(S_T)$: choose $N \geq 1$ and then
  1) generate $Z_i, i = 1, 2, \dots, N$ from the standard normal distribution;
  2) for each $i = 1, 2, \dots, N$, set
  $$S_T(i) = S_0 \exp\left(\left(r - \frac{\sigma^2}{2}\right) T + \sigma\sqrt{T} Z_i\right);$$
  3) for each $i = 1, 2, \dots, N$, set
  $$V_T(i) = g(S_T(i));$$
  4) finally, the Monte Carlo estimator of the option's price
  $$\hat{V}_0 = \mathrm{e}^{-rT} \times \frac{1}{N} \sum_{i=1}^{N} V_T(i).$$
- Monte Carlo pricing algorithm for the payoff $V_T = g(S_{t_1}, S_{t_2}, \dots, S_{t_n})$: set $h = t_j - t_{j-1} = T/n$ (equally spaced), choose $N \geq 1$ and then
  1) generate $Z_{i,j}, i = 1, 2, \dots, N, j = 1, 2, \dots, n$ from the standard normal distribution;
  2) for each $i = 1, 2, \dots, N$, set recursively, for each $j = 1, 2, \dots, n$,
  $$S_{t_j}(i) = S_{t_{j-1}}(i) \exp\left(\left(r - \frac{\sigma^2}{2}\right) h + \sigma\sqrt{h} Z_{i,j}\right);$$
  3) for each $i = 1, 2, \dots, N$, set
  $$V_T(i) = g\left(S_{t_1}(i), S_{t_2}(i), \dots, S_{t_n}(i)\right);$$
  4) finally, the Monte Carlo estimator of the option's price
  $$\hat{V}_0 = \mathrm{e}^{-rT} \times \frac{1}{N} \sum_{i=1}^{N} V_T(i).$$
- Monte Carlo pricing algorithm for the payoff $V_T = g(S_t, 0 \leq t \leq T)$: choose $N \geq 1$ and then choose $n \geq 1$ and g_n such that
  $$V_T = g(S_t, 0 \leq t \leq T) \approx g_n\left(S_0, S_{t_1}, S_{t_2}, \dots, S_{t_{n-1}}, S_T\right),$$
  Then apply the previous algorithm for $g_n(S_0, S_{t_1}, S_{t_2}, \dots, S_{t_{n-1}}, S_T)$.

Variance reduction techniques in the BSM model

- Variance reduction technique: define an estimator $\hat{\theta}^{\text{alt}}$ such that
  - $\hat{\theta}^{\text{alt}}$ is unbiased;
  - $\mathbb{V}\text{ar}(\hat{\theta}^{\text{alt}}) \leq \mathbb{V}\text{ar}(\hat{\theta})$.
- Stratified sampling: partitioning the interval $[0, 1]$ into m strata and sampling $n = N/m$ uniform random numbers from each stratum $(\frac{k-1}{m}, \frac{k}{m}]$, where $k = 1, 2, \dots, m$.

- Pricing algorithm with stratified sampling: choose $m \geq 1$ and $n \geq 1$ so that $N = n \times m$, and then
  1) simulate N uniform random numbers $U_i, i = 1, 2, \dots, N$;
  2) for each stratum, i.e. for each $k = 1, 2, \dots, m$, and for each $j = 1, 2, \dots, n$, set:

$$U_j^{(k)} = \frac{k-1}{m} + \frac{U_{(k-1)n+j}}{m};$$

  3) compute the N corresponding realizations of V_T using $U_j^{(k)}$;
  4) stratified sampling estimator of the option price:

$$\hat{V}_0^{\text{strat}} = \frac{1}{N} \sum_{k=1}^{m} \sum_{j=1}^{n} X_j^{(k)}.$$

- Antithetic variates: for each uniform random number u, sample V_T twice with the normal random numbers $z = N^{-1}(u)$ then $z' = N^{-1}(1-u)$.
- Pricing algorithm with antithetic variates: choose $N \geq 1$ and then
  1) generate a sample $Z_i, i = 1, 2, \dots, N$ from the standard normal distribution;
  2) for each $i = 1, 2, \dots, N$, set

$$S_T(i) = S_0 \exp\left(\left(r - \frac{\sigma^2}{2}\right) T + \sigma\sqrt{T} Z_i\right)$$
$$S_T'(i) = S_0 \exp\left(\left(r - \frac{\sigma^2}{2}\right) T - \sigma\sqrt{T} Z_i\right)$$
$$V_T(i) = g(S_T(i))$$
$$V_T'(i) = g\left(S_T'(i)\right);$$

  3) antithetic variates estimator of the option price:

$$\hat{V}_0^{\text{anti}} = \mathrm{e}^{-rT} \frac{1}{2N} \sum_{i=1}^{N} \left(V_T(i) + V_T'(i)\right).$$

- Control variates: approximating the error committed with $\hat{\theta}$ by computing the true error committed by estimating $\varphi = \mathbb{E}[Y]$ with $\hat{\varphi}$, where Y is the control variate.
- Pricing algorithm with control variates:
  1) choose $Y = \widetilde{V}_T$ and compute $\varphi = \mathbb{E}^{\mathbb{Q}}[\widetilde{V}_T]$;
  2) choose the number of sample paths $N \geq 1$ and the number of discretization time-points $n \geq 1$;
  3) for each $i = 1, 2, \dots, N$: generate a discretized path of the GBM (using its risk-neutral distribution)

$$(S_0, S_h(i), S_{2h}(i), \dots, S_{(n-1)h}(i), S_T(i))$$

  and compute

$$X_i = g(S_0, S_h(i), S_{2h}(i), \dots, S_{(n-1)h}(i), S_T(i))$$
$$Y_i = \tilde{g}(S_0, S_h(i), S_{2h}(i), \dots, S_{(n-1)h}(i), S_T(i)).$$

  4) compute $\overline{X}$, $\overline{Y}$ and $\hat{\beta}$.
  5) control variates estimator of the option price:

$$\hat{V}_0^{\text{cont}} = \mathrm{e}^{-rT}(\overline{X} - \hat{\beta}(\overline{Y} - \varphi)).$$

19.5 Exercises

For the following exercises, the random numbers in Table 19.1 have been generated with a computer. Note that U_i is drawn from a uniform distribution whereas $Z_i = N^{-1}(U_i)$. For exercises 1–3 and 5, assume the BSM model holds with the following parameters: $S_0 = 100$, $\mu = 0.08$, $r = 0.03$ and $\sigma = 0.25$.

Table 19.1 Random numbers

i	U_i	Z_i
1	0.0371	−1.7861
2	0.6421	0.3641
3	0.8604	1.0818
4	0.032	−1.8536
5	0.8116	0.8838
6	0.4274	−0.1831
7	0.2277	−0.7465
8	0.7707	0.7411
9	0.9266	1.4508
10	0.6797	0.4669
11	0.3265	−0.4499
12	0.3631	−0.3503

19.1 Compute the crude Monte Carlo estimate of the initial price for the following derivatives:
(a) a 105-strike 6-month call option;
(b) a 100-strike 3-month gap put option with $H = 97$;
(c) a 3-month asset-or-nothing call option with strike price $K = 108$;
(d) an investment guarantee maturing in 1 year that applies on 75% of an initial investment of 1,000.
For each contract, use the first five random numbers of Table 19.1.

19.2 In the preceding exercise, use simulations to compute the probability that each of these derivatives will mature in the money.

19.3 Compute the crude Monte Carlo estimate of the initial price of the following exotic derivatives:
(a) a 3-month floating-strike lookback call option monitored monthly;
(b) a 3-year CAR EIA with participation rate of 75% and an initial investment of 1000;
(c) a 3-month floating-strike Asian put option monitored monthly;
(d) a 3-month up-and-in barrier call option with $H = 102$ and $K = 99$.
For each contract, generate four paths using the first three random numbers of Table 19.1 for the first path, the next three random numbers for the second path, and so on.

19.4 Let X be an exponentially distributed random variable with mean 100.
(a) Using four strata with three random numbers each, compute a stratified sampling estimator of $\mathbb{E}[\log(X)]$. Use the first three random numbers to fill in the first stratum

$(0, 0.25]$, the next three random numbers to fill in the second stratum $(0.25, 0.50]$, and so on.

(b) Compute the antithetic variates estimator of $\mathbb{E}[\log(X)]$ using the first five random numbers.

(c) Compute the control variates estimator of $\mathbb{E}[\log(X)]$ using X as a control variate and the first five random numbers.

19.5 Consider a 1-year at-the-money call option.

(a) Compute the stratified sampling estimator of the initial price of that option. Use the first three random numbers to fill in the first stratum $(0, 0.25]$, the next three random numbers to fill in the second stratum $(0.25, 0.50]$, and so on.

(b) Compute the antithetic variates estimator of the initial price of that option using the first five random numbers.

(c) Compute the control variates estimator of the initial price of that option using the payoff of an asset-or-nothing call option with same strike as the control variate. Use only the first five random numbers.

19.6 For the antithetic variates estimator, show that:

(a) If $X \sim \mathcal{N}(m, s^2)$, then $\mathbb{V}\mathrm{ar}(\widehat{\theta}^{\mathrm{anti}}) = 0$.

(b) If $X = bZ^2$ with $Z \sim \mathcal{N}(0, 1)$, then $\mathbb{V}\mathrm{ar}(\widehat{\theta}^{\mathrm{anti}}) = \mathbb{V}\mathrm{ar}(\widehat{\theta})$.

19.7 Let Y be a random variable with mean m and variance s^2 (not necessarily normal) and X be such that

$$X = \begin{cases} 0, & \text{if } I = 0 \\ Y, & \text{if } I = 1 \end{cases}$$

where I is a Bernoulli random variable with distribution $\mathbb{P}(I = 1) = p = 1 - \mathbb{P}(I = 0)$ which is independent of Y.

(a) Show that $\frac{\mathbb{V}\mathrm{ar}(X)}{N\mathbb{E}[X]} = \frac{1}{N}\left(\frac{s^2}{m} + m(1 - p)\right)$.

(b) Provide an interpretation for $\frac{\mathbb{V}\mathrm{ar}(\widehat{\theta})}{\mathbb{E}[X]}$ and analyze its behavior when p increases.

20

Hedging strategies in practice

Insurance companies, pension plans and investment banks have important liabilities that stem from their main business and that are affected by the variations of many economic and financial variables. For example, life insurance policies and annuities generate liabilities with very long-term maturities and thus are highly exposed to variations in the level of interest rates, among other risks. Similarly, investment banks trade large volumes of options and other financial derivatives whose values are sensitive to the movements of the underlying assets' prices.

In most situations, these risks cannot be eliminated by simply taking an offsetting position in a similar financial product or derivative. For example:

- An insurance company sells a 15-year EIA based on the S&P 500 index. There are no financial derivatives with such long-term maturities available on exchanges.
- A pension plan will need to pay monthly amounts starting in 15 to 50 years from now but only 15-year and 30-year bonds are available.
- An investment bank sells a 2-year 47-strike call option on ABC inc. However, only calls and puts with maturity 6 months, 1 year and 3 years are available, with strike prices ranging from \$48 to \$52.

In those cases, the risk manager must set up an investment strategy to mitigate the risks coming from these liabilities.

As opposed to a replication strategy, which is a portfolio reproducing exactly (*in all possible scenarios*) a given payoff, a **hedging strategy** is an investment strategy having the more realistic objective of reproducing some features of an important liability.

In general, *hedging* is an important piece of an efficient **asset liability management (ALM)** strategy. Indeed, the objective of hedging could alternately be viewed as assets (investments) designed and managed with the specific goal of protecting one's financial position against adverse changes in liabilities.

The main objective of this chapter is to analyze various risk management practices, mostly hedging strategies used for interest rate risk and equity risk management. The specific objectives are to:

- apply cash-flow matching or replication to manage interest rate risk and equity risk;
- understand how Taylor series expansions can be used for risk management purposes;
- understand the similarities between duration-(convexity) matching and delta(-gamma) hedging;
- apply duration-(convexity) matching to assets and liabilities sensitive to interest rate risk;
- determine whether an investment strategy is a full or a Redington immunization;
- differentiate between the various *Greeks*;

Actuarial Finance: Derivatives, Quantitative Models and Risk Management, First Edition.
Mathieu Boudreault and Jean-François Renaud.

- apply delta(-gamma) hedging to assets and liabilities sensitive to changes in the underlying asset price;
- apply delta-rho hedging to assets and liabilities sensitive to changes in both the interest rate and the underlying asset price;
- apply delta-vega hedging to assets and liabilities sensitive to changes in both the underlying asset price and its volatility;
- update the hedging portfolio (rebalancing) as conditions in the market evolve.

In this chapter, we will consider two types of hedging strategies: (1) perfect hedging strategies (whenever possible) based on exactly matching *cash flows* from assets and liabilities, and (2) hedging strategies based on matching *sensitivities* of liabilities and assets.

20.1 Introduction

Let $\mathcal{L}_t(x)$ be the time-t value of a company's liabilities which is (mainly) influenced by a financial variable x. In other words, if we observe the value x, then the time-t value of the company's future obligations is $\mathcal{L}_t(x)$. There are many important examples we can think of: the liability from

- a life insurance policy or pension plan whose important driver x is the interest rate;
- a written option on a stock which is obviously affected by x the underlying asset price;
- an EIA or VA which is driven mostly by the value x of the investment/reference portfolio selected by the policyholder.

Of course, most liabilities are affected by more than one financial variable. In what follows, we will focus first on protecting a financial position against changes in one of them (and denoted by x). Later on (see Section 20.5.4), we will also consider two financial variables (x, y).

Example 20.1.1 *Liability tied to life insurance*
An insurance company has a *very large* number of contracts so that, according to the *law of large numbers*, it can reasonably approximate the *proportion* of policyholders that will die over the next 30 years.

Given the size of its portfolio, the company expects that 100 of its policyholders will die 5 years from now, 500 will die 10 years from now and 250 will die 20 years from now. Let us assume that this *average scenario* is the scenario that will occur. If the benefit paid upon death is 1, then in this specific scenario today's value for this liability is simply

$$\mathcal{L}_0(r) = \frac{100}{(1+r)^5} + \frac{500}{(1+r)^{10}} + \frac{250}{(1+r)^{20}}.$$

We would like to determine the sensitivity of this actuarial liability to an immediate change in the interest rate. Suppose the term structure of interest rates is flat at 3%. The initial value of this liability becomes

$$\mathcal{L}_0(0.03) = \frac{100}{(1+0.03)^5} + \frac{500}{(1+0.03)^{10}} + \frac{250}{(1+0.03)^{20}} = \$596.73.$$

Let us look at the effect of a sudden change of the interest rate by $\pm 1\%$. Computing $\mathcal{L}_0(0.02)$ and $\mathcal{L}_0(0.04)$, we see that

$$\begin{aligned}\mathcal{L}_0(0.02) &= \$668.99\\ \mathcal{L}_0(0.03) &= \$596.73\\ \mathcal{L}_0(0.04) &= \$534.07.\end{aligned}$$

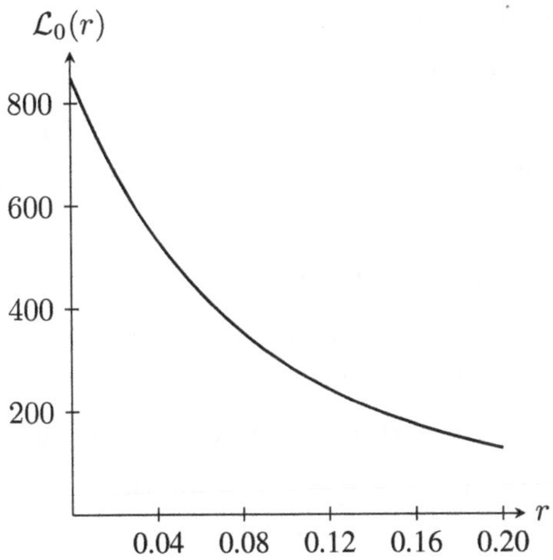

Figure 20.1 Liability tied to the life insurance business of example 20.1.1 as a function of the interest rate r

Of course, as shown in Figure 20.1, when the level of the interest rate increases (decreases), the value of the liability goes down (up). A change of 1% in interest rates provokes a change of about \$60–75 in the value of liabilities, more than 10% in relative terms, which is a significant change. Clearly, something must be done to mitigate this risk. ■

Similarly, let $\mathcal{A}_t(x)$ be the time-t value of this company's assets, where its relationship with the financial quantity x is emphasized. Most of the time, the assets correspond to a portfolio, i.e. a combination of investment products from the financial market. This portfolio can be composed of bonds, stocks, options, swaps, etc. Typically, these financial products, and consequently the portfolio, will also be influenced by the level of the variable x.

The main objective of asset liability management is to choose assets aligned with the company's liabilities. An important ALM strategy is hedging. The idea is that assets should be chosen such that whenever x moves by a quantity Δx, then $\Delta\mathcal{L}_t(x)$ and $\Delta\mathcal{A}_t(x)$ will be approximately the same. For example, when the interest rate goes down from 3% to 2%, as in example 20.1.1, then the value of the actuarial liabilities increases by more than 10%. Ideally, the insurer will build a portfolio so that its assets also increase when the interest rate goes down.

In the next section, we will have a look at cash-flow matching (or, equivalently, replication), which is the *strongest* risk management approach. Unfortunately, as we have seen before, it is not always possible to perfectly match the cash flows of a given liability/payoff. Other strategies need to be set up in order to diminish the exposure to various risks. This will be the topic of the following sections.

20.2 Cash-flow matching and replication

An intuitive ALM strategy is known as **cash-flow matching.** The idea is simple: for every liability cash flow, we will find a matching asset cash flow. This is equivalent to finding a replicating strategy. Most of the time, this strategy is static (buy and hold) as opposed to dynamic where rebalancing is needed.

Mathematically, cash flows are matched if

$$\mathcal{L}_t(x) = \mathcal{A}_t(x), \quad \text{for all } x \text{ and } t,$$

i.e. no matter the value of x, the value of assets and the value of liabilities are equal.
The following two examples illustrate cash-flow matching.

Example 20.2.1 *Cash-flow matching for derivatives*
Your investment bank has issued two derivatives:

- a 6-month forward contract on a stock for a delivery price of \$46;
- a 1-year 45-strike European call option on the same stock.

In other words, the bank is short a forward contract and short a call option. Its liabilities are:

- deliver a share of stock in 6 months in exchange for \$46;
- deliver another share of stock in a year, only if $S_1 > 45$.

Let us describe how to build an investment portfolio that matches the liability cash flows, where x is the stock price.
Of course, if the bank also takes the following positions:

- long a 6-month forward contract on the same stock with same delivery price;
- long a 1-year 45-strike European call option;

then all cash flows would be perfectly matched. It is important to understand that no matter what the values $S_{0.5}$ and S_1 turn out to be, the cash flows of the liabilities would always be matched.
Instead of taking the long position in another 6-month forward, the bank could also buy a share of stock and borrow the present value of the delivery price. This static replicating strategy was derived in Chapter 3 and it also matches the cash flows of the short forward.
Note that we cannot do the same thing for the long call option, i.e., set up a model-free replicating strategy. More assumptions are needed, such as assuming the underlying stock price evolves according to a given market model (e.g. a binomial tree, a trinomial tree or a BSM model). ■

Example 20.2.2 *Cash-flow matching for an actuarial liability*
A pension plan will terminate in 3 years and it has the following liability cash flows: 1M next year, 1.5M in 2 years and 4M 3 years from now.
In the financial markets, prices of corporate bonds for a highly-rated company are as follows:

Maturity	Coupon	Price
1	7%	104.39
2	6%	104.81
3	6.5%	107.09

The coupons are based on a face value of 100 and paid annually.
Let us find how many of each bond is required to match the liability cash flows, where x is the interest rate.

The following table summarizes the cash flows of each bond at each date:

Bond	$t = 1$	$t = 2$	$t = 3$
1	107	0	0
2	6	106	0
3	6.50	6.50	106.50

Let n_1, n_2 and n_3 be the number of the corresponding bond, chosen in order to replicate the liabilities. We must have that

$$\begin{aligned} 1 &= 107n_1 + 6n_2 + 6.50n_3 \\ 1.5 &= 106n_2 + 6.50n_3 \\ 4 &= 106.50n_3. \end{aligned}$$

In other words, we need to solve a system of three equations and three unknowns. This system is easy to solve: we first find $n_3 = 4/106.50 = 37558.69$, then we find $n_2 = 11847.82$ and $n_1 = 6399.828$.

Buying 6400 units of the first bond, 11848 units of the second one and 37559 of the third one will set up a portfolio exactly matching the pension obligations, no matter how interest rates evolve between now and the plan termination. ■

In the last example, for cash-flow matching to work, having access to bonds with those specific maturities, i.e. corresponding to the liability cash flows, was crucial. For insurance companies and pension plans, this is difficult, not to say impossible, because they have liabilities with very long maturities. Since the fixed-income market is illiquid for very long maturities, we cannot expect cash-flow matching to be an ALM strategy that we can always implement.

Credit and counterparty risks

The last two examples also demonstrate that cash-flow matching ensures the solvency of the plan sponsor or the investment bank. In other words, they will meet the obligations set forth by the contracts they have written if and only if the assets deliver the cash flows they promise. The loss resulting from such a broken promise is known as credit risk or counterparty risk in this specific context.

The application of cash-flow matching demonstrated above implicitly expects zero counterparty or credit risk. It is reasonable to assume that Treasury bonds will always make the required payments but the financial crisis of 2007–2009 showed that even highly-rated investment-grade bonds (or issuers of derivatives) can default. Cash-flow matching will fail in these situations as with any other ALM strategy if we do not account for counterparty or credit risk on the assets backing liabilities.

20.3 Hedging strategies

We saw that aiming for

$$\mathcal{L}_t(x) = \mathcal{A}_t(x),$$

for all x and t, is often impossible in practice, so instead of trying to meet this ultimate objective, we will now aim at having

$$\Delta \mathcal{L}_t(x) \approx \Delta \mathcal{A}_t(x), \quad \text{for all } \Delta x \text{ and } t.$$

In words, for a *variation* in x, the resulting variation in liabilities is *closely* (as opposed to *exactly*) matched by the corresponding variation in assets.

The mathematics backing up the hedging strategies presented in the remainder of this chapter is based on simple *Taylor series expansions*. We therefore briefly recall Taylor series and then we discuss how it can be used to design effective hedging strategies.

20.3.1 Taylor series expansions

It is possible to write most real-valued (infinitely differentiable) functions as a power series. In the neighborhood of x_0, we can rewrite a function $f(x)$ as

$$f(x) = f(x_0) + f'(x_0)(x - x_0) + \frac{f''(x_0)}{2!}(x - x_0)^2 + \frac{f'''(x_0)}{3!}(x - x_0)^3 + \cdots \tag{20.3.1}$$

where the equality is obtained at the limit. Written compactly, equation (20.3.1) becomes

$$f(x) = \sum_{k=0}^{\infty} \frac{f^{(k)}(x_0)}{k!}(\Delta x)^k$$

where

$$f^{(k)}(x) = \frac{\mathrm{d}^k f}{\mathrm{d}x^k}(x)$$

is the k-th derivative of f and $\Delta x = x - x_0$. This is called the *Taylor series expansion* of f.

A Taylor series expansion can be truncated if we consider only a finite number of terms/derivatives in which case the series becomes a sum. It can be used to approximate the function f.

The **first-order approximation** of f is given by

$$f(x) \approx f(x_0) + f'(x_0)\Delta x. \tag{20.3.2}$$

This is an approximation, not an equality. In a calculus course, we assess the quality of this approximation by quantifying the *error term*. Typically, if x is *close to* x_0, then the approximation is *good*.

Clearly, if the derivative of f at x_0 is positive, i.e. if $f'(x_0) > 0$, then if x is larger than x_0 but still close to x_0, $f(x)$ will be larger than $f(x_0)$. Said differently, if $\Delta x > 0$ then $\Delta f(x) = f(x) - f(x_0) > 0$. And vice versa if the derivative of f at x_0 is negative, i.e. if $f'(x_0) < 0$, and if $\Delta x > 0$ then $\Delta f(x) < 0$.

Similarly, the **second-order approximation** of f is given by

$$f(x) \approx f(x_0) + f'(x_0)\Delta x + \frac{f''(x_0)}{2}(\Delta x)^2. \tag{20.3.3}$$

Again, it is an approximation, not an equality.

Figure 20.2 shows the true price of a zero-coupon bond as a function of the interest rate along with its first-order and second-order approximations around $x_0 = 3\%$. We see that the first-order approximation can be interpreted as the affine function with a slope equal to the derivative of f at $x_0 = 3\%$. In the immediate surroundings of 3%, both approximations work

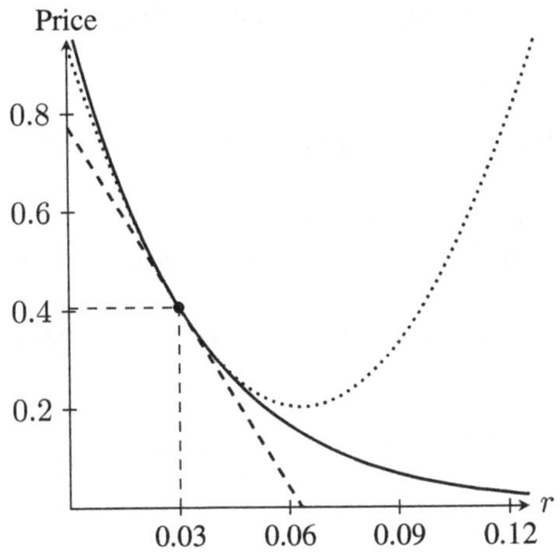

Figure 20.2 The price of a zero-coupon bond as a function of the interest rate (solid line) along with its first-order (dashed line) and second-order (dotted line) Taylor approximations around $x_0 = 3\%$

well. But when, for example, the interest rate goes down, the first-order approximation does not stay as close to the true value of the bond as the second-order approximation.

20.3.2 Matching sensitivities

Instead of matching cash flows, we will now look at hedging strategies matching the *sensitivities* of liabilities and assets.

Note that to avoid any confusion between the *mathematical derivative*, which results from differentiating a function, and a *financial derivative*, which is a financial security, we will use the word *sensitivity* as a synonym of mathematical derivative. We will see in the next sections that the first-order and second-order sensitivities will have their own specific names whether we manage interest rate risk or equity risk.

We will now drop the time index t and assume, without loss of generality, that hedging occurs at time 0. In other words, in what follows, $\mathcal{L} = \mathcal{L}_0$ and $\mathcal{A} = \mathcal{A}_0$.

From the Taylor series expansion around x_0 of the value of liabilities with respect to the financial variable x, we have

$$\Delta\mathcal{L}(x) = \mathcal{L}'(x_0)\Delta x + \frac{\mathcal{L}''(x_0)}{2!}(\Delta x)^2 + \frac{\mathcal{L}'''(x_0)}{3!}(\Delta x)^3 + \cdots \tag{20.3.4}$$

where $\Delta\mathcal{L}(x) = \mathcal{L}(x) - \mathcal{L}(x_0)$. Similarly, for the value of assets, we get

$$\Delta\mathcal{A}(x) = \mathcal{A}'(x_0)\Delta x + \frac{\mathcal{A}''(x_0)}{2!}(\Delta x)^2 + \frac{\mathcal{A}'''(x_0)}{3!}(\Delta x)^3 + \cdots \tag{20.3.5}$$

where $\Delta\mathcal{A}(x) = \mathcal{A}(x) - \mathcal{A}(x_0)$.

In a risk management context, x_0 is the current or initial value of an important economic or financial variable. Of course, the value of this quantity will eventually change, or suddenly move; more on this in Section 20.6.

Ideally, if we could find an investment strategy matching all the sensitivities, i.e. such that

$$\mathcal{L}'(x_0) = \mathcal{A}'(x_0), \quad \mathcal{L}''(x_0) = \mathcal{A}''(x_0), \quad \mathcal{L}'''(x_0) = \mathcal{A}'''(x_0), \ldots$$

then, from equations (20.3.4) and (20.3.5), we would find that

$$\mathcal{L}(x) - \mathcal{L}(x_0) = \mathcal{A}(x) - \mathcal{A}(x_0)$$

in the surroundings of x_0. This would be equivalent to cash-flow matching/replication as in this case $\mathcal{L}(x) = \mathcal{A}(x)$.

Of course, in practice, we cannot match infinitely many sensitivities. However, as discussed previously, whenever Δx is small, we can *reasonably* approximate $\Delta\mathcal{L}(x)$ and $\Delta\mathcal{A}(x)$ using only a finite number of sensitivities.

Assuming second-order terms and higher are *small enough* to be ignored, i.e. using first-order approximations, we can write

$$\Delta\mathcal{L}(x) \approx \mathcal{L}'(x_0)\Delta x$$

and

$$\Delta\mathcal{A}(x) \approx \mathcal{A}'(x_0)\Delta x.$$

A hedging strategy for which assets and liabilities are such that

$$\begin{cases} \mathcal{A}(x_0) &= \mathcal{L}(x_0) \\ \mathcal{A}'(x_0) &= \mathcal{L}'(x_0) \end{cases}$$

will be called a **first-order hedging strategy** and will hedge a financial position in the surroundings of x_0, i.e. for small values of Δx.

In the next sections, we will cover in more detail the cases for which x is the level of interest rates or the value of an underlying stock or index:

- Whenever the company wants to protect against an immediate variation in the interest rate r, this first-order hedging strategy is known as **duration matching**.
- When the company wants to protect against an immediate change in the stock price S_0, this first-order hedging strategy is known as **delta hedging**.

Assuming third-order and higher terms are *negligible*, i.e. using second-order approximations, we can write

$$\Delta\mathcal{L}(x) \approx \mathcal{L}'(x_0)\Delta x + \frac{\mathcal{L}''(x_0)}{2!}(\Delta x)^2$$

and

$$\Delta\mathcal{A}(x) \approx \mathcal{A}'(x_0)\Delta x + \frac{\mathcal{A}''(x_0)}{2!}(\Delta x)^2.$$

A hedging strategy for which assets and liabilities are such that

$$\begin{cases} \mathcal{A}(x_0) &= \mathcal{L}(x_0) \\ \mathcal{A}'(x_0) &= \mathcal{L}'(x_0) \\ \mathcal{A}''(x_0) &= \mathcal{L}''(x_0) \end{cases}$$

will be called a **second-order hedging strategy** and will hedge a financial position in the surroundings of x_0, i.e. for small values of Δx.

For interest rate risk management, this second-order hedging strategy is known as **duration-convexity matching** and for equity risk management, this hedging method is known as **delta-gamma hedging**.

Both first-order and second-order hedging strategies applied to interest rate risk and equity risk are the subject of the next two sections.

20.4 Interest rate risk management

We now focus on hedging strategies based on the matching of sensitivies for interest rate risk management. Therefore, the variable x will be the risk-free rate r and we will consider the case of a risk manager who wants to hedge immediate or sudden changes in that risk-free rate level. This means we assume the term structure of interest rates is flat and constant at r. This is the only modeling assumption we need.

20.4.1 Sensitivities

Let us start by setting up some notation. The 1-year discount factor, i.e. the present value of \$1 paid a year from now, is represented by $v(r)$. When the interest rate is compounded continuously, then $v(r) = \mathrm{e}^{-r}$, and when it is compounded annually, then $v(r) = (1+r)^{-1}$.

Assume that an asset or a liability has a stream of cash flows $\{c_1, c_2, \dots, c_n\}$ occurring at times $\{t_1, t_2, \dots, t_n\}$. The present value of these cash flows is then

$$\mathcal{P}(r) = \sum_{k=1}^{n} c_k v(r)^{t_k}.$$

The **duration** of those cash flows is defined as

$$\mathcal{D}(r) = \frac{\mathrm{d}\mathcal{P}}{\mathrm{d}r}(r) = \mathcal{P}'(r). \tag{20.4.1}$$

In the financial markets, $-\mathcal{D}(r)$ is known as the **dollar duration**.

We find that, under continuous compounding,

$$\mathcal{D}(r) = -\sum_{k=1}^{n} \left(t_k \times c_k v(r)^{t_k}\right) = -\mathcal{P}(r) \sum_{k=1}^{n} w_k t_k$$

whereas, under annual compounding,

$$\mathcal{D}(r) = -\sum_{k=1}^{n} \left(t_k \times c_k v(r)^{t_k+1}\right) = -\mathcal{P}(r) v(r) \sum_{k=1}^{n} w_k t_k = -\mathcal{P}(r) \frac{1}{1+r} \sum_{k=1}^{n} w_k t_k$$

where, in both cases,

$$w_k = c_k v(r)^{t_k} / \mathcal{P}(r).$$

Since $\sum_{k=1}^{n} w_k = 1$, then w_k is the weight of the k-th cash flow in the total present value $\mathcal{P}(r)$. The weighted average of the timing of those cash flows, given by $\sum_{k=1}^{n} w_k t_k$ is thus an important component of the duration.

Example 20.4.1 *Duration of a 3-year coupon bond*
A bond with annual coupons of 4% (on a face value of 100) matures in 3 years. The yield (to maturity) on that bond is 5.5% compounded annually. Calculate the duration of that coupon bond.

The following table computes some important quantities.

t_k	c_k	v^{t_k}	$c_k \times v^{t_k}$	w_k	$w_k \times t_k$
1	4	0.9478673	3.79146919	0.03951377	0.03951377
2	4	0.89845242	3.59380966	0.03745382	0.07490763
3	104	0.85161366	88.5678211	0.92303241	2.76909723
Sum		$\mathcal{P}(r)$	95.9530999	1	2.883518635

Therefore, the duration is

$$\mathcal{D}(0.055) = -2.883518635 \times 95.9530999 = -276.6825516.$$

■

Other definitions of duration

Durations can also be expressed *relatively* to the total present value of the cash flows. The **modified duration** is simply

$$\text{Mod}\mathcal{D}(r) = -\frac{\mathcal{D}(r)}{\mathcal{P}(r)}$$

so that, under continuous compounding,

$$\text{Mod}\mathcal{D}(r) = \sum_{k=1}^{n} w_k t_k$$

and, with annual compounding,

$$\text{Mod}\mathcal{D}(r) = \frac{1}{1+r} \sum_{k=1}^{n} w_k t_k.$$

Finally, the **Macaulay duration** is directly defined as

$$\text{Mac}\mathcal{D}(r) = \sum_{k=1}^{n} w_k t_k.$$

Thus, the *modified* duration under continuous compounding or the Macaulay duration (no matter what is the compounding frequency) really correspond to the weighted average of cash flow occurrence times.

Similarly, the **convexity** of the stream of cash flows $\{c_1, c_2, \ldots, c_n\}$ occurring at times $\{t_1, t_2, \ldots, t_n\}$ is defined as

$$\mathcal{C}(r) = \frac{\mathrm{d}^2\mathcal{P}}{\mathrm{d}r^2}(r) = \mathcal{P}''(r). \tag{20.4.2}$$

Again, using the properties of derivatives, we get

$$\mathcal{C}(r) = \mathcal{P}(r) \sum_{k=1}^{n} w_k t_k^2,$$

under continuous compounding and

$$C(r) = \mathcal{P}(r)\frac{1}{(1+r)^2}\sum_{k=1}^{n} w_k t_k (t_k + 1),$$

under annual compounding.

Example 20.4.2 *Convexity of a 3-year coupon bond*
Following from example 20.4.1, the convexity is

$$\begin{aligned} C(0.055) &= 95.9530999 \times 0.8984524 \\ &\quad \times (1 \times 2 \times 0.0395138 + 2 \times 3 \times 0.0374538 + 3 \times 4 \times 0.923032) \\ &= 95.9530999 \times 10.22451 \\ &= 981.0734295. \end{aligned}$$

■

Other definitions of convexity

As with duration, the convexity can also be expressed *relative* to the total present value of the cash flows, so that the **modified convexity** is

$$\mathrm{Mod}C(r) = \frac{C}{\mathcal{P}(r)}$$

and the **Macaulay convexity** is directly

$$\mathrm{Mac}C(r) = \sum_{k=1}^{n} w_k t_k^2.$$

20.4.2 Duration matching

The hedging method known as *duration matching* is a first-order hedging strategy that consists in making sure the duration of assets matches the duration of liabilities. Suppose that the current risk-free rate is r_0. For a first-order approximation, the following two conditions must hold:

$$\begin{cases} \mathcal{A}(r_0) = \mathcal{L}(r_0) \\ \mathcal{A}'(r_0) = \mathcal{L}'(r_0). \end{cases}$$

Let $\mathcal{P}_\mathcal{A}(r_0)$ be the current value of the assets and let $\mathcal{P}_\mathcal{L}(r_0)$ be the current value of the liabilities, both for an interest rate of r_0. Also, let $\mathcal{D}_\mathcal{A}(r_0)$ be the duration of the assets and $\mathcal{D}_\mathcal{L}(r_0)$ the duration of the liabilities, both when the current risk-free rate is r_0.

For **duration matching**, the following two conditions must hold:

$$\begin{cases} \mathcal{P}_\mathcal{A}(r_0) = \mathcal{P}_\mathcal{L}(r_0) \\ \mathcal{D}_\mathcal{A}(r_0) = \mathcal{D}_\mathcal{L}(r_0). \end{cases}$$

Example 20.4.3 *Duration matching with zero-coupon bonds*
The term structure of interest rates is currently flat at $r_0 = 3\%$ (continuously compounded). Two zero-coupon bonds, maturing in 1 year and 15 years respectively, are available. Your company owes \$100 due in 10 years. Find a portfolio of these two bonds that matches the duration of this liability. Then, assess the quality of this ALM strategy if the interest rate suddenly moves by $\pm 0.5\%$.

This is a typical situation where we cannot match the liability cash flow (occurring at time 10) using the assets available in the market. To find the hedging strategy that matches the duration of the liability, we need to determine the appropriate number of each bond in the portfolio. Let n_1 be the number of 1-year bonds and n_{15} the number of 15-year bonds. The initial values of assets and liabilities need to be equal, so we must have

$$100\,e^{-10\times0.03} = n_1 \times e^{-1\times0.03} + n_{15} \times e^{-15\times0.03}.$$

To match the durations, we must have

$$10 \times 100\,e^{-10\times0.03} = n_1 \times e^{-1\times0.03} + 15 \times n_{15} \times e^{-15\times0.03}$$

where the negative signs have been cancelled out on both sides. This is a system of two equations and two unknowns. We find that we need $n_1 = 27.2635534$ units of the 1-year bond and $n_{15} = 74.6893442$ units of the 15-year bond.

Next, we analyze the effect of an immediate shift of ±0.5% in the interest rate level on the performance of this hedging portfolio. The values are shown below.

r	Liability	1-year bond	15-year bond	Assets
2.50%	77.8800783	0.97530991	0.68728928	77.9235994
3%	74.0818221	0.97044553	0.63762815	74.0818221
3.50%	70.468809	0.96560542	0.59155536	70.5087171

For example, when the interest rate goes down by 0.5%, the liability increases from 74.0818221 to 77.8800783 and simultaneously the bond portfolio goes up from 74.0818221 to 77.9235994. This yields a small profit of 4 cents. The increase of $3.80 in the liability, which results in a loss for the company, is canceled out by an increase of $3.84 in the assets. A similar conclusion can be obtained if the interest rate goes up by 0.5%. In conclusion, small variations in the risk-free rate are hedged. ∎

Because we are ignoring second-order terms and other higher-order terms in the Taylor series expansion, duration matching will commit a more significant error for large changes in the interest rate. This is illustrated in the following example.

Example 20.4.4 *Performance of duration matching for large interest rate movements*
What if, in the previous example, the interest rate goes from 3% to either 10%, 0.5% or 0.001%? Let us calculate the mismatch (hedging error in absolute values) between the values of assets and liabilities.

If we update the table from the previous example by adding rows corresponding to the other interest rate levels and if we add a column for the corresponding mismatch, we obtain the following:

r	Liability	1-year bond	15-year bond	Assets	Mismatch
0.001%	99.9900	1.0000	0.9999	101.9414	1.9514
0.50%	95.1229	0.9950	0.9277	96.4201	1.2972
2.50%	77.8801	0.9753	0.6873	77.9236	0.0435
3%	74.0818	0.9704	0.6376	74.0818	0.0000
3.50%	70.4688	0.9656	0.5916	70.5087	0.0399
10%	36.7879	0.9048	0.2231	41.3345	4.5466

We see that when the interest rate level deviates significantly from $r_0 = 3\%$, which was used to set up the duration-matching portfolio, the mismatch is more important. For an extreme increase of the risk-free rate from 3% to 10%, the mismatch is greater than \$4.50, i.e. more than 10% of the current liability value. When the interest rate approaches 0, the mismatch is close to \$2 and, in relative terms, this is an impact of about 2%. This illustrates how the mismatch can quickly increase when r deviates from r_0. ■

20.4.3 Duration-convexity matching

The hedging method known as **duration-convexity matching** consists in making sure that the duration and convexity of assets match the duration and convexity of liabilities. As it is a second-order hedging strategy, the conditions for duration-convexity matching, if the interest rate level is r_0, are:

$$\begin{cases} \mathcal{A}(r_0) &= \mathcal{L}(r_0) \\ \mathcal{A}'(r_0) &= \mathcal{L}'(r_0) \\ \mathcal{A}''(r_0) &= \mathcal{L}''(r_0). \end{cases}$$

Example 20.4.5 *Duration-convexity matching with zero-coupon bonds*
In the context of example 20.4.3, if we want to match both duration and convexity, then we need a third asset in the hedging portfolio (acting as the third unknown for the duration-convexity matching system of equations). In this direction, we introduce a 7-year zero-coupon bond as another possible investment.

In order to match present values, durations and convexities of assets and liabilities, we need to solve the following system of equations (let n_7 be the number of 7-year bonds):

$$\begin{aligned} 100\,\mathrm{e}^{-10\times 0.03} &= n_1 \times \mathrm{e}^{-0.03} + n_7 \times \mathrm{e}^{-7\times 0.03} + n_{15} \times \mathrm{e}^{-15\times 0.03} \\ 10 \times 100\,\mathrm{e}^{-10\times 0.03} &= n_1 \times \mathrm{e}^{-0.03} + 7 \times n_7 \times \mathrm{e}^{-7\times 0.03} + 15 \times n_{15} \times \mathrm{e}^{-15\times 0.03} \\ 100 \times 100\,\mathrm{e}^{-10\times 0.03} &= n_1 \times \mathrm{e}^{-0.03} + 49 \times n_7 \times \mathrm{e}^{-7\times 0.03} + 225 \times n_{15} \times \mathrm{e}^{-15\times 0.03}. \end{aligned}$$

Solving for (n_1, n_7, n_{15}), we obtain the portfolio $(-13.63177668, 85.68104862, 28.00850407)$, in which we need to short 13.63177668 units of the 1-year bond.

Now, let us analyze the effectiveness of this hedging strategy. The next table shows the value of assets and the value of liabilities along with the corresponding mismatches (hedging errors in absolute value).

	Liability	1-year bond	7-year bond	15-year bond	Assets	Mismatch
0.001%	99.9900	1.0000	0.9999	0.9999	100.0477	0.0577
0.5%	95.1229	0.9950	0.9656	0.9277	95.1550	0.0321
2.5%	77.8801	0.9753	0.8395	0.6873	77.8803	0.0002
3%	74.0818	0.9704	0.8106	0.6376	74.0818	0.0000
3.5%	70.4688	0.9656	0.7827	0.5916	70.4686	0.0002
10%	36.7879	0.9048	0.4966	0.2231	36.4629	0.3250

Comparing with duration matching, we see that the mismatches are much lower for duration-convexity matching. This is also shown in Figure 20.3 where the value of the liability is depicted by the solid line, whereas the value of assets with both hedging strategies is given by either the dashed (duration matching) or dotted (duration-convexity matching) line. ■

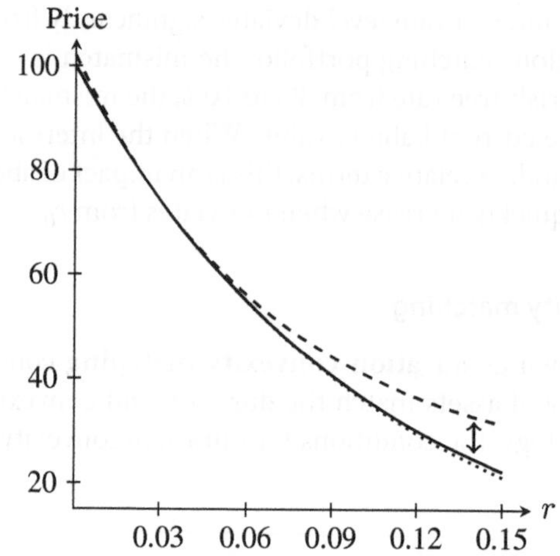

Figure 20.3 Quality of duration and duration-convexity matching hedging strategies as a function of the interest rate. Value of liability (solid line), value of assets with duration matching (dashed line) and value of assets with duration-convexity matching (dotted line)

20.4.4 Immunization

Immunization is the set of tools and techniques meant to protect (or immunize) a fixed-income portfolio against adverse changes in interest rates. Immunization theory was introduced in 1952 by the British actuary Frank Redington (1906–1984).

Immunization is an alternative word most often used specifically for interest rate risk management. In addition to duration matching and duration-convexity matching, immunization techniques comprise Redington and full immunization.

Redington immunization is a generalization of the duration-convexity matching technique where the convexity of the assets should be larger than the convexity of liabilities. We have Redington immunization of a position when the following three conditions hold:

- $\mathcal{A}(r_0) = \mathcal{L}(r_0)$;
- $\mathcal{A}'(r_0) = \mathcal{L}'(r_0)$;
- $\mathcal{A}''(r_0) \geq \mathcal{L}''(r_0)$.

For relatively small changes in the interest rates, Redington immunization should yield a profit (under the assumptions of the approach).

Full immunization is a technique that matches asset and liability durations in addition to making sure asset cash flows occur before and after each liability cash flow. Conditions for full immunization are:

- $\mathcal{A}(r_0) = \mathcal{L}(r_0)$;
- $\mathcal{A}'(r_0) = \mathcal{L}'(r_0)$;
- There are asset cash flows before and after each liability cash flow.

For any changes in the interest rates, full immunization always yields a profit (under the assumptions of the approach).

Redington and full immunization strategies are different only in the third condition. Due to the construction of the portfolio of assets, the strategy depicted in example 20.4.3 is also a full immunization, whereas the strategy of example 20.4.5 is also a Redington immunization as convexities match.

20.5 Equity risk management

As for interest rate risk, it is possible for (the risk manager of) an investment bank, an insurer or a pension plan to hedge its financial position with respect to the variations of a stock/an index price, if exposed to such an equity risk. For example, when issuing an option or selling a complex insurance policy, a company's position is sensitive to variations in the equity markets.

To perform *equity risk management*, one has to make a few market assumptions. In what follows, we will use the Black-Scholes-Merton model. This implies that the stock price (i.e. the underlying risky asset) follows a GBM, which means it has a constant volatility, and that the term structure of interest rates is flat and constant.

First, we will seek to protect the investor's position against an *immediate* and *sudden* change in the underlying asset price $x = S_0$. Then, we will extend the methodology to protect simultaneously against two risks.

20.5.1 Greeks

We know from Section 16.5 that the value of an option, or a complex derivative such as an ELIA, is affected by the level of various variables and parameters such as the:

- spot price S_0 of the underlying asset;
- volatility σ of the underlying asset;
- risk-free rate r;
- time to maturity T.

Let us consider a simple payoff $V_T = g(S_T)$, such as that of a forward, a call or a put, as well as a gap option or a binary option. This also includes investment guarantees, equity-indexed annuities and GMMBs. Recall that the time-t value of such a financial derivative is of the form $V_t = F(t, S_t; r, \sigma)$, a notation now emphasizing that V_t not only depends on time t and stock price S_t but also on the BSM model parameters r and σ.

When we performed option pricing and hedging in the BSM model (see Chapter 16), parameters such as r and σ were assumed to be known and fixed in advance. But as market conditions evolve, the value of these parameters will change. Later on, we will also want to protect against changes in these variables.

The sensitivities of an option price V_t with respect to various quantities, such as S_t, r and σ, are usually represented using Greek letters, thus the common name **Greeks**. The main Greeks are:

$$\Delta_t = \frac{\partial F}{\partial S_t}(t, S_t) \quad \text{(delta)} \tag{20.5.1}$$

$$\Gamma_t = \frac{\partial^2 F}{\partial S_t^2}(t, S_t) \quad \text{(gamma)} \tag{20.5.2}$$

$$\nu_t = \frac{\partial F}{\partial \sigma}(t, S_t) \quad \text{(vega}^1\text{)} \tag{20.5.3}$$

$$\rho_t = \frac{\partial F}{\partial r}(t, S_t) \quad \text{(rho)} \tag{20.5.4}$$

1 It is important to note that this sensitivity is called "*vega*" even if there is no "*vega*" in the Greek alphabet. The most commonly used Greek letter is ν which looks like a "*v*" for volatility.

$$\Theta_t = \frac{\partial F}{\partial t}(t, S_t) \quad \text{(theta}^2\text{)} \tag{20.5.5}$$

$$\psi_t = \frac{\partial F}{\partial \gamma}(t, S_t) \quad \text{(psi).} \tag{20.5.6}$$

For example, Δ_t is the sensitivity (at time t) of the option price with respect to S_t (underlying asset spot price) and ν_t is the sensitivity of the option price with respect to the parameter σ (volatility level in the BSM model).

In Section 16.5, we found explicit expressions for some of those Greeks when the payoff is that of a call or a put option. In Table 20.1, we provide the most important Greeks for both call and put options, computed at any time t and accounting for continuous dividends. Recall that d_1 and d_2 stand for $d_1(t, S_t; r, \sigma)$ and $d_2(t, S_t; r, \sigma)$, i.e. they are functions of time and the asset price, along with the BSM parameters r and σ. Moreover, we use $\tau = T - t$.

In Section 16.5 we analyzed the sensitivity of V_0 with respect to the *time to maturity*,[3] which is $\frac{\partial F}{\partial T}(0, S_0)$. This is not the same as Θ_t above, which measures the sensitivity of V_t with respect to the passage of time or *time decay*. Indeed,

$$\Theta_t = \frac{\partial F}{\partial t}(t, S_t) = -\frac{\partial F}{\partial T}(t, S_t).$$

In the BSM model, there is an important relationship between some of the sensitivities in Table 20.1. It is commonly known as the **Black-Scholes partial differential equation**:

$$\Theta_t + (r - \gamma)S_t\Delta_t + \frac{1}{2}\sigma^2 S_t^2 \Gamma_t = rV_t. \tag{20.5.7}$$

For more details on the Black-Scholes PDE, see Chapter 17.

Note that if we know the Greeks of a derivative, then from the Black-Scholes PDE we can retrieve its Black-Scholes price.

Table 20.1 Most important Greeks for call and put options

<table>
<tr><th></th><th>Call (C_t)</th><th>Put (P_t)</th></tr>
<tr><td>Δ_t</td><td>$e^{-\gamma(T-t)}N(d_1)$</td><td>$-e^{-\gamma(T-t)}N(-d_1)$</td></tr>
<tr><td>Γ_t</td><td colspan="2">$e^{-\gamma(T-t)}\dfrac{\phi(d_1)}{S_t\sigma\sqrt{T-t}}$</td></tr>
<tr><td>ν_t</td><td colspan="2">$S_t e^{-\gamma(T-t)}\phi(d_1)\sqrt{T-t}$</td></tr>
<tr><td>ρ_t</td><td>$K(T-t)e^{-r(T-t)}N(d_2)$</td><td>$-K(T-t)e^{-r(T-t)}N(-d_2)$</td></tr>
<tr><td>Θ_t</td><td>$-\dfrac{S_t\phi(d_1)\sigma}{2\sqrt{\tau}} - rKe^{-r\tau}N(d_2) + \gamma S_t e^{-\gamma\tau}N(d_1)$</td><td>$-\dfrac{S_t\phi(d_1)\sigma}{2\sqrt{\tau}} + rKe^{-r\tau}N(-d_2) - \gamma S_t e^{-\gamma\tau}N(-d_1)$</td></tr>
<tr><td>ψ_t</td><td>$-S_t(T-t)e^{-\gamma(T-t)}N(d_1)$</td><td>$S_t(T-t)e^{-\gamma(T-t)}N(-d_1)$</td></tr>
</table>

2 In order to be consistent with the financial literature, we use the Greek letter Θ_t exceptionally in this section, even though in the rest of the book, it represents the number of units invested in the risk-free asset.

3 At time 0, the maturity date T is the *time to maturity*. However, at time t, it is $T - t$ that represents *time to maturity*.

Example 20.5.1 *Manipulating the Greek letters*
You are given the following information about the Greeks of an option[4] issued on a non-dividend-paying stock:

- $\Theta_0 = -2.96713$;
- $\Delta_0 = 0.73279$;
- $\Gamma_0 = 0.01723$.

If the current stock price is \$54, its volatility is 25% and the risk-free rate is 5%, let us find the Black-Scholes price of this option.

Using the Black-Scholes PDE (20.5.7), we get

$$\begin{aligned}\Theta_0 + rS_0\Delta_0 + \frac{1}{2}\sigma^2 S_0^2\Gamma_0 \\ &= -2.96713 + 0.05 \times 54 \times 0.73279 + 0.5 \times 0.25^2 \times 54^2 \times 0.01723 \\ &= 0.58148675\end{aligned}$$

Dividing by the risk-free rate $r = 0.05$, we find that the option price is 11.629735. ∎

Option elasticity

The delta of an option can be re-expressed in terms of percentage changes. This is also known as the option elasticity, often denoted by Ω_t. We thus have

$$\Omega_t = \Delta_t \times \frac{S_t}{V_t}.$$

In words, the elasticity represents the percentage change in the option price for a 1% change in the underlying asset price.

20.5.2 Delta hedging

We know that issuing an option, or selling a variable annuity, is equivalent to owing its random future cash flows (usually its payoff V_T). This is a liability whose value today is denoted by

$$\mathcal{L}(S_0) = F(0, S_0; r, \sigma),$$

since $V_0 = F(0, S_0; r, \sigma)$.

As we saw already in Section 16.6.4 of Chapter 16, in a delta-hedging portfolio, the liability is *managed* with an investment portfolio composed of shares of the underlying asset S and of the risk-free asset B. It is a discretization (in time) of the Black-Scholes replicating portfolio.

Now, let δ_0^S and δ_0^B be the number of units of the risky asset and of the risk-free asset held in that portfolio,[5] respectively, at time 0. Then the time-0 value of these assets is given by

$$\mathcal{A}(S_0) = \delta_0^B \times B_0 + \delta_0^S \times S_0.$$

Note the emphasis on the fact that both assets and liabilities depend on the value of $x = S_0$. As for interest rate risk management, our objective is to match sensitivities with respect to this variable.

4 For the benefit of the reader, the Greeks provided below are those of a 51-strike 2-year call option.

5 We use a different notation than the usual couple (Δ, Θ) since we are setting up a *hedging* portfolio rather than a *replicating* portfolio.

Delta hedging is to equity risk management what duration matching is to interest rate management, i.e. it is a first-order hedging strategy. More precisely, **delta hedging** consists in making sure that

$$\begin{cases} \mathcal{A}(S_0) &= \mathcal{L}(S_0) \\ \mathcal{A}'(S_0) &= \mathcal{L}'(S_0). \end{cases}$$

Using the above assumptions and notation, we have

$$\mathcal{A}'(S_0) = \delta_0^S$$

and

$$\mathcal{L}'(S_0) = \frac{\partial F}{\partial S_0}(0, S_0) = \Delta_0.$$

Then the condition $\mathcal{A}'(S_0) = \mathcal{L}'(S_0)$ yields

$$\delta_0^S = \Delta_0$$

while $\mathcal{A}(S_0) = \mathcal{L}(S_0)$ yields

$$\delta_0^B = \frac{V_0 - \Delta_0 \times S_0}{B_0}.$$

The results we just found should be reminiscent of what we obtained in Chapter 16, where delta hedging was described as a *realistic* hedge for an option because trading occurred in discrete time, as opposed to the replicating portfolio (on which it is based) which requires continuous trading. Now, even though we are using a different approach, i.e. matching the first-order sensitivities of assets and the liability, we found the same time-0 quantities to hold in the portfolio: (δ_0^B, δ_0^S) is equal to (Θ_0, Δ_0) obtained in Chapter 16.

Below we will use the **relative hedging error (RHE)**, which is defined as follows:

$$\text{RHE} = \frac{|\mathcal{L}(S_0) - \mathcal{A}(S_0)|}{\mathcal{L}(S_0)}.$$

Example 20.5.2 *Delta hedging a call*
You need to hedge the issuance of a 1-year 45-strike call option on a stock worth \$43. The stock price evolves according to a BSM model with a volatility of 25%. The risk-free rate is 3%. Illustrate how to implement a delta-hedging strategy and analyze its performance in various scenarios.

We have $S_0 = 43$, $K = 45$, $T = 1$, $\sigma = 0.25$ and $r = 0.03$. First, we have $d_1(0, 43) = 0.063150504$ and $N(d_1) = 0.525176671$. Then, using the Black-Scholes formula, we find that $C_0 = 3.983989415$, which is also the value of the liability $\mathcal{L}(S_0) = \mathcal{L}(43)$.

Delta hedging this liability requires holding $\delta_0^S = \Delta_0 = N(d_1(0, 43)) = 0.525176671$ shares of the stock while the rest is invested at the risk-free rate, i.e.

$$\delta_0^B = \frac{C_0 - \Delta_0 \times S_0}{B_0} = 3.983989415 - 0.525176671 \times 43 = -18.59860743.$$

In other words, at time 0, we invest an amount of

$$\Delta_0 \times S_0 = 0.525176671 \times 43 = 22.58259684$$

in the stock, i.e. we buy $\Delta_0 = 0.525176671$ shares, which is financed by borrowing an amount of

$$-(C_0 - \Delta_0 \times S_0) = 18.59860743.$$

Suppose now that, a few milliseconds after setting up this hedging portfolio, there is a shock to the stock price. The value of the above hedging portfolio (set up for $S_0 = 43$) will change if the value of the underlying asset changes; one should think of S_{0+} as the new value of the underlying asset, immediately after having set up the hedging portfolio. Therefore, we will have

$$\mathcal{A}(S_{0+}) = 0.525176671 \times S_{0+} - 18.59860743.$$

On the other hand, the liability value, corresponding here to the call option price, will be equal to $\mathcal{L}(S_{0+}) = C(0, S_{0+})$.

The next table shows the values of assets and liabilities and the RHE.

S_0	$\mathcal{A}(S_0)$	$\mathcal{L}(S_0)$	RHE
38	1.35810606	1.83553306	26.01%
39	1.88328273	2.18789223	13.92%
40	2.4084594	2.57899886	6.61%
41	2.93363607	3.00896047	2.50%
42	3.45881274	3.47749967	0.54%
43	**3.98398942**	**3.98398942**	**0.00%**
44	4.50916609	4.52749415	0.40%
45	5.03434276	5.10681457	1.42%
46	5.55951943	5.72053374	2.81%
47	6.0846961	6.36706277	4.43%
48	6.60987277	7.04468455	6.17%

The value of assets and liabilities is also illustrated in Figure 20.4 for $S_0 = 38, 39, \ldots, 48$. The value of the delta-hedging portfolio is approximating the value of the liability with

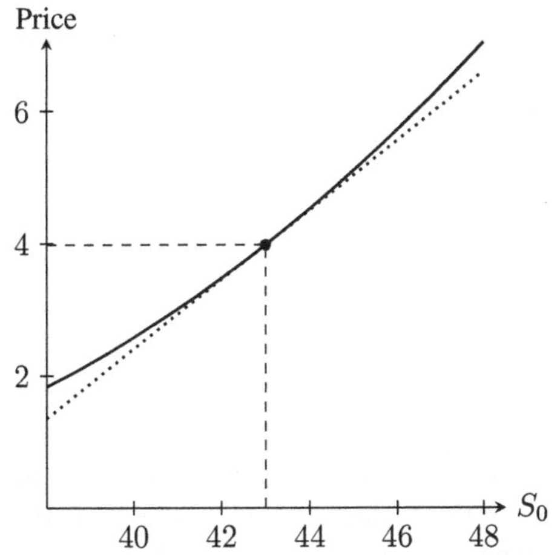

Figure 20.4 Value of the 1-year call option (liability, continuous line) and of the asset portfolio obtained with a delta hedging strategy (dotted line) as a function of the underlying asset price at $t = 0$

a linear function of S_0 around 43. Without surprise, we see the smallest relative hedging errors near 43, but when the shock is large, the hedging strategy struggles to capture the change in the liability value. ■

Let us now illustrate how to implement delta hedging for an investment guarantee written on a non-dividend-paying stock in the BSM model. We know from equation (18.2.2) in Chapter 18, that the initial value of an investment guarantee with payoff $V_T = \max(S_T, K)$ is related to the initial value of a put option:

$$V_0^{IG} = S_0 + P(0, S_0),$$

where $P(0, S_0)$ is the initial value of the corresponding put option as given by the Black-Scholes formula.

Here the liability is the issuance of the investment guarantee, so

$$\mathcal{L}(S_0) = S_0 + P(0, S_0)$$

and then

$$\mathcal{L}'(S_0) = 1 + \frac{\partial P}{\partial S_0}(0, S_0).$$

From Section 16.5 , we know that

$$\frac{\partial P}{\partial S_0}(0, S_0) = N(d_1(0, S_0)) - 1,$$

which means that

$$\mathcal{L}'(S_0) = N(d_1(0, S_0)).$$

This is exactly the delta of the corresponding underlying call option, i.e. with the same maturity date T and with strike price K. This means that delta hedging an investment guarantee requires the same portfolio of assets as for delta hedging a call option.

This should not come as a surprise if we recall that the initial value of an investment guarantee is also related to the initial price of a call option, as given by equation (18.2.1) in Chapter 18:

$$V_0^{IG} = \mathrm{e}^{-rT} K + C(0, S_0).$$

Consequently, we have

$$\frac{\partial}{\partial S_0} V_0^{IG} = 0 + \frac{\partial C}{\partial S_0}(0, S_0).$$

20.5.3 Delta-gamma hedging

We saw in example 20.5.2 that the *quality* of a delta-hedging strategy deteriorates quickly as the stock price changes. As for duration matching, this is because delta hedging ignores second-order and higher-order terms in the Taylor series expansions of $\mathcal{A}(S_0)$ and $\mathcal{L}(S_0)$.

Delta-gamma hedging is to equity risk management what duration-convexity matching is to interest rate management, i.e. a second-order sensitivity-matching hedging strategy. To delta-gamma hedge a financial position, we need to be allowed to invest in a third product which is sensitive to S in a non-linear fashion.[6] This means that the underlying asset itself or forward contracts on the underlying are not eligible candidates.

Let $G(t, S_t)$ be the time-t value of this other tradable asset. We need it to be such that

$$\Gamma_0^G = \frac{\partial^2 G}{\partial S_0^2}(0, S_0) \neq 0.$$

The asset portfolio now also includes δ_0^G units of asset G such that its value becomes

$$\mathcal{A}(S_0) = \delta_0^B \times B_0 + \delta_0^S \times S_0 + \delta_0^G \times G(0, S_0).$$

The value of the liability is still

$$\mathcal{L}(S_0) = F(0, S_0; r, \sigma).$$

Therefore, **delta-gamma hedging** is the hedging strategy that consists in making sure that

$$\begin{cases} \mathcal{A}(S_0) &= \mathcal{L}(S_0) \\ \mathcal{A}'(S_0) &= \mathcal{L}'(S_0) \\ \mathcal{A}''(S_0) &= \mathcal{L}''(S_0). \end{cases}$$

To find the delta-gamma hedging strategy corresponding to the liability $\mathcal{L}(S_0) = F(0, S_0)$, we first need to (re)compute the first-order derivatives of $\mathcal{A}(S_0)$ and $\mathcal{L}(S_0)$. We have, for the liability,

$$\mathcal{L}'(S_0) = \frac{\partial F}{\partial S_0}(0, S_0) = \Delta_0^F$$

$$\mathcal{L}''(S_0) = \frac{\partial^2 F}{\partial S_0^2}(0, S_0) = \Gamma_0^F$$

and, for the assets,

$$\mathcal{A}'(S_0) = \delta_0^S + \delta_0^G \times \Delta_0^G$$

$$\mathcal{A}''(S_0) = \delta_0^G \times \Gamma_0^G$$

where

$$\Delta_t^G = \frac{\partial G}{\partial S_t}(t, S_t)$$

$$\Gamma_t^G = \frac{\partial^2 G}{\partial S_t^2}(t, S_t).$$

The solution, i.e. the resulting hedging portfolio $(\delta_0^B, \delta_0^S, \delta_0^G)$, is thus:

$$\delta_0^G = \frac{\Gamma_0^F}{\Gamma_0^G}$$

$$\delta_0^S = \Delta_0^F - \frac{\Gamma_0^F}{\Gamma_0^G}\Delta_0^G$$

$$\delta_0^B = \frac{F(0, S_0) - \delta_0^S \times S_0 - \delta_0^G \times G(0, S_0)}{B_0}.$$

6 This is because delta hedging approximates the value of liabilities with a linear function of the stock price.

In other words, the number of units of the third asset G and the risky asset S are determined with Greeks, while the rest is invested at the risk-free rate.

It should be clear now, from the solution of the latter system of equations, that if $\Gamma_0^G = 0$ (when for example G is a linear function of S, like the asset itself or a forward on that asset), then delta-gamma hedging cannot be implemented.

Example 20.5.3 *Delta-gamma hedging a call*
In example 20.5.2 we sought to hedge the issuance of a 1-year call option. Assume now there exists a 3-month 45-strike put option that you intend to use to set up a delta-gamma hedging strategy. Build such a strategy at inception and compare its effectiveness with the delta hedge. You are given the following values:

	V_0	Δ_0	Γ_0
1-year call option	3.98398942	0.52517667	0.03703698
3-month put option	3.11608149	−0.59529955	0.07209392

Here, the added asset is the 3-month put option so $G(0, S_0) = P_0$. Note that the put does not depend linearly on S, i.e. P_0 is not a linear function of S_0. As usual, we will use the notation δ_0^P (instead of δ_0^G) for the number of units of the put (third asset) to hold in the delta-gamma portfolio.

For the delta-gamma hedging strategy, we get:

$$\begin{aligned}
\delta_0^P &= \frac{\Gamma_0^F}{\Gamma_0^P} = \frac{0.03703698}{0.07209392} \\
&= 0.513732439 \\
\delta_0^S &= \Delta_0^F - \frac{\Gamma_0^F}{\Gamma_0^P}\Delta_0^P = 0.52517667 - \frac{0.03703698}{0.07209392} \times -0.59529955 \\
&= 0.831001362 \\
\delta_0^B &= \frac{F(0,43) - \delta_0^S \times 43 - \delta_0^P \times P(0,43)}{B_0} \\
&= 3.98398942 - 0.831001362 \times 43 - 0.513732439 \times 3.11608149 \\
&= -33.34990131.
\end{aligned}$$

Therefore, the delta-gamma hedging asset portfolio's value is given by

$$\begin{aligned}
\mathcal{A}(43) &= -33.34990131 + 0.831001362 \times S_0 + 0.513732439 \times P_0 \\
&= 3.983989,
\end{aligned}$$

which is, as planned, equal to the call option's value of $C_0 = 3.98398942$.

The next table shows the values of the assets and the liability as functions of S_0, along with the relative hedging error.

S_0	$\mathcal{A}(S_0)$	$\mathcal{L}(S_0)$	RHE
38	1.773728	1.83553306	3.37%
39	2.15807094	2.18789223	1.36%
40	2.56731749	2.57899886	0.45%
41	3.00579739	3.00896047	0.11%
42	3.47714444	3.47749967	0.01%
43	**3.9839894**	**3.98398942**	**0.00%**
44	4.52776477	4.52749415	0.01%
45	5.10864177	5.10681457	0.04%
46	5.72559563	5.72053374	0.09%
47	6.3765756	6.36706277	0.15%
48	7.05874405	7.04468455	0.20%

Comparing the results of example 20.5.2 with the latter table, we see that delta-gamma hedging reacts much better to larger shocks in the stock price. The relative hedging errors with delta-gamma hedging are much lower and the mismatch is minimal as shown also in Figure 20.5. ■

20.5.4 Hedging with additional Greeks

According to the assumptions of the BSM model, only the underlying asset price evolves with time, while the volatility σ and the risk-free interest rate r are constant. However, in reality, the volatility of the risky asset and the risk-free rate also fluctuate with time and economic conditions.

For short-term options, changes in the risk-free rate have almost no impact on the option value while changes in the volatility level have a much bigger impact. Therefore, *hedging vega* is common for options with short maturity. For long-term contracts such as ELIAs, the

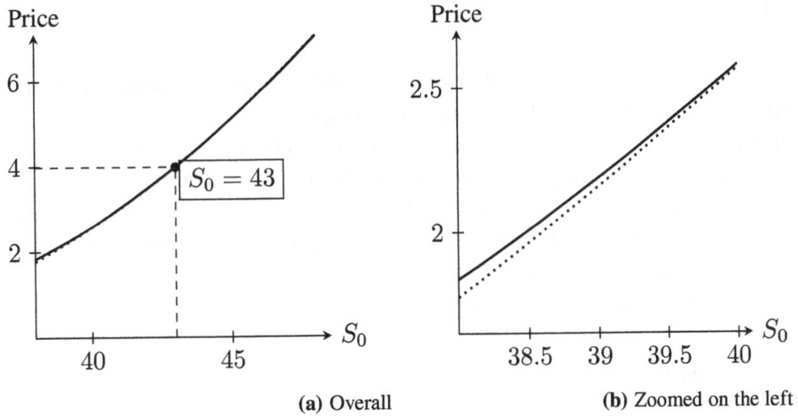

Figure 20.5 Value of the 1-year call option (liability, continuous line) and of the investment portfolio (assets) obtained with a delta-gamma hedging strategy (dotted line) as a function of the initial stock price

fluctuations of interest rates have a more significant impact, so *hedging rho* is very important. In what follows, we will extend delta hedging by considering an additional Greek, such as rho (interest rate) or vega (volatility), when building the hedging strategy.

Mathematically, for a function $f(x,y)$ of two variables x and y, the first-order Taylor series expansion of f around (x_0, y_0) is given by

$$f(x,y) - f(x_0,y_0) \approx \frac{\partial f}{\partial x}(x_0,y_0)\Delta x + \frac{\partial f}{\partial y}(x_0,y_0)\Delta y,$$

where $\Delta x = x - x_0$ and $\Delta y = y - y_0$. Note the similarity with the one-dimensional case considered in equation (20.3.1).

In what follows, we will derive **bivariate first-order hedging strategies** for the liability value $\mathcal{L}(x,y)$ and the asset value $\mathcal{A}(x,y)$, both depending now on two financial variables, x and y.

20.5.4.1 Delta-rho hedging

To illustrate the idea, let us first show how to *delta-rho hedge* an option. The objective is simple: we want to match the first-order sensitivities with respect to the stock price and to the risk-free rate. Now, the value of the liability is $\mathcal{L}(S_0, r) = F(0, S_0; r)$, i.e. $x = S_0$ and $y = r$, and the value of the investment portfolio (assets) is

$$\mathcal{A}(S_0, r) = \delta_0^B \times B_0 + \delta_0^S \times S_0 + \delta_0^G \times G(0, S_0; r).$$

As announced, the notation has changed to emphasize the dependence on both S_0 and r.

Note that, as for delta-gamma hedging, we need to be able to invest in a third asset with initial value $G(0, S_0; r)$, in addition to the basic assets B and S. To hedge rho, i.e. the sensitivity to changes in the level of the risk-free rate r, we need this third asset to be also sensitive to changes in the risk-free rate. It could be, for example, a bond, a swap or any other fixed-income security.

Delta-rho hedging is the hedging strategy that consists in making sure that

$$\begin{cases} \mathcal{A}(S_0, r) &= \mathcal{L}(S_0, r) \\ \frac{\partial}{\partial r}\mathcal{A}(S_0, r) &= \frac{\partial}{\partial r}\mathcal{L}(S_0, r) \\ \frac{\partial}{\partial S_0}\mathcal{A}(S_0, r) &= \frac{\partial}{\partial S_0}\mathcal{L}(S_0, r). \end{cases}$$

Therefore, we match the initial value of both assets and liabilities, in addition to matching the first partial derivatives with respect to S_0 and with respect to r.

It is important to understand how the risk-free rate affects $G(0, S_0; r)$. For example, in a bond or a swap, a sudden change in the risk-free rate affects the present value of all future cash flows, thus changing the current price. However, as B represents the accumulated value of \$1 at the risk-free rate, it has no future cash flows and only the balance is reinvested at the risk-free rate. A change in the level of interest rates will not affect its current value.

The above conditions for delta-rho hedging yield

$$\Delta_0^F = \frac{\partial}{\partial S_0}\mathcal{L}(S_0, r) = \frac{\partial}{\partial S_0}\mathcal{A}(S_0, r) = \delta_0^S$$

and

$$\rho_0^F = \frac{\partial}{\partial r}\mathcal{L}(S_0, r) = \frac{\partial}{\partial r}\mathcal{A}(S_0, r) = \delta_0^G \times \rho_0^G,$$

where

$$\Delta_0^F = \frac{\partial}{\partial S_0} F(0, S_0; r)$$

$$\rho_0^F = \frac{\partial}{\partial r} F(0, S_0; r)$$

$$\rho_0^G = \frac{\partial}{\partial r} G(0, S_0; r).$$

Therefore, the delta-rho hedging portfolio is given by

$$\begin{cases} \delta_0^S = \Delta_0^F \\ \delta_0^G = \frac{\rho_0^F}{\rho_0^G} \\ \delta_0^B = \frac{F(0,S_0;r) - \delta_0^S \times S_0 - \delta_0^G \times G(0,S_0;r)}{B_0}. \end{cases}$$

Example 20.5.4 *Delta-rho hedging a long-term put option*
An at-the-money 10-year put option, written on a stock index whose initial value is 100, is issued by an insurance company. If the current risk-free rate is 4% and if the volatility of the index is 17%, build a delta-rho hedging strategy using shares of the stock and a 15-year zero-coupon bond with a face value of 100.

The liability is the short put option, i.e. $\mathcal{L}(S_0, r) = P(0, S_0; r)$. Using the Black-Scholes formula for a put option, we obtain $P(0, 100; 0.04) = 5.712653$. Also, using the formulas for the delta and the rho of a put, as given in Table 20.1 (with $\gamma = 0$), we have

$$\begin{aligned} \Delta_0^P &= -N(-d_1(0, S_0)) \\ &= -N(-d_1(0, 100)) \\ &= -0.155563773 \\ \rho_0^P &= -KT\mathrm{e}^{-rT} N(-d_2(0, S_0)) \\ &= -100 \times 10 \times \mathrm{e}^{-0.04 \times 10} \times N(-d_2(0, 100)) \\ &= -212.690307. \end{aligned}$$

Recall that a negative rho means that when the interest rate increases, the put option price decreases.

In this situation, we have $G(0, S_0; r) = 100\,\mathrm{e}^{-r \times 15}$ (value of the bond), so its duration is given by

$$\rho_0^G = \frac{\partial}{\partial r} G(0, S_0; r) = -15 \times 100\,\mathrm{e}^{-r \times 15}.$$

Since $r = 0.04$, we have $G(0, 100; 0.04) = 54.88116$ and $\rho_0^G = -823.2175$.

Therefore, the positions in the delta-rho hedge are

$$\delta_0^S = \Delta_0^P = -0.155563773$$

$$\delta_0^G = \frac{\rho_0^P}{\rho_0^G} = \frac{-212.690307}{-823.2175} = 0.2583647$$

$$\delta_0^B = \frac{P(0, S_0; r) - \delta_0^S \times S_0 - \delta_0^G \times G(0, S_0; r)}{B_0}$$
$$= \frac{5.71265337 - (-0.155563773) \times 100 - 0.2583647 \times 54.88116}{1}$$
$$= 7.089676,$$

yielding an assets value of

$$\begin{aligned} \mathcal{A}(100, 0.04) &= 7.089676 \times B_0 + (-0.155563773) \times S_0 + 0.2583647 \times G(0, 100; 0.04) \\ &= 7.089676 \times 1 + (-0.155563773) \times 100 + 0.2583647 \times 54.88116 \\ &= 5.712653. \end{aligned}$$

For validation purposes, we see that the liability and assets values match.

If the risk-free rate *immediately* drops from 4% to 3.5%, the 10-year put option price becomes $P(0, 100; 0.035) = 6.861382$, while the 15-year bond price becomes $G(0, 100; 0.035) = 59.15554$. In this case, the value of the investment portfolio changes to

$$\begin{aligned} \mathcal{A}(100, 0.035) &= 7.089676 + (-0.155563773) \times 100 + 0.2583647 \times 59.15554 \\ &= 6.817002. \end{aligned}$$

The loss is

$$|\mathcal{A}(100, 0.035) - \mathcal{L}(100, 0.035)| = |6.81700 - 6.861382| = 0.044382$$

which is very small compared with the put option price. The delta-rho hedge performed well. ■

Figure 20.6 shows the price of an at-the-money 10-year put option for various values of r. In the neighborhood of $r = 0.04$ the quality of the delta-rho hedge is *very good* whereas the delta hedge does not account for changes in the risk-free rate.

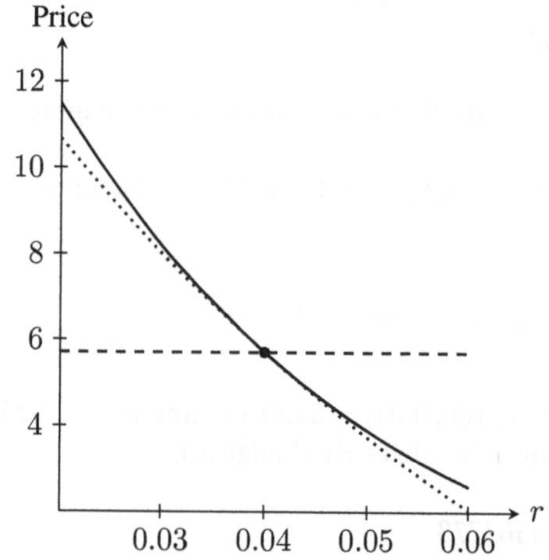

Figure 20.6 Initial value of a 10-year put option as a function of the risk-free rate r (solid line) along with a delta hedging strategy (dashed horizontal line) and a delta-rho hedging strategy (dotted line)

Example 20.5.5 *Delta-rho hedging vs. delta hedging*
In the context of the previous example, if the company uses only a delta-hedging strategy, then the asset portfolio is not sensitive to changes in the level of the risk-free rate.

Therefore, the asset portfolio is worth $\mathcal{A}(100, 0.035) = \mathcal{A}(100, 0.04) = 5.712653$ in both cases, which corresponds to the value of the 10-year put option with $r = 0.04$.

Under this hedging strategy, the company will suffer a much bigger loss from the decrease in the risk-free rate:

$$|\mathcal{A}(100, 0.035) - \mathcal{L}(100, 0.035)| = |5.712653 - 6.861382| = 1.148729.$$

Hedging rho has reduced the loss by more than 95% in this hypothetical scenario. ■

ELIAs have very long effective maturities and the previous examples illustrate that long-term put options, which are similar to GMMBs, are particularly sensitive to adverse changes in the level of interest rates. Rho hedging helps attenuate that kind of risk.

20.5.4.2 Delta-vega hedging

Since *delta-vega hedging* uses the same ideas as delta-rho hedging, we will proceed directly to the computations. The objective is to match the first-order sensitivities with respect to the stock price and to the volatility. The value of the liability is now $\mathcal{L}(S_0, \sigma) = F(0, S_0; \sigma)$, whereas the value of the investment portfolio (assets) is

$$\mathcal{A}(S_0, \sigma) = \delta_0^B \times B_0 + \delta_0^S \times S_0 + \delta_0^G \times G(0, S_0; \sigma).$$

Again, we need to be able to invest in a third asset, with initial value $G(0, S_0; \sigma)$, which is sensitive to changes in the volatility level σ. It could be, for example, another call or put option.

Delta-vega hedging is the hedging strategy that consists in making sure that

$$\begin{cases} \mathcal{A}(S_0, \sigma) &= \mathcal{L}(S_0, \sigma) \\ \frac{\partial}{\partial \sigma}\mathcal{A}(S_0, \sigma) &= \frac{\partial}{\partial \sigma}\mathcal{L}(S_0, \sigma) \\ \frac{\partial}{\partial S_0}\mathcal{A}(S_0, \sigma) &= \frac{\partial}{\partial S_0}\mathcal{L}(S_0, \sigma). \end{cases}$$

The above conditions for delta-vega hedging yield

$$\Delta_0^F = \frac{\partial}{\partial S_0}\mathcal{L}(S_0, \sigma) = \frac{\partial}{\partial S_0}\mathcal{A}(S_0, \sigma) = \delta_0^S,$$

and

$$v_0^F = \frac{\partial}{\partial \sigma}\mathcal{L}(S_0, \sigma) = \frac{\partial}{\partial \sigma}\mathcal{A}(S_0, \sigma) = \delta_0^G \times v_0^G,$$

where

$$\begin{aligned} \Delta_0^F &= \frac{\partial}{\partial S_0}F(0, S_0; \sigma) \\ v_0^F &= \frac{\partial}{\partial \sigma}F(0, S_0; \sigma) \\ v_0^G &= \frac{\partial}{\partial \sigma}G(0, S_0; \sigma). \end{aligned}$$

Therefore, the delta-vega hedging portfolio is given by

$$\begin{cases} \delta_0^S = \Delta_0^F \\ \delta_0^G = \frac{\nu_0^F}{\nu_0^G} \\ \delta_0^B = \frac{F(0,S_0;\sigma)-\delta_0^S\times S_0-\delta_0^G\times G(0,S_0;\sigma)}{B_0}. \end{cases}$$

20.6 Rebalancing the hedging portfolio

So far, we have seen that matching sensitivities at time 0 yields (static) hedging portfolios working reasonably well as long as the changes in the variables/parameters are relatively small, which can be assumed for short periods of time. But as time goes by, the interest rate or the stock price will make larger movements, further and further away from their initial values. The hedging strategies presented before will then become less and less efficient. To solve this, hedging portfolios need to be updated periodically. As we have seen in previous chapters, this is known as rebalancing.

In Section 16.6.4, we have shown how to periodically rebalance a delta-hedging strategy making sure it is self-financing. Recall that a *self-financing strategy* is a strategy that does not require any additional deposits or withdrawals until the maturity of the liability, which means that gains and losses in the investment portfolio (assets) are only reallocated between each individual asset.

Assume we set up the hedging strategy at time 0 and that it will be rebalanced at time 1. Assume also that the variable/parameter of interest evolves from x_0 to x_1 over that period. We define $\mathcal{A}_{1-}(x_1)$ as the value of assets *just before* rebalancing at time 1 and $\mathcal{A}_{1+}(x_1)$ as the value of assets *just after* rebalancing, with x_1 being the time-1 value of the financial variable of interest.

A self-financing hedging strategy based on the matching of sensitivities should be rebalanced to make sure that

$$\begin{cases} \mathcal{A}_{1-}(x_1) = \mathcal{A}_{1+}(x_1) \\ \mathcal{A}'(x_1) \;\; = \mathcal{L}'(x_1) \end{cases}$$

for a first-order strategy, and such that

$$\begin{cases} \mathcal{A}_{1-}(x_1) = \mathcal{A}_{1+}(x_1) \\ \mathcal{A}'(x_1) \;\; = \mathcal{L}'(x_1) \\ \mathcal{A}''(x_1) \;\; = \mathcal{L}''(x_1) \end{cases}$$

for a second-order strategy. The first common condition, i.e. $\mathcal{A}_{1-}(x_1) = \mathcal{A}_{1+}(x_1)$ is the *self-financing condition*, while the other ones are *hedging conditions*.

The following example illustrates the process with a duration-matching strategy. Otherwise, the reader can also go back to Section 16.6.4 for self-financing delta hedging.

Example 20.6.1 *Self-financing duration-matching strategy*
Let us continue example 20.4.3 and rebalance the hedge portfolio 1 year later, in a scenario where the risk-free rate drops to 2%. From time 0 to time 1, the duration-matching hedging portfolio is composed of 27.2635534 units of the 1-year bond and 74.6893442 units of the 15-year bond.

One year later, the 1-year bond has expired and, in the chosen scenario, the 15-year bond is now worth

$$\exp(-0.02 \times 14) = 0.755783741.$$

Thus, the value of assets 1 year later is

$$\mathcal{A}_{1-}(0.02) = 27.2635534 \times 1 + 74.6893442 \times 0.755783741 = 83.71254541.$$

We need to update the hedging portfolio. The 1-year bond has expired but we can enter into a new 1-year bond. To rebalance the portfolio, we need to match durations and maintain a value of 83.71254541. Therefore, we need to solve

$$\begin{aligned} 83.71254541 &= n_1\, e^{-0.02} + n_{15}\, e^{-14\times 0.02} \\ 9 \times 100\, e^{-9\times 0.02} &= n_1\, e^{-0.02} + 14\, n_{15}\, e^{-14\times 0.02}. \end{aligned}$$

Solving for n_1 and n_{15}, we obtain $n_1 = 32.97859261$ and $n_{15} = 67.99163552$, which are the portfolio quantities after rebalancing. ■

With a self-financing strategy, the hedging errors are rolled over until the company needs to meet its obligation, i.e. at the maturity of the liability. The value of assets at maturity $\mathcal{A}_T(x)$ needs to be sufficient to avoid losses. Instead of running the risk of having a huge profit or a big loss at maturity, the risk manager can deposit or withdraw money in the asset portfolio, thus attempting to smooth hedging errors (profits and losses) between inception and maturity. In that case, the strategy is no longer self-financing.

Mathematically, this is equivalent to matching sensitivities periodically, requiring the latter system of equations to hold at every time t chosen in a set of trading dates $\{t_1, t_2, \ldots, t_n\}$:

$$\begin{cases} \mathcal{A}_t(x_t) = \mathcal{L}_t(x_t) \\ \mathcal{A}'_t(x_t) = \mathcal{L}'_t(x_t) \\ \mathcal{A}''_t(x_t) = \mathcal{L}''_t(x_t), \end{cases}$$

where x_t is the time-t value of the financial quantity of interest. Note that the last condition is used only for second-order hedging strategies. For this approach to work, the risk manager needs to deposit/withdraw any amount necessary in between adjustments.

Example 20.6.2 *Periodic duration-matching*
In the preceding example, the value of liabilities after a year is the present value of a promise of \$100 to be paid 9 years in the future. The value is

$$\mathcal{L}_1(0.02) = 100 \exp(-0.02 \times 9) = 83.52702114.$$

This is slightly different from $\mathcal{A}_{1-}(0.02) = 83.71254541$. Therefore, reapplying a duration matching strategy at time 1 requires solving for

$$\begin{aligned} 83.52702114 &= n_1 e^{-0.02} + n_{15} e^{-14\times 0.02} \\ 9 \times 100 e^{-9\times 0.02} &= n_1 e^{-0.02} + 14 n_{15} e^{-14\times 0.02}. \end{aligned}$$

Notice how the left-hand side of the first condition changed only slightly. Since $\mathcal{A}_{1-}(0.02) = 83.71254541$, the company can withdraw $83.71254541 - 83.52702114 = 0.18552427$, cash it as a profit and perform a regular duration-matching with 83.52702114 instead. This strategy is no longer self-financing and the company can report the 19 cents as a profit resulting from hedging errors.

This yields a slightly different portfolio, i.e. solving the latter system of equations, we get that one should hold $n_1 = 32.77476111$ and $n_{15} = 68.01051804$ in the second year to match durations. ■

Model assumptions

The approaches depicted in this chapter will be successful in controlling risk coming from liabilities as long as hedging assumptions do not deviate too much from reality. In this case, the more often the asset portfolio is rebalanced, to account for changing market conditions, the smaller the hedging errors should be. In other words, at maturity the assets should be sufficient for the investor to meet its obligations.

To attain this important objective, there are three critical assumptions from the BSM model that do not hold in reality:

- The stock price is adequately modeled by a geometric Brownian motion and the risk manager knows the exact parameters.
- The term structure of interest rates is flat and moves in a parallel fashion over time.
- It is a frictionless market (no transaction costs, no taxes, no bid-ask spread, etc.).

In practice:

- it is widely accepted that geometric Brownian motions cannot adequately model the evolution of most stock prices and indices;
- the term structure of interest rates is often increasing. Moreover, short-term rates (say 3-month spot rate) usually are more volatile than long-term rates (say 10-year spot rates) so that parallel moves of the term structure are rare;
- the financial market is not entirely frictionless, thus rebalancing entails important transaction costs which reduce the benefit of rebalancing more often.

Whenever the model used to evaluate sensitivities deviates too much from the *real dynamics* of the market, the hedging strategy is exposed to **model risk.** In general, model risk is the possibility of loss resulting from not knowing what the *true model* is. In this context, model risk can seriously impede one's ability to hedge a liability and thus affect the *hedge effectiveness*.

20.7 Summary

Definitions and notation

- Financial variable/quantity of interest: x (and y).
- Value (at time t) of liabilities: $\mathcal{L}_t(x)$.
- Value (at time t) of assets/investment portfolio: $\mathcal{A}_t(x)$.
- Notation: $\mathcal{L} = \mathcal{L}_0$ and $\mathcal{A} = \mathcal{A}_0$.
- Asset liability management (ALM): choose assets aligned with the company's liabilities.
- Hedging: ALM strategy where assets are chosen such that they attenuate variations in the liabilities.
- Replication and cash-flow matching: ALM strategy where assets are chosen such that every liability cash flow is matched/offset by an equivalent asset cash flow (in all scenarios).

Cash-flow matching and replication

- Objective: design an investment strategy (choose assets) such that cash flows are matched:

$$\mathcal{L}_t(x) = \mathcal{A}_t(x), \quad \text{for all } x \text{ and } t.$$

- Difficult to implement in practice.

Hedging strategies and sensitivity matching

- Objective: design an investment strategy (choose assets) such that

$$\Delta\mathcal{L}_t(x) \approx \Delta\mathcal{A}_t(x), \quad \text{for all } \Delta x \text{ and } t.$$

- Mathematical tools: Taylor series expansions.
- First-order hedging strategy: assets are chosen such that

$$\begin{cases} \mathcal{A}(x_0) = \mathcal{L}(x_0) \\ \mathcal{A}'(x_0) = \mathcal{L}'(x_0). \end{cases}$$

- Second-order hedging strategy: assets are chosen such that

$$\begin{cases} \mathcal{A}(x_0) = \mathcal{L}(x_0) \\ \mathcal{A}'(x_0) = \mathcal{L}'(x_0) \\ \mathcal{A}''(x_0) = \mathcal{L}''(x_0). \end{cases}$$

Interest rate risk management or immunization

- Financial variable x: interest rate r.
- Assumption: term structure is flat and constant at r.
- One-year discount factor: $v(r)$.
- Cash flows: $\{c_1, c_2, \ldots, c_n\}$ at times $\{t_1, t_2, \ldots, t_n\}$.
- Present value of cash flows: $\mathcal{P}(r) = \sum_{k=1}^{n} c_k v(r)^{t_k}$.
- Duration of cash flows: $\mathcal{D}(r) = \mathcal{P}'(r)$.
- Convexity of cash flows: $\mathcal{C}(r) = \mathcal{P}''(r)$.
- Duration matching: first-order hedging strategy around r_0 such that

$$\begin{cases} \mathcal{P}_{\mathcal{A}}(r_0) = \mathcal{P}_{\mathcal{L}}(r_0) \\ \mathcal{D}_{\mathcal{A}}(r_0) = \mathcal{D}_{\mathcal{L}}(r_0). \end{cases}$$

- Duration-convexity matching: second-order hedging strategy around r_0 such that

$$\begin{cases} \mathcal{A}(r_0) = \mathcal{L}(r_0) \\ \mathcal{A}'(r_0) = \mathcal{L}'(r_0) \\ \mathcal{A}''(r_0) = \mathcal{L}''(r_0). \end{cases}$$

- Immunization: set of tools/techniques meant to protect/immunize a fixed-income portfolio against adverse changes in interest rates. Alternative word for interest rate risk management.
- Redington immunization:

$$\begin{cases} \mathcal{A}(r_0) = \mathcal{L}(r_0) \\ \mathcal{A}'(r_0) = \mathcal{L}'(r_0) \\ \mathcal{A}''(r_0) \geq \mathcal{L}''(r_0). \end{cases}$$

- Full immunization:

$$\begin{cases} \mathcal{A}(r_0) = \mathcal{L}(r_0) \\ \mathcal{A}'(r_0) = \mathcal{L}'(r_0) \end{cases}$$

There are asset cash flows before and after each liability cash flow.

Greeks in the Black-Scholes-Merton model

- Greeks: sensitivities of a derivative's price $V_t = F(t, S_t; \sigma, r)$ with respect to S_t, σ, r and more.
- Main Greeks:

$$\Delta_t = \frac{\partial F}{\partial S_t}(t, S_t) \quad \text{(delta)}$$
$$\Gamma_t = \frac{\partial^2 F}{\partial S_t^2}(t, S_t) \quad \text{(gamma)}$$
$$\nu_t = \frac{\partial F}{\partial \sigma}(t, S_t) \quad \text{(vega)}$$
$$\rho_t = \frac{\partial F}{\partial r}(t, S_t) \quad \text{(rho)}$$
$$\Theta_t = \frac{\partial F}{\partial t}(t, S_t) \quad \text{(theta)}$$
$$\psi_t = \frac{\partial F}{\partial \gamma}(t, S_t) \quad \text{(psi).}$$

- Black-Scholes PDE: $\Theta_t + (r - \gamma)S_t\Delta_t + \frac{1}{2}\sigma^2 S_t^2 \Gamma_t = rV_t$.

Equity risk management

- Assumption: Black-Scholes-Merton model.
- Financial variable x: initial price of risky asset S_0.
- Liability: $\mathcal{L}(S_0) = V_0 = F(0, S_0; r, \sigma)$
- Assets (investment portfolio): $\mathcal{A}(S_0) = \delta_0^B \times B_0 + \delta_0^S \times S_0$.
- Delta hedging: first-order hedging strategy around S_0 given by:

$$\begin{cases} \delta_0^S = \Delta_0 \\ \delta_0^B = \frac{V_0 - \Delta_0 \times S_0}{B_0}. \end{cases}$$

- Delta-gamma hedging: second-order hedging strategy around S_0.
  - Requires a third tradable asset with initial price $G(0, S_0; r, \sigma)$.
  - Assets' value becomes $\mathcal{A}(S_0) = \delta_0^B \times B_0 + \delta_0^S \times S_0 + \delta_0^G \times G(0, S_0)$.
  - Delta-gamma hedging strategy is given by:

$$\begin{cases} \delta_0^G = \frac{\Gamma_0^F}{\Gamma_0^G} \\ \delta_0^S = \Delta_0 - \frac{\Gamma_0^F}{\Gamma_0^G}\Delta_0^G \\ \delta_0^B = \frac{F(0, S_0) - \delta_0^S \times S_0 - \delta_0^G \times G(0, S_0)}{B_0}. \end{cases}$$

- Delta-rho hedging: bivariate first-order hedging strategy around (S_0, r_0).
  - Requires a third tradable asset with initial price $G(0, S_0; r, \sigma)$.
  - Delta-rho hedging strategy is given by:

$$\begin{cases} \delta_0^S &= \Delta_0^F \\ \delta_0^G &= \frac{\rho_0^F}{\rho_0^G} \\ \delta_0^B &= \frac{F(0,S_0;r)-\delta_0^S \times S_0-\delta_0^G \times G(0,S_0;r)}{B_0}. \end{cases}$$

- Delta-vega hedging: bivariate first-order hedging strategy around (S_0, σ_0) similar to delta-rho hedging and given by:

$$\begin{cases} \delta_0^S &= \Delta_0^F \\ \delta_0^G &= \frac{\mathcal{V}_0^F}{\mathcal{V}_0^G} \\ \delta_0^B &= \frac{F(0,S_0;\sigma)-\delta_0^S \times S_0-\delta_0^G \times G(0,S_0;\sigma)}{B_0}. \end{cases}$$

Rebalancing the hedging portfolio

- Financial quantity's time-1 value: x_1.
- Self-financing condition (at time 1): $\mathcal{A}_{1-}(x_1) = \mathcal{A}_{1+}(x_1)$.
- Hedging conditions (at time 1): match sensitivities.
- Repeat at each trading/rebalancing date $t_2, t_3, \ldots, t_n$.

20.8 Exercises

20.1 You owe \$50 to be paid in a year, \$40 in 2 years, \$30 in 3 years and \$20 in 4 years. Four coupon bonds are available in the market (each having a face value of 100):

Maturity	Coupon	Price
1	0%	97.561
2	0%	94.719
3	4.5%	104.424
4	3.7%	102.284

Coupons are paid annually. You seek to exactly replicate the cash flows of your liability using the bonds available in the market.
(a) Find the replicating strategy.
(b) What is the initial cost of this strategy?

20.2 Using the conventional actuarial notation (such as $v^n, a_{\overline{n}|}, Ia_{\overline{n}|}$, etc.), write the:
(a) duration of an annuity $a_{\overline{n}|}$;
(b) duration of a coupon bond with a face value of F, annual coupon rate of r and YTM of i (compounded annually).

20.3 Assuming that third and higher order effects can be ignored, using Taylor series, show that Redington immunization always yields a profit (assets over liabilities) for small interest rate changes.

20.4 You owe $100 to be paid in 3 years and seek to replicate (cash-flow match) this liability. However, there are no 3-year zero-coupon bonds available in the market. There is a 2-year bond and a 10-year coupon bond, with annual coupons of 3% and 7% respectively, traded on the market.

(a) How many of each bond do you need in order to match the duration of your assets and liabilities given that the term structure of interest rates is flat at 5.5%?

(b) Calculate the mismatch between assets and liabilities of the strategy obtained in (a) if the interest rate immediately drops by 0.5%.

20.5 Your valuation actuary informs you that an actuarial liability is such that:

- current value: 100 million;
- duration: 4.6 times 100 million;
- convexity: 14.9 times 100 million;
- 0.98 invested today is worth 1 in a year;
- term structure of interest rates is flat and interest rates are compounded annually.

Calculate the approximate variation in the actuarial liability if interest rates immediately decrease by 0.75%.

20.6 Your insurance company owes $100 to be paid in 2 years and another $100 in 4 years. The term structure of interest rates is flat at 3% (compounded annually). Three zero-coupon bonds are available with maturities of 1, 3 and 5 years (all with a face value of 100), respectively.

(a) How many of each bond do you need in order to match the duration and convexity of the assets and liabilities?

(b) Calculate the mismatch between assets and liabilities of the strategy obtained in (a) if the interest rate immediately drops by 1%.

20.7 For a single payment owed at time T, show that full immunization involving zero-coupon bonds implies that the liability is also Redington immunized.

20.8 Using the chain rule of calculus, compute ρ_t for a call option.

20.9 Using the put-call parity and the corresponding Greek letter for a call option, derive Δ_t, Γ_t and ρ_t for a standard put option.

20.10 For a 3-month 70-strike put option, you are given:

- $\Delta_0 = -0.178082$;
- $\Gamma_0 = 0.038613$;
- $S_0 = 75, r = 0.04, \sigma = 0.18$.

You seek to delta-gamma hedge this option with a 9-month 78-strike put option whose Greeks are:

- $\Delta_0 = -0.492504$;
- $\Gamma_0 = 0.034117$.

Describe your hedging strategy at time 0.

20.11 You know that one share of a non-dividend-paying stock trades for \$75 whereas coupon bonds of any maturity are available. You seek to delta-rho hedge an at-the-money 6-month call option written on this stock. If the risk-free rate is flat at 4%, the volatility of the stock return is 20%, describe your delta-rho hedge strategy at time 0 using a 5-year bond paying annual coupons of 6%.

20.12 For a 53-strike call option on a non-dividend-paying stock, you are given:
- $\Delta_0 = 0.621228$;
- $\Gamma_0 = 0.055343$;
- $\Theta_0 = -3.362835$;
- continuously compounded risk-free rate is 2.5%;
- $\sigma = 0.18$;
- the current price of that call option is \$3.60772.

Find the current price of the stock.

20.13 You need to delta hedge an at-the-money call option issued on a stock paying a continuous dividend with a yield of 3% per year. Moreover, you are given that $S_0 = 100$, $r = 0.03$ and $\sigma = 0.25$. The option matures in 3 months and you delta hedge every month.

(a) How many shares of stock and units of the risk-free asset do you need at time 0 if you need to delta hedge this call option?

(b) In the scenario that 1 month later the stock price is down to \$95, rebalance your hedging portfolio, making sure it is self-financing.

(c) In the scenario that 1 month prior to maturity the stock price is up to \$107, rebalance your hedging portfolio, making sure it is self-financing.

(d) In the scenario that at maturity the stock price is \$110, compute the hedging error.

References

1 Baxter, M. and Rennie, A. (1996) *Financial Calculus: An Introduction to Derivative Pricing*, Cambridge University Press, 1st edn.
2 Björk, T. (2009) *Arbitrage Theory in Continuous Time*, Oxford University Press, 3rd edn.
3 Cvitanic, J. and Zapatero, F. (2004) *Introduction to the Economics and Mathematics of Financial Markets*, MIT Press.
4 Lamberton, D. and Lapeyre, B. (2007) *Introduction to Stochastic Calculus Applied to Finance*, CRC Financial Mathematics Series, Chapman & Hall, 2nd edn.
5 Mikosch, T. (1998) *Elementary Stochastic Calculus – With Finance in View*, World Scientific.
6 Musiela, M. and Rutkowski, M. (2011) *Martingale Methods in Financial Modelling*, Springer, 2nd edn.
7 Shreve, S. (2004) *Stochastic Calculus for Finance I – The Binomial Asset Pricing Model*, Springer, 2nd edn.
8 Shreve, S. (2004) *Stochastic Calculus for Finance II – Continuous-Time Models*, Springer, 2nd edn.
9 Boyle, F. and Boyle, P.P. (2001) *Derivatives: The Tools That Changed Finance*, Risk Books.
10 Hull, J.C. (2017) *Options, Futures and Other Derivatives*, Pearson, 10th edition.
11 McDonald, R.L. (2013) *Derivatives Markets*, Pearson, 3rd edition.
12 Wilmott, P. (2007) *Paul Wilmott Introduces Quantitative Finance*, John Wiley & Sons, 2nd edn.
13 Devroye, L. (1986) *Non-Uniform Random Variate Generation*, Springer, 1st edn.
14 Glasserman, P. (2003) *Monte Carlo Methods in Financial Engineering*, Springer, 1st edn.
15 Hardy, M.R. (2003) *Investment Guarantees: Modelling and Risk Management for Equity-Linked Life Insurance*, Wiley.
16 Kalberer, T. and Ravindran, K. (2009) *Variable Annuities: A Global Perspective*, Risk Books, 1st edn.
17 Poitras, G. (2009) "The early history of option contracts," in Hafner, W. and Zimmermann, H. (eds.), *Vinzenz Bronzin's Option Pricing Models*, Berlin: Springer, pp. 487–518.
18 Boudreault, M. (2012) Pricing and hedging financial and insurance products. part 1: Complete and incomplete markets. *Risk & Rewards (Society of Actuaries)*.
19 Black, F. and Scholes, M. (1973) The pricing of options and corporate liabilities. *The Journal of Political Economy*, **81**, 637–654.
20 Merton, R.C. (1973) Theory of rational option pricing. *Bell J. Econom. and Management Sci.*, **4**, 141–183.
21 Harrison, J.M. and Kreps, D.M. (1979) Martingales and arbitrage in multiperiod securities markets. *J. Econom. Theory*, **20** (3), 381–408.

Actuarial Finance: Derivatives, Quantitative Models and Risk Management, First Edition.
Mathieu Boudreault and Jean-François Renaud.

22 Harrison, J.M. and Pliska, S.R. (1981) Martingales and stochastic integrals in the theory of continuous trading. *Stochastic Process. Appl.*, **11** (3), 215–260.
23 Harrison, J.M. and Pliska, S.R. (1983) A stochastic calculus model of continuous trading: complete markets. *Stochastic Process. Appl.*, **15** (3), 313–316.
24 Klugman, S.A., Panjer, H.H. and Willmot, G.E. (2013) *Loss Models – Further Topics*, John Wiley & Sons.

Index

Actuarial Finance: Derivatives, Quantitative Models and Risk Management, First Edition.
Mathieu Boudreault and Jean-François Renaud.